# Lecture Notes in Biomathematics

# Lecture Notes in Biomathematics

Managing Editor: S. Levin

## 45

Competition and Cooperation
in Neural Nets

Proceedings of the U.S.-Japan Joint Seminar
held at Kyoto, Japan
February 15–19, 1982

Edited by S. Amari and M.A. Arbib

Springer-Verlag
Berlin Heidelberg New York 1982

**Editors**

S. Amari
University of Tokyo
Dept. of Mathematical Engineering and Instrumentation Physics
Bunkyo-ku, Tokyo 113, Japan

M.A. Arbib
Center for Systems Neuroscience
Computer and Information Science, University of Massachusetts
Amherst, MA 01003, USA

AMS Subject Classifications (1980): 34 A 34, 34 D 05, 35 B 32, 35 Q 20, 58 F 13, 58 F 14, 60 G 40, 60 J 70, 68 D 20, 68 G 10, 92-06, 92 A 15, 92 A 27

ISBN-13: 978-3-540-11574-8    e-ISBN-13: 978-3-642-46466-9
DOI: 10.1007/978-3-642-46466-9

## Preface

The human brain, with its hundred billion or more neurons, is both one of the most complex systems known to man and one of the most important. The last decade has seen an explosion of experimental research on the brain, but little theory of neural networks beyond the study of electrical properties of membranes and small neural circuits. Nonetheless, a number of workers in Japan, the United States and elsewhere have begun to contribute to a theory which provides techniques of mathematical analysis and computer simulation to explore properties of neural systems containing immense numbers of neurons. Recently, it has been gradually recognized that rather independent studies of the dynamics of pattern recognition, pattern formation, motor control, self-organization, etc., in neural systems do in fact make use of common methods. We find that a "competition and cooperation" type of interaction plays a fundamental role in parallel information processing in the brain.

The present volume brings together 23 papers presented at a U.S.-Japan Joint Seminar on "Competition and Cooperation in Neural Nets" which was designed to catalyze better integration of theory and experiment in these areas. It was held in Kyoto, Japan, February 15-19, 1982, under the joint sponsorship of the U.S. National Science Foundation and the Japan Society for the Promotion of Science. Participants included brain theorists, neurophysiologists, mathematicians, computer scientists, and physicists. There are seven papers from the U.S. (plus two not presented at the Seminar), twelve from Japan, and one each from England, Finland, Italy and Mexico. The order of the papers herein is somewhat different from the order of the oral presentations, grouping the papers into six sections: An Opening Perspective; Reaction-Diffusion Equations; Single-Neuron and Stochastic Models; Oscillations in Neural Networks; Development and Plasticity of the Visual System; and Sensori-Motor Transformations and Learning. The remainder of the preface provides a brief overview of the papers in each section.


## I. An Opening Perspective

At a physiological level, we may speak of excitation and inhibition between neurons. At a functional level, we may speak of competition and cooperation between functionally defined groups of neurons. In the first stage of their work together at the Center for Systems Neuroscience in 1975-76, the Editors proposed a primitive neural model of competition and cooperation which could be adapted to model prey-selection mechanisms in the frog brain, and stereopsis mechanisms in the mammalian brain. In setting the theme for many of the papers in this volume, Amari devotes Chapter 1 to a survey of "Competitive and Cooperative Aspects in Dynamics of Neural Excitation and Self-Organization" which builds upon such work to study the dynamics of neuron pools, and the dynamics of self-organization of neural systems. Neural pools can function as analog devices or switches, depending on the strength of connections. Systems of neuron pools can exhibit oscillations. Neural fields can exhibit local excitations, function as maximum detectors, spontaneously form spatial patterns, and exhibit temporal patterns. Issues in self-organization include the formation of pattern recognizers, and the development of topographic maps between layers, with more active regions in the first layer having wider representation in the second layer.


## II. Reaction-Diffusion Equations

One contribution of Amari's paper is to show that the dynamics of neural fields can be treated from the competition and cooperation point of view, since neural fields can in general be regarded as continuum versions of the basic model of competition and cooperation. The present section is devoted to the more general study of interacting populations (chemical substances in a reaction-diffusion system; interacting biological species; neurons in a network) of which the study of neural fields is but one case. We thus offer the brain theorist a variety of general mathematical methods, and a number of non-neural systems for comparison. Note, however, that these models do not usually assess the effect of spatiotemporal variations in input, variations which are crucial in any study of the way in which the brain monitors changing sensory stimulation to control behavior.

In Chapter 2, "Sigmoidal Systems and Layer Analysis," Fife shows how reaction-diffusion equations exhibit patterns and waves suggestive of neural net phenomena, and presents the theory of such waves, which is moderately well-developed, as a potentially fruitful source of methods for the mathematical analysis of neural models. (Warning: the words 'layer analysis' do not denote the study of interacting layers, as in brain theory, but refer to the technique of expanded space scale as used to study boundary layer behavior in hydrodynamics.) Special attention is given to the way in which travelling wave solutions may develop over time from various initial conditions. In Chapter 3 on "Asymptotic Behavior of Stationary Homogeneous Neuronal Nets," Ermentrout applies related techniques to the study of small-amplitude phenomena in the asymptotic (i.e. long-term) behavior of neuronal nets with spatially and temporally uniform input. He starts with a spatially uniform solution and applies perturbation and bifurcation theory to obtain non-trivial spatio-temporally patterned solutions. The theory is applied to the study of epilepsy and flicker phosphenes. Mimura's Chapter 4,

"Aggregation and Segregation Phenomena in Reaction-Diffusion Equations," also analyzes stable spatially inhomogeneous solutions, but this time the biological motivation is provided by the aggregation and segregation of the individuals of competing species in population biology.


## III. Single-Neuron and Stochastic Models

The papers in Section III focus on the behavior of single neurons. As such, they fall outside the focal theme of "competition and cooperation." Nonetheless, they are important in helping us assess the propriety of the simplifications involved in specifying the elements of the large neural networks that preoccupy us in most of this volume. The papers in this section may be seen as falling into four groups: those by Scott and by Koch, Poggio and Torre discuss the interaction of electrical events within axonal and dendritic, respectively, "micronetworks." Sato reviews the role and use of noise in biological systems. Holden and Ricciardi use stochastic processes to model the behavior of the single neuron. Lastly, both Yoshizawa and Yamaguti and Hata offer a mathematical account of a phenomenon found by L.D. Harmon in experimental studies with his transistor model of a neuron.

Scott, in Chapter 5, studies "Nerve Pulse Interactions." Three modes of interaction are considered: Interaction of pulses as they travel along a single fiber, resulting in velocity dispersion; propagation of pairs of pulses through a branching region, leading to quantum pulse code transformations, and interaction of pulses on parallel fibers through which they may form a "pulse assembly"--analogous to Hebb's "cell assembly," but on a lower level of the neural hierarchy. In Chapter 6, "Micronetworks in Nerve Cells," Koch, Poggio and Torre analyze the electrical properties of cat retinal ganglion cells using passive cable theory to conclude that these neurons need not be equipotential, mainly because of their extensive branching. It is suggested that the different dendritic structures of the various cell types yield different electrical properties, and that these are relevant for their physiological function.

S. Sato's Chapter 7, "Role and Use of Noise in Biological Systems," suggests that the destabilizing effect of noise in systems described by nonlinear differential equations may play an essential role in making such a system switch from one stable region of behavior to another. He then discusses applications of Wiener's analysis of nonlinear systems to the use of noise in studying biological systems.

Holden's "Stochastic, Quantal Membrane Conductances and Neuronal Function," Chapter 8, is centered on the stochastic kinetics of a number of models of how ions cross membranes. Holden argues that there is little direct evidence for cooperativity within the neuron; rather, he suggests that the behavior of the neuron results from nonlinear membrane properties that are produced by systems of stochastic quantal channels. However, inter-channel cooperativity can be used to explain a major component of membrane noise, namely flicker noise. Where Holden focusses on membrane noise, Ricciardi's Chapter 9, "Diffusion Approximations and Computational Problems for Single Neuron's Activity," obtains diffusion equations to represent the stochastic behavior of a neuron's membrane potential by analyzing the random effects of the very many excitatory and inhibitory inputs (posited to be uncorrelated, as a simplifying assumption) which the neuron receives from the large network in which it is embedded. Consideration is then given to computing stochastic descriptions of neural firing by solving the first crossing problem for the resultant diffusion process.

In experiments on an analog neuron model built from transistors, Harmon in 1961 demonstrated a very complicated step-function relation between the amplitude of the input pulse and the firing rate of the neuron model. In Chapter 10, "Periodic Pulse Sequences Generated by an Analog Neuron Model," Yoshizawa presents a neuron model, due to Nagumo et al., based on a tunnel-diode, and shows that the periodic output sequences belong to a special class of periodic sequences generated by a simple algorithm. In a special case, the relation between the firing rate and the strength of the input (pulse width or pulse height) becomes an extended Cantor function. In Chapter 11, "On a Mathematical Neuron Model," Yamaguti and Hata summarize (proofs omitted) the theorems of a related mathematical study, and show that there exists a Cantor attractor if and only if the average firing rate is irrational.


## IV. Oscillations in Neural Networks

Oscillations form an important class of dynamic behaviors of neural networks, as in the control of respiration, heartbeat, and locomotion. R. Suzuki, S. Majima and H. Tatumi devote Chapter 12 to the study of "Control of Distributed Neural Oscillators." In particular, they use the Wilson-Cowan model of neural oscillators to explain some, but not all, the phase entrainment effects they see in the coordinated rhythmical behavior of crayfish swimmerets. In Chapter 13, "Characteristics of Neural Network with Uniform Structure," Noguchi and Araki depart from the continuous-time neural models of most of this volume to give a graph-theoretic analysis of the cycle modes of a fully-synchronized net of formal discrete-time neurons.

## V. Development and Plasticity of Visual Systems

The papers of Section V fall naturally into three groups: the first three offer specific models tuned to neurophysiological data on the development of visual systems; the next three offer formal models of the self-organization of pattern-recognition networks; while Grossberg's paper goes well beyond the visual system to offer an overview of his general theory, which embraces developmental models, network adaptation, and psychological learning theory.

One of the most notable properties of many parts of the visual system is retinotopy--the presence of layers of neurons in two-dimensional (but not metric-preserving) correspondence with the visual axes of the retina. A fruitful experimental locus for the study of such topographic mappings has been the projection from retina to tectum in such lower vertebrates as goldfish, frog and toad. In Chapter 14, Overton and Arbib offer "Systems Matching and Topographic Maps: The Branch-Arrow Model (BAM)," which builds on earlier models to show how competition of branches of retinal fibres for space on the tectum may account for some but not all of the relevant experimental data. With Chapter 15, "Differential Localization of Plastic Synapses in the Visual Cortex of the Young Kitten: Evidence for Guided Development of the Visual Cortical Networks" by Toyama and Komatsu, we turn from retino-tectal connections in fish and amphibia to geniculo-cortical connections in the kitten. These authors present experimental evidence for differential plasticity in synapses in kitten visual cortex, with plastic synapses mostly confined to layers I and IV. They speculate that this differential location may provide the basis for guided organization of neural networks. Sawada and Sugie's Chapter 16, "Self-Organization of Neural Nets with Competitive and Cooperative Interaction" also studies geniculo-cortical connections in the kitten, offering a computer simulation of a model which can acquire binocular stereopsis through self-organization in response to presentation of a sequence of visual stimuli.

In Chapter 17, "A Simple Paradigm for the Self-Organized Formation of Structured Feature Maps," Kohonen offers a simple principle of self-organization which can generate ordered feature maps that are directly or indirectly related to sensory or somatic information. The paradigm is illustrated by computer simulation of the organization of a network in response to inputs encoding phonemes from the Finnish language. The model has the important "magnification property" that the more frequent is the presentation of a pattern during the training phase, the more widespread are units responsive to the pattern. "The Neocognitron: A Self-Organizing Neural Network Model for a Mechanism of Visual Pattern Recognition" is presented by Fukushima and Miyake in Chapter 18. Self-organization of the network progresses by "learning without a teacher," so that patterns are clustered together into similar classes on the basis of shape, without an external input specifying the category to which an input pattern belongs. The network is layered, and the feature extraction from the input pattern is performed step by step in each layer of the network: Local features are initially extracted, and these features are combined into more global features. In each stage of feature extraction, a small amount of positional error is tolerated. Thus the cells of the deepest layer of the network are relatively resistant to shift in position and deformation in shape of the input pattern. Harth, in Chapter 19, offers two general principles "On the Spontaneous Emergence of Neuronal Schemata": One is the flow of information from the micro- to the macroscale; the other is a stochastic feedback process, which he calls Alopex, capable of generating patterns under the guidance of a scalar evaluation signal. Harth has applied Alopex to the computer determination of visual receptor fields, but he speculates that an analogous process in the brain may be involved in perception and "higher nervous activity."

Grossberg devotes Chapter 20, "Associative and Competitive Principles of Learning and Development," to a survey of principles, mechanisms and theorems from his work over the past 20 years. The key questions are: how are short-term memory (STM) and long-term memory (LTM) patterns formed, how do they unfold over time, and how are they stabilized? Grossberg provides his answers in terms of mathematically-defined networks which may be interpreted at either the neural or the psychological level.


## VI. Sensori-Motor Transformations and Learning

The final five papers return us to neural models developed in close interaction with experimentation, and bring us to a question which has been all too seldom addressed in the previous papers: How does sensory input affect behavior? The first two papers offer models of visuo-motor transformations in the amphibian brain and the third paper models transformations performed by the cerebellum. The last two suggest mechanisms for motor learning, in the cerebellum and in the brainstem respectively.

Chapter 21, by Arbib, provides a general perspective on neural modelling, and an introduction to the 'style of the brain' in models of prey-selection, stereopsis, distributed motor control, and cerebellar parameter setting. This is then followed by an exposition of three 'evolutionary stages' in "Modelling Neural Mechanisms of Visuomotor Coordination in Frog and Toad". This serves as background for Lara, Cervantes and Arbib's "Two-Dimensional Model of Retinal-Tectal-Pretectal Interactions for the Control of Prey-Predator Recognition and Size Preference in Amphibia" in Chapter 22. The tectum is modelled as an array of columns each comprising a glomerulus and four types of neurons, receiving retinal input, and exciting the pretectum from which it in turn receives inhibition. The resultant competition and cooperation accounts, e.g. for the toad's increasing preference for worm-shaped stimuli as their length increases, and also accounts for a number of other behavioral data in a manner consistent with known anatomy and physiology.

In Chapter 23, "Tensor Theory of Brain Function: The Cerebellum as a Space-Time Metric," Pellionisz and Llinas argue that the cerebellum embodies a metric tensor for transforming covariant vectors representing motor intention to contravariant vectors representing motor execution. They cite data that some Purkinje cells act like differentiators to argue that the cerebellum embodies "lookahead" modules based on a truncated Taylor series to compensate for delays within the CNS. Here we enter an area of current controversy, for where Pellionisz and Llinas see little if any plasticity in the cerebellum, Ito argues, in Chapter 24 on "Mechanisms of Motor Learning," for the Marr-Albus model of the self-organization of the cerebellar cortical network through synaptic plasticity. He supports the applicability of this model to the floccular control of the vestibulo-ocular reflex by neuronal circuit analysis, lesion experiments and unit recording from Purkinje cells. However, he notes that the plasticity of parallel fibre-Purkinje cell synapses has not yet been shown stable enough for permanent memory. Finally, in Chapter 25 on "Dynamic and Plastic Properties of the Brain Stem Neuronal Networks as the Possible Neuronal Basis of Learning and Memory," Tsukahara and Kawato suggest that motor learning may not rest on plasticity of the cerebellar cortex but rather on plasticity of the red nucleus, for Tsukahara has demonstrated that red nucleus exhibits a form of plasticity called "sprouting," forming new synaptic connections. They argue that the "internal model" necessary for motor control can be realized by the loop involving cerebral cortex, red nucleus, inferior olive and cerebellum. They predict that if this loop is destroyed, the internal model would be destroyed and motor performance disturbed, and they note that this prediction is in accord with the "tremor" frequently reported after lesions anywhere in the loop.

Tokyo, February 26, 1982.                Shun-Ichi Amari

Michael A. Arbib

# TABLE OF CONTENTS

U.S., Mexican & European Participants

in the U.S.-Japan Joint Seminar on

COMPETITION AND COOPERATION IN NEURAL NETS

February 15-19, 1982
Kyoto, Japan

Michael A. Arbib  (Co-Organizer)
Center for Systems Neuroscience,
Department of Computer and
 Information Science
University of Massachusetts
Amherst, MA 01003, USA

G. Bard Ermentrout
Mathematical Research Branch
NIADDK, Bldg. 31, Rm. 4B-54
National Institutes of Health
Bethesda, MD 20205, USA

Paul C. Fife
Department of Mathematics
University of Arizona
Tucson, AZ 85721, USA

Stephen Grossberg
Department of Mathematics
Boston University
264 Bay State Road
Boston, MA 02215, USA

Erich Harth
Department of Physics
Syracuse University
Syracuse, NY 13210, USA

Arun V. Holden
Department of Physiology
University of Leeds
Leeds LS2 9NQ, ENGLAND

Teuvo Kohonen
Department of Technical Physics
Helsinki University of Technology
SF-02150 Espoo, FINLAND

Rolando Lara
Centro de Investigaciones en
 Fisiología Celular
Universidad Nacional Autónoma de
 México
Apartado postal 70-600
04510 México, D.F., MÉXICO

Andras Pellionisz
Department of Physiology
New York University Medical Center
550 First Avenue
New York, NY 10016, USA

Luigi Ricciardi
Instituto di Matematica
 dell 'Universita'
Napoli, ITALY

Alwyn C. Scott
Los Alamos National Laboratory
Center for Nonlinear Systems
Mail Stop B258
Los Alamos, NM 87545, USA

Japanese Participants

in the U.S.-Japan Seminar on

COMPETITION AND COOPERATION IN NEURAL NETS

February 15-19, 1982
Kyoto, Japan

Shun-ichi Amari  (Co-Organizer)
Department of Mathematical
 Engineering
University of Tokyo
Tokyo, 113 JAPAN

Kunihiko Fukushima
NHK Broadcasting Science
Research Laboratory
Kinuta, Setagaya
Tokyo, 157 JAPAN

Masao Ito
Department of Physiology
University of Tokyo
Tokyo, 113 JAPAN

Masayasu Mimura
Department of Mathematics
Hiroshima University
Nakaku, Hiroshima
730 JAPAN

Shoichi Noguchi
Electrocommunication Laboratory
Tohoku University
Sendai, 980 JAPAN

Shunsuke Sato
Department of Biological Engineering
Osaka University
Toyonaka, 560 JAPAN

Noboru Sugie
Department of Information Processing
Nagoya University
Nagoya, 464 JAPAN

Ryoji Suzuki
Department of Biophysical
 Engineering
Osaka University
Toyonaka, 560 JAPAN

Keisuke Toyama
Department of Physiology
Kyoto Prefecture College of Medicine
Sakyoku, Kyoto, 606 JAPAN

Nakaakira Tsukahara
Department of Biophysical Engineering
Osaka University
Toyonaka, 560 JAPAN

Masaya Yamaguti
Department of Mathematics
Kyoto University
Kyoto, 606 JAPAN

Shuji Yoshizawa
Department of Mathematical
 Engineering
University of Tokyo
Tokyo, 113 JAPAN

## Observers

Masahiko Fujita
Nagasaki Institute of Applied
 Science
Aba-cho, Nagasaki-shi
851-01 JAPAN

James Greenberg
Applied Mathematics Program
National Science Foundation
Washington, DC 20550, USA

Yuzo Hirai
Division of Electronics and
 Information
Tsukuba University
Niibari-gun
Ibaraki, 305 JAPAN

Mitsuo Kawato
Department of Biophysical
 Engineering
Osaka University
Toyonaka, 560 JAPAN

Kenjiro Maginu
Department of Mathematical
 Engineering
University of Tokyo
Tokyo, 113 JAPAN

Hiroshi Matano
Department of Mathematics
Faculty of Science
University of Tokyo
Tokyo, 113 JAPAN

Sei Miyake
NHK Broadcasting Science Research
 Laboratory
Kinuta, Setagaya
Tokyo, 157 JAPAN

Yasushi Miyashita
Department of Physiology
University of Tokyo
Tokyo, 113 JAPAN

Toshi Nagano
Electrotechnical Laboratory
Niibari-gun
Ibaraki, 305 JAPAN

Kazuhisa Niki
Electrotechnical Laboratory
Niibari-gun
Ibaraki, 305 JAPAN

Hidenosuke Nishio
Department of Biophysics
Kyoto University
Kyoto, 606 JAPAN

Akio Tanaka
Fujitsu International Institute
 of Social Information Science
Numazu, 410-03 JAPAN

Ei Teramoto
Department of Biophysics
Kyoto University
Kyoto, 606 JAPAN

Toyoshi Torioka
College of Engineering
Yamaguchi University
Ube, 755 JAPAN

Shigehiro Ushiki
Department of Mathematics
Faculty of Science
Kyoto University
Kyoto, 606 JAPAN

# CHAPTER 1

# COMPETITIVE AND COOPERATIVE ASPECTS IN DYNAMICS OF NEURAL EXCITATION AND SELF-ORGANIZATION

Shun-ichi Amari

Mathematical Engineering, University of Tokyo
Tokyo , 113 Japan

**Abstract**. The nervous system seems to process information in parallel through the competitive and cooperative interactions of neurons. Dynamical processes of competition and cooperation are formulated by the use of the dynamics of excitations in neuron networks. Dynamical behaviors of neuron pools and neural fields are analyzed from this point of view. It is shown that self-organization of neural systems can also be represented by a dynamics of competition and cooperation in an abstract field of a signal space. A field theory of self-organization is proposed and the behaviors of self-organizing nerve nets are analyzed from this unified point of view. The formation of topographic structures is elucidated by this method.

## 1. Introduction

The brain is indeed a hierarchical system which processes information in a parallel manner. Although modern neurophysiology has clarified the functioning of a single neuron with the aid of the microelectrode techniques, little has been known about the functioning of neuron networks, in which parallel information processing takes place. In order to elucidate possible neural mechanisms of parallel information processing, a number of neural models have been proposed and analyzed mathematically. It has gradually been recognized through these researches that the neural system processes information in parallel through the dynamics of competitive and cooperative interactions of neurons (Montalvo, 1975).

Amari and Arbib (1977) proposed a primitive neural model of competition and cooperation to explain a neural mechanism of parallel information processing. They not only analyzed the dynamical behavior of the model, but also explained the mechanism of bug detection in the frog and that of stereoscopic depth perception by using versions of the primitive model. Their model of streopsis perception consists of a neural field in which disparities of the signals from both eyes are fused through the dynamics of competition and cooperation of neurons. This model is similar to those independently proposed by Marr and

Poggio (1976) and Sugie and Sawa (1977). It is mathematically tractable and the analysis shows that the model accounts for the hysteresis characteristics of perception and the behavior under ambiguous inputs.

It has been recognized through these theoretical researches that the idea of competition and cooperation provides not only mathematical method for analyzing parallel information processing in neuron net models but also a unified principle in modeling various aspects of neural information processing. The present paper elucidates the competitive and cooperative interactions of neurons existing in various levels of neural systems, by developing further the idea of Amari and Arbib (1977). This shows that a wide range of dynamical behaviors of neural systems can be understood from the unified point of view of competition and cooperation.

We summarize in §2 the dynamics of neuron pools (Amari, 1971, 1972, 1974; Anninos et al. 1970; Wilson and Cowan, 1972) from the cooperative point of view, and give a primitive neural model of competition and cooperation. The dynamics of neural fields (Wilson and Cowan, 1973; Amari, 1977a; Ellias and Grossberg, 1975; Ermentrout and Cowan, 1979a, b, 1980; see also Grossberg, 1982; Ermentrout, 1982) is treated in §3 from the competition and cooperation point of view, since neural fields can in general be regarded as continuum versions of the above primitive model. Finally, we study dynamics of self-organization of neural systems (cf. Amari, 1977b; Amari and Takeuchi, 1978; Amari, 1980; Takeuchi and Amari, 1979; Grossberg, 1976; Grossberg, 1982; Sugie, 1982; Fukushima, 1982). It will be proved that the process of self-organization can be represented by a competitive and cooperative dynamics of excitations on an abstract field of a signal space. We thus propose a field theory of self-organization. This shows that competition and cooperation play also a fundamental role even in self-organization.

Finally, we apply the field theory to the formation of topographic structures in the nerve field (Amari, 1980; Kohonen, 1982). We have in this case a dynamics of competition and cooperation type on the product space of a signal space and a neuron field. We have a uniform continuous solution of a topographic map. However, it becomes unstable under a certain condition, and a spatial pattern appears. It is hence suggested that some microstructures are formed in the topographic map. They are compared with the columnar microstructures, microzones, barrel structures, etc. (Amari, 1980; Takeuchi and Amari, 1979).

## 2. Primitive Aspects of Competition and Cooperation

### 2.1. Cooperation in neuron pool

Let us consider a simple neuron model which receives n input signals $x_1$, $x_2$, ..., $x_n$ and emits an output signal z. We represent the state of the neuron at time t by a quantity $u(t)$. It increases in proportion to the total stimulus or the weighted sum s of the inputs

$$s = \sum_{i=1}^{n} w_i x_i \qquad (2.1)$$

where $w_1$, $w_2$, ..., $w_n$ are the synaptic efficiencies or weights of these inputs $x_i$, and decays at the same time with a time constant $\tau$ toward -h. Then, we have the equation

$$\tau \dot{u} = - u + s - h \ , \qquad (2.2)$$

where $\cdot$ denotes the time derivative $\dot{u} = du/dt$ . The quantity u may be compared to the short-time average of the membrane potential of the neuron and -h to the resting potential. Input and output quantities $x_i$ and z are analog quantities representing the pulse frequencies. The output pulse rate $z(t)$ at time t is determined as a function of the potential $u(t)$ as

$$z = f(u) \ , \qquad (2.3)$$

where f is called the output function. It is a monotonically non-decreasing function satisfying

$$f(u) = 0 \qquad (u \leq 0) \ , \qquad 0 \leq f(u) \leq 1 \ .$$

When the input signals $x_i$ are kept constant for a while (or their changes are sufficiently slow compared with the time constant $\tau$), the potential u quickly converges to

$$u = \sum w_i x_i - h = s - h \qquad (2.4)$$

so that the input-output relation is given by

$$z = f(s - h) \ . \qquad (2.5)$$

Hence, the output z is a monotone function of the total input stimulus s.

Let us consider a pool of n mutually connected neurons. The behavior of the neuron pool is quite different from the mere average of the behaviors of the component neurons, because of cooperation in the pool. Let $s_i$ be the total stimulus applied to the i-th neuron from the outside, and let $u_i$ be its potential. When the output of the j-th neuron $f(u_j)$ enters the i-th neuron with the synaptic weight $w_{ij}$, the dynamical behavior of the i-th neuron is described by

$$\tau \dot{u}_i = - u_i + \sum_j w_{ij} f(u_j) + s_i - h. \qquad (2.6)$$

We regard the average of the total stimuli $s = \sum s_i/n$ as the overall input to the neuron pool, and the average $z = \sum z_i/n$ as the overall output from the neuron pool. The overall behavior of the neuron pool is shown by the relation between s and z. When the synaptic weights $w_{ij}$ are identical and independently distributed random variables, by taking the average over all the neurons, we have the equation for the average $u = \sum u_i/n$ of the potentials $u_i$. If we can assume the relation

$$z = f_p(u)$$

between the average u and z, the dynamical behavior of the neuron pool is described by the equation

$$\tau \dot{u} = - u + w f_p(u) + s - h , \qquad (2.8)$$

where $f_p$ is the output function of the pool and w is the average of all $w_{ij}$.

We encounter a difficult but interesting theoretical problem in deriving the above macroscopic equation rigorously. The same difficulty is involved in the Boltzmann equation in statistical mechanics. Statistical neurodynamics studies such problems, which we do not treat here. (See, e.g., Amari, 1974, Amari et al. 1977, Geman, 1980.)

When the input s is kept constant, the average potential u converges to the equilibrium state of the equation (2.8). However, this equation is in general multi-stable, having a number of equilibria, so that u converges to one of them. It has been shown by Amari (1971) (see also Amari, 1972, Holden, 1976) that the dynamics of the above neuron pool is monostable or bistable depending on w and s. There are two kinds of dynamical behaviors of the neuron pool,

depending on s and w. When w is larger than a certain value $w_0$ and s
is limited in a certain range determined by w, the dynamics is
bistable, having two stable equilibria $\bar{u} = u_1$ and $\bar{u} = u_2$ and one
unstable equilibrium $\bar{u} = u_0$. Otherwise, the dynamics is monostable
with one stable equilibrium $\bar{u}$. This situation can clearly be shown by
the type of the catastrophe of the dynamics. It is given by

$$\bar{u} = \bar{u}(s, w),$$

denoting the equilibria of the dynamics as a (multi-valued) function
of s and w. When the pool is monostable,
$\bar{u}$ is unique, while when the pool is
bistable, $\bar{u}$ takes three values $u_1$, $u_2$ and
$u_0$. Fig.1 shows the catastrophe of the
dynamics (2.8), showing $\bar{u}(s, w)$, where
the hatched part is the range of (s, w)
in which the dynamics is bistable.

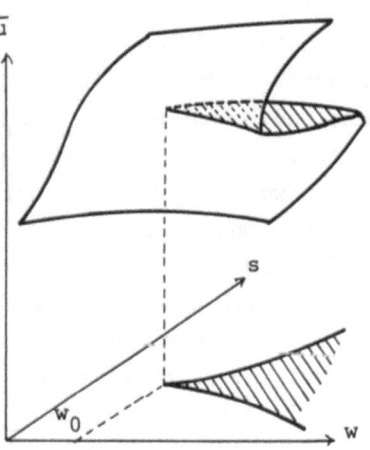

Fig.1. Catastrophe $\bar{u}(s, w)$

The equilibrium $\bar{u}$ is uniquely
determined as a function of the input
stimulus s, $\bar{u} = \bar{u}(s)$, in a monostable
neuron pool whose average mutual
connection w is smaller than a certain
value $w_0$. The relation $\bar{u} = \bar{u}(s)$ is
obtained by cutting Fig.1 with a plane w
= const., as is shown in Fig.2. The
equilibrium $\bar{u}$ is a continuous monotone
function of s, and so is the output z =
$f_p(\bar{u})$. Hence, the behavior of this pool
seems to be similar to that of a
component neuron. However, the
convergence rate is different. By
putting $u(t) = \bar{u} + \delta u(t)$, we have the
linearized variational equation of (2.8),

$$\tau(\delta u) = -\delta u + f_p'(\bar{u})\delta u$$

which represents the dynamical behavior
of the deviation $\delta u$ around the
equilibrium $\bar{u}$. Hence, the time constant
of $\delta u$ is

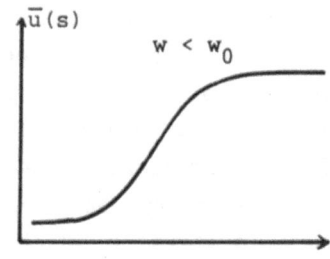

Fig.2. Monstable net

$$\tau' = \tau/[1 - f_p'(\bar{u})] \; ,$$

which is quite different from $\tau$.

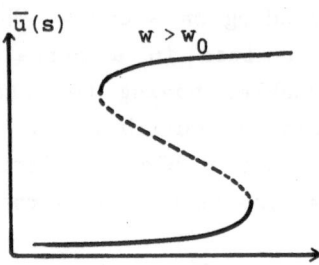

Fig.3. Bistable net

When w is larger than $w_0$, the equilibrium $\bar{u}$ is not necessarily uniquely determined by s. The relation between $\bar{u}$ and s has a hysteresis characteristic as is shown in Fig.3, which is also obtained by cutting Fig.1 with a plane w = const. > $w_0$ (Amari, 1971, 1972). In this case, the average potential $\bar{u}$ is either in the excited state $u_1$ or in the resting state $u_2$ in the equilibrium and the intermediate state $u_0$ is unstable. Hence, the pool behaves like a switch with hysteresis controlled by s. The output $z = f_p(u_1)$ is nearly equal to 1 in the excited state and $z = f_p(u_2)$ is nearly equal to 0 in the resting state.

Two typical behaviors of simple neuron pools have thus been shown. One is an analog signal processor whose time constant is $\tau'$ different from that of component neurons. The other behaves like a switch with a hysteresis characteristic. We can analyze the behaviors of more complex systems in which various neuron pools are connected (Amari, 1971, Harth et al. 1970). They are in general multi-stable or oscillatory. It has been shown (Amari 1971, Wilson and Cowan, 1972) that a system of mutually coupled excitatory and inhibitory neuron pools behaves as a neural oscillator, and its characteristics were studied in detail (Amari, 1971, 1972). Another type of neural oscillators is obtained by connecting more than three neuron pools in a ring form, where the connections are unilateral. When an odd number of inhibitory neuron pools are included in it, the system can behave as an oscillator.

These results demonstrate primitive aspects of cooperation in neuron pools. Neuron pools indeed behave quite differently from those of the components elements. They are elementary building-blocks of neural systems in which competition and cooperation take place in a more elaborate manner.

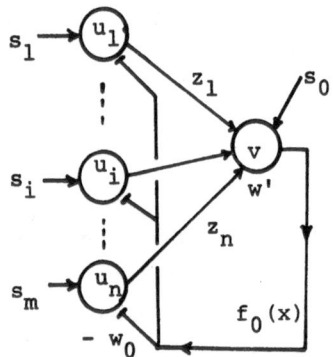

Fig.4. Primitive competition

## 2.2. Primitive model of competition and cooperation

Let us consider a simple competitive and cooperative system composed of m excitatory neuron pools $E_i$ $(i = 1,\ldots, m)$ and one inhibitory neuron pool I. Each excitatory neuron pool $E_i$ receives input of average intensity $s_i$ from the outside. It emits output $z_i$. These outputs $z_i$ excite the inhibitory neuron pool I and its outputs in turn inhibit the excitatory neurons (Fig. 4). This is similar to the primitive competition model proposed by Amari and Arbib (1977).

Let $u_i$ and $v$ be the average potentials of $E_i$ and I, respectively. Then, the dynamical equation of the system is written as

$$\tau \dot{u}_i = - u_i + w f_p(u_i) - f_0(v) + s_i - h \tag{2.9}$$

$$\tau \dot{v} = - v - w_0 f_0(v) + w' \sum_i f_p(u_i) + s_0 - h_0 \tag{2.10}$$

where $z_i = f_p(u_i)$ and $f_0(v)$ are the average outputs of $E_i$ and I, w and $-w_0$ are the average weights of connections within $E_i$ and I, respectively, and $w'$ is the average weight of connections from $E_i$ to I. The activity of I is directly controlled by input $s_0$ to I. When w is large, each $E_i$ is bistable. Once it is excited, it can retain the excitation even after the stimulus is removed, because of cooperation of the neurons inside it. On the other hand, an excitation of $E_i$ brings about inhibition to all the $E_j$'s through the common inhibitory neuron pool I, so that $E_i$'s compete with one another. A typical process of competition and cooperation takes place in the model. We show its behavior, by simplifying the equations without destroying the intrinsic structures.

Since $E_i$ is bistable, it becomes eventually either in the excited state $z_i = 1$ or in the resting state $z_i = 0$, approximately. Hence, we may replace the output function $f_p$ by the step function $1(u)$,

$$1(u) = \begin{cases} 1, & u > 0 , \\ 0, & u \leq 0 . \end{cases}$$

On the other hand, the inhibitory pool I is monostable, so that its output is a continuous function of the input $z_i + s_0$. Hence, we may approximate $f_0(v)$ by the semi-linear function $v^+$,

$$v^+ = \begin{cases} v, & v > 0 , \\ 0, & v \leq 0 . \end{cases}$$

We further assume $w_0 = 0$. The system then reduces to the Amari-Arbib primitive competition model, whose behaviors have been studied in detail (Amari and Arbib, 1977). Their results can be generalized as follows.

When the system receives m inputs $s_1$, $s_2$, ..., $s_m$ ($s_i$ to $E_i$), the neurons cooperate in each $E_i$ to get excited and retain the aroused excitation. On the other hand all the $E_i$ compete with each other for excitations, so that only a limited number of $E_i$'s can remain in the excited state. When the intensities of the stimuli are limited to $0 < s_i < 1$, the neuron pool $E_i$ is excited when input $s_i$ is larger than h. However, because of the competition through the inhibitory pool I, only a limited number of the neuron pools can retain the excitations in the equilibrium. The maximum number k of excited neuron pools is limited to be smaller than $(w + 1 + h_0 - h - s_0)/w'$. Hence, the system selects the k largest stimuli among m stimuli $s_1, ..., s_m$ and retain persistently the k excitations. Even after the stimuli are removed, the system can retain at most k' excitations, where k' is the integar part of $(w + h_0 - h - s_0)/w'$. Hence, the system behaves as a detector of the k largest stimuli, and remembers the positions of the k' largest stimuli even after the stimuli are removed. The numbers k and k' can be controlled by varying the intensity $s_0$ of the direct input to I. When $k = k' = 1$, the system chooses the neuron pool which receives the largest stimulus and keeps that neuron pool excited.

This gives a prototype of competitive and cooperative information processing in neural systems. We shall see versions of the behaviors of this primitive model not only in the dynamics of excitations in neural fields but also in self-organization of neural systems.

## 3. Pattern Formation in Neural Fields

### 3.1. Dynamics in neural fields of lateral inhibition

Let us consider a neural field consisting of one excitatory layer and one inhibitory layer (Fig.5). Let u(y, t) and v(y, t) be, respectively, the average potentials of the neurons at around the spatial position y at time t of the excitatory layer and

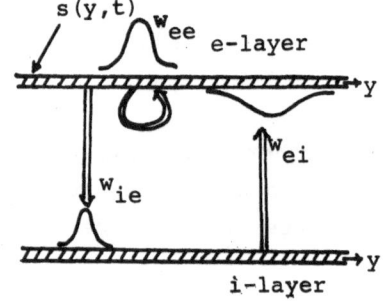

Fig.5. Neural field

inhibitory layer. Let $w_{ee}(y, y')$ be the average weight of the connections from the neurons at position $y'$ to the neurons at position $y$ within the e-layer (excitatory layer). Similary, we denote by $w_{ie}(y, y')$ the average weight from position $y'$ of the e-layer to position $y$ of the i-layer (inhibitory layer). We also denote similarly by $-w_{ei}(y, y')$ and $w_{ii}(y, y')$ the average inhibitory weights from the i-layer.

The dynamical equation of this field can be written as

$$\tau \partial u(y, t)/\partial t = - u(y, t) + w_{ee}*f_e(u)$$

$$- w_{ei}*f_i(v) + s(y, t) - h_e , \qquad (3.1)$$

$$\tau \partial v(y, t)/\partial t = - v(y, t) + w_{ie}*f_e(u)$$

$$- w_{ii}*f_i(v) - h_i , \qquad (3.2)$$

where * denotes the convolution such as

$$w_{ee}*f_e(u) = \int w_{ee}(y, y')f_e[u(y', t)]dy' , \qquad (3.3)$$

$s(y, t)$ is the average intensity of stimuli entering directly to the neurons at position $y$ at time $t$ from the outside, and $h_e$ and $h_i$ are constants.

In order to analyze the competitive and cooperative aspects in neural fields, the following mathematical simplifications are assumed. 1) The field is one-dimensional, 2) homogeneous, and 3) isotropic (symmetric). 4) The spread of the inhibitory connections $w_{ei}$ is wider than the spread of the excitatory connections $w_{ee}$. 5) There are no self-inhibitory connections in the i-layer, i.e., $w_{ii} = 0$. 6) The spread of the connections from the e-layer to the i-layer is so narrow that we can approximate $w_{ie}(y, y')$ by the delta-function $\delta(y - y')$.

The effective connections from the neurons at $y'$ to those at $y$ are given by the sum of $w_{ee}(y, y')$ and $-w_{ei}(y, y')$,

$$w(y - y') = w_{ee}(y, y') - w_{ei}(y, y') , \qquad (3.4)$$

which depends only on the distance $|y - y'|$ of the two points. This type of connections is called lateral inhibition, where a neuron excites its neighboring neurons directly and inhibits distant neurons via the inhibitory neurons. Hence, neighboring neurons cooperate to retain common excitations and distant neurons compete with one another for excitations. Hence, the neurons in the e-layer are either in the excited state or in the resting state in the equilibrium. This enables us to approximate the output functions $f_e(u)$ and $f_i(v)$ by the step function $1(u)$.

Under the above simplifications, the dynamical equations of the field reduce to

$$\tau \partial u / \partial t = - u + w_{ee} * 1(u) - w_{ei} * 1(v) + s - h_e , \qquad (3.5)$$

$$\tau \partial v / \partial t = - v + 1(u) - h_i . \qquad (3.6)$$

The neural field of this type is regarded as a continuous version of the primitive competition model. The e-layer is indeed a continuum version of the primitive model in which neighboring $E_i$'s are connected with positive weights. The inhibitory neuron pool I is expanded to form the i-layer in the present field model.

We look for the excitation patterns which the field can retain in the equilibrium under the constant distribution

$$s(y, t) = s = const.$$

of the input stimuli. We call this uniform s the basic stimulus level of the field. We show that non-homogeneous excitation patterns are formed in a uniform field. The neural field seems to have a richer ability of pattern formation than the reaction-diffusion system has, waiting for the further analysis.

3.2.  Existence of local excitation pattern

An excitation pattern $u(y)$ of the field is called a local excitation pattern of length b, when the set of the positions of the excited neurons is an interval of length b, i.e., when $1[u(y)]$ is equal to 1 in the interval and 0 outside it. Since the field is homogeneous, when $u(y)$ is an equilibrium solution of (3.5), (3.6), so is its parallel shift $u(y + y_1)$ for any $y_1$. Hence, we search for the condition that the local excitation $u(y)$ of length b

$$1[u] = \begin{cases} 1 , & \text{for} \quad 0 < y < b , \\ 0 , & \text{otherwise} , \end{cases}$$

exists as a stable equilibrium solution of (3.5) and (3.6).

When (u, v) is an equilibrium solution, they satisfy $\partial u/\partial t = 0$, $\partial v/\partial t = 0$ in (3.5) and (3.6). Hence, we have

$$v(y) = 1[u(y)] - h_i$$

or

$$1[u(y)] = 1[v(y)] ,$$

where $0 < h_i < 1$ is assumed. Therefore, the equilibrium solution u(y) satisfies

$$u(y) = w*1(u) + s - h_e$$

or

$$u(y) = \int_0^b w(y - y')dy' + s - h_e$$

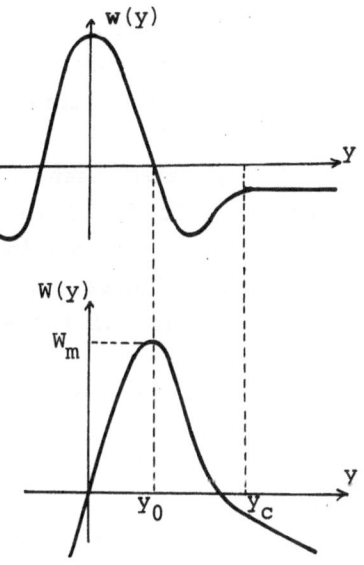

where $w = w_{ee} - w_{ei}$. From the relation $u(0) = 0$ and $u(b) = 0$, we easily have the condition

$$W(b) + s - h_e = 0 , \qquad (3.7)$$

where

Fig. 6. w(y) and W(y)

$$W(y) = \int_0^x w(z)dz . \qquad (3.8)$$

We show the graphs of W(y) and w(y) in Fig.6. The curve W(y) is monotonically increasing for $y > 0$ until $y = y_0$ where $w(y_0) = 0$ and W(y) attains its maximum $W(y_0) = W_m$ at $y_0$. Then, it decreases monotonically. It becomes constant $w(y) = -w_0$ for $y > y_c$. Hence, the equation (3.7) has no solution, when the basic stimulus level s is smaller than $-W_m + h_e$. In this case, the field cannot retain any local excitation patterns. However, when the level s becomes larger than $-W_m + h_e$

$$h_e > s > -W_m + h_e ,$$

there are two solutions $b_1$ and $b_2$ ($b_1 < b_2$) in (3.7). We can show that the local excitation of length $b_1$ is unstable, but that of length $b_2$ is stable. Hence, the field can retain persistently a local excitation of length $b_2$ at any position in this case. When s becomes larger, spontaneous excitation patterns are aroused. (See Amari, 1977a for details).

When distributed input stimuli $s(y)$ arrive at the field, excitations occur at various positions. However, after the input stimuli are removed, all the excitations will be eliminated except the largest local excitation, if the field can retain only one local excitation pattern. Hence, this field behaves as a detector of the position of the maximum stimulus and keep its position. Amari and Arbib (1977) used this model to explain the mechanism of the bug detection in the tectum of the frog.

It has also been shown that, when an input distribution $s(y)$ arrives to the field in which a local excitation is retained, the local excitation moves in the direction in which the intensity of the input stimuli increases. Hence, the local excitation moves searching for the position of the maximum stimulus and stops at the position of the maximum.

We have so far discussed a single local excitation pattern. However, the field can retain a number of local excitation patterns at the same time in some case. To show this, we should discuss the mutual interaction of two local excitations (Amari, 1977a). When the distance between two local excitations is smaller than a constant (which is nearly equal to $y_0$, the distance within which two neurons are connected with a positive weight), they attract each other. Hence, they become closer and join into one single local excitation. When their distance is larger than $y_0$ but smaller than $y_c$, they repulse each other, moving in the opposite directions. Two local excitations cannot coexist within a distance smaller than $y_c$. However, when their distance is larger than $y_c$, neither attraction nor repulsion takes place. There is no direct interaction between two local excitations whose distance is larger than $y_c$. However, one local excitation of length b inhibits the neurons of the other excitation by the amount of $-w_0 b$, because the weight of the inhibitory connection $w_{ei}(y)$ is equal to $-w_0$ when $y > y_c$. This shows that a local excitation has an effect of attraction, repulsion or decreasing the basic stimulus level s by $-w_0 b$ to another excitation, depending on their distance.

We now show the condition that the field can retain k local excitations persistently. Presume that there coexist k local excitation patterns of length b. Then, one local excitation receives inibitory stimuli from the k - 1 other local excitations, amounting to reduce the basic stimulation level from s to s - (k - 1)$w_0$b.  Hence, we have the condition

$$W(b) + s - (k - 1)w_0b - h_e = 0 \qquad\qquad (3.9)$$

from the fact that u(y) should be equal to 0 at the boundaries of every local excitation.  This condition reduces to (3.7) when k = 1.

The solution of (3.9) is given by the intersection of the graph W(b) and the linear graph (k - 1)$w_0$b - s + $h_e$.  When it has two solutions $b_1$, $b_2$ ($b_1 < b_2$), the bigger one is stable.  Given a basic stimulation level s, let K(s) be the set of the numbers of local excitations which can exist in the field at the same time.  When (3.9) has a stable solution b(k, s) for k, the k is included in K(s).  We define that k = 0 is included in K(s), when the resting state in which no neurons are excited is stable.  As is shown in Fig.7, K(s) can easily be obtained by writing the graphs.  In the case of Fig.7,

$$K(s) = \{0, 1, 2, 3\} .$$

When s < - $W_m$ + $h_e$, K(s) = {0}. Hence, the field can retain no excitation patterns persistently.  As the basic stimulation level s increases, the set K(s) becomes larger.  When

$$K(s) = \{0, 1, 2, \ldots, k \},$$

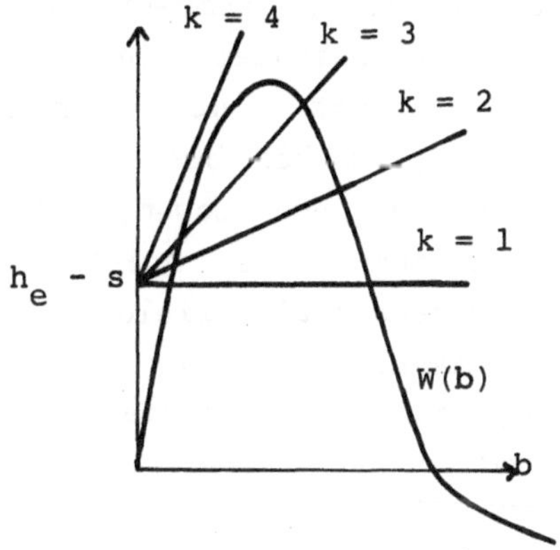

Fig.7. Number of excitations

the field can retain at most k local excitations persistently.  Hence, when more than k excitations are aroused by input stimuli, the field selects the k largest stimuli and keep their positions excited even after the stimuli are removed.  It is not possible for this field to keep more than k patterns.  Such a phenomenon is given rise to by the competition and cooperation mechanism in the neural field.  It is known in psychology as the law of "magical number seven", that one can remember about seven randomly chosen items in the short term memory. The above type of competitive and cooperative interaction is suggested to underlie in the short term memory.

When s becomes larger than $h_e$, the resting state becomes
unstable. In this case, the set $K(s)$ is of the form

$$K(s) = \{k_1, k_1+1, k_1+2, \ldots, k_2\},$$

and at least $k_1$, and at most $k_2$, excitations are spontaneously aroused
even no stimuli are applied. This is a kind of non-homogeneous
pattern formation in the homogeneous field. A periodic excitation
pattern is obtained when $w_0 = 0$.

We have thus shown how the gloval dynamical behavior of pattern
formation changes as the constant stimulus level s of the field
increases, without using the technique of bifurcation, which does not
account for the gloval behavior of the field.

## 3.3. Further results on the dynamics of nerve fields

We have shown the ability of pattern formation in the simple
neural field. Competitive and cooperative interactions of neurons
play a fundamental role in information processing in the neural field.
We can treat a more complex dynamical aspect of the field. For
example, it has been shown (Wilson and Cowan, 1973, Amari, 1977b,
Ermentrout and Cowan, 1971b, 1980, Ermentrout, 1982) that the
travelling wave solution exists in the above field. The
two-dimensional field (in which position parameter y is
two-dimensional) can be treated similarly, but the results are more
complex. More elaborate mathematical techniques such as functional
analysis and bifurcation theory are also useful (Kishimoto and Amari,
1979, Oguzutörelei, 1975, Ermentrout and Cowan, 1980, Ermentrout,
1982).

Amari and Arbib (1977) proposed a neural model of stereopsis
perception, similar to that proposed by Marr and Poggio (1976), and
Sugie and Sawa (1977) independently. The competitive and cooperative
interactions are essential in all of the above models of the
stereopsis perception. Their model consists of a two-dimensional
neural field consisting of depth and space dimensions. The
competitive aspect is dominant in the connections along the depth
dimension, and the cooperative aspect is dominant in the direction of
spatial dimension. They gave a mathematical analysis of the model,
and shown that the model has desired hysteresis characteristics. The
model also works well under ambiguous input patterns. This shows that
the present mathematical method is applicable to a more complex
competition and cooperation model, which can successfully explain the

essential features of neural information processing at least qualitatively.


4. Competition and Cooperation in Self-Organization

4.1. Average learning equation

A neural system has an ability of modifying its behavior to adapt itself to the varying environment. This is carried into effect by changing the synaptic weights of connections, and is called self-organization of the neural system. Competitive and cooperative interactions play a fundamental role in self-organization, too. Let us first treat a very simple rule of learning that a neuron, receiving a bunch of inputs $x_i$, automatically modifies the synaptic weights $w_i$. We assume a generalized Hebbian law that the synaptic weight $w_i$ increases in proportion to the product of the output $z = f(u)$ and the input $x_i$ when it is excited, and decreases with a time constant T. The equation of this learning is written as

$$T\dot{w}_i = - w_i + cf(u)x_i , \qquad\qquad (4.1)$$

where c is a constant and the potential u is determined from the input as $u = \Sigma w_i x_i - h$. It is convenient to adopt the vector notation

$$\underline{x} = (x_1, x_2, \ldots, x_n) , \qquad \underline{w} = (w_1, w_2, \ldots, w_n) .$$

Then, (4.1) can be rewritten as

$$T\dot{\underline{w}} = -\underline{w} + cf(\underline{w}\cdot\underline{x} - h)\underline{x} . \qquad\qquad (4.2)$$

where $\underline{w}\cdot\underline{x}$ is the inner product of $\underline{w}$ and $\underline{x}$.

Since the synaptic weight vector $\underline{w}$ is determined from the differential equation, its present value depends on the history of the past inputs $\underline{x}(t)$, which represents the nature of the environment in which the neuron is put. We consider the case where the inputs $\underline{x}$ are emitted from a stationary information source S representing the environment of the neuron. The information source S is specified by the probability distribution $p(\underline{x})$ of the signals x in it. The source S selects a signal $\underline{x}$ with probability $p(\underline{x})$ and provides it to the neuron for a fixed short time and then selects another signal $\underline{x}$ independently with probability $p(\underline{x})$, repeating these processes. Then, S is an ergodic information source, and an input sequence $\underline{x}(t)$

includes every signal $\underline{x}$ with relative frequency $p(\underline{x})$ for a long time. Here the probability distribution $p(\underline{x})$ represents the property of the environment from which the neuron receives a signal $\underline{x}$ with relative frequency $p(\underline{x})$.

Whichever sequence $\underline{x}(t)$ the neuron receives, it is expected that $\underline{w}(t)$ converges to the same value in some sense of approximation, provided $\underline{x}(t)$ is a typical sequence from an ergodic source S. This is also a version of the ergodic theorem. The probability $p(\underline{x})$ of S defines a measure P in the space of signals $S = \{\underline{x}\}$,

$$dP(\underline{x}) = p(\underline{x}) d\underline{x} . \tag{4.3}$$

Let $\langle \ \rangle$ denotes the expectation with respect to $p(\underline{x})$, i.e.,

$$\langle A(\underline{x}) \rangle = \int A(\underline{x}) dP(\underline{x}) . \tag{4.4}$$

By averaging the right-hand side of (4.2), we have the equation

$$T\dot{\underline{w}} = - \underline{w} + c \langle f(\underline{w} \cdot \underline{x} - h)\underline{x} \rangle , \tag{4.5}$$

which is a deterministic equation in $\underline{w}$. We call this the average learning equation under the information source S. Amari (1977b) has proposed the above type of average learning equations in more general situations and shown the various interesting behaviors of learning or self-organizing neural systems (see also Amari and Takeuchi, 1978, Amari, 1980). It has also been shown that the solution $\underline{w}(t)$ of the average learning equation gives a sufficient approximation to the solution of stochastic equation (4.2) for almost all input sequences $\underline{x}(t)$ from S. Geman (1979) gave a rigorous mathematical treatment of the above average equation.

## 4.2. Field theory of self-organization

Amari and Takeuchi (1978) have proposed a simple model to explain the mechanism that the nerve system forms detectors of signals included in the environment by self-organization (cf. Malsburg, 1978, Grossberg, 1976). Their model consists of a pool of neurons which

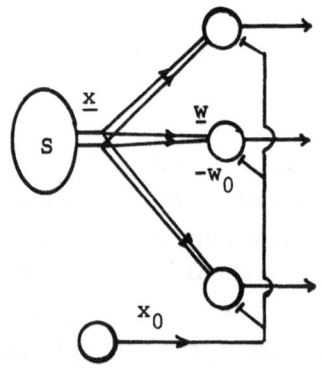

Fig.8. Self-organization

receive common signals $\underline{x}$ from an information source S and modify their synaptic weights. They also receive an inhibitory input of constant intensity $x_0$ (Fig.8). The inhibitory synaptic weights are also modifiable. Let $\underline{w}$ and $-w_0$ be the synaptic weight of a neuron in the pool for the inputs $\underline{x}$ and $x_0$, respectively. Then, the average learning equations are given by

$$T\dot{\underline{w}} = -\underline{w} + c \left\langle f(\underline{w} \cdot \underline{x} - w_0 x_0) \underline{x} \right\rangle , \qquad (4.6)$$

$$T\dot{w}_0 = -w_0 + c' x_0 \left\langle f(\underline{w} \cdot \underline{x} - w_0 x_0) \right\rangle , \qquad (4.7)$$

which are the same for all the neurons in the pool.

A neuron is called a detector of a set A of signals, when it is excited by any signals in A and it is not excited by any signals outside A. In other words, a neuron with synaptic weights $\underline{w}$, $w_0$ is a detector of A, when

$$\underline{w} \cdot \underline{x} - w_0 x_0 > 0 \qquad \text{for} \qquad \underline{x} \in A , \qquad (4.8)$$

$$\underline{w} \cdot \underline{x} - w_0 x_0 \leq 0 \qquad \text{for} \qquad \underline{x} \notin A . \qquad (4.9)$$

Conversely, the above set A is called the receptive signal set of the neuron.

It has been shown that the equations (4.6), (4.7) are in general multi-stable, having a number of stable equilibrium solutions $(\underline{w}, w_0)$ depending on the property of S. The receptive signal set of a neuron, whose weights are given by an equilibrium solution of (4.6) and (4.7), is called a stable receptive set under S. When the synaptic weights converge to an equilibrium solution $(\underline{w}, w_0)$, the neuron becomes a detector of the corresponding stable set A. Hence, by receiving a signal sequence from S, each neuron converges to a detector of one of the stable receptive sets under S. If the initial synaptic weights $\underline{w}$, $w_0$ of the neurons are randomly distributed, the neurons differentiate even by receiving the same input sequence to become detectors of all the stable receptive sets under S.

What are the stable receptive sets under S ? This important problem can be studied from the competitive and cooperative point of view, leading us to the study of the field theory of self-organization. Let us introduce a quantity

$$U(\underline{x}; \underline{w}, w_0) = \underline{w} \cdot \underline{x} - w_0 x_0 \qquad (4.10)$$

which is a function of $\underline{x}$ depending on $(\underline{w}, w_0)$. Obviously, $U(\underline{x})$ is positive on the receptive set A of the neuron with weights $(\underline{w}, w_0)$, and hence $1[U(\underline{x})]$ is the indicator of A. We regard $U(\underline{x})$ as a hypothetical potential distribution over points $\underline{x}$ of the signal space S. Obviously, the potential distribution $U(\underline{x})$ is determined by $\underline{w}$, $w_0$ of the neuron. When the weights $\underline{w}$, $w_0$ change by self-organization, so does the potential distribution $U(\underline{x})$ or the receptive set. The time derivative of the potential U can be obtained from

$$\dot{U}(\underline{x}, t) = \underline{\dot{w}}(t) \cdot \underline{x} - \dot{w}_0(t) x_0 , \qquad (4.11)$$

by substituting (4.6), (4.7), as

$$T\dot{U}(\underline{x}, t) = -U(\underline{x}, t) + c \left\langle f(\underline{w} \cdot \underline{x} - w_0 x_0) \underline{x} \right\rangle \cdot \underline{x}$$

$$-c' x_0^2 \left\langle f(\underline{w} \, \underline{x} - w_0 x_0) \right\rangle .$$

The equation can be rewritten as

$$T\dot{U}(\underline{x}, t) = -U(\underline{x}, t) + \int w(\underline{x}, \underline{x}') f[U(\underline{x}', t)] dP' , \qquad (4.12)$$

where we put $dP' = dP(\underline{x}')$ and

$$w(\underline{x}, \underline{x}') = c\underline{x} \cdot \underline{x}' - c' x_0^2 . \qquad (4.13)$$

This gives a dynamical equation of competition and cooperation on the signal field S consisting of all the input signals. The dynamics describes the change in the receptive set of a neuron by self-organization under S. The field equation (4.12) shows that the receptive region or "excitation" $U(\underline{x})$ at point $\underline{x}$ is determined by competitive and cooperative interactions with "excitations" at the other points in the field S. Two points $\underline{x}$ and $\underline{x}'$ in the field S are considered to be mutually connected with weight $w(\underline{x}, \underline{x}')$. Hence, two neighboring points $\underline{x}$ and $\underline{x}'$ cooperate to retain common excitations, because $c\underline{x} \cdot \underline{x}'$ is large. On the other hand, every point $\underline{x}$ receives inhibition in proportion to the total sum of the excitations in the field because of the term $- c' x_0^2$. Hence, all the points compete to each other through this inhibition mechanism.

We have thus obtained the field equation of self-organization. The equilibrium solution $\bar{U}(\underline{x})$ gives the stable receptive sets whose detectors can be obtained by self-organization under S.  Since the field has a structure of lateral inhibition, a local excitation pattern $\bar{U}(\underline{x})$ exists in the equilibrium.  Here, neighboring signals cooperate to form a stable set, while different signals compete with one another for forming detectors.

Let A be a stable set corresponding to an equilibrium solution $\bar{U}(\underline{x})$.  Let us define P(A) by

$$P(A) = \int f[\bar{U}(\underline{x}')]dP' , \qquad (4.14)$$

which is a measure of A.  When f is equal to the step function 1, P(A) is the probability of a signal belonging to A.  We also define the weighted average $\underline{x}_A$ of the signals in A by

$$\underline{x}_A = \int f[U(\underline{x}')]\underline{x}'dP'/P(A) . \qquad (4.15)$$

The equilibrium solution is then written in terms of P(A) and $\underline{x}_A$ as

$$\bar{U}(\underline{x}) = P(A)(c\underline{x}_A \cdot \underline{x} - c'x_0^2) . \qquad (4.16)$$

This should be positive for $\underline{x} \in A$ and should not be positive for $\underline{x} \notin A$.  Hence, a stable receptive set A satisfies

$$A = \{\underline{x} \mid \underline{x}_A \cdot \underline{x} > \lambda\} , \qquad (4.17)$$

where $\lambda = c'x_0^2/c$.  This proves the theorem (cf. Amari and Takeuchi, 1978) that a set A is stable, if and only if

$$\inf_{\underline{x} \in A} \underline{x} \cdot \underline{x}_A > \lambda \geq \sup_{\underline{x} \notin A} \underline{x} \cdot \underline{x}_A . \qquad (4.18)$$

Let us consider the case where signals $\underline{x}$ are distributed in S uniformly on the surface of the unit sphere satisfying

$$\underline{x} \cdot \underline{x} = 1 .$$

Since the distribution $p(\underline{x})$ is homogeneous, if A is a stable set, then OA is also a stable set, where O is any orthogonal transformation. Let us consider a stable set A whose average is $\underline{x}_A$.  Then, A is given by the signals $\underline{x}$ satisfying $\underline{x}_A \cdot \underline{x} > \lambda$.  Since $\underline{x}_A \cdot \underline{x}_A = 1$ approximately holds, $\underline{x}_A \cdot \underline{x}$ is the cosine of the angle $\theta$ between $\underline{x}$ and $\underline{x}_A$. Hence, we see that A consists of the signals $\underline{x}$ such that the angles of $\underline{x}$ and $\underline{x}_A$ are smaller than $\cos^{-1}\theta$. We thus proves that every stable set is of a disc shape on the sphere, whose radius is $\cos^{-1}\theta$ and vice versa. The parameter $\lambda$ designates the size of the stable set or resolution of detectors.

When the signals $\underline{x}$ are not uniformly distributed but the function $p(\underline{x})$ has a number of peaks, the situation is different. As can be understood from (4.17), a stable set is a disc with radius $\cos^{-1}\theta$ on the unit sphere in S. A disc A is a stable set, when and only when its center coincides with the average of the signals in A with respect to the measure $P(\underline{x})$. This coincidence never occurs unless the set A includes a peak of the distribution $p(\underline{x})$. On the other hand, we can prove that there exists a stable set around a peak, if the peaks of $p(\underline{x})$ are sufficiently separated. This shows that the stable sets, and hence their detectors, are formed around the peaks of the distribution $p(\underline{x})$ in S. This elucidates the ability of formation of signal detectors in the simple model of the neural system.

## 4.3. Self-organization of neural system and field

We have so far treated learning of a single neuron. Self-organization of a neural system is more complicated, because there exist mutual interactions among the component neurons. Amari (1982) has formulated the equation of self-organization of neural systems in the following way. Let us consider a neural system, which receives signals $\underline{x}$ from an ergodic information source. The system includes a number of modifiable synapses. Let W be a matrix whose components $w_{ij}$ are the weights of all the modifiable synapses in the system. When an input $\underline{x}$ is applied to the system, the state of the system changes according to the dynamics of neural excitation. We assume that the state converges to the unique equilibrium, which depends on the input $\underline{x}$, the modifiable synaptic weights W and other fixed synaptic weights in the net. Let

$$U_i = U_i(\underline{x}, W) \qquad (4.19)$$

be the potential of the i-th neuron in the equilibrium where $\underline{x} \in S$ is applied. These functions $U_i$ can be obtained by solving the dynamical equations governing the potentials $u_i$'s. Since the time constant T of the synaptic modification is much larger than the time constant $\tau$ of neural excitation, we can regard W as fixed parameters when we solve the dynamical equations of neural excitation.

The synaptic weight $w_{ij}$ changes slowly. We can describe its change by the average learning equation

$$T\dot{w}_{ij} = - w_{ij} + c \left\langle f[U_i(\underline{x}, W)]y_{ij} \right\rangle, \qquad (4.20)$$

where $y_{ij}$ is the input signal arriving at the synapse $w_{ij}$ in the equilibrium of the dynamics of neural excitation when $\underline{x}$ is applied. It is determined depending on the input $\underline{x}$ and W. Since the time constant T is much larger than $\tau$, we presumed that the system is always in the equilibrium state determined by the input $\underline{x}$ at that time.

We can analyze the behavior of a revised model of formation of signal detectors in which all the neurons are connected through a common inhibitory neuron pool. The inhibitory neuron pool is excited by the outputs of the neurons and it in turn inhibits all the neurons. Each neuron, in this model, has a dynamical equation of self-organization on the signal field S. Hence, the self-organization of the total system is described by a dynamical equation on the multi-layer signal fields, one layer for each neuron. Competition and cooperation take place within one field as well as among the layers.

Self-organization of a neural field can be treated in a similar manner. We consider a one-dimensional homogeneous neural field Y, in which neurons are recurrently connected with lateral-inhibitory weight function r, i.e., the synaptic weight of connection from the neuron at position y' to the neuron at position y is r(y - y'). The neurons in Y receive a common input $\underline{x}$ from an information source S. Let $\underline{w}(y)$ be the synaptic weight vector for the input $\underline{x}$ of the neuron at position y. Let $- w_0(y)$ be the inhibitory synaptic weight for a constant inhibitory input $x_0$ of the neuron at y. (See Fig. 9). The synaptic weights $\underline{w}(y)$ and $- w_0(y)$ are assumed to be modifiable, but r(y) is fixed.

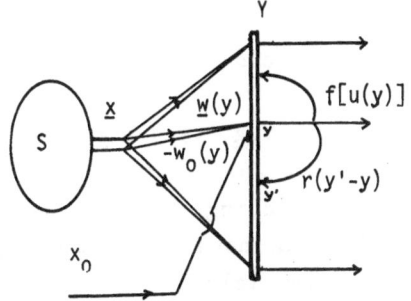

Fig. 9 Self-organization of nerve field

The dynamics of excitation in the field, when a signal $\underline{x}$ is applied, is written as

$$\tau\dot{u}(y) = - u(y) + \underline{w}(y)\cdot\underline{x} - w_0(y)x_0$$

$$+ \int r(y - y')f[u(y')]dy' , \tag{4.21}$$

where $u(y)$ is the potential of the neuron at position $y$. Let $U(\underline{x}, y)$ be the equilibrium solution of the above equation, which depends on the input $\underline{x}$. It satisfies

$$U(\underline{x}, y) = \underline{w}(y)\cdot\underline{x} - w_0x_0 + r*f[U] , \tag{4.22}$$

where $*$ implies the covolution operator over Y, i.e.,

$$r*f[U] = \int r(y - y')f[U(\underline{x}, y')]dy' .$$

This $U(\underline{x}, y)$ represents the hypothetical potential distribution over the product space $S \times Y$, designating that the neuron at $y \in Y$ is responsive in the equilibrium for signal $\underline{x} \in S$ when $U(\underline{x}, y) > 0$. Hence, $U(\underline{x}, y)$ denotes the receptive region of the neuron at $y$.

Since $U(\underline{x}, y)$ depends on the synaptic weight distributions $\underline{w}(y)$ and $-w_0(y)$, it changes as the field self-organizes by receiving inputs $\underline{x}$ from S. The average learning equations for $\underline{w}(y)$ and $w_0(y)$ are

$$T\dot{\underline{w}}(y) = - w(y) + c \langle f[U(\underline{x}, y)]\underline{x} \rangle , \tag{4.23}$$

$$T\dot{w}_0(y) = - w_0(y) + c'x_0 \langle f[U(\underline{x}, y)] \rangle . \tag{4.24}$$

By introducing

$$V(\underline{x}, y) = U(\underline{x}, y) - r*f[U] , \tag{4.25}$$

differentiating it with respect to t and substituting (4.23) and (4.24), we have the following equation governing the change in the receptive field,

$$T\dot{V}(\underline{x}, y) = V + w\otimes f[U] , \tag{4.26}$$

where

$$w(\underline{x}, \underline{x}') = c\underline{x}\cdot\underline{x}' - c_0x_0{}^2$$

as before and ⊗ denotes the convolution operator over the signal space S with respect to the measure $P(\underline{x})$, i.e.,

w⊗f[U]

$$= \int w(\underline{x}, \underline{x}')f[U(\underline{x}', y)]dP'$$

The above equation (4.26) shows that competitive and cooperative interactions take place in the field S × Y in the self-organizing process (Fig.

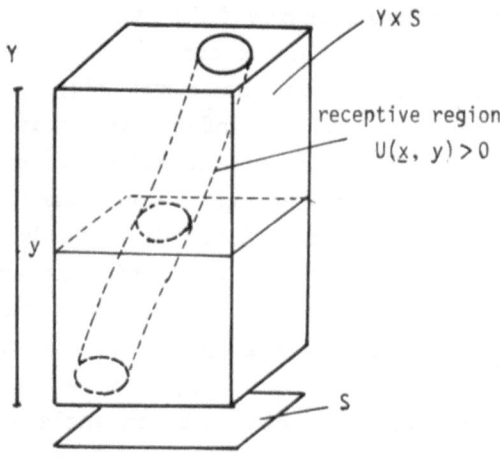

Fig. 10 Receptive region on Y × S

10). The equilibrium solution $\bar{U}(\underline{x}, y)$ of self-organization is given from

$$\bar{U}(\underline{x}, y) = r*f[\bar{U}] + w⊗f[\bar{U}] .\tag{4.27}$$

## 4.4. Formation of topographic map

It is known that many neural tissues have topographic organization such that continuous projection is formed from a presynaptic neural field to a postsynaptic field. There are many interesting experimental facts on the formation of such a topographic map, and there have been proposed a number of theoretical models (e.g., Willshaw and von der Malsburg, 1979, Amari, 1980, Takeuchi and Amari, 1981, Kohonen, 1982, Overton and Arbib, 1982). It seems that there are two stages of processes in the formation of the map: A rough and inaccurate map is formed in the first stage. The rough map is modified in the much slower second stage to adapt the environmental information structure, so that the map becomes accurate having fine resolution.

The first and second stages may be based on different mechanisms, and it is the second stage that might be analyzed by our field theory of self-organization. We give a simple one-dimensional example. In the field model of the previous subsection, we consider the case when the information source S has a one-dimensional structure. Let $\underline{x}(q)$ be a continuous curve in S, where $q \in Q$ is a scalar parameter specifying the position in the curve. When the probability measure $P(\underline{x})$ of S is concentrated on the curve $\underline{x}(q)$, only the signals on the curve are produced from the information source S. The source S can be said in this case to have a one-dimensional structure, with a one-dimensional coordinate q. We give some examples. When the signals $\underline{x}$ represent the local excitations of a one-demensional presynaptic neural field, $\underline{x}(q)$ being a local excitation at around position q, the source S is one-dimensional if it produces signals $\{\underline{x}(q)\}$ alone. On the other hand, consider the case when the signals $\underline{x}$ represent patterns on the retina, and when $\underline{x}(q)$ is a bar pattern on the retina of the orientation specified by an angle q. The source S is one-dimensional, if S produces bar patterns alone. In this case, $\{\underline{x}(q)\}$ forms a circle in S.

By putting

$$U[\underline{x}(q), y] = U(q, y) ,$$

$$V[\underline{x}(q), y] = V(q, y) ,$$

$$w[\underline{x}(q), \underline{x}(q')] = w(q, q') ,$$

$$p[\underline{x}(q)] = p(q) ,$$

the field equation of self-organization reduces to

$$V(q, y) = U(q, y) - r*f[U] , \tag{4.28}$$

$$\tau \dot{V}(q, y) = - V + w \circledS f[U] , \tag{4,29}$$

where $*$ and $\circledS$ are the convolutions over Y and Q, respectively. This is a competition and cooperation equation on the field $Q \times Y$, and the topographic map from Q to Y can be formed subject to this equation.

The equation has already been analyzed in some special cases by approximating f by the step function 1. Amari (1980) treated the case where the correlation between $\underline{x}(q)$ and $\underline{x}(q')$ decreases rapidly to 0 as $|q - q'|$ becomes large. The quantity $\underline{x}(q) \cdot \underline{x}(q')$ was approximated by the delta function $\delta(q - q')$, and the effect of an unequal probability distribution p(q) was studied. It was proved mathematically that signals $\underline{x}(q)$ which are applied more frequently (i.e., p(q) are relatively large), have a wide area representing it in the postsynaptic field Y and fine resolution in Y.

Takeuchi and Amari (1979) treated the case where the field Q is uniform so that $p(q)$ is constant over Q and $w(q, q') = w(q - q')$ holds. Since the field is homogeneous in this case, it is easy to show that the equilibrium solution of the type

$$U(q, y) = a(q - y) , \tag{4.30}$$

where $a(q)$ is a unimodal function, exists, where we chose some suitable scales in Q and Y. This solution corresponds to the continuous topographic map. It is surprising that this solution becomes unstable in some case. The variational equation of (4.48) and (4.29) is written as

$$\tau(I - R*)^{-1}\dot{\delta U} = - \delta U + (W\circledast + R*)\delta U , \tag{4.31}$$

where I is the identity operator and

$$R*\delta U = \int r(y - y')f'[a(y' - q)]\delta U(y', q)dy' ,$$

$$W\circledast\delta U = \int w(q - q')f'[a(y - q')]\delta U(y, q')dq' .$$

Hence, the stability of the uniform solution depends on the eigenvalues of the sum of the operators $W\circledast$ and $R*$. Generally speaking, when the correlation function $w(q - q')$ becomes wide compared to the lateral-inhibitory connection or when the relative frequency $p(q)$ becomes small, the uniform solution has a tendency of becoming unstable.

When the uniform solution becomes unstable, a spatially periodical pattern emerges. This seems to explain the mechanism of stability of the microstructures in the brain such as columns, barrels and microzone. Competition and cooperation, and related abilities of pattern formation seem to play a fundamental role in self-organization.


Conclusion. We have shown that competitive and cooperative interactions play an important role in the dynamical behaviors not only of neuron pools and neural fields but also of self-organization of nerve systems. We have proposed a field theory of self-organization and analyzed some characteristics of self-organizing neural systems. The principle of competition and cooperation seems to play a fundamental role in modeling and analyzing neural systems.

## References

Amari, S., 1971: Characteristics of randomly connected threshold element networks and network systems. Proc. IEEE 59, 35-47

Amari, S., 1972: Characteristics of random nets of analog neuron-like elements. IEEE Trans. SMC-2, 643-657

Amari, S., 1974: A method of statistical neurodynamics. Kybernetik 14, 201-215

Amari, S., 1977a: Dynamics of pattern formation in lateral-inhibition type neural fields. Biol. Cybernetics 27, 77-87

Amari, S., 1977b: Neural theory of association and concept-formation. Biol. Cybernetics 26, 175-185

Amari, S., 1980: Topographic organization of nerve fields. Bull. Math. Biology 42, 339-364

Amari, S., 1982: A mathematical theory of self-organizing nerve systems. In Proceedings of Biomathematics: Current Status and Perspectives (Ricciardi, L.M., Scott, A., ed.), North-Holland

Amari, S., Arbib, M.A., 1977: Competition and cooperation in neural nets. In Systems Neurosciences (Metzler, J. ed.) Academic Press, 119-165

Amari, S., Takeuchi, A., 1978: Mathematical theory on formation of category detecting nerve cells. Biol. Cybernetics 29, 127-136

Amari, S., Yoshida, K., Kanatani, K., 1977: A mathematical foundation for statistical neurodynamics. SIAM J. on App. Math. 33, 95-126

Anninos, P.A. et al., 1970: Dynamics of neural structures. J. Theor. Biol. 26, 121-148

Ellias, S.A., Grossberg, S., 1975: Pattern formation, contrast control, and oscilations in the short term memory of shunting on-center off-surround networks. Biol. Cybernetics, 20, 69-98

Ermentrout, G.B., 1982: Asymptotic behavior of stationary homogeneous neuronal nets. In Competition and Cooperation in Neural Nets (Amari, S. and Arbib, M.A. ed.), Lecture Notes in Biomathematics (This volume), Springer-Verlag

Ermentrout, G.B., Cowan, J.D., 1979a: A mathematical theory of visual hallucination patterns. Biol. Cybern. 34, 137-150

Ermentrout, G.B., Cowan, J.D., 1979b: Temporal oscilations in neuronal nets. J. Math. Biol. 7, 265-280

Ermentrout, G.B., Cowan, J.D., 1980: Secondary bifurcation in neuronal nets. SIAM J. Appl. Math. 39, 323-341

Fukushima, K., Miyake, S., 1982: Neocognitron: A self-organizing neural network model for a mechanism of visual pattern recognition. In Competition and Cooperation in Neural Nets (Amari, S. and Arbib, M.A. ed.), Lecture Notes in Biomathematics (This volume), Springer-Verlag

Geman, S., 1979: Some averaging and stability results for random differential equations. SIM J. App. Math. 36, 86-105

Geman, S., 1980: Almost sure stable oscillations in a large system of randomly coupled equations. Rep. Pattern Analysis No. 97, Brown University

Grossberg, S., 1976: Adaptive pattern classification and universal recording, I. Biol. Cybern. 23, 121-134

Grossberg, S., 1982: Associative and competitive principles of learning and development. In Competition and Cooperation in Neural Nets (Amari, S. and Arbib, M.A. ed.), Lecture Notes in Biomathematics (This volume), Springer-Verlag

Harth, E.M., et al., 1970: Brain functions and neural dynamics. J. Theor. Biol. 26, 93-120

Holden, A.V., 1976: Models of Stochastic Activity of Nurones. Lecture Notes in Biomath. 12, Springer

Kishimoto, K., Amari, S., 1979: Existence and stability of local excitations in neural fields. J. Math. Biol. 7, 303-318

Kohonen, T., 1982: A simple paradigm for the self-organized formation of structured feature maps. In Competition and Cooperation in Neural Nets (Amari, S. and Arbib, M.A. ed.), Lecture Notes in Biomathematics (This volume), Springer-Verlag

Malsburg, Ch. von der, 1973: Self-organization of orientation sensitive cells in the striate cortex. Kybernetik 14, 85-100

Marr, D., Poggio, T., 1976: Cooperative computation of stereo disparity. Science, 194, 283-287

Montalvo, F.S., 1975: Consensus vs. Competition in neural networks. Int. J. Man-Machine Studies 7, 333-346

Oguztöreli, M.N., 1975: On the activities in a continuous neural network. Biol. Cybern. 18, 41-48

Overton, K.J., Arbib, M.A., 1982: Systems matching and topographic maps: The branche-arrow model (BAM). In Competition and Cooperatio in Neural Nets (Amari, S. and Arbib, M.A. ed.), Lecture Notes in Biomathematics (This volume), Springer-Verlag

Sawada, R., Sugie, N., 1982: Self-organization of neural nets with competitive and cooperative interaction. In Competition and Cooperation in Neural Nets (Amari, S. and Arbib, M.A. ed.), Lecture Notes in Biomathematics (This volume), Springer-Verlag

Sugie, N., Sawa, M., 1977: A scheme for binocular depth perception suggested by neurophysiological evidence. Biol. Cybern. 26, 1-15

Takeuchi, A., Amari, S., 1979: Formation of topographic maps and columnar microstructures. Biol. Cybern. 35, 63-72

Willshaw, D.J., Malsburg, C. von der, 1979: A marker induction mechanism for the establishment of ordered neural mappings: its application to the retinotectal connections. Phil. Trans. Proc. R. Soc. Lond. B, 287, 203-243

Wilson, H.R., Cowan, J.D., 1972: Excitatory and inhibitory interactions in localized populations of model neurons. Biophys. J. 12, 1-24

Wilson, H.R., Cowan, J.D., 1973: A mathematical theory of the functional dynamics of cortical and thalamic nervous tissues. Kybernetik 13, 55-80

# Chapter 2

## SIGMOIDAL SYSTEMS AND LAYER ANALYSIS

Paul C. Fife

Mathematics Department, University of Arizona

### 1.  Introduction

My purpose here is to describe a variety of wave-like
phenomena which depend, for their existence, on the nonlinear
nature of the underlying laws governing the dynamics of the
medium.  It will be assumed that these underlying laws can be
expressed in terms of field equations of reaction-diffusion type,
although some of the methods which have been developed for such
systems have also been extended to other types of evolutionary
laws.  I make no claim that reaction-diffusion systems have a
direct bearing on acceptable models for large scale neural nets.
But they exhibit patterns and waves suggestive of phenomena
expected in such nets, and it is hoped that their theory, which is
moderately well developed, may provide a fruitful source of
methods for the mathematical analysis of the neural models.
Indeed, some parts of the theory, dealing for example with small-
amplitude phenomena, have already been extended to the latter
context (Ermentrout 1980; Ermentrout and Cowan 1979, 1980a,
1980b).

A preliminary word on the role of nonlinearity is in order.
Field equations of hyperbolic type exhibit wave-like solutions,
i.e., solutions with features which propagate in space with finite
velocities. This has been known for centuries, and is true
whether the equations are linear or nonlinear. Reaction-diffusion
systems, on the other hand, are parabolic in nature, and were very
seldom used in modelling waves prior to the late 1960's. Linear
parabolic systems (with more than one equation) do at times have
bounded propagating wave solutions (Turing 1953, for example),
although this fact is not too well known or used. Far more
important, however, are the wave phenomena which occur in
nonlinear systems; and in fact the wave solutions which I shall
emphasize in this talk depend on the nonlinearity; they would not
exist were it absent.

Let me be more specific about the evolution problem I wish to
discuss. Consider a medium distributed in space (the entire
space $\mathbb{R}^n$, for simplicity), and describable, at each point
$(x,t)$ in space-time , by a state vector $u \in \mathbb{R}^m$. This state
function $u: \mathbb{R}^n \times I \rightarrow \mathbb{R}^m$ (where I is the time interval under
consideration), and the restrictions to which it is subject, are
the objects of this lecture. It will be assumed that the state
$u$ must satisfy a reaction-diffusion equation

$$u_t = D\Delta u + f(u), \tag{1}$$

where  D  is an  n × n  matrix,  $\Delta = \sum \dfrac{\partial^2}{\partial x_i^2}$ ,  and  f  (the

reaction term) is a smooth function,  $\mathbb{R}^m \rightarrow \mathbb{R}^m$ .

A wave could loosely be defined as a solution  u(x,t)  with
the property that some localized feature of the function at one
time (or perhaps several such features) propagate to other
locations at later times.  But I shall generally use a tighter
definition and deal only with bounded strict traveling waves,
namely solutions

$$u(x,t) = U(x \cdot \nu - ct) \tag{2}$$

for some unit vector  ν  and constant velocity  c,  with the
profile  U(z)  bounded.  Traveling waves are apparently
commonplace for nonlinear systems although existence theorems for
them do not enjoy such abundance.  A vast amount of numerical
simulation has been done, which I shall not mention further.  As
indicated before, waves are quite rare for linear systems.  In
fact, suppose the reaction term is linear:  f(u) = Au  for some
matrix  A.  If it should happen that traveling waves do exist for
a particular matrix  A,  then  A  can always be altered by an
arbitrarily small amount so as to make the existence of such waves
impossible.

Being undirectional in the (arbitrary) direction  ν,  the
waves (2) are a one-dimensional phenomenon, and so we may as well

consider space to be one-dimensional, with  x  a scalar.  Then
u = U(x - ct) = U(z) ∈ R^m  satisfies

$$DU'' + cU' + f(U) = 0, \quad (3)$$

the "traveling wave" equation associated with (1).

In the case  m = 1,  the existence and stability theory of
traveling waves is almost complete.  It is the subject of Section
2.  These results have a bearing on the theory, dealing with
higher order systems, to be examined in the later sections.  After
some general remarks in Sections 3 and 4, sigmoidal systems,
with  m = 2,  will be defined and analysed by asymptotic methods
in Sections 5-7.  Other results will be discussed in Section 8.

## 2.  Scalar Nonlinear Diffusion

The case when  m = 1,  so that the state vector is a single
number, has been thoroughly investigated.  Some of the relevant
references are Kolmogorov, Petrovskii and Piskunov (1937), Kanel'
(1962), Fife and McLeod (1977), Aronson and Weinberger (1978),
Uchiyama (1978).  See also the references in Fife (1979).  Salient
facts about traveling wave solutions are the following:

(1)  For  $c \neq 0$,  all nonconstant waves have the property
     that  U(z)  approaches distinct limits  $U_\pm$  as
     $z \to \pm \infty$,  and these are zeros of  f:  $f(U_\pm) = 0$.  We

call these waves "fronts".

(ii)   c  necessarily has sign opposite that of  $J = \int_{U_-}^{U_+} f(u)du$
(and must vanish when  $J = 0$).

(iii)   If  f  is nonzero and of  constant sign for  U  between
$U_-$  and  $U_+$ ,  there is a whole continuum of possible
values of  c  associated with monotone profiles  U.  If
$f'(U_-) < 0$, $f'(U_+) < 0$,  and  f  has only one zero  $U_0$
between the other two, then there is a unique  c  with its
unique associated profile.  In the latter case, a global
stability result has been proved.  Specifically, if
$u(x,0)$  is bounded away from  $U_0$  for large  $|x|$,  lies
between  $U_-$  and  $U_+$,  and lies on opposite sides of  $U_0$
for large negative and positive values of  x,  then
$u(x,t)$  converges uniformly to a traveling front as
t → ∞    (Fife and McLeod 1977).  A weaker convergence
result had previously been proved by Kanel' (1962) by a
different method under the assumption that  $u(x,0)$  is
monotone and identically equal to  $U_-$  or  $U_+$  for large
$|x|$.

(iv)   If  $c = 0$,  solutions of (3) periodic in  z  may exist, as
well as a unimodular solution approaching the same limit
as  z → ±∞.   These solutions are unstable in a certain

sense (Fife 1979). If $J = 0$, a stable stationary front also exists.

### 3. Classification of Methods

The case when $m > 1$ is of course more interesting and challenging. To distinguish waves from stationary configurations, I shall always assume that $c \neq 0$. Most of the existing results are perturbative in nature (there are certainly exceptions to this rule; one will be mentioned in Section 8). In turn, the most important kinds of perturbation methods can be classified into four categories:

(a)  layer-type singular perturbation methods,

(b)  Conley index methods, useful in proving the validity of the formal constructions,

(c)  bifurcation methods, and

(d)  long fast wave extensions of homogeneous oscillations.

Because of time limitations, I shall limit my remarks to methods of type (a). The powerful methods of type (b) were illustrated recently in Conley-Gardner (1981), Gardner-Smoller (1981), and Conley-Fife (1982). Bifurcation approaches (type (c)) are very popular in continuum mechanics and in biological applications. Earlier work in connection with reaction-diffusion problems was surveyed in Fife (1978). Very general recent

treatments have been given by Ermentrout (see his paper in these
Proceedings). Wave trains of type (d) were constructed in
Ortoleva-Ross (1973) and in Howard-Kopell (1974). Their rigorous
justification is found in the latter paper.

### 4. Layer Analysis: Earlier Work

This approach has proved to be exceedingly fruitful for the
problem of propagating waves in systems of reaction-diffusion
equations. Layer analysis for ordinary and partial differential
equations in general has long been a standard technique. As
applied in reaction-diffusion contexts, the technique has a flavor
of its own.

Roughly speaking, layer analysis proceeds by recognizing that
the solutions of interest may have natural length and/or time
scales which depend on location in space-time. Typically, the
relative magnitude of these differing scales is a function of some
small or large parameter occuring in the differential equations.
In the case of travelling waves discussed here, the wave profile
U may have a steep gradient at certain places, resulting in
rather abrupt, but smooth, jumps at those locations, where the
natural length scale will be shorter than it is between the
jumps. Since the profile moves, the time scale will change in a
similar manner. The classical example of layer analysis has to do
with hydrodynamical problems involving the flow of a slightly
viscous fluid near a stationary boundary to which the fluid

adheres. In this case the steep-gradient regime is adjacent to
the boundary and has thickness proportional to the square root of
the viscosity. It was termed a "boundary layer", which accounts
for the present-day uses of the term "layer" in other connections.

In the present section, I shall mention some of the first
work in layer analysis of reaction-diffusion systems, deferring a
detailed description of the mechanics of the method to later
sections.

The problem first systematically handled this way was
apparently one in combustion theory (Bush and Fendell 1970), but I
shall omit further mention of combustion-type problems, under the
assumption that they bear the least similarity to the theme of
this conference.

In 1971, H. Cohen (1971) described a layer analysis of a
problem more in the spirit of the conference:  that of steady
pulses for the FitzHugh-Nagumo equations, in which a small
parameter $\varepsilon$ appears. These equations constitute a model for
the propagation of signals along a homogeneous nerve axon, and the
pulse constructed is meant to represent a single nerve impulse.
The pulse constructed in the above reference is relatively flat
topped, with more abrupt drop-offs (layers) in the front and the
rear. The construction Cohen described was performed by Casten,
Cohen, and Lagerstrom, and appeared in detail in their paper in
1975.

This involved only formal arguments.  The existence of

pulses, as well as of periodic traveling waves, for FitzHugh-
Nagumo and Hodgkin-Huxley equations was proved a little later by
various people using various methods; see Conley (1975), Conley
and Smoller (1976), Hastings (1974, 1976a,b), Carpenter (1979a,b).

The FitzHugh-Nagumo model is a system of two equations whose
nonlinear term  f  has a certain "sigmoidal" feature which I shall
describe in a moment.  Pairs of equations with this characteristic
arise as models in other connections as well.  They provide a rich
array of dynamic as well as static pattern phenomena.

Another example of the use of sigmoidal equations has been in
the analysis of exotic chemical systems.  This was possibly first
done by Ortoleva and Ross (1975), where layer analysis was used to
construct pulses, and single fronts as well.  In that paper, the
use of a sigmoidal function in a chemical context was made
credible by exhibiting a model chemical reaction which leads to
reaction functions of that sort.  In all of these layer analyses,
the layer equations (those using the short space scales) reduce to
the one-dimensional scalar nonlinear diffusion equation considered
above.  In Ortoleva and Ross, it was assumed that this equation
had a unique associated velocity, which was found by numerical
integration.

The analyses of all the authors mentioned above were confined
to the structure of a steady traveling pulse or front.  In
Ostrovskii and Yahno (1975), on the other hand, descriptive
analysis was made of the "initiation" process by which a pulse of

this type might evolve in a natural way from a fairly arbitrary initial disturbance in a medium whose state is otherwise uniform. The analysis, though formal, clearly shows that the pulse is a very stable entity. Ostrovskii and Yahno worked with equations of sigmoidal type with the essential features of the FitzHugh-Nagumo equations; but the applications they had in mind were more wide-ranging than to signal propagation along an axon. By that time, it had become clear that a number of biophysical and chemico-physical phenomena had excitable properties analogous to those of the nerve membrane. In particular, they had been observed in cardiac tissue and in the Belousov-Zhabotinskii reagent. Sigmoidal reaction-diffusion systems had also been proposed as a phenomenological model to account for some of the observed phenomena, and had been analyzed by numerical methods. Ostrovskii and Yahno's advance was to show the applicability of layer analysis in the pulse initiation process. At the same time, of course, this analysis provided information on the structure of steady pulses, as did that of the previous authors mentioned.

The paper of Fife (1976b), written independently of that of Ostrovskii-Yahno, contains, among other things, an initiation analysis in another context. Again, a sigmoidal system was used; this time it was meant to model the dynamics of a chemically reacting and diffusing medium, and a hypothetical reaction mechanism was provided (different from that in Ortoleva-Ross), which yields a sigmoidal nonlinearity. The medium was allowed to

be higher-dimensional, but bounded and subject to boundary
conditions. The object was to exhibit stationary pattern
formation by means of the initiation of traveling wave fronts,
their migration through the medium, and their final attainment of
a permanent stationary configuration. The fronts in this case
were not planar, and their velocity was not constant.

Rigorous existence results on static layer configurations had
been obtained previously for a two-dimensional inhomogeneous
medium with a single sigmoidal differential equation (Fife and
Greenlee 1974), and for a pair of equations with internal layers
by Fife (1976).

All the above papers show (1) the great variety of exotic
phenomena capable of being explained by sigmoidal systems, and (2)
the relative ease with which layer analysis, both dynamic and
static, can be performed with these systems. There have since
been a number of papers further exploring these directions. Let
me merely mention Fife (1976c,1977) and Pismen (1978) where a
variety of types of fronts were catalogued and analyzed with layer
methods, Fife (1976c) where it was shown that they may be used to
describe the various stages in the evolving dynamics of exotic
systems, Rinzel and Terman (1981) where phenomena like front
reversal were analyzed, and Tyson-Fife (1980), which contains an
extensive layer-type analysis of target patterns for the
Oregonator equations. I have completely ignored a lot of good
work on other exotic reaction-diffusion phenomena, like that on

spirals and stationary patterns.

## 5.   Sigmoidal Systems.

For the present purposes, I define them as systems

$$u_t = D_1 u_{xx} + f(u,v),$$

$$v_t = D_2 v_{xx} + g(u,v),$$

where the curve  $f(u,v) = 0$  in the  u-v  plane bends back:

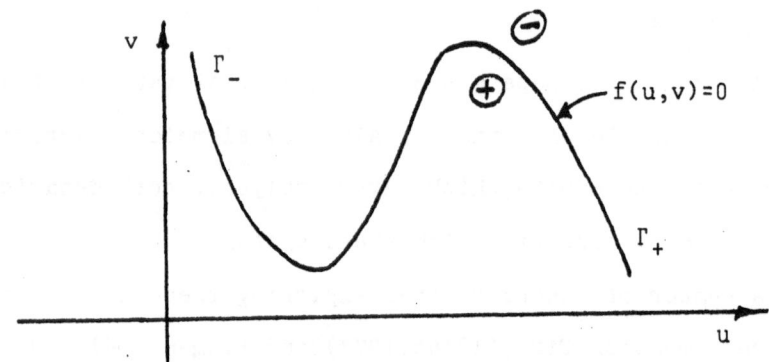

Fig. 1

with  f  negative above the curve, positive below.  As mentioned
before, this type of configuration arises in models from many
fields.  In particular, I mention nerve signal propagation,
chemical reactors, chemical kinetics and population biology.

The type of dynamics to be expected depends also on the
function  g  and the relative magnitude of  $D_1$,  $D_2$,   f  and  g.

For layer analysis to be possible, of course, we need to have at least one small parameter in the system.  Examples follow.

### 6.  Construction of Traveling Waves.

The basis of the following construction is the assumption that  f  is larger than  g.  This assumption provides the needed small parameter  $\varepsilon$; we simply replace  f  by  $(1/\varepsilon)f$.  In addition, the spatial domain is unbounded, so  x  may be rescaled at will:  we do this in such a way that  $D_1 = \varepsilon$.  A further, completely unnecessary, assumption is that  $D_2 = 0$.

The following traveling wave equations (from (3)) result. Using  $u = U(z) = U(x-ct)$,  $v = V(z)$, we have

$$\varepsilon^2 U'' + \varepsilon cU' + f(U,V) = 0$$
$$cV' + g(U,V) = 0. \qquad (4)$$

At this stage, not only is the pair  $(U(z), V(z))$  unknown, but the velocity  c  as well.

A necessary assumption is that  $g > 0$  on at least part of the right hand (descending) branch  $\Gamma_+$  of the nullcurve in Fig. 1, and that  $g < 0$  on  $\Gamma_-$.  For the moment, assume these inequalities hold everywhere on the respective branches.  We shall construct solutions of (4), periodic in  z,  whose image in the U-V plane is a closed orbit like the one in dotted lines:

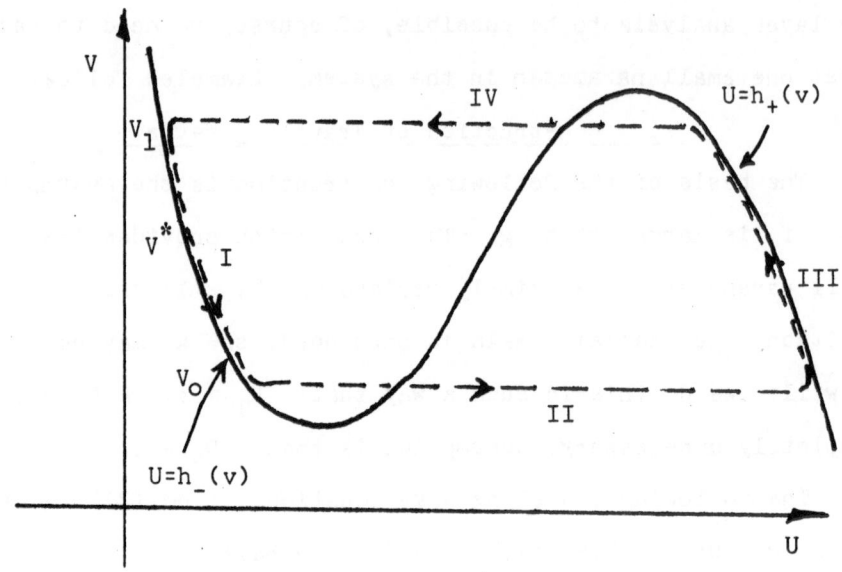

Fig. 2

The portions I and III are the easiest to analyze: in (4) one naively disregards the terms which are $O(\varepsilon)$, and so obtains $f(U,V) = 0$ and $cV' + g(U,V) = 0$. On either I or III, one solves for $U = h_{\pm}(V)$ and thus obtains

$$V' = \frac{g(h_{\pm}(V),V)}{c} \qquad (5)$$

The more difficult part comes in attempting to connect I and III by means of II and IV. This is where layer analysis saves the day. These transitions are supposed to be rapid, so that $U$ and $V$ have steep gradients. A new length scale is needed at the location of each of these jumps. For example, let $z_0$ be the location of a transition II, and define $\zeta = \frac{z-z_0}{\varepsilon}$. With this new scale, the equation (4) becomes

$$U_{\zeta\zeta} + cU_{\zeta} + f(U,V) = 0, \tag{6a}$$

$$cV_{\zeta} + \varepsilon g(U,V) = 0. \tag{6b}$$

Again disregarding the $\varepsilon$ term, one concludes that approximately $V_{\zeta} = 0$, so that $V = \text{const} = V_0$. This has been anticipated by depicting the transitions II and IV as horizontal lines. The first equation (6a) is now a self-contained scalar equation with $V$ an input parameter; we can refer to the results in sec. 2. On the line II, $f(U,V_0)$ is such that $f_u < 0$ at both of the endpoints $U_{\pm} = h_{\pm}(V_0)$, and has a single internal zero. Therefore according to fact (iii) in sec. 2, there exists a unique stable front solution of (6a) with $U(\pm\infty) = h_{\pm}(U_0)$, and a unique velocity $c$. To emphasize that $c$ depends on $V_0$, we write $c = c_0(V_0)$. Now $J = \int_{U-}^{U_+} f(U,V_0)dU > 0$ for transitions near the bottom of the curve, so we assume it is positive on II.

Therefore $c < 0$, according to fact (ii).

An entirely similar analysis (but now with $U_+$ and $U_-$ reversed) regarding transition IV yields another relation $c = c_1(V_1) < 0$. For a given $V_0$ with $c(V_0) < 0$, two possibilities arise:

(a) There exists a $V_1$ such that $c_1(V_1) = c_0(V_0)$. Let $V^*$ be the value of $v$ such that $\int_{h_-(V^*)}^{h_+(V^*)} f(U,V^*)\,dU = 0$; it is clear that such a $V^*$ exists. As $V_1$ increases from $V^*$ to its maximum possible value, $|c_1(V_1)|$ increases from zero (in typical cases monotonically) to a maximum value $|c_m|$. If $|c_m| >$ $|c_0(V_0)|$, then case (a) holds. Then the construction of the traveling waves is complete: its velocity is $c = c_0(V_0) = c_1(V_1)$, and its U-profile has been pieced together from four elements as follows:

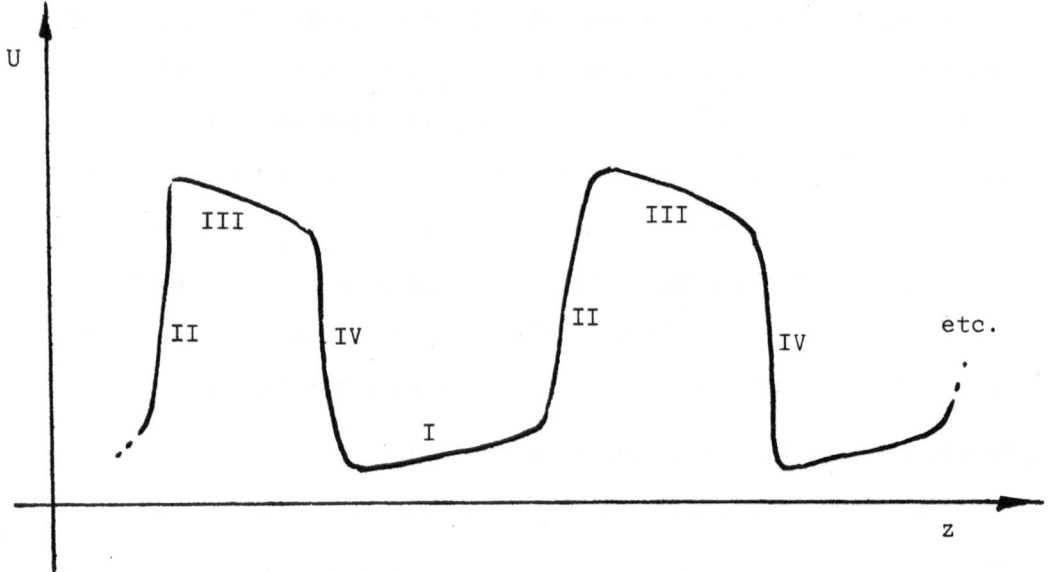

Fig. 3

(b) There exists no such $V_1$. This is the case when $|c_m| < |c_0(V_0)|$. Although the above construction fails, there still exists a periodic traveling wave. It can be obtained by moving the segment IV to the top of the nullcline:

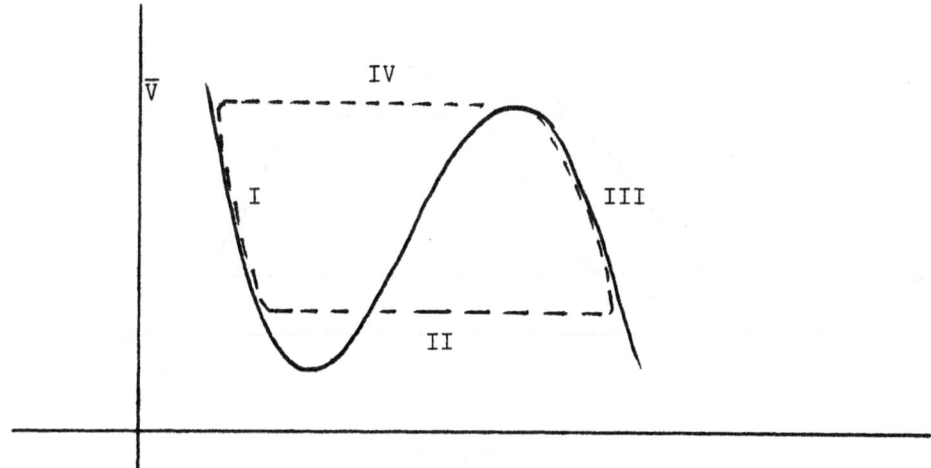

Fig. 4

For this value of $V_1 = \bar{V}$, $f(U,\bar{V})$ is of constant sign on IV and fact (iii) of sec. 2 implies the existence of a continuum of possible velocities $c$ with $|c| >$ some positive minimum. It turns out that this minimum equals $\lim_{V \to \bar{V}} c_1(V)$, and so if the continuum is included, all possible negative velocities may be obtained by choosing $V_1$ appropriately. So in case (b), we take $V_1 = \bar{V}$ and choose the front solution of (6a) whose velocity equals $c_0(V_0)$.

From the above, it is clear that a one-parameter family of periodic wave trains exists. Going back to (5), we see that the assumed sign of $g$ on $\Gamma_\pm$, together with the fact that $c < 0$, gives the information that the directions of traversal of III and I are compatible with the analysis.

This argument really only assumes the inequalities on $g$ to hold on I and III, not on all of $\Gamma_\pm$. For example, if $g = 0$ on the curve as shown below, then there exists a stable equilibrium at $(U_e, V_e) = P$. Periodic wave trains exist as before, provided only that $V_0 > V_e$.

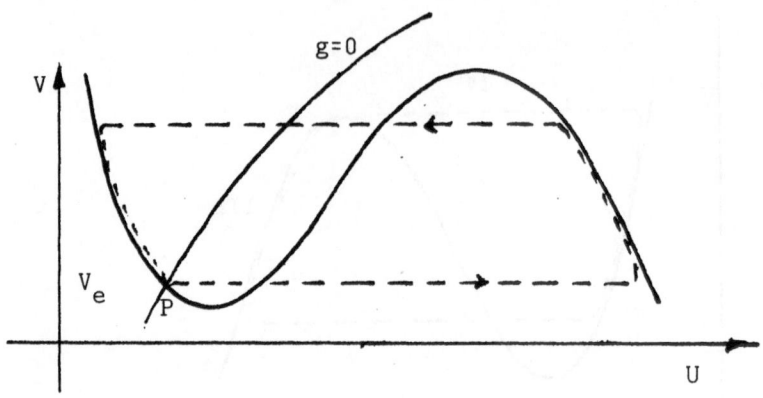

Fig. 5

But in addition, there now exists a nonperiodic solution of (4) with orbit shown in Fig. 5. It leaves the equilibrium P at z = -∞, and returns to it as z → ∞. This is a typical pulse solution, as described before the FitzHugh-Nagumo equation. Here also, case (a) or (b) may arise (as noted by Ostrovskii-Yahno 1975).

## 7. Initiation Problems

An important question is whether given initial data will evolve into one of the traveling wave solutions discussed above. Layer analytical arguments, easily employed to construct such steady waves, are also easily used to decide the answer to such transient questions. This is most easily illustrated for pulses, as I shall do below. The question of how periodic wave trains are initiated is more involved; possible mechanisms are given in Tyson-Fife (1980) and Fife (1981).

Consider the partial differential equation whose traveling wave equation is (4):

$$\varepsilon u_t = \varepsilon^2 u_{xx} + f(u,v) \tag{7a}$$
$$v_t = g(u,v), \tag{7b}$$

with nullcurves as shown:

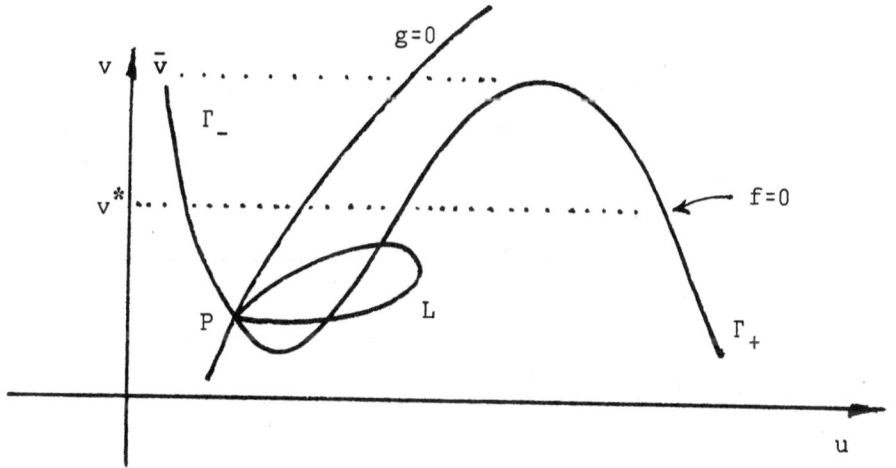

Fig. 6

imagine given an initial state $(u(x,0), v(x,0))$ which differs from the stable uniform state P only within the confines of a bounded interval. With x as parameter, suppose the initial data trace out the curve L in Fig. 6, beginning and ending at P.

Assuming $\varepsilon \ll 1$ , we can infer the subsequent evolution of the solution $u(x,t)$:

First, in a time interval $= 0(\varepsilon)$, the curve is attracted to the two branches $\Gamma_+$ and $\Gamma_-$. This is because, from (7a),

$$u_t = \frac{1}{\varepsilon} f(u,v) + 0(\varepsilon),$$

as long as $u_{xx} = 0(1)$. The part of L to the right of the intermediate branch is attracted to $\Gamma_+$ , and vice versa.

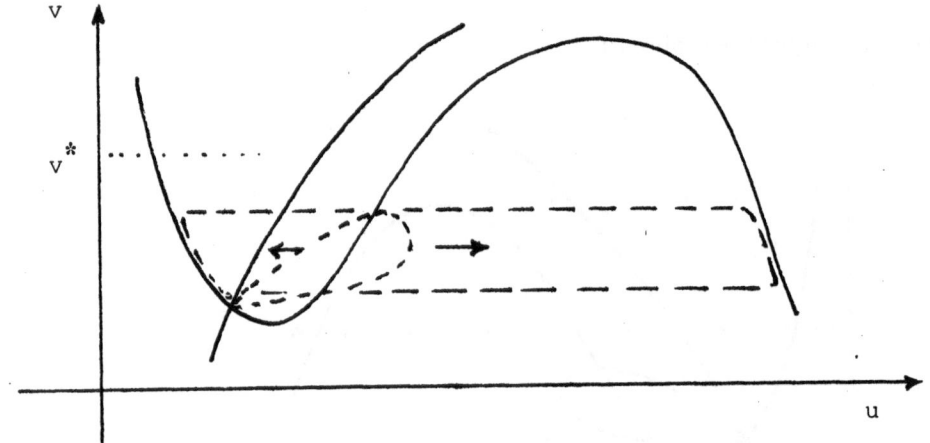

Fig. 7a

In the x-u plane, this results in a flat-topped configuration .

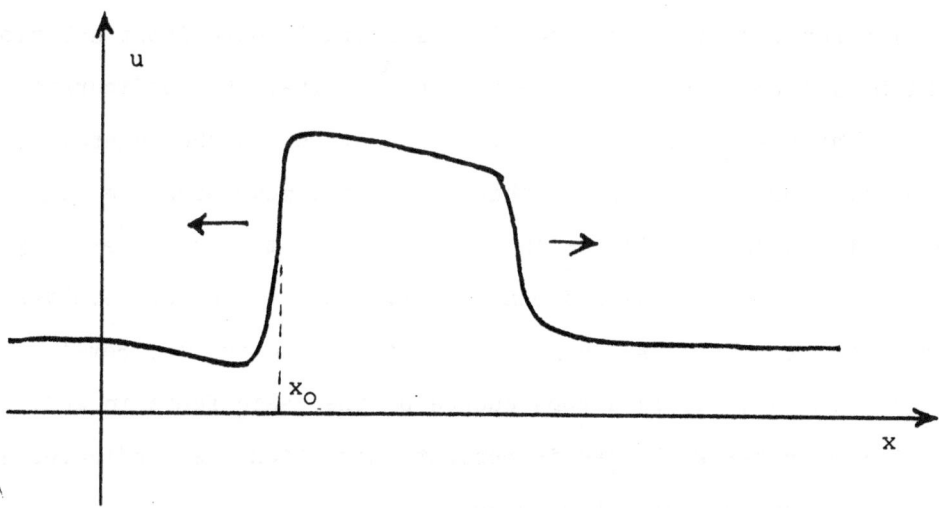

Fig. 7b

The function $v$ does not change much during this initial stage.

It is clear that the $u$ profile, as a result of this equilibration, will steepen at the two locations corresponding to the two horizontal dotted lines in Fig. 7a; these steep parts are shown in Fig. 7b. Eventually, $u_{xx}$ will no longer be $O(1)$ there, and the diffusion term $u_{xx}$ in (7a) becomes important. At that point, one introduces scaled local variables $\xi = \dfrac{x-x_0}{\varepsilon}$ , $\tau = \dfrac{t}{\varepsilon}$ , at the location $x_0$ where such a transition is made, in order to obtain the system

$$u_\tau = u_{\zeta\zeta} + f(u,v),$$
$$v_\tau = \varepsilon g(u,v) \sim 0.$$

In this same system, $v$ is approximately constant, so that a

scalar equation is obtained. It has a stable wave front solution
which will be approached, due to the fact that the configuration
of u at the end of the equilibration stage has the general
character of such a front. This front will move with the same
velocity in the (x,t) as in the (ζ,τ) system; this velocity
will be 0(1), and will depend on v. And v in turn evolves
according to (7b). As long as v, at the two fronts, remains
less than $v^*$, it is a consequence of the basic facts in sec. 2
that the fronts will move in opposite directions, as indicated in
Fig. 7b, thus widening the pulse.

Between the two fronts, (u,v) lies on $\Gamma_+$, where g > 0,
so that v increases. Eventually, v will attain its maximum
possible value on $\Gamma_+$, , namely $\bar{v}$ (see Fig. 4), at some point
between the two diverging fronts. Then (u,v) will be forced off
the nullcurve of f, and at that location will be strongly
attracted to $\Gamma_-$. This is illustrated below.

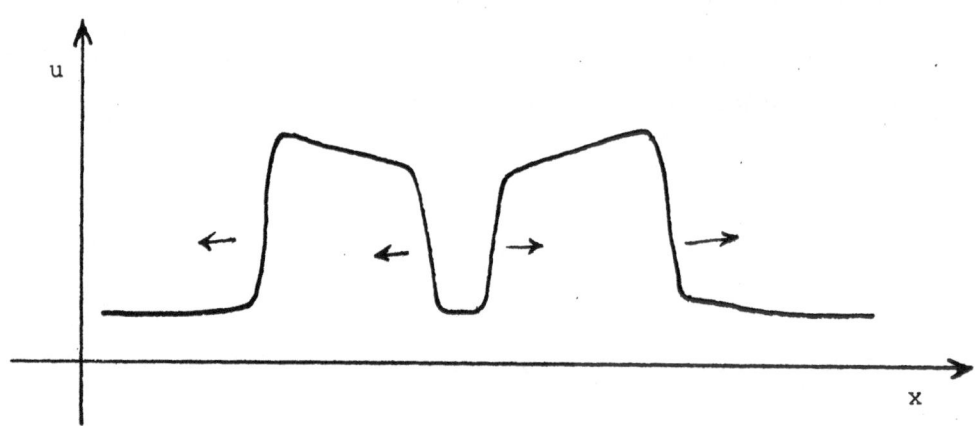

Fig. 8

In the same manner as before, this generates two new fronts which "follow the leader" and diverge. The picture now begins to appear like a pair of pulses moving in opposite directions.

A continuation of the same type of analysis shows that indeed two diverging pulses with standard shape and velocity do form, and everywhere else the medium approaches the equilibrium state.

This process of double-pulse formation was described in Ostrovskii-Yahno. Other transient phenomena were described in Fife (1976b,c), Keener (1980), Tyson-Fife (1980). The crucial assumptions on the initial perturbation (L, in Fig. 6) were (1) a part of L lies to the right of the ascending intermediate branch of the nullcline of f, and (2) all of L lies below the line $v = v^*$.

If one of these two conditions is violated, something else happens. It is left as an exercise to show that (a) if (1) is violated (and L is situated above the minimum point of $\Gamma_-$, as Dr. Yoshizawa pointed out), no pulse forms, and the medium returns uniformly to P; (b) if (1) holds but (2) is violated, so that the picture in Fig. 9 holds, then a single pulse forms, rather than two.

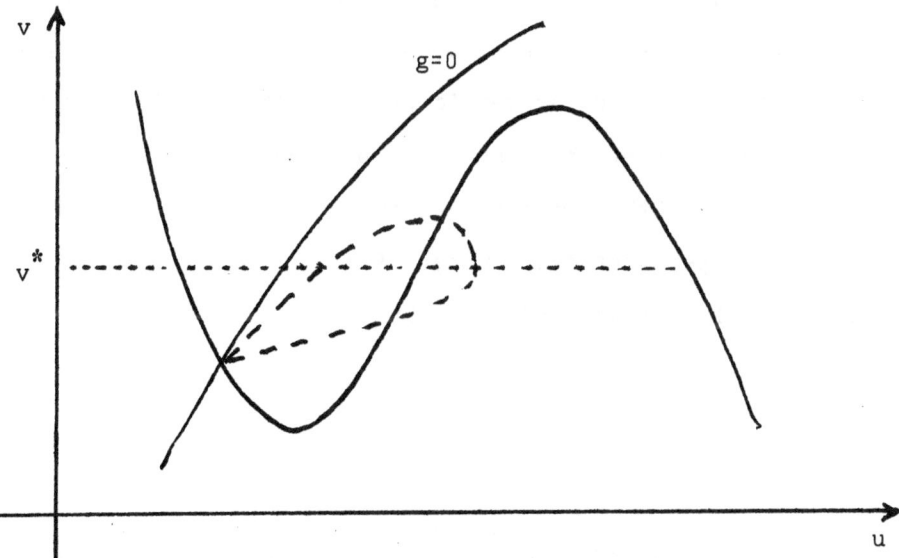

## 8. Discussion

Sigoidal nonlinearities, though apparently rather special, are practically ubiquitous in models for large-amplitude waves and patterns in reaction-diffusion systems. When, in addition, the variable with the sigmoidal nonlinearity is "fast" relative to the other, then simple lowest-order analysis, as described here, is applicable in both transient and steady-state problems. One can then understand these patterns and their dynamics without resorting to numerical simulation. This is the reason why Tyson's (1979) discovery that the Oregonator equations, when scaled properly, result in a 2-component sigmoidal system, was a breakthrough which led to the ability to analyze, much more completely than before, target patterns in the Belousov-Zhabotinsky reagent (Tyson-Fife 1980). It should be noted that Tyson (1982) has recently gone one step further and shown how the Field-Koros-Noyes mechanism can be reasonably reduced to the Oregonator model.

Applications in population dynamics have scarcely been mentioned here, but the methods are definitely applicable. This was brought out in, among other places, the splendid recent papers by Fujii, Mimura and Nishiura (1981) and Gardner and Smoller (1981). In the latter, the existence of traveling waves in a diffusive predator-prey system is proved. This system has the essential features of the singularly perturbed sigmoidal systems discussed here. Other recent rigorous results have been obtained which are not perturbative (Dunbar 1981, Conley-Gardner 1981)) or involve a different kind of perturbation (Conley-Fife 1982). The

former establish certain kinds of wave fronts in diffusive
Volterra-Lotka systems, and the latter uses the Conley index to
establish their existence in a system of three reaction-diffusion
equations occuring in a selection-migration model in population
genetics.

## References

1. D. G. Aronson and H. F. Weinberger 1978, Multidimensional nonlinear diffusion arising in population genetics, Advances in Math. 30, 33-76.

2. W. R. Bush and F. E. Fendell 1970, Asymptotic analysis of laminar flame propagation of general Lewis numbers, Combustion Science and Technology 1, 421-428.

3. G. Carpenter 1977, Periodic solutions of nerve impulse equations, J. Math. Anal. Appl. 57, 152-173.

4. G. Carpenter 1977, A geometric approach to singular perturbation problems with applications to nerve impulse equations, J. Differential Equations 23, 335-367.

5. R. Casten, H. Cohen, and P. Lagerstrom 1975, Perturbation analysis of an approximation to Hodgkin-Huxley theory, Quart. Appl. Math 32, 365-402.

6. H. Cohen 1971, Nonlinear diffusion problems, Studies in Appl. Math. 7, Math. Assoc. of America and Prentice Hall, 27-63.

7. C. Conley 1975, On travelling wave solutions of nonlinear diffusion equations, Dynamic Systems Theory and Appl. (J. Moser, Editor), Lecture Notes in Physics, Vol. 38, Springer-Verlag Berlin and New York, 498-510.

8. C. Conley and P. Fife 1982, Critical manifolds, travelling waves, and an example from population genetics, J. Math. Biology, to appear.

9. C. Conley and R. Gardner 1981, An application of the generalized Morse index to travelling wave solutions of a competitive reaction-diffusion model, preprint.

10. C. C. Conley and J. A. Smoller 1976, Remarks on travelling wave solutions of nonlinear diffusion equations, Structural Stability, The Theory of Catastrophes, and Applications in the Sciences (P. Hilton, Ed.), Lecture Notes in Math., Vol. 525, Springer-Verlag, Berlin and New York.

11. S. R. Dunbar 1981, Travelling wave solutions of diffusive Volterra-Lotka interaction equations, Ph.D. Thesis, University of Minnesota.

12. G. B. Ermentrout 1980, Stationary homogeneous media I. One dimensional, isotropic media, preprint.

13. G. B. Ermentrout and J. D. Cowan 1979, Temporal oscillations in neuronal nets, J. Math. Biol., 7, 265-280.

14. G. B. Ermentrout and J. D. Cowan 1980, Secondary bifurcation in neuronal nets, SIAM J. Appl. Math., 39, 323-340.

15. G. B. Ermentrout and J. D. Cowan 1980, Large scale spatially organized activity in neural nets, SIAM J. Appl. Math., 38, 1-21.

16. P. C. Fife 1981, On the question of the existence and nature of homogeneous-center target patterns in the Belousov-Zhabotinskii reagent. pp. 45-56 in Analytical and Numerical Approaches to Asymptotic Problems in Analysis, O. Axelsson, L. S. Frank, A. Van der Sluis, eds., Mathematics Studies 47, North-Holland, Amsterdam.

17. P. C. Fife 1979, Mathematical Aspects of Reacting and Diffusing Systems, Lecture Notes in Biomathematics 28, Springer-Verlag.

18. P. C. Fife 1976a, Boundary and interior transition layer phenomena for pairs of second order differential equations, J. Math. Anal. and Appls. 54, 497-521.

19. P. C. Fife 1976b, Pattern formation in reacting and diffusing systems, J. Chem. Phys. 64, 854-864.

20. P. C. Fife 1976c, Singular perturbation and wave front techniques in reaction-diffusion problems, in: SIAM-AMS Proceedings, Symposium on Asymptotic Methods and Singular Perturbations, New York, 23-49.

21. P. C. Fife 1977, Asymptotic analysis of reaction-diffusion wave fronts, Rocky Mountain J. Math., 7, 389-415.

22. P. C. Fife and W. M. Greenlee 1974, Interior transition layers for elliptic boundary value problems with a small parameter, Usp. Matem. Nauk SSSR, 24, 103-130; Russ. Math. Surveys 29, 103-131.

23. P. C. Fife and J. B. McLeod 1977, The approach of solutions of nonlinear diffusion equations to travelling front solutions, Arch Rational Mech. Anal. 65, 335-361. Also: Bull. Amer. Math. Soc. 81, 1075-1078 (1975).

24. H. Fujii, M. Mimura and Y. Nishiura 1981, A picture of global bifurcation diagram in ecological interacting and diffusing systems, preprint.

25. R. Gardner and J. Smoller 1982, The existence of periodic travelling waves for singularly perturbed predator-prey equations via the Conley index, preprint.

26. S. P. Hastings 1974, The existence of periodic solutions to Nagumo's equation, Quart. J. Math. Oxford, Ser. 25, 369-378.

27. S. P. Hastings 1976, On travelling wave solutions of the Hodgkin-Huxley equations, Arch. Rational Mech. Anal., 60, 229-257.

28. S. P. Hastings 1976, On the existence of homoclinic and periodic orbits for the FitzHugh-Nagumo equations, Quart, J. Math. Oxford, Ser. 27, 123-134.

29. Ya. I. Kanel' 1962, On the stabilization of solutions of the Cauchy problem for the equations arising in the theory of combustion, Mat. Sbornik 59, 245-288.

30. J. P. Keener 1980, Waves in excitable media, SIAM J. Appl. Math. 39, 528-548.

31. A. N. Kolmogorov, I. G. Petrovskii, and N. S. Piskunov 1937, A study of the equation of diffusion with increase in the quantity of matter, and its application to a biological problem, Bjul. Moskovskovo Gos. Univ. 17, 1-72.

32. N. Kopell and L. N. Howard 1973, Plane wave solutions to reaction-diffusion equations, Studies in Appl. Math. 52, 291-328.

33. P. Ortoleva and J. Ross 1975, Theory of propagation of discontinuities in kinetic systems with multiple time scales: fronts, front multiplicity, and pulses, J. Chem. Phys. 63, 3398-3408.

34. L. A. Ostrovskii and V. G. Yahno 1975, The formation of pulses in an excitable medium, Biofizika 20, 489-493.

35. L. M. Pismen 1978, Multiscale propagation phenomena in reaction-diffusion systems, preprint.

36. J. Rinzel and D. Terman 1981, Propagation phenomena in a bistable reaction diffusion system, preprint.

37. A. M. Turing 1953, The chemical basis of morphogenesis Phil. Trans. Roy. Soc. Lond. B237, 37-72.

38. J. Tyson 1982, On scaling and reducing the Field-Koros-Noyes mechanism of the Belousov-Zhabotinskii reaction, preprint.

39. J. Tyson 1979, Oscillations, bistability, and echo waves in models of the Belousov-Zhabotinskii reaction, Ann. N. Y. Acad. Sci. 36, 279-295.

40. J. Tyson and P. C. Fife 1980, Target patterns in a realistic model of the Belousov-Zhabotinskii Reaction, J. Chem. Physics 73, 2224-2237.

41. K. Uchiyama 1978, The behavior of solutions of some nonlinear diffusion equations for large time, J. Math. of Kyoto University 18, 453-508.

CHAPTER 3

## ASYMPTOTIC BEHAVIOR OF STATIONARY

## HOMOGENEOUS NEURONAL NETS

G. B. Ermentrout
National Institutes of Health
Bldg. 31, Rm. 4B-54
Bethesda, MD 20205/USA

## 1. Introduction

There are many approaches available for the explanation of the behavior of nervous systems. These range from the physiological models of nerve membranes [Rinzel, 1978] to abstract models of learning [Grossberg, 1980]. In an attempt to connect these disparate approaches, one is forced into a somewhat intermediate level of modelling. Over the last five years, we have studied the interactions of simplified neurons in order to deduce more global behavior; such models are called neuronal nets.

Neuronal nets are aggregates of elements each of which lump together the properties of a single nerve cell. Each neuron is capable of producing trains of action potentials which propagate to other regions of the net. The influence of one neuron on another is a function of their spatial positions. At the synaptic site the presynaptic impulses are converted into postsynaptic potentials, (PSP's) either excitatory or inhibitory and these are spatially and temporally summed to form the somatic membrane potential. This graded potential causes the generation of action potentials (≡ impulses), starting the whole cycle again.

Neuronal nets have been used to study various psychological [Grossberg, 1980] and psychophysical [Wilson and Cowan, 1973] functions in the brain. They are capable of producing a surprising range of dynamic behaviors including waves, oscillations, spatial patterns, and chaos. In a recent book [an der Heiden, 1980] the properties of neuronal nets have been discussed in a fairly general fashion. In this paper, some of the asymptotic properties of neuronal nets (i.e. longtime behavior) in the absence of inhomogeneous inputs will be examined. When possible the results will be reported for completely general media. Simple neural nets will serve to clarify the underlying principle. This approach is plausible because homogeneity is assumed. That is, the dynamic properties of the net are assumed to be independent of spatial position. Furthermore, these properties are also time independent so that we do not consider nets with temporally or spatially varying inputs. Because of this assumed symmetry, we are able to construct a variety of new, more complicated, behaviors from simpler behavior. Our approach is through the use of perturbation and bifurcation theory. Generally, we will start with a spatially uniform solution and obtain non-trivial spatially patterned solutions. For example, if the neuronal net admits a bulk periodic solution, it is possible to construct propagating wavetrains in some instances. Our results allow us to give some rudimentary explanations for certain pathological physiological findings. Two of those discussed in this paper are epilepsy and flicker phosphenes.

In the remainder of this section we formulate a mathematical model for a typical neuronal net. The next two sections describe some of the behavior possible in general neuronal nets and present two examples. Finally, we relate these mathematical solutions to physiological phenomena.

To motivate the generality of the next two sections, and to give the reader a taste of the complexity of these models, we derive a typical neuronal net. There are N layers of cells, each distributed in some region in $\mathbb{R}^1$ or $\mathbb{R}^2$; each layer corresponds to a population of cells of a specific type (e.g. excitatory, inhibitory, pyramidal, etc.). Let $U(x,t) \epsilon \mathbb{R}^N$ be the vector of instantaneous firing rates (impulses per second) of these cells at each point x and time t. Each cell in each layer sends axons out to the dendrites of other cells which are synaptically coupled. The fraction of impulses a cell in layer j at a point x receives from a cell in layer k at a point x' is a function of distance and potentials:

$$P_{jk}(x,x',t) = S_{jk}(x-x',U_j(x,t), U_k(x',t)$$

or in matrix form:

$$P(x,x',t) = S(x-x', U(x,t), U(x',t)) .$$

For most neuronal nets, $S_{jk}$ is linear:

$$P_{jk}(x,x',t) = S_{jk}(x-x') U(x',t) .$$

An impulse results in transmitter release and a concomitant change in the post-synaptic potential. This is modeled by the impulse response function (IRF) which of course depends on past time. Thus, the post-synaptic potential contributed by the above impulse is:

$$PSP_{j,k}(x,x',t) = H'_{jk}[U_j(x,t), \int_0^\infty dt' \ H^2_{jk}(P_{jk}(x,x',t-t'),t')] .$$

In most systems $H'_{jk}(u,v) = v$ and $H^2_{jk}(w) = h_{jk}(t') w(t-t')$, i.e., linear temporal summation. The total synaptic contribution to the cell in layer j at $(x,t)$ is the sum over each layer and over each x':

$$POT_j(x,t) = \sum_{j=1}^N \int_\Omega dx' \ P_{jk}(x,x',t) .$$

Here, $\Omega$ is the domain of the net and in this paper will always be the real line. By assuming an infinite cortical layer, we eliminate the boundary effects. Since cortical interactions are only of the order of 100μ-500μ and the cortical nets are of the order of centimeters, this approximation is not unreasonable near the center of the net. To close the system, we must relate the firing rate $U_j(x,t)$ to the membrane potential, $P_j(x,t)$:

$$U_j(x,t) = N_j(P_j(x,t)) .$$

$N_j$ is typically S-shaped or asymptotically linear. In general, this is the only nonlinearity in most neuronal nets. Most models [Amari, 1977, Ellias and Grassberg, 1975, Wilson and Cowan, 1973, and Maxwell and Renninger, 1980] have the form:

$$U(x,t) = N(H(t) * S(x) \otimes U(x,t)) ,$$

a particular case of the model derived here. Note that by * we mean the temporal convolution:

$$U(t) * V(t) \equiv \int_0^\infty U(t') V(t-t')dt'$$

and by $\otimes$ we mean the spatial convolution:

$$U(x) \otimes V(x) \equiv \int_{-\infty}^\infty U(x') V(x')dx' .$$

## 2.   General Theory and Example

In a series of papers [Ermentrout, 1981 a, b, c] and in [Ermentrout, 1979] a general multiple scaling theory coupled with symmetry arguments is developed for the study of one dimensional multicomponent system which are homogeneous and stationary (their physical properties do not vary in space or with time). We generally consider a problem of the form:

(2.1)    $L(\gamma)u + F(u) = 0 , \quad U(x,t) \in \mathbb{R}^N$

where L is linear, $\gamma$ is a parameter, and F is quadratic plus higher order terms. $L(\gamma)$ can be any spatio-temporal operator. The solution to (2.1) is $u \equiv 0$, so we ask whether it is a stable solution. Since $L(\gamma)$ is homogeneous and stationary, the linearized problem (2.1 without F) has solutions of the form $\chi e^{ikx+\nu t}$, $\chi \in \mathbb{C}^N$, $k \in \mathbb{R}$, $\nu \in \mathbb{C}$, which satisfy:

(2.2)    $\hat{L}(k,\nu,\gamma)\chi = 0 .$

The N X N complex matrix $\hat{L}(k,\nu,\gamma)$ is defined as:

(2.3)    $\hat{L}(k,\nu,\gamma)\chi = \lim_{\substack{x \to 0 \\ t \to 0}} L(\gamma)\chi e^{ikx+\nu t} .$

Since $\hat{L}$ is a matrix, (2.1) has a solution if and only if

(2.4)    $f(\nu,k,\gamma) = \det[\hat{L}(k,\nu,\gamma)] = 0 .$

(2.4) gives a relationship between $\nu$ and the two numbers $(k,\gamma)$. Clearly if Re $\nu > 0$

then u = 0 is unstable since small perturbations will exponentially increase as $e^{\nu t}$.

We suppose there is a pair $(\gamma_0, k_0)$ such that for any $(\gamma, k)$ in a neighborhood, $\eta$, of $(\gamma_0, k_0)$, (2.4) has a solution $\nu(k,\gamma)$ with maximal real part (unique up to complex conjugates). For $\gamma < \gamma_0$, Re $\nu(k,\gamma) < 0$ for all $(\gamma, k) \in \eta$. At $\gamma = \gamma_0$, Re $\nu(k_0, \gamma_0) = 0$, but for $|k| \neq |k_0|$, Re $\nu(k, \gamma_0) < 0$. Finally, for $\gamma > \gamma_0$, Re $\nu(k_0, \gamma) > 0$ and there is a small band of values of k near $k_0$ such that Re $\nu(k, \gamma) > 0$. These assumptions state that as the parameter $\gamma$ is increased, the solution u = 0 loses stability at a wavenumber $k_0$. Let $\omega_0 = $ Im $\nu(k_0, \gamma_0)$. There are four distinct cases: $(k_0, \omega_0) = (0,0)$; $(k_0, \omega_0) = (k_0, 0)$; $(k_0, \omega_0) = (0, \omega_0)$; $(k_0, \omega_0) = (k_0, \omega_0)$ (i.e. $k_0 \neq 0$, $\omega_0 \neq 0$). Furthermore, the final case can be subdivided into two cases: (i) $f(i\omega_0, k_0, \gamma_0) = f(-i\omega_0, k_0, \gamma_0) = 0$ (ii) $f(i\omega_0, k_0, \gamma_0) \neq f(-i\omega_0, k_0, \gamma_0)$. Case (i) occurs in an isotropic medium where spatial interactions are only functions of distance. Case (ii) occurs when there is anisotropy. The models cited at the end of section 1 are all isotropic.

In [Ermentrout, 1981 a,b,c] we use multiple scales and bifurcation theory to construct new nonzero solutions to (2.1) when $\gamma \neq \gamma_0$. This is done by introducing a scaling, $\gamma = \gamma_0 + \epsilon^2 \hat{\gamma}$, and a new variable, u = $\epsilon \nu$. Two additional space-time scales are introduced and are generally of the form $\xi = \epsilon x$ and $\tau = \epsilon^2 t$. We seek solutions of the form

$$v(x,t) = V(x,t,\xi,\tau;\epsilon) .$$

V is expanded in a power series in $\epsilon$ and the lowest order solution is always of the form:

$$(2.5) \qquad V_0(x,t,\xi,\tau) = \sum_{j=1}^{d} z_j(\xi,\tau) \chi_j(x,t) .$$

$\chi_j(x,t)$ is any of the functions $\chi_0 e^{\pm i\omega_0 t \pm i k_0 x}$, $\hat{L}(i\omega_0, k_0, \gamma_0)\chi_0 = 0$, and d is the dimension of the nullspace of $L(\gamma_0)$. It is either 1, 2 or 4. The behavior of the solutions is determined by the complex scalar functions $z_j(\xi,\tau)$ which satisfy $z_1 = \bar{z}_2$, $z_3 = \bar{z}_4$. These functions obey universal parabolic partial differential equations the form of which depends only on the type of bifurcation (i.e. whether or not $\omega_0 = 0$, $k_0 = 0$). The coefficients in these universal equations are the only problem dependent parameters. Table I contains the three different equations. Case (i) leads to stable propagating wave fronts with amplitudes of order $\epsilon$ and velocities of order $1/\epsilon$. Stability occurs as long as $\nu$ is small and q > 0. Case (ii) is the Turing bifurcation and one obtains a one parameter family stationary spatially periodic solutions. These are stable when b > 0. Case (iii) is a construction of long periodic wave trains near an Hopf bifurcation. The velocity of these waves is $0(1/\epsilon)$ and some waves are stable as long as Re b > 0. Case (iv) corresponds to periodic wave trains which travel in only one direction with velocities of the order $\omega_0/k_0$. These are "short" waves in the sense that as $\epsilon \to 0$ their wave number and velocity remain finite. Stability occurs for some of them when Re b > 0. Finally case (v) corresponds to $\omega_0 \neq 0$, $k_0 \neq 0$ in an isotropic medium. Solutions for which $z_1 \neq 0$ and $z_2 = 0$ or vice versa lead to a one

TABLE I

| | d | $\omega_0$ | $k_0$ | parameters | Equation |
|---|---|---|---|---|---|
| i | 1 | 0 | 0 | $r,D,q,\nu \in \mathbb{R}$ | $r_\tau = Dr_{\xi\xi} + r(\hat{\gamma}-qr^2+\nu r)$ |
| ii | 2 | 0 | $\neq 0$ | $z \in \mathbb{C}, D,a,b \in \mathbb{R}$ | |
| iii | 2 | $\neq 0$ | 0 | $z,D,a,b \in \mathbb{C}$ | $z_\tau = Dz_{\xi\xi} + z(a\hat{\gamma}-bz\bar{z})$ |
| iv | 2 | $\neq 0$ anisotropic | $\neq 0$ | $z,D,a,b \in \mathbb{C}$ | |
| v | 4 | $\neq 0$ isotropic | $\neq 0$ | $z_1,z_2,D,a,b,c \in \mathbb{C}$ $\xi = \epsilon x + \bar{m}\epsilon t$ $\eta = \epsilon x - \bar{m}\epsilon t$ $\tau = \epsilon^2 t$ | $z_{1\tau} = Dz_{1\eta\eta}+z_1(a\hat{\gamma}-bz_1\bar{z}_1-cz_2\bar{z}_2)$ $z_{2\tau} = Dz_{2\xi\xi}+z_2(a\hat{\gamma}-bz_2\bar{z}_2-cz_1\bar{z}_1)$ $z_{1\xi} = z_{2\eta} = 0$ |

parameter family of traveling waves which propagate in either direction. The velocity is $\approx \omega_0/k_0$ and they are stable as long as Re $c$ > Re $b$ > 0. Solutions of the form $z_1 = z_2 = e^{i\sigma\tau}$ exist and correspond to standing spatially periodic oscillations. They are stable when Re $h$ > Re $c$ > 0.

EXAMPLE In this example we show how certain physical properties of the spatial connectivity can influence the behavior of the neural net. In [Maxwell & Renninger, 1980] a version of the following has been proposed as a model of the Limulus retina:

$$(2.6) \quad r(x,t) = m(\gamma + \int_{-\infty}^{\infty} S(x-x')dx' \int_0^{\infty} h(t')dt' \, r(x',t-t')) - m(\gamma) \; .$$

$r(x,t)$ represents the response of a single omnatidium at $(x,t)$, $m$ is a nonlinear function, $\gamma$ is the background illumination, and $h$, $S$ are the temporal and spatial transfer functions. $h(t)$ takes on the differential-delay form:

$$(2.7) \quad h(t) = 0 \qquad t \leq d$$
$$h(t) = \frac{1}{\mu} \exp(-(t-d)) \quad t \geq d \; .$$

Thus, an impulse causes an abrupt rise to the maximum PSP after a short delay. This is followed by exponential decay. For the problem:

$$(2.8) \quad L(\gamma)u = u(x,t) - m'(\gamma)h(t)* S(x)\otimes u(x,t) \; ,$$

and

$$(2.9) \quad \hat{L}(k,\nu,\gamma) = [1 - m'(\gamma)\hat{\ell}(k)e^{-\nu d}/(1+\mu\nu)] \equiv f(\nu,k,\gamma) \; .$$

$\hat{\ell}(k)$ is the Fourier transform of $S(x)$. This is a scalar equation and the next Lemma characterises $\nu(k,\gamma)$.

LEMMA 1 Re $\nu(k,\gamma)$ < 0 if and only if

(2.10)  (a)  $1 > m'(\gamma)\hat{\ell}(k) > -|\sec a|$ ,    where

        (b)  $a = -\frac{d}{\mu} \tan a$ ,    $a \epsilon (0,\pi)$ .

If for some $(k_0, \gamma_0)$:

(2.11)  (a)  $m'(\gamma_0)\hat{\ell}(k_0) = 1$    then    $\nu(k_0, \gamma_0) \equiv 0$

        (b)  $m'(\gamma_0)\hat{\ell}(k_0) = -|\sec a|$    then    $\nu(k_0, \gamma_0) = \pm ia/d$ .

If (2.11) occurs there is a loss of stability at $\gamma = \gamma_0$ with $(k_0, \omega_0) = (k_0, \nu(k_0, \gamma_0))$. Typically, $m$ is monotone so $m'(\gamma_0) > 0$. Thus for (2.11a) to occur $\ell(k_0) > 0$ and for

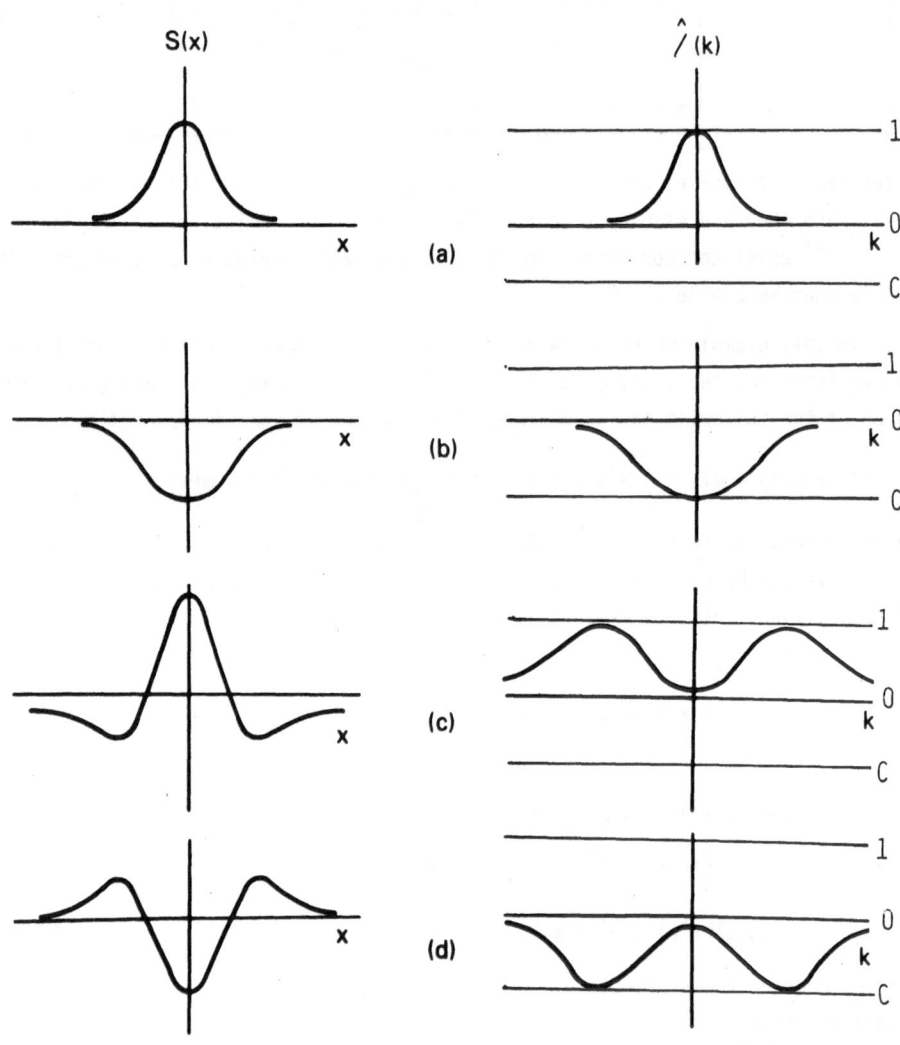

FIGURE 1

(2.11b) to occur $\hat{\ell}(k_0) < 0$. In Fig. 1 we depict the lines $y = 1$, $y = -|\sec a| \equiv c$, and the functions $y = \hat{\ell}(k_0)$, $S(x)$, for 4 qualitatively different cases: (a) pure excitation, (b) pure inhibition, (c) lateral inhibition/self excitation (LI/SE), (d) lateral excitation/self inhibition (LE/SI). Purely excitatory nets lead to $(k_0, \omega_0) = 0$ and propagating wave fronts. Inhibitory nets lead to long waves and oscillations since $\omega_0 \neq 0$ and $k_0 = 0$. LI/SE nets cause stationary periodic patterns since $\omega_0 = 0$, $k_0 \neq 0$. Finally LE/SI nets lead to propagating wave trains and since the medium is isotropic, the possibility of standing spatially periodic oscillations.

The lesson learned here is that the spatial interactions between networks of neurons can lead to important qualitative differences in their behavior. Local inhibition leads to a tendency toward time periodic behavior when there are delays, while local excitation tends to lead to spatial patterning. Lateral excitation is necessary to obtain spatio-temporal periodic solutions with order 1 velocities. The solutions constructed here can be given a certain amount of physiological meaning. Wave fronts are a means of switching the cortical net from one state to another - they also occur as a lowest order approximation to propagating pulse waves [see Burns, 1950 for a physiological example]. In section 4, we discuss the possible physiological meaning of the periodic patterns.

## 3. Waves

In this section we use the spatial homogeneity of the medium to construct large amplitude traveling waves (i.e. far from bifurcation). Through the use of a simple long wave number approximation, we arrive at a Burger-type equation which gives information on dispersion and stability of the resultant solutions. This approach has been exploited in chemical reaction diffusion equations [Ortoleva and Ross, 1974]. Our goal is to demonstrate that these waves are a natural property of any homogeneous medium. We introduce the WC [Wilson and Cowan, 1973] system as an example of the qualitative results.

We consider for simplicity the model system:

$$(3.1) \quad u(x,t) = F(h(t)* S(x) \otimes u(x,t))$$

where $h(t)$ and $S(x)$ are the temporal and spatial transfer functions, the domain, $\Omega$, is the whole line and F is nonlinear. $u(x,t) \in \mathbb{R}^n$, i.e. there are n neuron types. We define the matrix $\tilde{S}$ as:

$$(3.2) \quad \tilde{S} = \int_{-\infty}^{\infty} dx \, S(x) dx \ .$$

We suppose that the system:

$$(3.3) \quad u(t) = F(h(t)* \tilde{S}u(t))$$

admits a unique asymptotically stable limit cycle,

$$(3.4) \quad u(t) = P_0(t)$$

with period, $T \equiv 2\pi/\omega_0$. We define the inner product of two vector functions, $(u(t), v(t))$ as:

$$\frac{1}{T}\int_0^T u(t)\cdot v(t)dt \ .$$

We seek slowly varying waves of (3.1), that is solutions dependent on t, $\xi = \epsilon x$, and $\tau = \epsilon^2 t$ where $\epsilon \ll 1$ is small. We thus seek

$$u(x,t) = U(t,\xi,\tau,\epsilon)$$

U is expanded in $\epsilon$,

$$U = U_0 + \epsilon^2 U_1 + 0(\epsilon^4) \ ,$$

and substituted into (3.1). This leads to:

(3.5)　$U_0(t,\xi,\tau) = F(h(t)* \tilde{S} U_0(t,\xi,\tau)) \ .$

The solution to this is

$$U_0(t,\xi,\tau) = P_0(t + \psi(\xi,\tau))$$

where $\psi(\xi,\tau)$ is an arbitrary (at this stage!) function of $(\xi,\tau)$. The next equation is:

(3.6)　$L_0 U_1 \equiv [U_1 - A(t) \{h(t)*\tilde{S} U_1(t)\}] =$

$$A(t)\{-[th(t)]*\tilde{S} P_0'(t+\psi)\psi_\tau$$

$$+ h(t)*[S_1 P_0'(t+\psi)\psi_{\xi\xi} + S_1 P_0''(t+\psi)\psi_\xi^2]\} \ .$$

$A(t)$ is the Jacobian of F evaluated at $P_0(t+\psi)$. Here, ' denotes $\partial/\partial t$ and $S_1 =$

$.5 \int_{-\infty}^{\infty} x^2 S(x)dx$. (3.6) is of the form $L_0 v = w$. Since $L_0$ has a one dimensional null-space (spanned by $P_0'(t+\psi)$), (3.6) is solvable if and only if the right hand side is orthogonal to the left eigenfunction of $L_0$, i.e. the solution to $P_0^* L_0 = 0$. This generally exists and we can scale it to satisfy

$$<P_0^*(t), A(t)[(th(t))*\tilde{S}P_0'(t)]> = 1 \ .$$

Thus applying $P_0^*(t+\psi)$ to both sides yields the equation:

(3.7)　$\psi_\tau = a(S_1) \psi_{\xi\xi} + b(S_1) \psi_\xi^2$

where

(3.8)　$a(S_1) = <P_0^*(t), A(t)[h(t)*S_1 P_0'(t)]>$

$$b(S_1) = <P_0^*(t), A(t)[h(t)*S_1 P_0''(t)]> \ .$$

A traveling wave is a $2\pi$-periodic function of $\omega t + kx$, thus we seek solutions to (3.7) of the form:

$$\psi(\tau,\xi) = \sigma\tau + \alpha\xi \ .$$

This leads to $\hat{\omega} = (\omega_0 + \epsilon^2\sigma)$, $k = \epsilon\alpha$ where from (3.7)

(3.9)　$\sigma = b(S_1)\alpha^2 \ ,$　　so

(3.10)   $\omega = \omega_0 + b(S_1)k^2$ .

To test stability of the solution $\psi = \sigma\tau + \alpha\xi$, we linearize (3.7):

(3.11)   $\phi_\tau = a(S_1)\phi_{\xi\xi} + 2b(S_1)\alpha\phi_\xi$ .

Solutions to this are $\phi = e^{\lambda\tau}e^{ikx}$ so that:

(3.12)   $\lambda = - a(S_1)\kappa^2 + 2i\kappa\alpha$ .

If $a(S_1) > 0$ then these waves are stable since Re $\lambda < 0$. Alternatively if $a(S_1)$ is negative, no waves are stable, including the bulk oscillation $\sigma = \alpha = 0$. This last fact follows from the fact that non constant spatial perturbations grow in time. As long as $b(S_1) \neq 0$ there will be traveling waves, but they are stable only if $a(S_1) > 0$. Thus, while spatial interactions exert little effect on existence of waves, they are important in the physical manifestation; unstable waves will not appear.

As a concrete example, we present the Wilson-Cowan system and the functions $a(S_1)$, $b(S_1)$. Let $E(x,t)$ and $I(x,t)$ denote the firing rates of an excitatory and an inhibitory population of neurons. Let $h(t) = e^{-t}$ be the IRF and let $S_{jk}(x) = S(x/\sigma_{jk})/\sigma_{jk}$ be the spatial transfer functions of interactions between neuron type $j$ and neuron type $k$. $S(x)$ is even and monotone for $x > 0$, e.g. $S(x) = \exp(-|x|)/2$. The variables $\sigma_{jk}$ describe the spread of various types of interactions; there are three of them, $\sigma_{ee}$ - the excitatory to excitatory interaction, $\sigma_{ei}$ - the excitatory to inhibitory interactions, and $\sigma_{ie}$ - the inhibitory to excitatory interactions. So if $\sigma_{ei}$, $\sigma_{ie} \gg \sigma_{ee}$ there is lateral inhibition because inhibition is reinjected at further points than the self excitation. Similarly if $\sigma_{ee} \gg \sigma_{ie}$, $\sigma_{ie}$ there is dominant excitation (see Fig. 2).

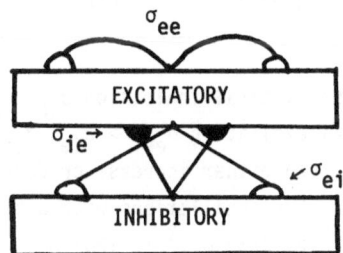

FIGURE 2

Let $\alpha_{ee}$, $\alpha_{ie}$, $\alpha_{ei}$ denote the amplitudes of the post synaptic potentials, e.g. $\alpha_{ee}$ is the amplitude of the EPSP on an excitatory cell, $\alpha_{ei}$ is the amplitude of the EPSP on an inhibitory cell, and $\alpha_{ie}$ is the amplitude of the IPSP on an excitatory cell. For simplicity, we include no inhibitory-inhibitory coupling. Finally, the nonlinear function is sigmoid of the form:

$$F(u) = (1 + \exp(-u))^{-1} .$$

Let $\theta_e$, $\theta_i$ be the thresholds of E and I. The WC model is:

(3.11)  $E(x,t) = F(\alpha_{ee} h(t)* S_{ee}(x) \otimes E(x,t) - \alpha_{ie} h(t)* S_{ie}(x) \otimes$

$$I(x,t) - \theta_e) - F(-\theta_e)$$

$$I(x,t) = F(\alpha_{ei} h(t)* S_{ei}(x) \otimes E(x,t) - \theta_i) - F(-\theta_i) \ .$$

When $\theta_e = 2$, $\theta_i = .85$, $\alpha_{ei} = 20$, $\alpha_{ie} = 10$, we find (3.11) admits oscillations for $\alpha_{ee} \in (9.52, 16.8)$. At $\alpha_{ee} = 9.52$ there is an Hopf bifurcation to stable oscillations; at $\alpha_{ee} = 16.8$ the limit cycle disappears through a homoclinic orbit.

For this model, we find

$$a(S_1) = \beta[J_1 \sigma_{ee}^2 - \sigma_{ie}^2 J_2 + \sigma_{ei}^2 J_3]$$
$$b(S_1) = \beta[I_1 \sigma_{ee}^2 - \sigma_{ie}^2 I_2 + \sigma_{ei}^2 I_3]$$

where $\beta = \int_{-\infty}^{\infty} x^2 S(x)dx$ and $I_k$, $J_k$ are constants depending on the period of the oscillation. Their values are shown in the table below.

TABLE 2

| $\alpha_{11}$ | $T_0$ | $J_1$ | $J_2$ | $J_3$ | $I_1$ | $I_2$ | $I_3$ |
|---|---|---|---|---|---|---|---|
| 12 | 74 | 2.16 | 6.2 | -4.986 | -3.648 | 6.08 | -.558 |
| 13 | 84 | 4.55 | 15.4 | -13.68 | -30.03 | 10.8 | 3.456 |
| 14 | 108 | 11.9 | 40.8 | -33.12 | -84.00 | 23.8 | 10.116 |
| 15 | 141 | 47.7 | 155.0 | -131.22 | -330.3 | 83.2 | 42.3 |
| 16 | 217 | 1296.16 | 3915 | -3281.4 | -8528.00 | 2050 | 894.06 |

Immediately evident is the fact that lateral inhibition $\sigma_{ie}$, $\sigma_{ei} \gg \sigma_{ee}$ destabilizes the propagating waves while lateral excitation $\sigma_{ee} \gg \sigma_{ie}$, $\sigma_{ei}$ results in stability. The dispersion relation, $\omega = F(k)$, can either increase or decrease for $k > 0$; this is dependent on the sign of $b(S_1)$. For oscillations with a period of $\approx 100$, we find the condition for stability is $\sigma_{ee}^2 > .2833 \sigma_{ei}^2 + .1204 \sigma_{ie}^2 \equiv L_1(\sigma_{ei}, \sigma_{ie})$. The frequency increases with wave number, $k$, if $\sigma_{ee}^2 < 3.42 \sigma_{ei}^2 + 2.783 \sigma_{ie}^2 \equiv L_2(\sigma_{ei}, \sigma_{ie})$. For $\sigma_{ee}^2 > L_2(\sigma_{ei}, \sigma_{ie})$, the frequency is lower for higher wave numbers. Thus, stable waves can have either type of dispersion relation and stronger lateral excitation results in a decreasing dispersion relation. This is in qualitative agreement with experimental results [Petsche, 1973] for dispersion relations on epileptic rabbit cortex. It is tempting to suggest that the reason that epilepsy is rare in sensory cortex is the strong lateral inhibition.

4.  Discussion

In the previous two sections we computed nontrivial spatio-temporal patterns for an infinite one-dimensional multilayer neuronal net. The cortex is inherently a two-dimensional multilayered structure. In order to understand the relationships of calculated patterns to real physiological phenomena, one has to realize that the one-dimensional patterns have two-dimensional analogues. Without describing the actual mathematical analysis, we discuss two of these two-dimensional patterns and their relationship to cortical pathologies.

In the last two sections, we saw that there were two types of spatially periodic activity: traveling or time-periodic standing waves and stationary spatial patterns. Both patterns are global in that they involve large amounts of tissue and both attain a high degree of synchrony. In a normal functioning cortex, one does not expect to see this type of activity - it has the effect of disrupting normal cortical processing. Global spatial patterns should be expected to occur only in pathological or abnormal situations. In several articles [Amari, 1978 and Wilson and Cowan 1973], localized peaks of spatial activity are conjectured to represent a simple form of short term memory. But in these patterns, only a local patch of tissue is active, all other regions remain normal and receptive to incoming stimuli. Thus, these local patterns contrast with the global activity calculated in this paper. We now describe two distinct pathological conditions which appear to result in global synchronous, spatial-temporal periodic patterns.

When penicillin or other agents are uniformly applied to the surface of exposed cortical tissue, shortly thereafter, large amplitude regular oscillations can be recorded intracellularly, extracellularly, and on the cortical surface. The oscil-lations have a frequency in the range of 8-10 cycles per second and have been used as an experimental model for clinical epilepsy. In many respects they mimic the onset of the clinical seizure [Ayala, et al., 1973]. Most investigators have recorded only from select neurons so that a complete spatio-temporal record is not generally obtained. Recently, a square surface electrode has been developed that allows one to record from 16 evenly spaced points simultaneously [Petsche, et al., 1974]. With the use of a computer, one can trace out equipotential contours as a function of time. What is found is that the oscillations occur simultaneously with a rotating potential field - i.e. rotating waves. In Figure 3A, we depict a typical contour plot taken from data on the rabbit cortex. The potential contours rotate clockwise and do not change sub-stantially over the whole period. Thus during experimental epilepsy, there exist rotating periodic waves. Far from the center, the wave fronts lose their curvature and begin to resemble plane waves (two-dimensional analogues of the periodic waves computed in sections 2 and 3). The wave trains computed analytically could then in some sense model the genesis and spread of epileptic activity. That the physio-logical syndrome spreads through synaptic interactions has been demonstrated by slicing across the cortex and abolishing the wave propagation [Petsche and Rappels-berger, 1970]. The ultimate mechanism underlying the epileptic oscillation is not known although there are many conflicting theories [see Ayala, 1973]. For this reason, the general model is good since it allows one to get to the basic qualitative

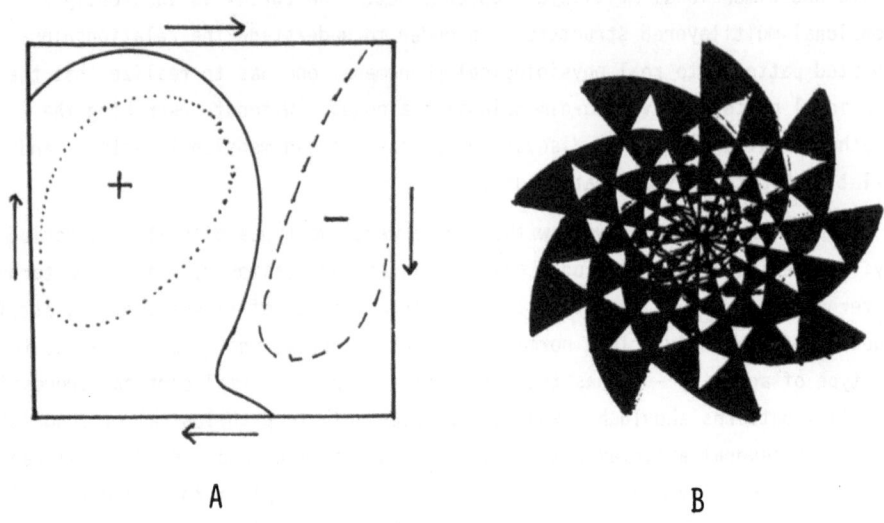

A                                       B

FIGURE 3

mechanisms underlying the spread of activity. As an example, we have seen that there
are two different routes to traveling waves: (1) the local region oscillates and
spatial interactions lead to waves (case (ii) in section 2 and section 3) and (2) the
actual spatial interactions lead to the oscillations although the space independent
medium is stable (case (iv) in section 2). So far there is not enough experimental
evidence to allow one to distinguish between these two.

There are various types of stimuli that can give rise to beautiful geometric
periodic patterns in the visual field. Among these are the application of binocular
deep pressure, [Tyler, 1978], binocular exposure to diffuse rapidly flickering light
[Remole, 1978], and ingestion of various hallucinogenic compounds [Siegel, 1977].
These geometric patterns seem to be limited to only a few different forms [Kluver,
1967]. A typical one is illustrated in Figure 3B. The most obvious property of
these patterns is their spatial periodicity. A cross section in one dimension yields
a pattern of activity analogous to the stationary spatial structures described in
section 2. The basic mechanism appears to be an enhanced cortical excitation with
long range lateral inhibition to prevent wave propagation and to induce the spatial
periodicity. Details of the theory of hallucinogenesis and a description of the two-
dimensional patterns can be found in [Ermentrout and Cowan, 1979]. More recently
(in preparation) a general theory for flicker phosphenes has been developed which
takes into account the time dependence of the solutions. Despite this additional
complication the main mechanism is the underlying lateral inhibition of the network.
Since deep pressure phosphenes require binocular stimuli, it seems likely that these
too originate in the visual cortex. Since it is difficult to record from humans

during these spontaneous visual disturbances, one can only base the theory on the general mechanisms described above: enhanced excitability and lateral inhibition.

In this discussion, we have attempted to assign some physiological interpretations to the patterns described in sections 2 and 3. A neuronal net is capable of eliciting many other types of behavior including transients, stimulus response, etc., all mimicking various physiological phenomena. Because the details of cortical organization are not well understood, we believe the best approach is one which does not rely on the exact details of the model but rather on its qualitative properties. For this reason, we have studied the general properties of arbitrary stationary, homogeneous media. We end by remarking that many other biological phenomena fall into the same category - reaction diffusion equations [Fife, 1979], population models [Murray, 1976], epidemiological systems [Diekmann, 1978], etc. Consequently, insight can be gained for many biological patterns within the context of these stationary, homogeneous systems.

## REFERENCES

Amari, S., 1977: Dynamics of pattern formation in lateral-inhibition type neural fields. Biol. Cybern. 27, 77-87.

Heiden, V., 1980. Analysis of Neural Networks, Lecture Notes in Biomathematics 35, Springer-Verlag, New York.

Ayala, G. F., M. Dichter, R. J. Gumnit, H. Matsumoto, and W. A. Spencer, 1973. Genesis of epileptic interictal spikes. New knowledge of cortical feedback systems suggests a neurophysiological explanation of brief paroxysms. Brain Res. 52, 1-17.

Burns, B. D., 1950. Some properties of the cats' isolated cerebral cortex. J. Physiol. 111, 50-68.

Diekmann, O. L., 1978. Thresholds and traveling waves for the geographical spread of infection. J. Math. Biol. 6, 109-130.

Ellias, S. A., and S. Grossberg, 1975. Pattern formation, contrast control, and oscillations in the short term memory of shunting on-center off-surround networks. Biol. Cyber. 20, 69-98.

Ermentrout, G. B., 1979. Symmetry Breaking in Homogeneous Stationary, Isotropic Neuronal Nets. Dissertation, University of Chicago, August.

Ermentrout, G. B., 1981a. Stationary, homogeneous media. I. Preprint.

Ermentrout, G. B., 1981b. Stationary, homogeneous media. II. Preprint.

Ermentrout, G. B., 1981c. Stationary, homogeneous media. III. In preparation.

Ermentrout, G. B., and J. D. Cowan, 1979. A mathematical theory for visual hallucination patterns. Biol. Cyber. 34, 137-150.

Fife, P. C., 1979. Mathematical Aspects of Reacting and Diffusing Systems, Lecture Notes in Biomathematics 28, Springer-Verlag, New York.

Grossberg, S., 1980. How does the brain build a cognitive code? Psycholog. Rev. 87, 1-51.

Kluver, H., 1967. Mescal and the Mechanisms of Hallucination. University of Chicago Press, Chicago.

Maxwell, J. A., and G. H. Renninger, 1980. On the theory of synchronization of lateral optic-nerve responses in "Limulus". I. Uniform excitation of the homogeneous retina. Math. Biosci. 52, 117-129.

Murray, J. D., 1976.  Spatial structures in predator-prey communities - a nonlinear time delay diffusional model.  Math. Biosci. 31, 73-86.

Ortoleva, P., and J. Ross, 1974.  On a variety of wave phenomena in chemical oscillations.  J. Chem. Phys. 60, 5090-5107.

Petsche, H., O. Prohaska, P. Rappelsberger, R. Vollmer, A. Kaiser, 1974.  Cortical seizure patterns in the multidimensional view:  The information content of equipotential maps.  Epilepsia 15, 439-463.

Petsche, H., and P. Rappelsberger, 1970.  Influence of cortical incisions on synchronization pattern and traveling waves.  Electroenceph. Clin. Neurophys. 28, 592-600.

Petsche, H., and P. Rappelsberger, 1973.  The problem of synchronization in the spread of epileptic discharges leading to seizures in man.  In:  M.A.B. Brazier (Ed.) Epilepsy, its Phenomena in Man.  UCLA Forum in Medical Sciences, #17, Acad. Press, N.Y. 121-151.

Remole, A., 1978.  Subjective patterns in a flickering field:  Binocular vs. monocular observation.  J. Opt. Soc. Amer. 63, 745-748.

Rinzel, J., 1978.  Integration and propagation of neuroelectric signals.  In: Studies in Mathematical Biology 15 (S. A. Levin, ed.), Math. Assoc. America, Washington, DC.

Siegel, R. K., 1977.  Hallucinations.  Scient. Amer. Oct., pp. 132-139.

Tyler, C. W., 1978.  Some new entoptic phenomena.  Vis. Res. 18, 1633-1639.

Wilson, H. R., and J. D. Cowan, 1973.  A mathematical theory of the functional dynamics of cortical and thalamic nervous tissue.  Kybernetik 13, 55-80.

# CHAPTER 4

## AGGREGATION AND SEGREGATION PHENOMENA
## IN REACTION-DIFFUSION EQUATIONS

M. Mimura

Department of Mathematics
Hiroshima University, Hiroshima, Japan

## 1. Introduction

Recently there has been considerable interest in a model of system where the components are interacting and biodiffusing. Such systems are described by

$$u_t + \text{div}\, J = f(u) \tag{1}$$

where the variable $u = (u_1, u_2, \ldots, u_n)$ represents quantities such as the temperature, chemical concentrations, and biomass density, $J = (J_1, J_2, \ldots, J_n)$ is the flux, $f = (f_1, f_2, \ldots, f_n)$ means the kinetics of the processes which are explicitly independent of the spatial variables. From the theory of biodiffusion, Okubo(1980) classified the flux $J_i$ into three types:

(i) (Fickian expression)

$$J_i^f = -\, d_i \text{grad}\, u_i \, ,$$

(ii) (Repulsive expression)

$$J_i^r = -\, \text{grad}(d_i u_i),$$

(iii) (Attractive expression)

$$J_i^a = -\, d_i^2 \text{grad}(u_i/d_i),$$

where $d_i (> 0)$ is called the coefficient of diffusion. If we assume that $d_i$ is an isentropic and constant diffusion, it follows that $J_i^f = J_i^r = J_i^a$ and then (1) is reduced to a system of reaction-diffusion equations

$$u_t = D\Delta u + f(u), \tag{2}$$

where $D$ is a diagonal constant matrix with $\{d_i\}$ and $\Delta$ is the Laplace operator in the spatial coordinates. On the other hand, if $d_i$ depends on $x$, then, although $J_i^f$ is always directed from high density to low one as we may expect, $J_i^r$ and $J_i^a$ is not necessarily directed down the density gradient. Okubo(1980) has noted that certain characteristics of animal diffusion and taxis can be explained with the use of flux $J_i^r$ or $J_i^a$.

---

A part of this paper is reported on a number of results which have been obtained jointly with H. Fujii and Y. Nishiura of Kyoto Sangyo University.

The asymptotic behavior of solutions to (1) exhibits a variety of interesting phenomena such as oscillation, wave propagation, stationary states, and so on (Fife (1979) and its bibliography, for instance). If we consider (1) in a bounded domain with the zero flux boundary condition on the boundary, then the asymptotic states may be classified into four categories: "spatially uniform stationary states or oscillations", "spatially non-uniform stationary states", "spatially non-uniform oscillations", and "spatio-temporal chaos".

In this report, we take as examples of (1) some specific equations describing the dynamics of populations whose individuals move out by biodiffusion in a population reservoir, and concentrate on the second asymptotic state of (1), which is obtained from the stationary system of (1),

$$\text{div } J = f(v) \tag{3}$$

subject to the zero flux boundary condition. If $J$ or $f$ depends on $x$, then it is not surprising that (3) can admit spatially inhomogeneous solutions (Matano (1980)). However, if $J$ and $f$ are independent of $x$, it is not intuitively obvious that there will exist *stable* spatially inhomogeneous solutions. This problem is motivated by pattern formation such as morphogenesis in developmental biology and aggregation or segregation in population biology.

This report is not concerned with the experimental aspect of biological pattern formation, but only with the mathematical analysis and includes several numerical simulations of these phenomena.

## 2. Activator-inhibitor dynamics

A plenty of attention has been focused on a class of diffusion systems with activator-inhibitor reaction dynamics described by (2). The main tools to study pattern formation of such systems are, in one hand, the local bifurcation analysis at simple or multiple eigenvalues, and on the other hand, regular and/or singular perturbation methods when some biological parameters externally controlled. These researches have been extensively clarified questions on the onset and existence of spatial patterns, their stability, existence of large amplitude patterns, and so on.

In this section, we take as an example of stationary activator-inhibitor systems a Lotka-Volterra prey-predator model with simple constant diffusion ($J = J^f$) in one dimensional space, which can be written as

$$\begin{cases} - d_1 u_{xx} = f_0(u)u - uv \equiv f(u,v), \\ \\ - d_2 v_{xx} = - g_0(v)v + uv \equiv g(u,v), \end{cases} \quad x \in I \equiv (0,1), \tag{4}$$

where  u  is a prey (activator) and  v  is its predator (inhibitor).  Here we state
the following assumtions on  (f,g)  in (4):

   (A.1)   There exists a unique positive constant solution  $(\bar{u},\bar{v})$  of  $f = g = 0$.

   (A.2)   At  $(u,v) = (\bar{u},\bar{v})$,  $f_u > 0$, $f_v < 0$, $g_u > 0$, $g_v < 0$, $f_u + g_v < 0$

         and  $f_u g_v - f_v g_u > 0$.

   (A.3)   The curve of  $f = 0$  is S-like shaped in the (u,v)-space.

(Figure 1).  The second assumption is well known as the requirement of occurence of
bifurcations in *Turing*'s sense.  By using the Fourier series expansion, it is easy
to know the stable and unstable regions of  $(\bar{u},\bar{v})$  in the $(d_1,d_2)$-space (Figure 2).

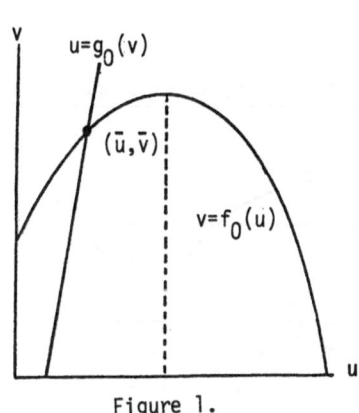

Figure 1.

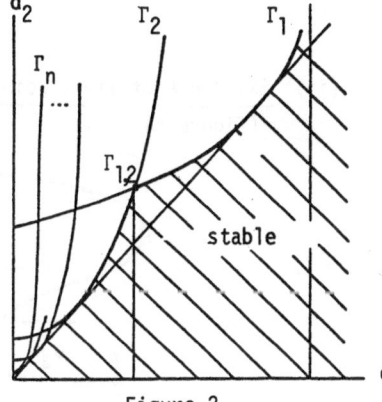

Figure 2.

    Our aim is, by both analytical and numerical approaches, to present a picture
of *global dependency* of non-uniform patterns on the diffusion coefficients  $(d_1,d_2)$
$\in \mathbf{R}_+^2$.  More precisely, the global branches bifurcating from  $(\bar{u},\bar{v})$  are studied
in  $\mathbf{R}_+^2 \times X$, where  X  is a suitable solution space.  Recently Fujii et al (1981)
has observed a variety of phenomena of bifurcating solutions, including appearance
of limit points, of secondary and tertiary bifurcation points, disappearance of
bifurcated branches, global existence of bifurcated branches, and so on.  Among
these, coexistence of several stable patterns as a consequence of successive
recovery of stability is the most interesting phenomenon.  Here we only show some
remarkable results without the mathematical discussion.

   (i)  <u>The local structure near one of the double singularities</u> $\Gamma_{1,2}$
       (Figure 3)

           ——— stable branch,   ----- unstable branch

---

In the computation, we set  $f_0(u) = (35+16u-u^2)/9$  and  $g_0(v) = 1+2v/5$.

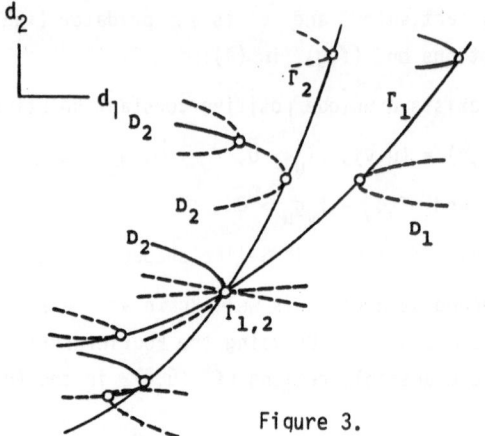

Figure 3.

(ii) __Existence of limit points in $D_1$-branch and hysteresis phenomena__
    (Figure 4)

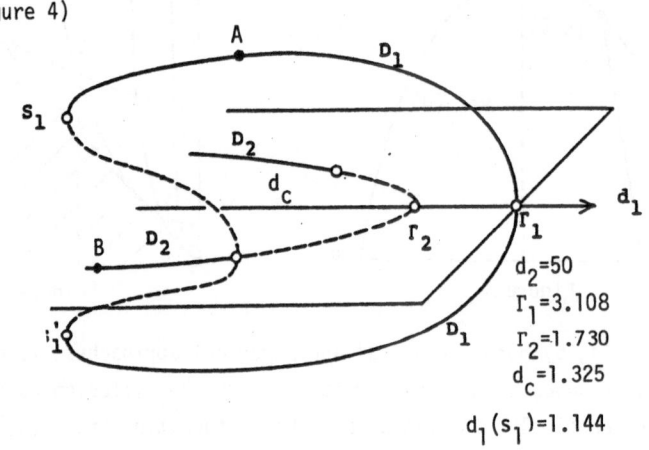

$d_2 = 50$
$\Gamma_1 = 3.108$
$\Gamma_2 = 1.730$
$d_c = 1.325$

$d_1(s_1) = 1.144$

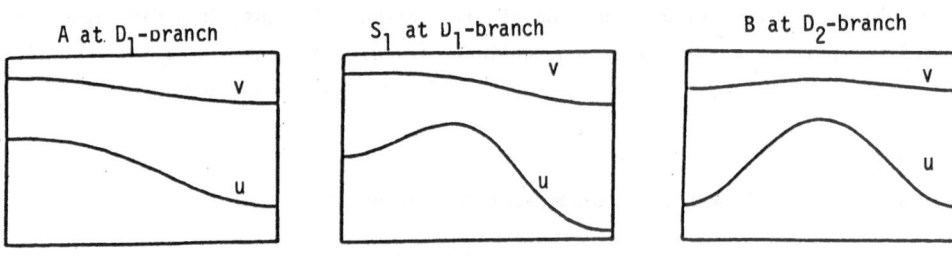

(iii) __Global branch of $D_1$-branch without loss of stability__
    (Figure 5)

When $d_2 = 200$, $D_1$-branch hits the wall($d_1 = 0$).

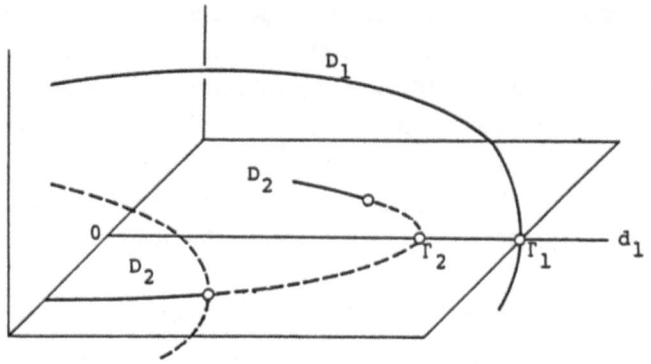

Figure 5.

(iv)  Global diagram of bifurcated branches in the $(d_1, d_2)$-space
(Figure 6)

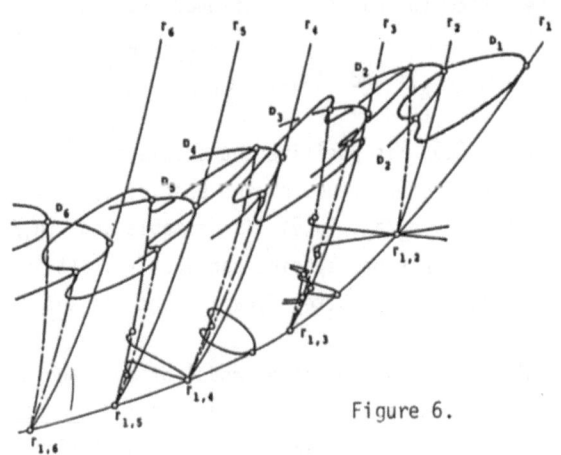

Figure 6.

From these figures, we conclude that when $d_2$ is sufficiently large, the global behavior of solutions is *close* to that of scalar problems(Nishiura(1980)).  On the contrary, if $d_2$ is not large, Figure 6 clearly shows the complexity and crucial difference of global behavior of branches.  I would like to say here that, though the equation (4) seems to be quite simple, there exist multiple stable and unstable non-uniform solutions.    This fact makes the mathematical study of pattern formation of reaction-diffusion equations difficult.

3. Competitive dynamics

In the natural environment, populations of many species exist and experience the struggle for existence. According to the competitive exclusion principle, (i) if two non-interbreeding populaitons have similar needs and habits, (ii) if they live in the same habitat, then one of the two populations becomes extinct (Gause (1934)). On the other hand, certain populations competing with populations of the other species can reduce the interspecific competition by evolving into forms which permit ecological separation by changing their food requirements in space. From a theoretical point of view, coexistence of two species by means of spatial segregation has been of interest.

The Lotka-Volterra type dynamics for two competing species is described by

$$\begin{cases} u_t = (R_1 - a_1u - b_1v)u, \\ v_t = (R_2 - b_2u - a_2v)v, \end{cases} \qquad t > 0, \qquad (5)$$

where u and v are the population densities of the species, $a_i$ is the coefficient of intraspecific competition and $b_i$ is that of the interspecific competition (i=1,2). The system (5) has four stationary points

$$(0,0), \ (R_1/a_1,0), \ (0,R_2/a_2) \ \text{and} \ (\bar{u},\bar{v}) = (\frac{R_1a_2 - R_2b_1}{a_1a_2 - b_1b_2}, \frac{R_2a_1 - R_1b_2}{a_1a_2 - b_1b_2}).$$

When $a_1/b_2 > R_1/R_2 > b_1/a_2$, $(\bar{u},\bar{v})$ is positive and stable and others are all unstable. This indicates that the coexistence of two species is realizable. On the other hand, when $a_1/b_2 < R_1/R_2 < b_1/a_2$, $(\bar{u},\bar{v})$ is unstable and $(R_1/a_1,0)$ and $(0,R_2/a_2)$ are stable, that is, only one species can survive. It is noted that the kinetics of (5) is not an activator-inhibitor dynamics but, in some sense, an inhibitor-inhibitor one.

Shigesada et al(1979) proposed a population model of two species which are moving outward by repulsively dipersive forces($J = J^r$) including population pressures and interacting through dynamics of the Lotka-Volterra type of (5). The resulting system in one dimensional space is of the form

$$\begin{cases} u_t - [(d_1 + \gamma_1u + \alpha_1v)u]_{xx} = (R_1 - a_1u - b_1v)u, \\ v_t - [(d_2 + \alpha_2u + \gamma_2v)v]_{xx} = (R_2 - b_2u - a_2v)v, \end{cases} \qquad t > 0, x \in I \ (6)$$

subject to the initial and boundary conditions

$$u(0,x) = u_0(x), \ v(0,x) = v_0(x), \ x \in \bar{I} \qquad (7)$$

and

$$u_x = v_x = 0, \ t > 0, \ x \in \partial I, \qquad (8)$$

respectively, where $d_i$ is a positive constant and $\gamma_i$ and $\alpha_i$ are non-negative constants(i=1,2). First we consider the system (6) in the absence of $\alpha_i$(i=1,2). The resulting system is

$$\begin{cases} u_t - [(d_1 + \gamma_1 u)u]_{xx} = (R_1 - a_1 u - b_1 v)u, \\ v_t - [(d_2 + \gamma_2 v)v]_{xx} = (R_2 - b_2 u - a_2 v)v, \end{cases} \quad t > 0, \ x \in I. \quad (9)$$

For (7)-(9), Kishimoto(1980) has shown that inhomogeneous steady state solutions which exist are *unstable*. This result claims that spatial segregation of two species can never occur.

We next consider (6) in the presence of cross-population pressures, i.e., at least one $a_i \neq 0$ (i=1,2). For simplicity only, we deal with a simplified version of (6),

$$\begin{cases} u_t - [(d_1 + \alpha_1 v)u]_{xx} = (R_1 - au - bv)u, \\ v_t - d_2 v_{xx} = (R_2 - bu - av)v, \end{cases} \quad (10)$$

where $a_1 = a_2 = a$ and $b_1 = b_2 = b$. Ecologically this situation implies that only the population pressure of species (v) is exerted on the species (u) and raises the dispersive force on the species (u). We may rewrite (10) as

$$\begin{cases} u_t - [(1 + \alpha v)u]_{xx} = \beta(R_1 - au - bv)u, \\ v_t - dv_{xx} = (R_2 - bu - av)v, \end{cases} \quad (11)$$

where $\alpha$, $d$ and $\beta$ are positive constants. For the parameters $R_1$, $R_2$, a, b and $\alpha$ in (11), we make two conditions (A.4) and (A.5):

(A.4)       $a/b > R_1/R_2 > b/a$,

which indicates that $(\bar{u}, \bar{v})$ is asymptotically stable in (5).

(A.5)       $\alpha > a/(b\bar{u} - a\bar{v}) > 0$

which states that the effect of the cross-diffusivity is not so weak.

It is convenient to transform (u,v) into (w,v) through $(1 + \alpha v)u = w$. Then the system resulting from (11) is

$$\begin{cases} (\frac{w}{1 + \alpha v})_t - w_{xx} = \beta(R_1 - \frac{aw}{1 + \alpha v} - bv)\frac{aw}{1 + \alpha v} \equiv \beta k(w,v), \\ v_t - dv_{xx} = (R_2 - \frac{bw}{1 + \alpha v} - av)v \equiv h(w,v). \end{cases} \quad (12)$$

The stationary system of (12) is

$$\begin{cases} - w_{xx} = \beta k(w,v), \\ - dv_{xx} = h(w,v), \end{cases} \quad x \in I \quad (13)$$

subject to the zero flux boundary conditions

$$w_x = v_x = 0, \quad x \in \partial I. \quad (14)$$

The curves of $k = h = 0$ under (A.4) and (A.5) are drawn in Figure 7, where $k = 0$ has a *humped* effect from (A.5). Here we find that (13) is of the similar type to the activator-inhibitor system (4), and that the nonlinearities (f,g) and (k,h) are also quite similar. In fact, though (13) does not fall within the framework of activator-inhibitor systems, one can find that the *cross-diffusion-instability* occurs when d and β are adjustable parameters(Figure 8).

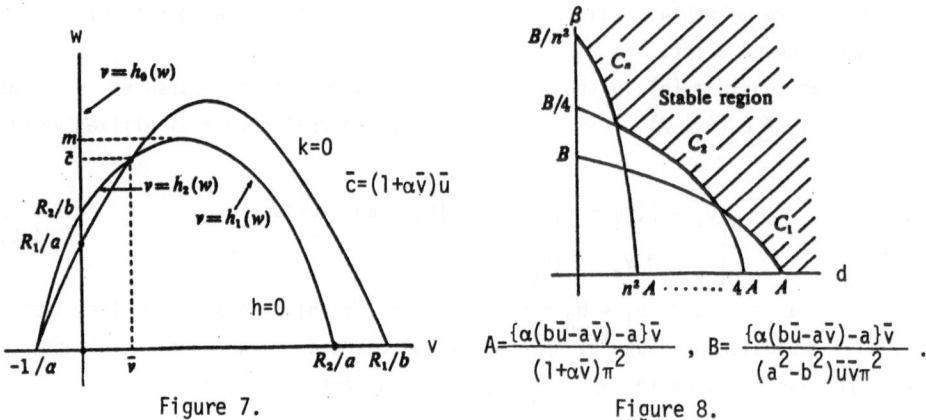

Figure 7.                                                Figure 8.

$$A = \frac{\{\alpha(b\bar{u}-a\bar{v})-a\}\bar{v}}{(1+\alpha\bar{v})\pi^2}, \quad B = \frac{\{\alpha(b\bar{u}-a\bar{v})-a\}\bar{v}}{(a^2-b^2)\bar{u}\bar{v}\pi^2}.$$

Using the Lyapunov-Schmidt method, we obtain an ε-family of bifurcating solutions $(d(\sigma(\varepsilon)), \beta(\sigma(\varepsilon)), w(\varepsilon), v(\varepsilon))$ from $(d(\sigma(0)), \beta(\sigma(0)), (1+\alpha\bar{v})\bar{u}, \bar{v})$ $(0 < |\varepsilon| < \varepsilon_0)$ for some $\varepsilon_0$. Here σ is a parameter which arbitrarily determines a path $(d(\sigma), \beta(\sigma))$ in such a way that it starts from the stable region and goes into the unstable one at σ = 0. Thus $(w(\varepsilon), v(\varepsilon))$ leads to a solution $(u(\varepsilon), v(\varepsilon))$ of the stationary problem of (11) through $(1+\alpha v)u = w$. This is of the form

$$\begin{pmatrix} u(\varepsilon) \\ v(\varepsilon) \end{pmatrix} = \begin{pmatrix} \bar{u} \\ \bar{v} \end{pmatrix} + \varepsilon\xi_n \begin{pmatrix} dn^2\pi^2 + a\bar{v} \\ -b\bar{v} \end{pmatrix} \cos(n\pi x) + o(\varepsilon),$$

where $\xi_n$ is a normalized constant such that $|\xi_n| = \sqrt{2}$(Mimura and Kawasaki(1980)). Therefore, if ε is sufficiently small, one knows that $(u(\varepsilon), v(\varepsilon))$ exhibits spatial segregation. Thus, we conclude that the cross-population pressure plays an important role for segregation processes. Furthermore, from the previous section, we may expect existence of many large amplitude solutions for fixed parameters (d,β) far away from primary bifurcation curves(see Figures 6 and 8). We can study spatial structures of the solutions in the special cases β << 1 and/or d << 1 by using regular and/or singular perturbation approaches.

### 3. Singular perturbation problem (0 < d << 1)

In this section, we look for non-solutions of (13), (14) when d is a sufficiently small parameter for fixed β > 0. First consider the reduced problem (d = 0) of (13), (14),

$$\begin{cases} - w_{xx} = \beta k(w,v), \\ 0 = h(w,v), \end{cases} \qquad x \in I \qquad (15)$$

and

$$w_x = 0, \quad x \in \partial I. \qquad (16)$$

Once a function $v = H(w)$ can be obtained from the second of (15), we have a scalar equation with respect to $w$,

$$- w_{xx} = \beta k(w,H(w)) \qquad (17)$$

From Figure 7, we have three different functions, say $v = h_0(w)(= 0)$, $v = h_1(w)$ and $v = h_2(w)$ ($h_0 < h_2 < h_1$). Here we define $H(w)$ by

$$H(w;s) = \begin{cases} h_0(w), & w \in \mathbb{R}_+\backslash(s,m), \\ h_1(w), & w \in (s,m), \end{cases}$$

for any fixed $s \in (R_2/b,m)$. The validity of this construction was discussed in Fife(1976b). Since $H(w;s)$ is discontinuous at $w = s$, $K(w;s) = k(w,H(w;s))$ is a discontinuous function(Figure 9).

Therefore we must define a weak solution $w(x) \in H^1(I)$ of (16), (17) by

$$(w_x, \phi_x) = \beta(K(w;s),\phi)$$

for all $\phi \in H^1(I)$. For the problem (16), (17), Mimura et al(1980) proved the existence of non-uniform solution $w_s(x)$ satisfying $R_1/a < w_s(x) < m$ for $0 < \beta < \beta_0$, where $\beta_0$ is some constant. Using this function $w_s(x)$, we construct a solution $(w_s(x),v_s(x))$ of (15) through $v = H(w;s)$ and then, through $(1+\alpha v)u = w$, obtain a solution $(u_s(x),v_s(x)) \in L^2(I) \times L^2(I)$,

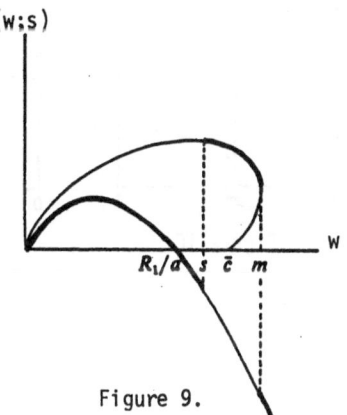

$H(w;s)$

Figure 9.

$$\begin{cases} - [(1+\alpha v)u]_{xx} = \beta(R_1 - au - bv)u, \\ 0 = (R_2 - bu - av)v, \end{cases} \qquad x \in I \qquad (18)$$

subject to the zero flux boundary conditions

$$u_x = v_x = 0, \quad x \in \partial I. \qquad (19)$$

Let us mention the spatial structures of the solution $(u_s(x),v_s(x))$ in relation to segregating pattern. For brevity, we consider the simple case when $w_s(x)$ satisfying $w_s(x^*) = s$ is monotone increasing on $I$. Then it turns out that the solution of (18), (19), say $(u(x),v(x))$, has only one point of discontinuity $x = x^*$ ($\epsilon$ I) such that

$$u(x) = w_s(x), \quad v(x) \equiv 0, \quad x \in (0,x^*)$$

and

$$u(x) = w_s(x)/(1+\alpha H(w_s(x);s)), \quad v(x) = H(w_s(x);s), \quad x \in (x^*,1).$$

Thus, one easily finds that in the subregion $(0,x^*)$ one species $(u)$ is not zero and that in $(x^*,1)$, $u(x)$ is monotone decreasing, because $h_1(w_s(x))$ is monotone decreasing in this region(Figure 10). This structure shows spatial segregation between two species.

When $d$ is not zero but sufficiently small, the discontinuity of $(u(x),v(x))$ at $x = x^*$ suggests us that an internal transition layer appears in both arguments. Thus, this case falls into the framework of singular perturbation problems(Fife (1976), Mimura et al(1979)). We only show the spatial structure of a solution $(u(x;d),v(x;d))$ in Figure 11. The details are discussed in Mimura(1981). Furthermore, for an arbitrarily fixed $d$, we can obtain large amplitude solutions of (13), (14) in the case when $\beta$ is sufficiently small(Nishiura(1981), Mimura(1981)). In this case, spatial structure of $(u(x),v(x))$ is drawn in Figure 12.

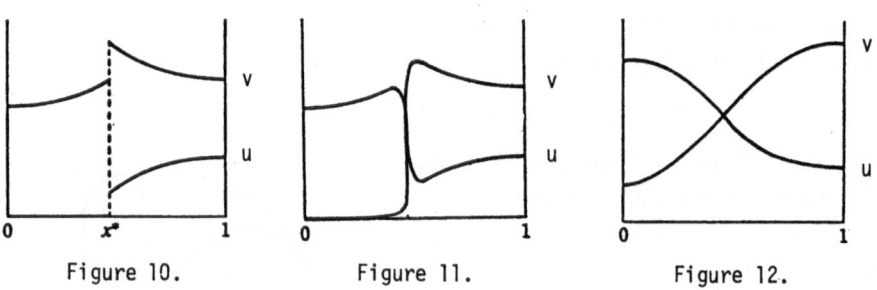

Figure 10.                    Figure 11.                    Figure 12.

## 4. Concluding remark

In this report, we have concluded that for the Lotka-Volterra type dynamics for two competing and biodiffusing species, certain cross population pressure is essential for the realization of spatial segregation. However, Kishimoto(1981) has recently reported that when 3-competing species models described by

$$\begin{cases} u_t = d_1 u_{xx} + (R_1 - a_1 u - b_1 v - c_1 w)u \\ v_t = d_2 v_{xx} + (R_2 - c_2 u - a_2 v - b_2 w)v \\ w_t = d_3 w_{xx} + (R_3 - b_3 u - c_3 v - a_3 w)w \end{cases}$$

are considered, Turing's diffusion-induced instability can occur for suitable parameters $d_i$, $R_i$, $a_i$, $b_i$ and $c_i (i=1,2,3)$. He has also shown stable bifurcating solutions exhibiting spatial segregation. Furthermore, If we consider a 4-competing species model with simple diffusions, it is found that there appears spatio-temporal segregation in this model. This study will be reported in a forthcoming paper.

## References

P. C. Fife(1976a) Boundary and interior transition layer phenomena for pairs of second-order differential equations, J. Math. Anal. Appl., 54, 497-521.

P. C. Fife(1976b) Pattern formation in reacting and diffusing systems, J. Chem. Phys., 64, 554-564.

P. C. Fife(1979) Mathematical aspects of reacting and diffusing systems, Lecture Notes in Biomath., 28, Springer-Verlag, Berlin.

H. Fujii, M. Mimura and Y. Nishiura(1981) A picture of global bifurcation diagram in ecological interacting and diffusing systems, Research Report, KSU-ICS, 81-05, Kyoto Sangyo Univ..

G. F. Gause(1934) The struggle for existence, Baltimore William and Wilkins Co..

K. Kishimoto(1980) Instability of non-constant equilibrium solutions of a system of competition-diffusion equations, J. Math. Biol.(in press).

K. Kishimoto(1981) The diffusive Lotka-Volterra system with three species can have a stable non-constant equilibrium solution, manuscript.

H. Matano(1981) Nonincrease of the lap-number of a solution for a one-dimensional semilinear parabolic equation, manuscript.

M. Mimura, M. Tabata and Y. Hosono(1980) Multiple solutions of two-point boundary value problems of Neumann type with a small parameter, SIAM J. Math. Anal., 11, 613-631.

M. Mimura and K. Kawasaki(1980) Spatial segregation in competitive interaction-diffusion equations, J. Math. Biol., 9, 49-64.

M. Mimura(1981) Spatial pattern of some density-dependent diffusion system with competitive dynamics, Hiroshima Math. J., 11, 621-633.

Y. Nishiura(1980) Global structure of bifurcating solutions of some reaction-diffusion systems, to appear in SIAM J. Math. Anal..

A. Okubo(1980) Diffusion and ecological problems: Mathematical models, Biomath., 10, Springer-Verlag, Berlin.

N. Shigesada, K. Kawasaki and E. Teramoto(1979) Spatial segregation of interacting species, J. Theor. Biol., 79, 89-99.

# CHAPTER 5

NERVE PULSE INTERACTIONS

Alwyn C. Scott
Center for Nonlinear Studies
Los Alamos National Laboratories
Los Alamos, New Mexico    87545   USA

## INTRODUCTION

Traditionally the neuron has been viewed as a linear threshold unit which generates an output when some weighted sum of the inputs exceeds a certain value. Over the past two decades, however, several suggestions have been made for mechanisms by which the neuron might perform more sophisticated forms of information processing, such as:   Boolean logic at dendritic branches [1-8], Time code to space code translations on the axonal tree [9-14], and Dendrodendritic interactions [15]. Such suggestions have led Waxman to propse the concept of a "multiplex neuron" [16] which bears about the same relationship to a linear threshold unit as does a "chip" to a "gate" in modern integrated circuit technology.

Certainly an important aim of neuroscience is to understand the true functional roles played by neurons.  A first step in this direction is to ask what it is that a neuron can do.  This is a question of biophysics.  From this spectrum of possible modes of neuronic behavior, one can then ask the biological question:  What is it that a particular neuron does do?

Here I review some recent experimental and theoretical results on mechanisms through which individual nerve pulses can interact.  Three modes of interactions are considered:   1) Interaction of pulses as they travel along a single fiber which leads to velocity dispersion, 2) Propagation of pairs of pulses through a branching region leading to quantum pulse code transformations, and 3) Interaction of pulses on parallel fibers through which they may form a pulse assembly.  This notion is analogous to Hebb's concept of a "cell assembly" [17], but on a lower level of the neural hierarchy.  It may help to explain the extreme sensitivity of neural systems to nonionizing electromagnetic fields [18,19].

## VELOCITY DISPERSION

Consider the "twin pulse" experiment sketched in Fig. 1.  The times $t_1$ and $t_3$ are when the first and second pulses, respectively, pass the first electrode.  The times $t_2$ and $t_4$ are similarly related to the second electrode.  Thus the average

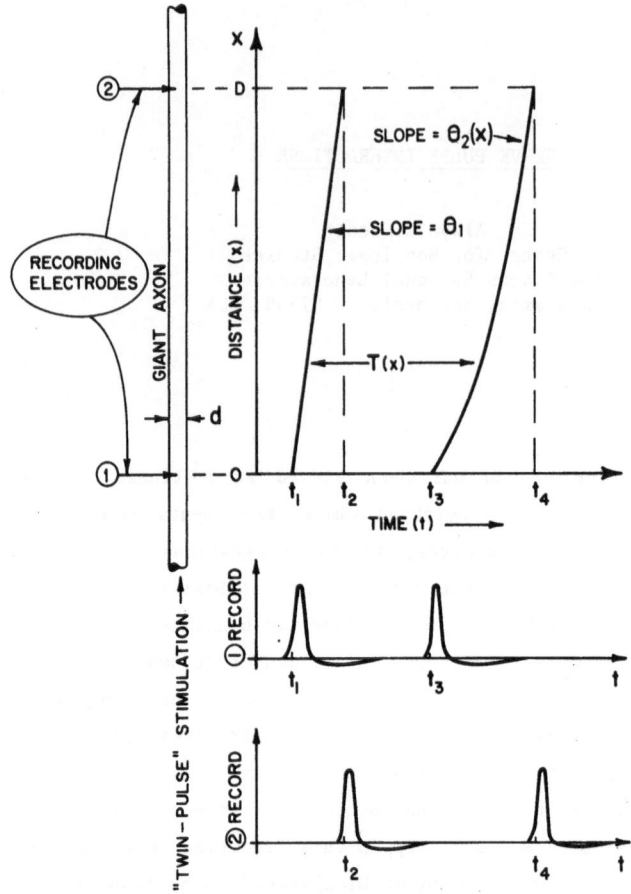

Fig. 1. Experimental details of
twin pulse experiment on giant axon of squid.

time interval between the two pulses is measured as

$$T = \frac{1}{2}(t_4 + t_3 - t_2 - t_1) ,$$ (1)

and the ratio of the speed of the second pulse, $\theta_2$, to that of the first pulse, $\theta_1$, is

$$R \equiv \theta_2/\theta_1$$

$$= \frac{t_2 - t_1}{t_4 - t_3} .$$ (2)

Actually (2) is not exact because the speed of the second pulse is not constant. A small correction to account for this effect is discussed in [20]. Since the time differences for a squid giant axon are a few milliseconds and can be measured with an rms accuracy of about 0.003 milliseconds, (1) and (2) determine the <u>velocity dispersion function</u>, R(T), with an accuracy of a few per cent.

A calculation of R(T) can be made by noting that the leading edge of the second pulse propagates into the tail of the first pulse. Since all parameters that describe the axon are identical for the two pulses, the velocity dispersion function is readily calculated as (see [20] for details)

$$R(T) = \left[ \frac{1 + K}{\left(1 - \frac{V(T)}{V_T}\right) 1 + K\left[\left(\frac{1 - V(T)/V_T}{1 + V(T)/V_+}\right)\right]} \right]^{\frac{1}{2}} \tag{3}$$

where $V_T$ is the threshold voltage of an isolated pulse, and $V_+$ and $V(T)$ describe the tail of an isolated pulse as is indicated in Fig. 2.

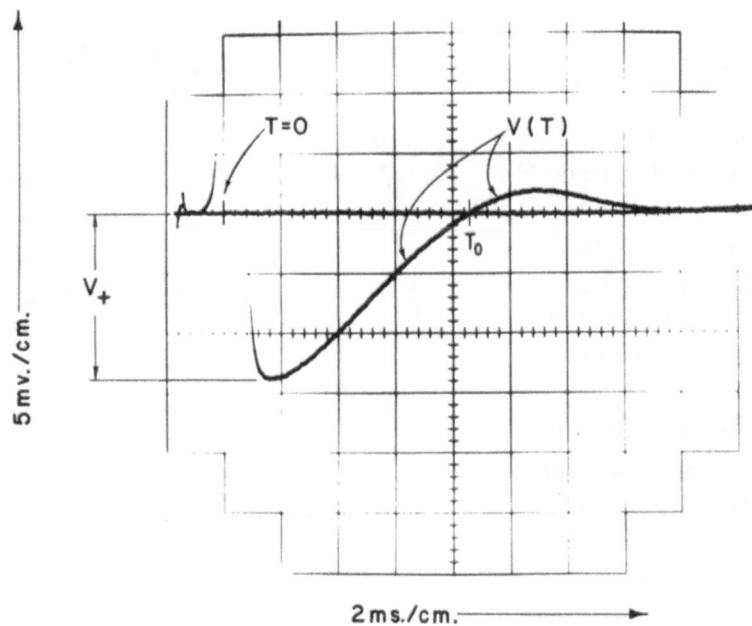

Fig. 2. The tail of a typical isolated pulse on a squid giant axon. The time T = 0 indicates when the pulse voltage passes through threshold (dv/dT = maximum valu

In (3) the parameter

$$K \equiv g_o \, \tau/c \qquad\qquad\qquad (4)$$

where $\tau$ is the time constant for exponential rise of the leading edge below threshold and $c/g_o$ is the time constant of a resting membrane ($= 1$ millisecond).

In Fig. 3 a measurement of R(T), from (1) and (2), is compared with calculations from (3). There are four calculations because the tail of an isolated pulse was measured on both electrodes before and after the measurements of R(T).

The measurements displayed in Fig. 3 confirm previous measurements by Donati and Kunov [21] and are in agreement with recent numerical studies of the Hodgkin-Huxley equations by Miller and Rinzel [22]. Particularly interesting is the "overshoot" when R < 1 and the second pulse is actually going faster than the first.

From the measurements one can define $T_1$ as the pulse spacing at which R(T), determined from (1) and (2), is equal to unity. Likewise from the calculations one can define $T_o$ as the pulse spacing at which R(T), determined from (3), is equal to unity. In Fig. 4 a comparison of $T_1$ with $T_o$ is presented for

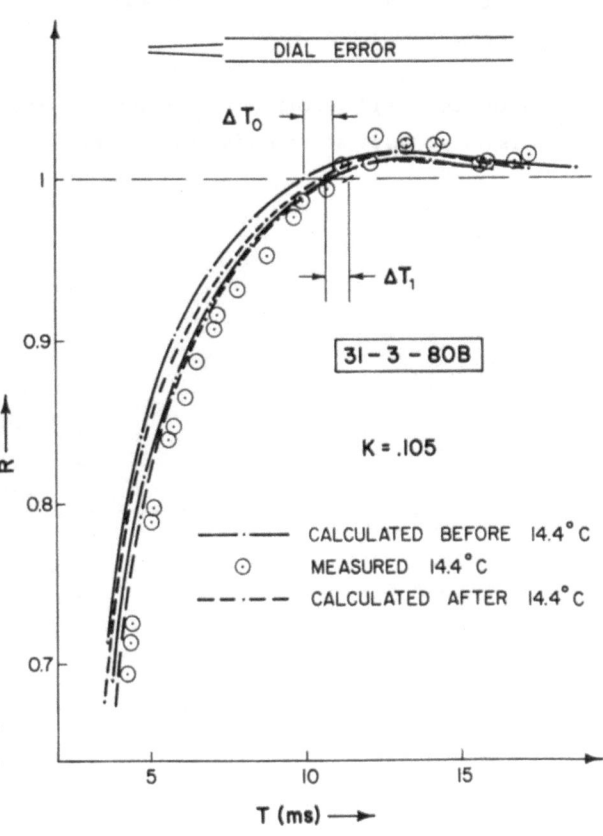

Fig. 3. Comparison of measurements and calculations of the velocity dispersion function R(T).

measurements on twelve giant squid axons where the error bars are defined as indicated on Fig. 3. Figure 4 indicates

$$T_1 \doteq T_0 \tag{5}$$

within the accuracy of the measurements and calculations. However (3) implies that
$T_0$ is the time at which
the tail crosses through
zero; i.e., $V(T_0) = 0$ (see
also Fig. 2). However calculations by John Rinzel
for the Hodgkin-Huxley equations do not confirm this
point. He finds [23]

| Temp (°C) | $T_1$ (ms.) | $T_0$ (ms.) |
|-----------|-------------|-------------|
| 18.5      | 4.77        | 6.14        |
| 14.0      | 7.25        | 8.53        |
| 6.3       | 16.0        | 17.2        |

These calculations are plotted as "⊘" on Fig. 4 and
appear to lie below the
experimental data although
this distinction is not
quite clear. This dis-
crepancy may arise because
the Hodgkin-Huxley equations
do not account for potassium
build up in the periaxonal
space [24,25].

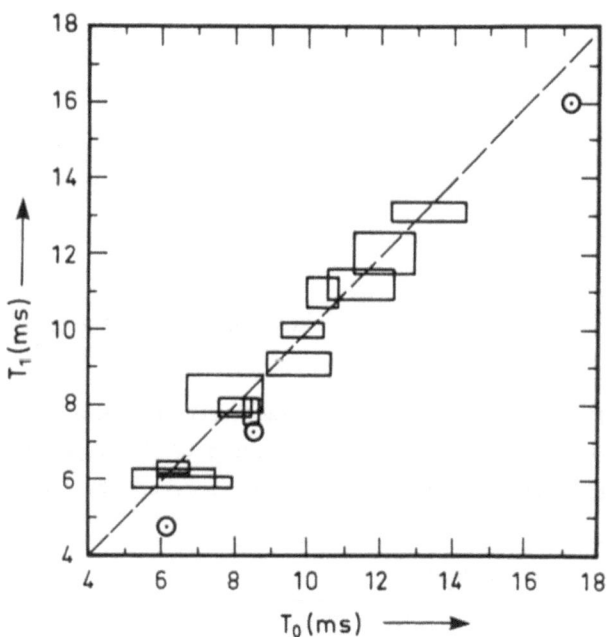

Fig. 4. Comparison of measured $T_1$ and calculated
$T_0$ for twelve giant squid axons. The points ⊘ are
calculated [23] from the Hodgkin-Huxley equations.

Apart from theoretical descriptions, it is interesting to consider all the
measurements of R vs T normalized to $T_1$ (made at the Stazione Zoologica between
January and June of 1980) on the "flyspeck" diagram of Fig. 5. Within a few per
cent, the data of Fig. 5 can be represented by the empirical relation

$$R = 1 + 0.105 \left( \frac{T}{T_1} - 1 \right) \exp \left[ 2.4 \left( 1 - \frac{T}{T_1} \right) \right]. \tag{6}$$

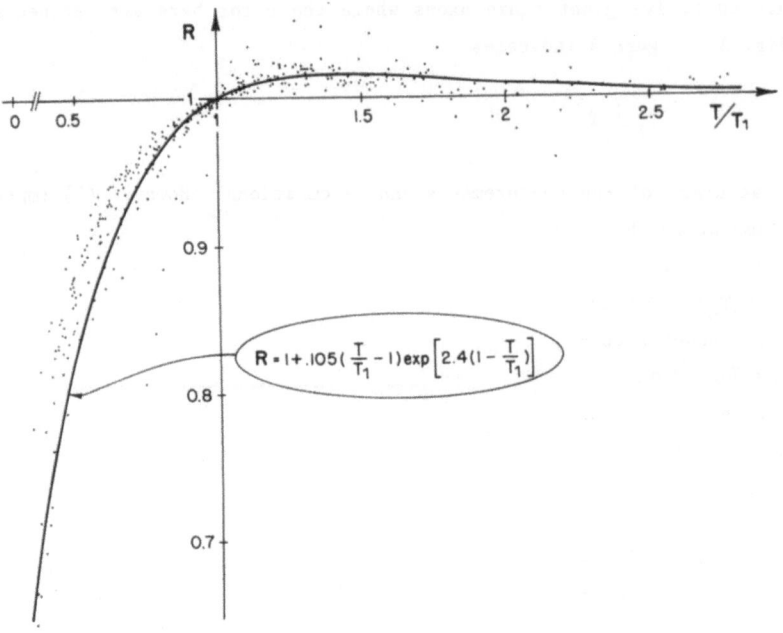

Fig. 5. Measured values of R vs $T/T_1$. Note that $T_1$ is defined such that $R(T_1) = 1$.

It should perhaps be emphasized that the velocity dispersion implied by (6) is not a negligible effect. The measurements displayed in Fig. 6, for example, show considerable "frequency smoothing" between the upstream record A and the downstream

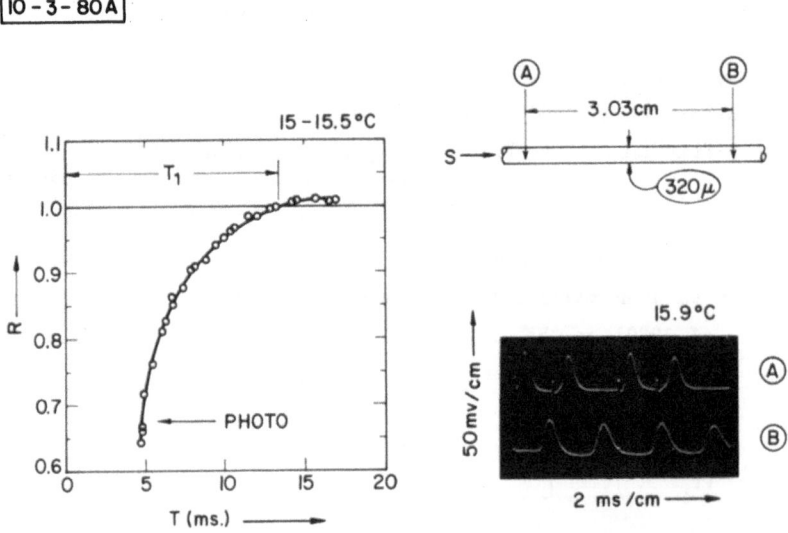

Fig. 6. Frequency smoothing on a squid giant axon.

## QUANTUM PULSE CODE TRANSFORMATIONS

We turn now to experimental situations in which a nerve pulse disappears suddenly during the course of its propagation. I term such effects "quantum" pulse code transformations to distinguish them from differential transformations related to velocity dispersion. An obvious place to seek quantum transformations is at the points where fibers branch (or "bifurcate"). Branches of the squid giant axon are found at the locations indicated in Fig. 7.

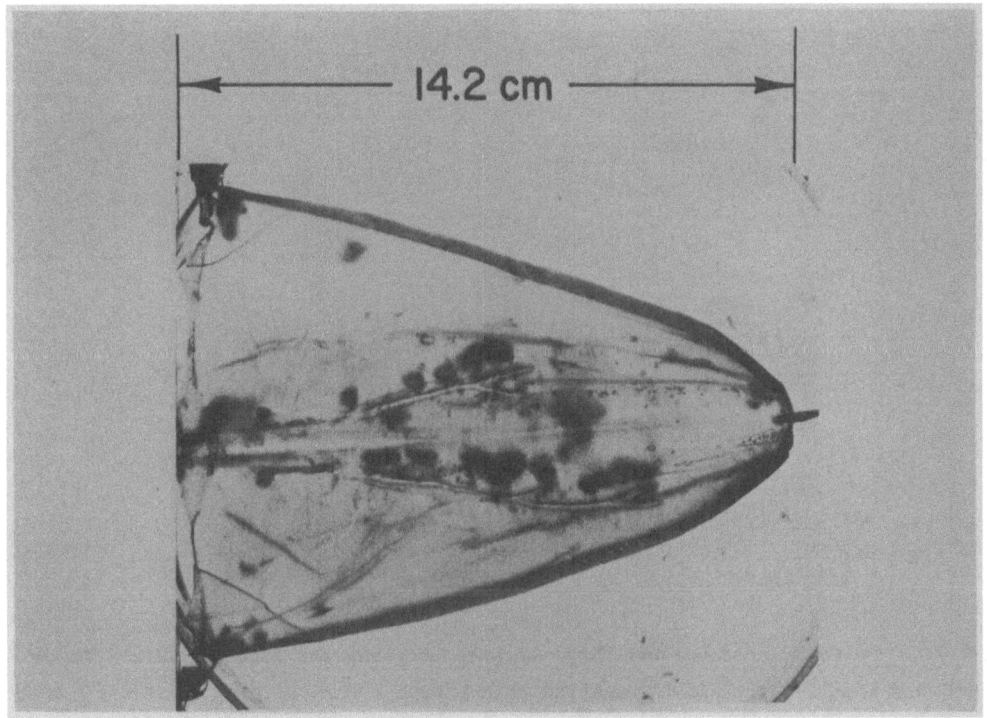

Fig. 7. Mantle of squid (Loligo vulgaris) showing
scars where giant axons and branches have been removed.

Some typical branches of the squid giant axon are displayed in Fig. 8 from which a large variation in the ratios of fiber diameters is observed. A measure of the difficulty experienced by a pulse as it propagates through a branching region is expressed by the geometrical ratio (GR) [26] which is essentially the total characteristic admittance of outgoing fibers divided by that carrying the incoming pulse. In terms of fiber diameters (d) it takes the form

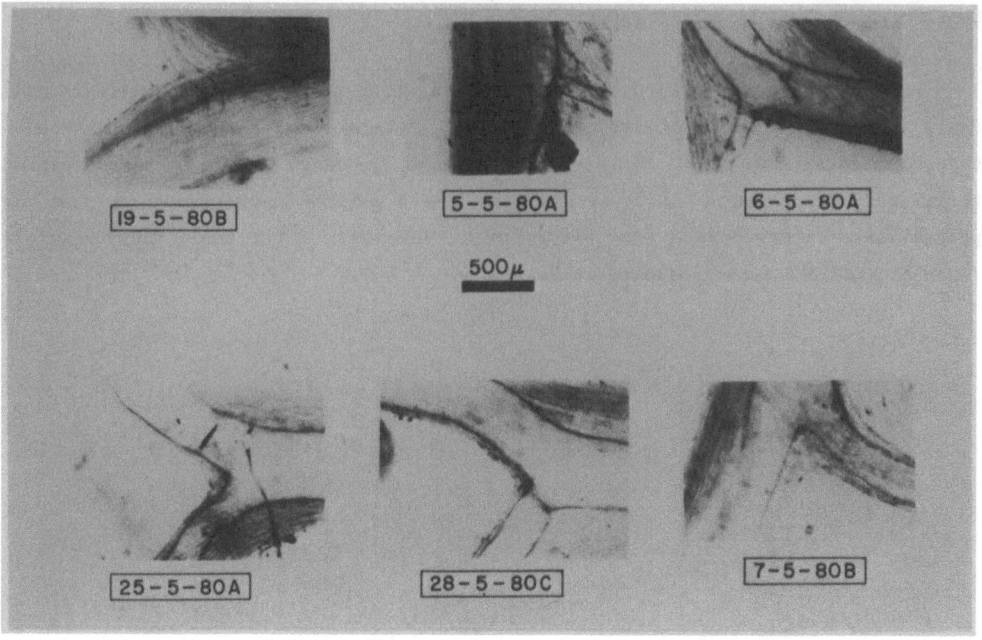

19-5-80B  5-5-80A  6-5-80A

500μ

25-5-80A  28-5-80C  7-5-80B

Fig. 8. Typical branches of squid giant axon.

$$GR = \frac{\sum d_{out}^{3/2}}{d_{in}^{3/2}} \quad . \tag{7}$$

If GR = 1, the "impedance matching" between outgoing and incoming fibers is perfect and a pulse proceeds through the branch with a minimum of difficulty. As GR becomes progressively greater than one, difficulties increase. Numerical calculations based on the Hodgkin-Hyxley model axon indicate that a solitary pulse will fail to propagate though a bifurcation with GR greater than about 10 [27,28].

A histogram for the GR's observed on branchings of squid giant axons under the assumption of orthodromic stimulation (i.e. incoming pulse on the main fiber) is presented in Fig. 9. It should be emphasized that these bifurcations include rather equal representations of all the different diameter ratios shown in Fig. 8; thus it seems that in the squid (Loligo vulgaris) nature is attempting to match impedances for orthodromic conduction under a variety of geometrical constraints. I have carefully examined sixteen bifurcations under orthodromic stimulation seeking evidence for a quantum change in a temporal pulse code before and after the bifurcation. Although such a change was sometimes observed, I failed to find any case for which surgical damage or the effects of tiring the axon during the search could be eliminated. This is consistent with the observations reported in Refs. [13 and 14].

It is interesting, therefore, to turn to antidromic stimulation where the incoming pulse is on one of the daughter branches and geometrical ratios greater than unity are readily

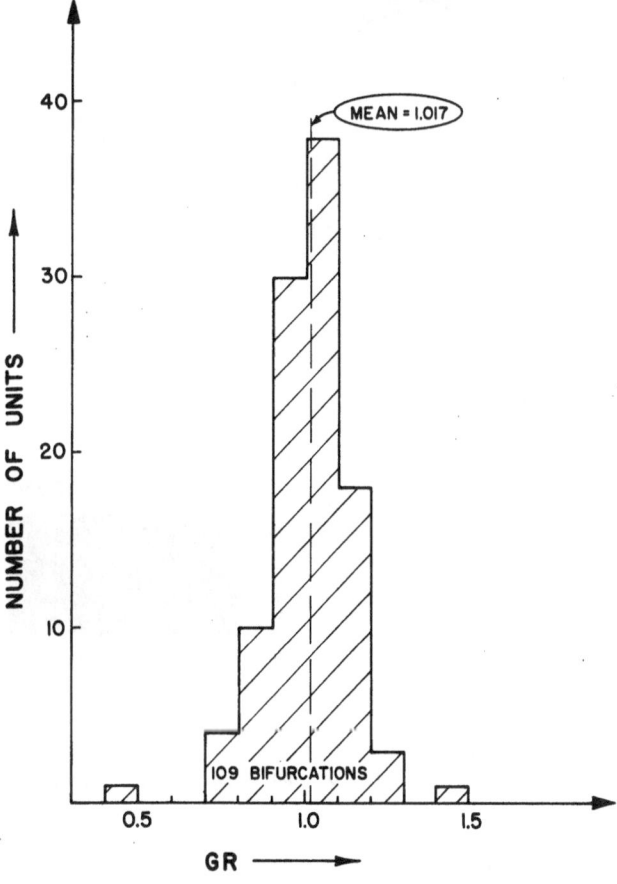

Fig. 9. Histogram of orthodromic GR's for 109 branches of the squid giant axon.

obtained. A typical example is bifurcation #20-3-80A which is shown in Fig. 10. With stimulation (as shown) on the 340 micron daughter, the GR = 1.71. In Fig. 10b are displayed the records of a single pulse on both the upstream electrode (B) and the downstream electrode (A). Figure 11 shows that upon stimulation of #20-3-80A with a pulse rate of 160 pps. in a single burst of one half second, I observed a gradual change in the outgoing pulse train until, after 200 milliseconds, the output pulse rate was 80 pps. This effect was observed on several preparations and, for a particular preparation was quite stable and reproducible. It is likely that the mechanism here is related to potassium accumulation in the periaxonal space [13,14,24,25,29].

An example that demonstrates the failure of two pulses to propagate through a branch is shown for #29-2-80B in Figs. 12 and 13. From Fig. 12a the GR = 2.14 and the state of health for isolated pulses is displayed in Fig. 12b. Figure 13 shows a critical value of pulse interval at which the second pulse just fails (upper) or

20-3-80A

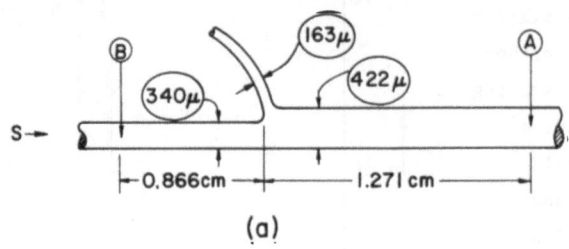

(a)

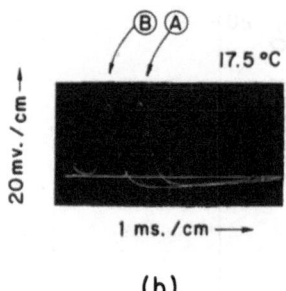

(b)

Fig. 10.   a) Bifurcation.   b) Action potentials.

20-3-80A

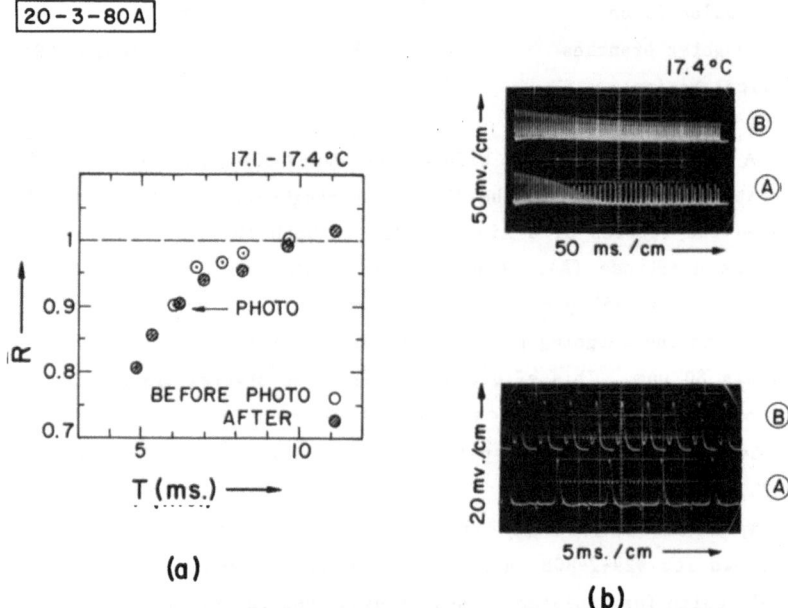

(a)

(b)

Fig. 11.   a) Dispersion for bifurcation of Fig. 10.
b) Quantum pulse code translation after 200 ms. of stimulation.

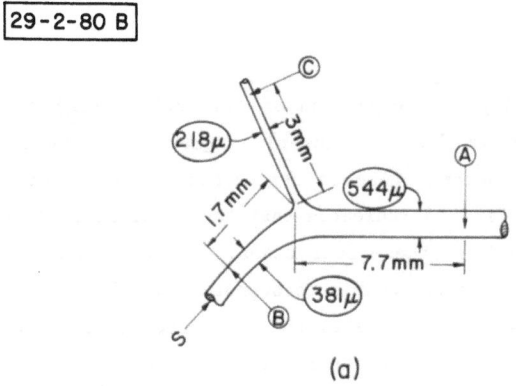

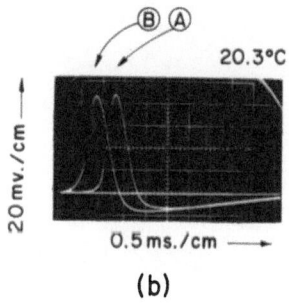

Fig. 12. a) Bifurcation. b) Action potentials.

succeeds (lower) to negotiate the bifurcation. The "hump" that appears in record (B)
of the lower photo in Fig. 13 is an artifact of the appearance of the second pulse.
This observation is important because it permits one to fix the position of the
critical dynamic event that reproduces the second pulse. This location proceeds as
follows. During the course of these experiments it was determined that conduction
velocity (θ) depends upon axon diameter (d) and temperature as

$$\theta = \left[\frac{d}{476}\right]^{\frac{1}{2}} [2.03 + 0.078 \text{ (Temp - 18.5)] cm/ms.} \tag{8}$$

where d is measured in microns. From (10) one can calculate the time delay for a
solitary pulse to go from the crotch of the bifurcation to electrode (A); this is
called $T_A$. Likewise $T_B$ is calculated as the time for a pulse to travel backward
from the crotch to electrode (B). The total time delay, $T_D$, between the second
pulse and the hump is greater than $T_A - T_B$ because the second pulse experiences
velocity dispersion as discussed in the previous section. Thus, assuming that
the second pulse is triggered at the bifurcation, one expects

$$T_D = \frac{T_A - T_B}{R} \quad .$$

(9)

In Fig. 14 is displayed a comparison of time delays ($T_D$) calculated from (9) with those measured, as in Fig. 13 (lower). Agreement between measurements and calculations implies the second pulse is regenerated at the crotch of the bifurcation. The agreement in Fig. 14 indicates that the critical event took place within a millimeter of the crotch in these four cases. Since no surgical damage was observed in this region, it is reasonable to suppose that the second pulse is blocked at the bifurcation. Furthermore three of the four observations indicated in Fig. 14 show agreement with calculations of the relation between GR and pulse interval from the Hodgkin-Huxley model [30]. See [29] for details.

29 - 2 - 80 B

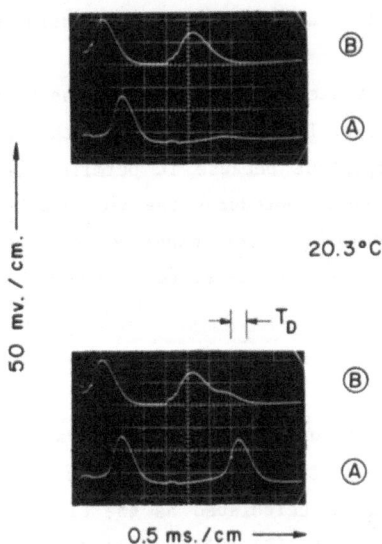

Fig. 13. The critical time interval between two pulses such that the second pulse just fails (upper) or just succeeds to pass through the branch.

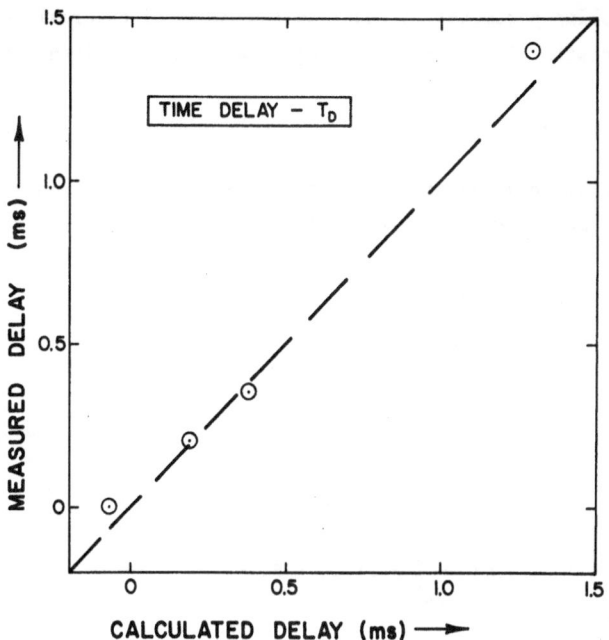

Fig. 14. A comparison of calculated and measured time delays $(T_o)$
assuming the second is initiated at the bifurcation.

One general statement about my observations is the following: Quantum pulse code transformations were only observed in the dispersive region where $T < T_1$ and $R < 1$.

## PULSE INTERACTIONS ON PARALLEL FIBERS

An electromagnetic analysis of two parallel fibers enclosed within an insulating sheath leads to a pair of coupled nonlinear diffusion equations [8,31-33]. In the context of a FitzHugh-Nagumo approximation [34], these take the form

$$V_{1,t} = (1 - \alpha)V_{1,xx} - \alpha V_{2,xx} - F(V_1) - R_1$$

$$R_{1,t} = \varepsilon(V_1 - bR_1)$$

$$V_{2,t} = (1 - \alpha)V_{2,xx} - \alpha V_{1,xx} - F(V_2) - R_2 \tag{10}$$

$$R_{2,t} = \varepsilon(V_2 - bR_2)$$

where the small coupling parameter is equal to the ratio of external to internal resistance per unit length. One seeks traveling wave solutions of the form

$$V_k(x,t) = V_k(\xi) = V_k(x - ut); \quad k = 1,2 \tag{11}$$

where u is the propagation speed of two pulses traveling in synchronism. Under this constraint, (10) becomes a set of ordinary differential equations

$$-u \frac{dV_1}{d\xi} = (1 - \alpha) \frac{d^2V_1}{d\xi^2} - \alpha \frac{d^2V_2}{d\xi^2} - F(V_1) - R_1$$

$$-u \frac{dR_1}{d\xi} = \varepsilon V_1$$

$$-u \frac{dV_2}{d\xi} = -\alpha \frac{d^2V_1}{d\xi^2} + (1 - \alpha) \frac{d^2V_2}{d\xi^2} - F(V_2) - R_1 \tag{12}$$

$$-u \frac{dR_2}{d\xi} = \varepsilon V_2$$

where, for conveninece, it has been assumed that b = 0.

A solution of (12) will be two pulses, one on each fiber, moving with the same velocity. Since $\alpha \ll 1$, this solution is written as a series

$$V_k = V_{ko} + \alpha V_{k1} + \ldots; \quad k = 1, 2$$

$$u(k) = u_o + \alpha u_1^{(k)} + \ldots \tag{13}$$

where it is provisionally assumed that solutions of (12) will have different velocities for k = 1, 2. Eliminating the R's yields

$$\frac{d^3V_{ko}}{d\xi^3} + u_o \frac{d^2V_{ko}}{d\xi^2} - F'(V_{ko})\frac{dV_{ko}}{d\xi} + \frac{\varepsilon}{u_o} V_{ko} = 0 \tag{14}$$

and

$$\frac{d^3V_{k1}}{d\xi^3} + u_o \frac{d^2V_{k1}}{d\xi^2} - F'(V_{ko}) \frac{dV_{k1}}{d\xi} - F''(V_{ko}) \frac{dV_{ko}}{d\xi} - \frac{\varepsilon}{u}_o V$$

$$= u_1^{(k)} \frac{\varepsilon}{u_o^2} V_{ko} - \frac{d^2V_{ko}}{d\xi^2} - \frac{d^3V_{10}}{d\xi^3} - \frac{d^3V_{20}}{d\xi^3} \quad . \tag{15}$$

The perturbation expanson (13) has reduced (12) to the uncoupled nonlinear equations (14) and the linear equations (15) for the first order corrections. Equations (15) are uncoupled in the $V_{k1}$ and the inhomogeneous parts involve only zero order solutions and the first order velocity perturbations. Thus each of (15) can be written

$$L_k V_{1k} = f_k \tag{15'}$$

for which a solvability condition is

$$(y_k, f_k) = 0 \tag{16}$$

where $y_k$ is a solution of

$$L_k^\dagger y_k = 0 \tag{17}$$

and $L_k^\dagger$ is the adjoint of $L_k$ under the inner product employed in (16). With the conventional definition

$$(v,w) \equiv \int_{-\infty}^{\infty} vw d\xi$$

(17) becomes

$$\frac{d^3y_k}{d\xi^3} - u_o \frac{d^2y_k}{d\xi^2} - F'(V_o) \frac{dy_k}{d\xi} - \frac{\varepsilon}{u_o} y_k = 0 \quad . \tag{18}$$

From solutions of (18) one can compute inner products with the right-hand sides of (15) and obtain useful expressions for the first order velocity perturbations, the

$u_1^{(k)}$. To effect this calculation it is convenient to take

$$F(V) = V - H(V - a) \tag{19}$$

where $H(\cdot)$ is the Heaviside step function. Choosing $a = 0.3$ and $\varepsilon = 0.1$, $V_{ko}(\xi)$ is the pulse shown in Fig. 15. Comparison with Fig. 10b or 12b shows that this choice of parameters is physiologically reasonable. The corresponding solution of (17) is shown in Fig. 16.

It is now assumed that $V_{20}$ differs from $V_{10}$ by a translation $\delta$ in $\xi$. Thus

$$V_{20}(\xi) = V_{10}(\xi - \delta)$$

$$y_2(\xi) = y_1(\xi - \delta) . \tag{20}$$

The solvability condition (16) requires

$$Nu_1^{(k)} = - \int_{-\infty}^{\infty} y_k \left( \frac{d^3 V_{10}}{d\xi^3} + \frac{d^3 V_{20}}{d\xi^3} \right) d\xi \tag{21}$$

where

$$N \equiv \int_{-\infty}^{\infty} y_1 \left( \frac{d^2 V_{10}}{d\xi^2} - \frac{\varepsilon}{u_o^2} V_{10} \right) d\xi = \int_{-\infty}^{\infty} y_2 \left( \frac{d^2 V_{20}}{d\xi^2} - \frac{\varepsilon}{u_o^2} V_{20} \right) d\xi . \tag{22}$$

To first order in $\alpha$, the condition for a traveling wave solution is

$$u(1) = u_o + \alpha u_1^{(1)} = u(2) = u_o + \alpha u_1^{(2)}$$

or

$$u_1^{(1)} = u_1^{(2)} . \tag{23}$$

In Fig. 17 these first order velocity corrections are plotted as functions of $\delta$, the displacement of pulse #2 with respect to pulse #1. Five solutions of (23) (i.e. intersections) are observed, but only three of these (at $\delta = \delta_1$, 0, and $- \delta_1$ denoted by open circles) are stable in the following sense. An increase in $\delta$

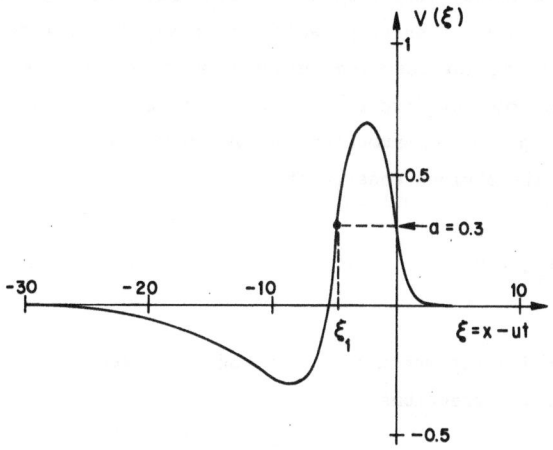

Fig. 15. Solution of (14)

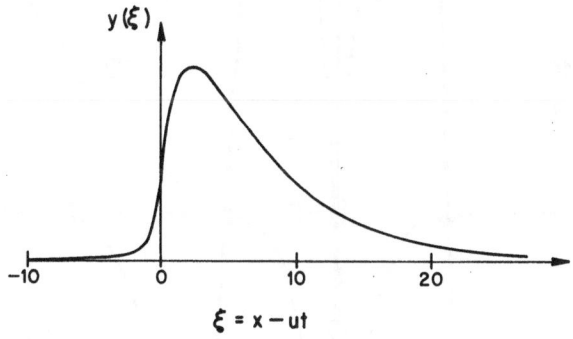

Fig. 16. Solution of (17)

(implying that pulse #2 advances with respect to pulse #1) causes the speed of pulse #1 to become greater than that of pulse #2 which, in turn, causes δ to decrease. By a corresponding argument the intersections denoted by closed circles are unstable. This effect has been confirmed by direct integration of the original pde's (10) [35].

For n weakly coupled fibers, numbered in any order, the relative pulse displacements must satisfy the obvious constraint

$$\delta_{12} + \delta_{23} + \ldots + \delta_{n1} = 0 \quad . \tag{24}$$

For each choice of n - 1 independent δ's, one can calculate n - 1 differences of the corresponding velocity corrections as

$$D_i = u_1^{(i)} - u_1^{(i+1)}; \; i = 1, 2, \ldots, n - 1 \quad . \tag{25}$$

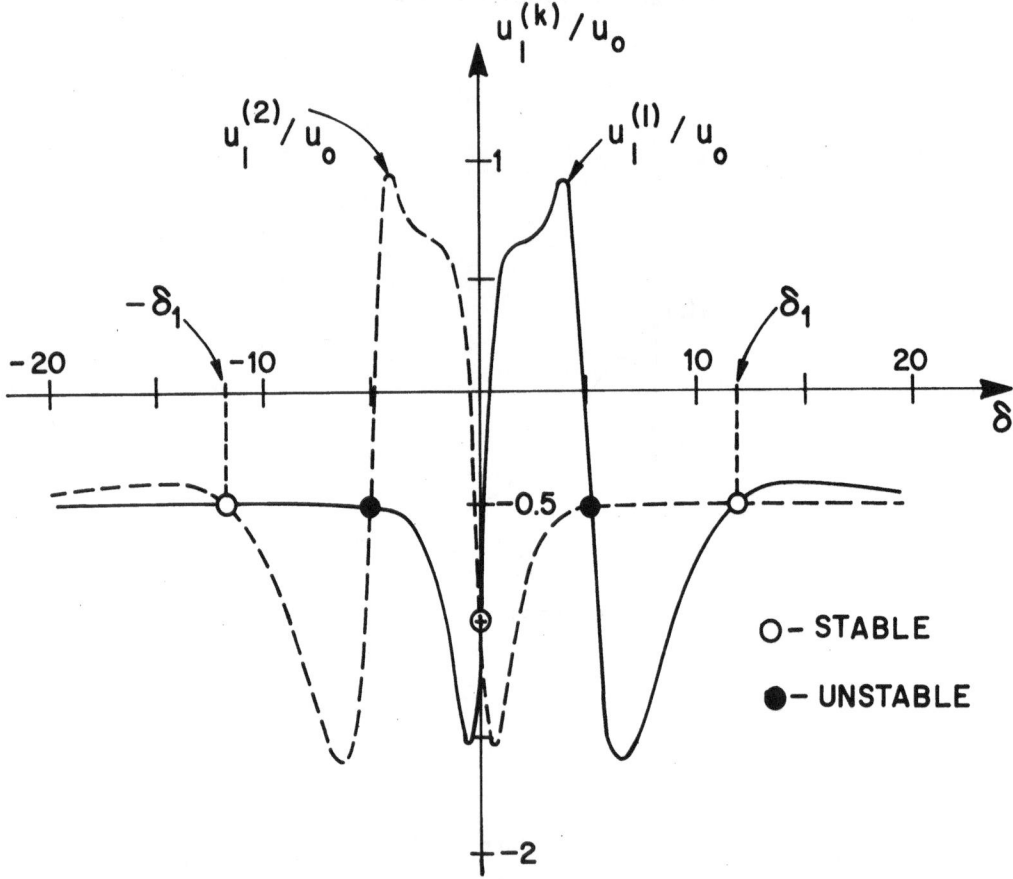

Fig. 17.  Stability diagram from (23).

Then $\bar{D} \equiv (D_1, D_2, \ldots, D_{n-1})$ is an $n - 1$ dimensional vector field in the $n - 1$ dimensional "$\delta$-space" where $\delta \equiv (\delta_{12}, \delta_{23}, \ldots, \delta_{n-1,n})$. Condensed pulse states are defined as stable zeros of

$$\bar{D} = \bar{D}(\bar{\delta}) = \bar{0} \ . \tag{26}$$

If $\delta_{i,i+1}$ denotes the position of a pulse on fiber $(i + 1)$ with respect to that of a pulse on fiber $i$, stability (in the sense described above) requires

$$\left. \frac{\partial D_i}{\partial \delta_{i,i+1}} \right|_{\bar{D} = \bar{0}} > 0 \tag{27}$$

for all $i$. For more than two coupled fibers ($n \geq 3$), this stability condition is satisfied only for the root of (26) that lies at the origin of $\delta$-space. Markin [31] has shown that if

$$\alpha n \approx 1 \tag{28}$$

this state can "recruit" additional pulses by stimulating neighboring fibers above threshold.

## THE PULSE ASSEMBLY

The recent reviews by Adey [18,19] indicate a surprising sensitivity of living tissue to nonionizing electromagnetic fields. Brain tissue, for example, responds to gradients as low as $10^{-7}$ volts/cm in the frequency range from 6 to 20 cycles per second. In speculating about mechanisms for this effect, it is necessary to consider how such weak fields can be recognized above thermal noise. More precisely one must suppose

$$\left[ \sqrt{(\pi f \mu_o \rho)} \ E^2 \right] \cdot \sigma \geq kT\Delta f \tag{29}$$

where the square-bracketed term on the left-hand side is the power per unit area of an incident electromagnetic wave and $\sigma$ is the absorption cross-section for some (unknown) dynamical variable. On the right-hand side, $\Delta f$ is the reciprocal of the memory time for the dynamical variable. Guided by Fig. 4, one can take $\Delta f \sim T^{-1} = 100 \ sec^{-1}$. Then with $\mu_o = 4\pi \times 10^{-7}$ henrys/meter, $\rho = 1$ ohm-meter, and $E = 10^{-7}$ volt/cm (29) implies

$$\sigma \gtrsim 10^{-5} \text{ meter}^2$$

$$\gtrsim (3 \text{ mm})^2 .$$

What neural mechanism could have such a large absorption cross-section? One possibility might be the "pulse assembly" sketched in Fig. 18. Here we suppose that the longitudinal pulse locking at time interval $T_1$ (see Fig. 5) is acting between pulses B and D and between pulses D and E. Also we assume a transverse pulse locking, determined by (26), is acting between pulses A, B, and C and between pulses E and F.

   This mechanism is highly speculative, but it would deliver information with an established time synchronism between the component pulses as is required by the information processing mechanisms that have been proposed for Waxman's multiplex neuron [16]. Furthermore it is not inconsistent with the observations of Scheibel and Scheibel [36] who have compared stained sections from newborn and mature cats and conclude that:

> During the process of maturation, dendrite shafts have been found to rearrange themselves into bundles in various parts of the nervous system including the ventral horn of the spinal cord, brain stem reticular core, nucleus reticularis thalami, cerebral cortex, and possibly in basal ganglia and certain cranial nerve nuclei. In some cases, the appearance of bundle complexes seems closely time-locked to the initial development of discrete items of motor performance.

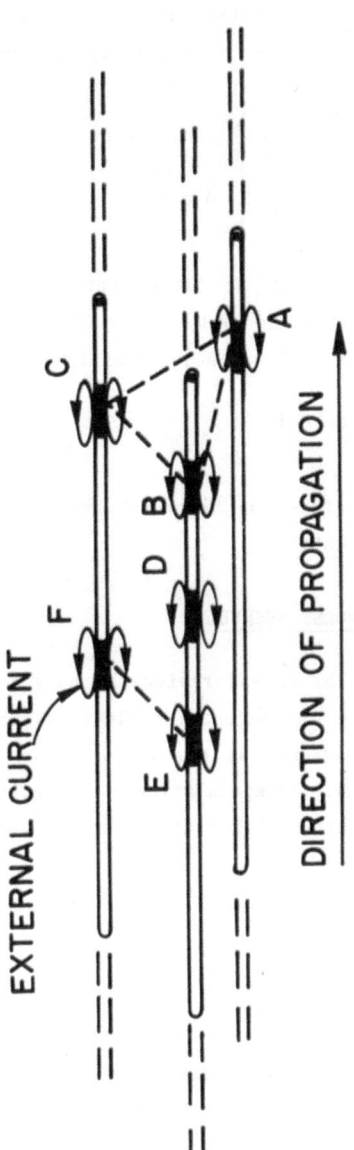

Fig. 18.   Structure of a "pulse assembly."

REFERENCES

1. R. Lorente de Nó, "Decremental conduction and summation of stimuli delivered to neurons at distant synapses," in Structure and Function of the Cerebral Cortex (Tower and Schade, eds) American Elsevier (New York, 1960) 278-281.

2. Yu. I. Arshavskii, et al., "The role of dendrites in the functioning of nerve cells," Dokl. Akad. Nauk SSSR 163 (1965) 994-997.

3. R. Llinás, C. Nicholson and W. Precht, "Preferred centripital conduction of dendrite spikes in alligator Purkinje cells," Sciences 163 (1969) 184-187.

4. V. F. Pastushenko, V. S. Markin and Yu. A. Chizmadzhev, "Branching as a summator of nerve pulses," Biophysics 14 (1969) 1130-1138.

5. A. M. Gutman, "Further remarks on the effectiveness of the dendrite synapses," Biophysics 16 (1971) 131-138.

6. M. B. Berkinblit, et al., "Interaction of the nerve impulse in a node of branching (investigation of the Hodgkin-Huxley model)," Biophysics 16 (1971) 105-113.

7. A. C. Scott, "Information processing in dendritic trees," Math. Biosci 18 (1973) 153-160.

8. A. C. Scott, Neurophysics, Wiley-Interscience (New York, 1977) Ch. 6.

9. K. Krnjević and R. Miledi, "Presynaptic failure of neuromuscular propagation in rats," J. Physiol. 149 (1959) 1-22.

10. S. H. Chung, S. A. Raymond and J. Y. Lettvin, "Multiple meaning in single visual units," Brain Behav. Evol. 3 (1970) 72-101.

11. Y. Grossman, M. E. Spira and I. Parnas, "Differential flow of information into branches of a single axon," Brain Res. 64 (1973) 379-386.

12. D. O. Smith and H. Hatt, "Axon conduction block in a region of dense connective tissue in crayfish," J. Neurophysiol. 39 (1976) 794-801.

13. Y. Grossman, I. Parnas and M. E. Spira, "Differential conduction block in branches of a bifurcating medium," J. Physiol. 295 (1979) 283-305; "Ionic mechanisms invloved in differential conduction of action potentials at high frequencies in a branching axon," ibid. 307-322.

14. D. O. Smith, "Morphological aspects of the safety factor for action potential propagation at axon branch points in the crayfish," J. Physiol. 301 (1980) 261-269; "Mechanisms of action potential propagation failure at sites of axon branching in the crayfish," ibid. 243-259.

15. F. O. Schmitt, P. Dev and B. H. Smith, "Electrotonic processing of information by brain cells," Science 193 (1976) 114-120.

16. S. G. Waxman, "Regional differentiation of the axon: a review with special reference to the concept of a multiplex neuron," Brain Res. 47 (1972) 269-288.

17. D. O. Hebb, Organization of Behavior Wiley (New York, 1949).

18. W. R. Adey, "Frequency and power windowing in tissue interactions with weak electromagnetc fields," Proc. IEEE 68 (1980) 119-125.

19.  W. R. Adey, "Tissue interactions with nonionizing electromagnetic fields," Physiol. Rev. 61 (1981) 435-514

20.  A. C. Scott and U. Vota-Pinardi, "Velocity variations on unmyelinated axons," J. Theoret. Neurobiol. (to appear).

21.  F. Donati and H. Kunov, "A Model for studying velocity variations on unmyelinated axons," IEEE Transactions on Biomedical Engineering BME-23 (1976) 23-23.

22.  R. N. Miller and J. Rinzel, "The dependence of impulse propagation speed on firing frequency, dispersion, for the Hodgkin-Huxley model," Biophys. J. 34 (1981) 227-259.

23.  J. Rinzel, private communication.

24.  B. Frankenhaeuser and A. L. Hodgkin, "The after-effects of impulses in the giant never fibers of Loligio," J. Physiol. 131 (1956) 341-376.

25.  W. J. Adelman, Jr. and R. FitzHugh, "Solutions of the Hodgkin-Huxley equations modified for potassium accumulation in a periaxonal space," Federation Proc. 34 (1975) 1322-1329.

26.  W. Rall, "Branching dendritic trees and motoneuron membrane resistivity," Exp. Neurol. 1 (1959) 491-527.

27.  M. B. Berkinblit, et al., "Computer investigation of the features of conduction of a nerve impulse along fibers with a different degree of widening," Biophysics 15 (1970) 1121-1130.

28.  I. Parnas and I. Segev, "A mathematical model for conduction of action potnetials along bifurcating axons," J. Physiol. 295 (1979) 323-343.

29.  A. C. Scott and U. Vota-Pinardi, "Pulse code transformations on axonal trees," J. Theoret. Neurobiol. (to appear).

30.  B. I. Khodorov, et al., "Conduction of a series of impulses through a portion of the fiber with increased diameter," Biophysics 16 (1971) 96-104.

31.  V. S. Markin, "Electrical interaction of parallel nonmyelinated nerve fibers," Biophysics 15 (1970) 122-133 and 713-721; ibid. 18 (1973) 324-332 and 539-547.

32.  A. C. Scott and S. D. Luzader, "Coupled solitary waves in neurophysics," Physica Scripta 20 (1979) 395-401.

33.  S. D. Luzader, "Neurophysics of parallel nerve fibers," Ph.D. Thesis, University of Wisconsin (1979).

34.  J. Nagumo, S. Arimoto and S. Yoshizawa, "An active pulse transmission line simulating nerve axon," Proc. IRE 50 (1962) 2061-2070.

35.  J. C. Eilbeck, S. D. Luzader, A. C. Scott, "Pulse evolution on coupled nerve fibers," Bull. Math. Biol. 43 (1981) 389-400.

36.  M. E. Scheibel and A. B. Scheibel, Intern. J. Neuroscience 6 (1973) 195.

# CHAPTER 6

# Micronetworks in nerve cells

C. Koch[$], T. Poggio and V. Torre[‡]

Artificial Intelligence Laboratory and Department of Psychology, Cambridge, Mass. 02139

[$] Max-Planck-Institut fur biologische Kybernetik
Spemannstrasse 38, D 7400 Tubingen, FRG

[‡] Universita' di Genova, Istituto di Fisica
Genova, Italy

## Introduction

A common belief is that the dendritic tree of neurons whose average length ($l$) is less that its estimated electronic space constant ($\lambda$) is virtually equipotential. As a consequence, the geometry of the dendritic tree just merely reflects the spatial convergence of synaptic inputs. The above reasoning is strictly correct for a single cylindrical cable. In branched structures, even those satisfying the equivalent cylinder condition, the argument can be totally wrong.[1,2] When a pulse of current is injected at one of the terminal tips, the observed voltage alternation between the tip and the soma can be orders of magnitude higher than that between the soma and the tip, when the same current is injected at the soma. Therefore, even in a small neuron, synaptic inputs can interact locally in a non-linear way, implementing elementary information processing operations.[3,4] In particular, it has been recently suggested that direction selectivity of some retinal ganglion cells is due to a non-linear interaction of excitatory conductance changes and of inhibitory inputs with an equilibrium potential near the resting membrane potential (shunting inhibition) on the dendritic membrane of the cell.[5]

We have analysed the electrical properties of the different type of retinal ganglion cells in the cat retina on the basis of passive cable theory. It is concluded that these neurons need not be equipotential despite their small dimensions, mainly because of their extensive branching. It is suggested that their dendritic architecture reflects characteristically different electrical properties, which are likely to be relevant for their physiological function and their information processing role. A more extensive report on this work can be found elsewhere.[12,13,11]

## Methods

We used data (kindly provided by B. Boycott and H. Wassle[6]) taken from two Golgi-Cox whole mounted retinae of an adult cat. Each analyzed cell was traced at x400 to x1000 magnification using a Zeiss drafting apparatus. Branching structure, length and diameters were coded into a list which served as input to a program (NEURON) which computed the characteristic electrical properties of the cell. The main electrical quantity computed by NEURON, under the assumption of sealed end terminations, is the complex transfer resistance $\tilde{K}_{ij}(\omega)$. If a current $I_j(t)$ is injected at location $j$, the voltage at location $i$ is:

$$V_i(t) = K_{ij}(t) * I_j(t) \tag{1}$$

where $*$ indicates convolution and $\tilde{K}_{ij}(\omega)$ is the Fourier transform of $K_{ij}(t)$. The D.C. value $\tilde{K}_{ij}(0)$ is the transfer resistance for D. C. inputs. We have assumed a value of $2\mu F cm^{-2}$ for the membrane capacitance, $70\Omega cm$ for the intracellular resistivity and $2500\Omega cm^2$ for the membrane resistivity.[1,7,8] We have checked our results for a broader range of values and only some of our results would change if the membrane resistivity is much higher than $8000\Omega cm^2$.

## Results

*Subunits*

Six out of the nine retinal ganglion cells that we have analyzed were located at about 3 mm from the center of the area centralis. They were one alpha cell, two beta cells, two gamma cells and one delta cell. We analyzed also one beta-ON and one beta-OFF at 3 mm of eccentricity and a delta cell in the periphery.

Of these nine cells, only the beta cells at 3 mm eccentricity were reasonably equipotential in agreement with its small dimensions (diameter about $180\mu m$) and an average short electrotomic length[12] of about .25. All the other cells showed substantial deviations from equipotentiality. In the $\alpha$ cell, a signal generated by a synapse in a distal dendrites could be attenuated $10 \div 25$ times in the soma. The voltage attenuation in gamma and delta cells were between 4 and 10. Cells in peripheral regions of the retina become larger and less equipotential.

107

If a neuron is not equipotential, its electrical properties are nonuniform in the sense that they strongly depend on input-output location. Therefore, it is natural to try to divide the dendritic tree into electrically homogeneous regions, each one being rather isolated from the other ones and rather equipotentive within. We propose the term "subunits" for such regions of the dendritic tree. A subunit is a region of the dendritic tree within which the transfer resistance is much higher than the transfer resistance from any location of the subunit to the soma.

Figure 1 shows a partition of the dendritic tree of a delta cell in distinct subunits. Using the same criterium, the alpha cell showed 32 sizable subunits while the $\beta$ cells in the area centralis behaved essentially as a single subunit. The $\delta$ cell of Figure 1, despite an overall dendritic field area which is less than half the area of the $\alpha$ cell, has 23 subunits. The type of dendritic tree, with daughter branches of almost the same diameter as the mother branches, is the main reason for this. $\gamma$ cells have a much smaller number of subunits.

Figure 1:

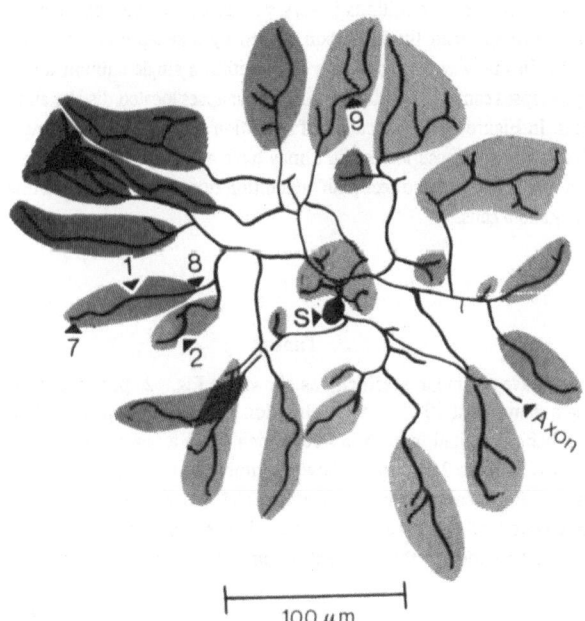

100 $\mu$m

Partition of the dendritic tree of a delta cell in subunits at 3 mm eccentricity. The shaded regions of the dendritic tree represent distinct subunits. A subunit is a region of the dendritic tree in which the transfer resistance between two arbitrary locations $i$ and $j$ is much higher than the transfer resistance from an arbitrary location of that region and the soma, that is

$$\frac{K_{ij}}{K_{is}} > C \quad C > 1$$

The subunits shown in the figure have been obtained with a $C$ equal to 4. The maximal set of non-overlapping subunits thus defined is clearly not unique. Because the solutions of the cable equations are continuous along branches, it formally impossible to define subunits as disjoint regions with distinct boundaries. (From ref. 12.)

A possible function of subunits is that nonlinear saturation at the synapses may be effectively reduced by spreading the same (conductance) input among several subunits on the dendritic field. The existence of subunits depends critically on the value of $R_m$: if $R_m$ were much higher than $R_m = 2500\Omega cm^2$ all subunits would effectively disappear (for steady state signals but not for transients).

## The finer structure or the direct path condition

The combination of an excitatory and an inhibitory conductance input may give rise to strong nonlinear interactions.[3—5,8,9] Let us consider the case of an excitatory synapse modulating the conductance $g_e$ to an ionic species with equilibrium potential $E_e > V_{rest}$ in location $e$ and an inhibitory synapse modulating the conductange $g_i$ to an ionic species with equilibrium potential $E_i \leq V_{rest}$ at location $i$. It is possible to prove (for DC signals) that for any arbitrary values of $g_e$ and $g_e$ the location where inhibition is maximally effective is always on the direct path from the location of the excitatory synapse to the soma.[10—12]

For $\gamma$ and $\delta$ cells, the above conditions is very strong, that is, the inhibition on the direct path may be 5-10 times stronger than the inhibition caused by a synapse on the same subunit but not on the direct path. In this way, $\gamma$ and $\delta$ cells have within a single subunit a finer organization in which inhibitory synapses can veto selectively only the synapses located distally and are uneffective on proximal synapses. In Figure 1, if the location of excitation is at 1, inhibitory synapses in 1 and 2 are rather uneffective, while a synapse located in 8 may have a strong control on the signal transmitted to the soma. In $\alpha$ and $\beta$ cells, "the direct path" condition does not determine a finer structure to the same extent as in $\gamma$ and $\delta$ cells.

## Timing

Transient inputs have a similar specificity as shown in Fig. 2, provided that the conductance changes are larger than about $10^{-8}$ mho. Inhibition on the direct path is then strong, whereas inhibition located behind excitation or on a side branch (even a few micron further away) is almost completely ineffective. Figure 2 also shows how the timing of the excitatory versus the inhibitatory input influences the effectiveness of the interaction. The sharpness of the timing curves of Figure 2 reflects the time course of the input conductance changes and the cable properties, whereas the optimal delay is essentially due to the propagation time from excitation to the location of shunting inhibition.

## Conclusion

Because of the strength and specificity of such nonlinear interactions, we wish to propose that they may perform characteristic information processing operating in passive dendritic trees. Since inhibi-

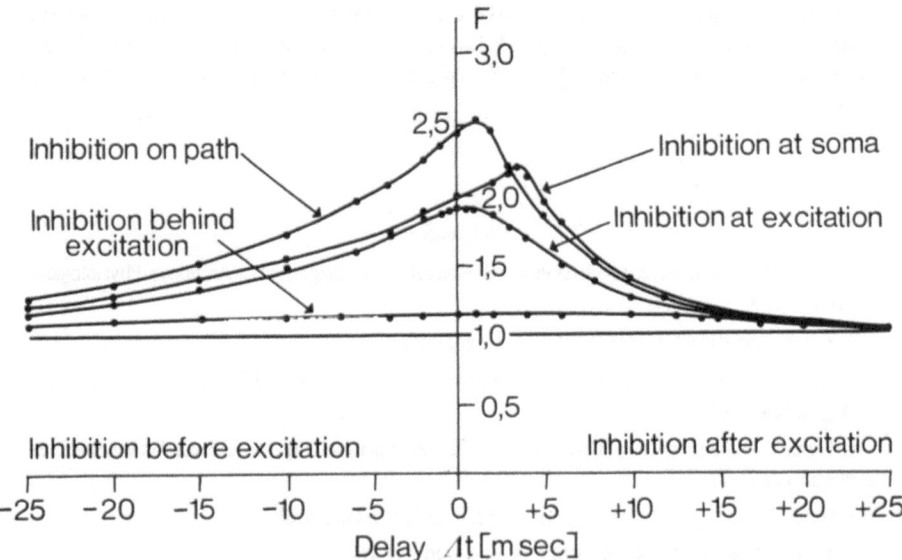

**Figure 2.** Inhibitory effect of a synapse as a function of delay between transient excitation and inhibition for the cell of Fig. 1. The equilibrium potential of the inhibitory synapse has been assumed to be equal to $V_{rest} = 0$. The conductance inputs are $\alpha$ functions [12] with a time to peak of 25 msec and peak values of $10^{-7} mho$. $F$ is the ratio between the somatic depolarization in the absence and in the presence of the inhibitory input. $F$ equal to 1 means that the soma depolarization is not affected by inhibition. Notice that the optimal timing for $e = i$ is not for $\Delta t = 0$ (Segev, pers. comm.) the deviation is, however, very small. The charge factor follows the $F$ factor, but is somewhat more broadly tuned. In these calculations, we have assumed $C_m = 2\mu F cm^{-2}$, $R_i = 70\Omega cm$, $R_m = 2500\Omega cm^2$. The basic results hold over over a wide range of parameters, in particular $R_m$ can be increased by several order of magnitudes without a significant change in the "direct path" property, which depends critically on branching geometry and effective intracellular resistance. From Koch & Poggio.[13]

tion vetoes effectively more distal excitatory inputs only when it is on the direct path to the soma, a variety of local operations can be performed, exploiting the branching geometry of a dendritic tree with a suitable location of excitatory and inhibitory inputs. Timing of inputs provides an additional important control variable. The $\delta$ cell with its many distinct subunits and with its fine structure within each subunit, seems to represent the optimal morphological substrate for direction selective ganglion cells.

**Acknowledgements.** Part of this work was done at the Max-Planck-Institut fur Biologische Kybernetik, Tubingen, West Germany. T.P. is presently at MIT Dept. of Psychology/Artificial

Intelligence Laboratory, Cambridge, Mass., USA. C.K. is at the Max-Planck-Institute fuer Biologische Kybernetik in Tuebingen, West Germany. V.T. is currently at the Istituto di Fisica, Genova, Italy. The authors would like to acknowledge support in their collaboration under NATO grant number 237.81.

## References

1. Rall, W. in Handbook of Physiology, E. Kandell & S. Geiger, eds. American Physiological Society, pages 39-97 (1977).

2. Rall, W. and Rinzel, J. 1973 Biophys. J., 13, 648-688.

3. Poggio, T. and Torre, V. in Theoretical approaches of neurobiology. MIT Press, W. Reichardt & T. Poggio, eds. (1981).

4. Torre, V. and Poggio, T. in Theoretical approaches of neurobiology. MIT Press, W. Reichardt & T. Poggio, eds. (1981).

5. Torre, V. and Poggio, T. Proc. R. Soc. Lond. B., 202, 409-416 (1978).

6. Boycott, B. B. and Wassle, H. 1974 J. Physiol., 240, 397-419.

7. Barrett, J. H. and Crill W. E. 1974 J. Physiol., 239, 301-324.

8. Barrett, J. H. 1975 Fed. Proc., 34, 1398-1407.

9. Redman, S. J. 1976 in International review of physiology, II, vol. 10, University Park Press.

10. Poggio, T. and Torre, V., in preparation.

11. Koch, C. 1982 Nonlinear information processing in dendritic trees. Ph.D. Thesis, Tubingen, Germany.

12. Koch, C., Poggio, T. and Torre, V. 1982 Trans. R. Soc., in press.

13. Koch, C., Poggio, T., Nonlinear interactions in a dendritic tree, in preparation.

# CHAPTER 7

ROLE AND USE OF NOISE IN BIOLOGICAL SYSTEMS

Shunsuke Sato

Department of Biophysical Engineering, Faculty of Engineering Science,
Osaka University, Toyonaka, Osaka, Japan

## 1. Introduction

Cooperative or competitive phenomena are not observed exclusively
in the brain, but widely in nature including ecosystems.  These
phenomena are often interpreted in mathematical biology according to
the behavior of the solutions of equations describing non-linear
dynamical systems that involve the variables of concern.  A phenomenon
is thus understood qualitatively and quantitatively in a very simplified
way in order to construct a manageable mathematical model.  For
instance, although one single prey and one single predator species do
not constitute an ecosystem, several prey and predator species interact
with each other in a sufficiently complicated manner to constitute an
ecosystem in the real world.  However, when we pay attention to some
relation existing between a prey and a predator species, we simplify
the relation itself by neglecting the influence of the remaining
species or by representing such influence just by means of some
parameters.

Let us consider two mutually interacting periodic phenomena
characterized by highly different periods.  If we are interested in
the shorter period phenomenon, it is expedient to account for the
influence of the longer one by means of a very slowly varying parameter
or by a constant set in the equation formulated to describe the
phenomenon itself.  While if we are interested in the longer period
phenomenon, we sometimes regard the effect of the quickly varying
variable as some kind of noise.  Thus we can sometimes reduce the
number of equations and obtain the necessary mathematical tractability
without loosing the essence.

While,  for instance, it is certainly reasonable to neglect the
random fluctuation of air molecules when studying the motion of a
pendulum, there exist numerous situations where we must take into
account random external forces.  The motion of pollen in water provides
the probably best known example in support of this statement.  Being
small and light enough, pollen particles indeed abruptly change the
direction of motion as a result of the collisions with the water
molecules.  This yields the much celebrated zig-zag motion of the

pollen particles in the water in which they are suspended.  In this
example the motion of the particles can not be inferred by means of
a deterministic model constructed by neglecting or averaging the
microscopic effect of the individual collisions.

In the sequel, we shall study the influence of noise on properties
of the solution of a deterministic dynamical equation.

## 2. Destabilizing effect of noise

Let the deterministic equation

$$\frac{dx}{dt} = f(x,t) \tag{1}$$

be given.  Is its stability property affected by the presence of noise?
The solution of eq. (1) can be affected by noise in various manners.
Noise itself has various statistical properties.  For the sake of
mathematical brevity, we restrict our considerations to a situation
where eq. (1) is perturbed by white Gaussian noise.  Namaly, we
consider the behaviour of the solution process of an equation of the
type:

$$dX = f(X,t)\,dt + g(X,t)\,dB(t) \tag{2}$$

and compare it with that of eq. (1).  Here $B(t)$ is referred to as the
Brownian motion process whose formal derivative is the white Gaussian
noise with mean 0 and variance 1.  We assume that f and g are vector
valued functions which satisfy some appropriate regularity conditions.
Eq. (2) is called the Ito type stochastic differential equation (See
Jazwinski, 1970, p.105).  One of the main interests is concerned with
the stability property of the solution of a dynamical system as a
mathematical model of a phenomenon under consideration.  One should
note that there are several different concepts or definitions of
stability of the solution process of eq. (2) as in the deterministic
case (See Ladde and Lakshmikantham, 1980, p.69).  We deal with the
moment stability.  Moment stability concerns the behaviour of $\|X\|_p$,
where

$$\|X\|_p = (E|X|^p)^{\frac{1}{p}} \tag{3}$$

The state X=0 is said to be stable, uniformly stable, and asymptotically stable, etc., in the p-th mean according to the corresponding behaviours of $\|x\|_p$.

Let us consider a simple example. Take the stochastic differential equation

$$dX = aXdt + \sigma XdB(t), \qquad X(0) = X_0 > 0, \ \sigma > 0 \qquad (4)$$

The solution process (see, for instance, McKean, 1969) reads.

$$X(t) = X_0 \exp((a - \frac{\sigma^2}{2})t + \sigma B(t)) \qquad (5)$$

It is easy to show that (5) is the solution of eq. (4), by computing the differential $dX(t)$ from (5) under the rules:

$$dB(t)^2 = dt, \quad dt^2 = dt \cdot dB(t) = 0 \qquad (6)$$

Now let us evaluate the p-th moment of $X(t)$ given by (5). Since $B(t)$ is Gaussian with mean 0 and variance t,

$$EX(t)^p = X_0^p \exp((a - \frac{\sigma^2}{2})pt) \int_{-\infty}^{\infty} e^{p\sigma b} \frac{1}{\sqrt{2\pi t}} e^{-\frac{b^2}{2t}} db$$

$$= X_0^p \exp((a + \frac{\sigma^2}{2}(p - 1))pt) \qquad (7)$$

Thus if $a + \frac{\sigma^2}{2}(p - 1) < 0$, X=0 is asymptotically stable in the p-th mean. We know that the mean of the solution process $X(t)$ will converge to the equilibrium X=0 as time goes on if $a < 0$, while the variance will increase infinitely unless $a < -\frac{\sigma^2}{2}$ (kozin, 1972).

Another simple example is offered by a non-linear dynamical system which is sometimes called a logistic equations:

$$\frac{dx}{dt} = \alpha(1 - \frac{x}{N})x \qquad (3)$$

Clearly x = N is a stable equilibrium point. Eq. (8) is also renamed as the Malthusian equation, in which density effect of population is taken into account when specifying the intrinsic fertility a:

$$a = \alpha (1 - \frac{x}{N}) \tag{9}$$

Assuming that the Malthusian coefficient a is perturbed by an additive
noise, we have

$$dX = \alpha (1 - \frac{X}{N}) Xdt + \sigma XdB(t) \tag{10}$$

A solution starting at positive $X_0$ remains positive for ever. We can
compute mean and variance of the steady state distribution obtained
from the associated Fokker-Planck eq:

$$\frac{\partial p}{\partial t} = -\frac{\partial}{\partial x} \{\alpha (1 - \frac{x}{N}) xp\} + \frac{\sigma^2}{2} \frac{\partial^2 x^2 p}{\partial x^2} \tag{11}$$

where $p = p(x,t|x_0)$ is the transition probability density of $x(t)$. A
simple computation yields the finite mean and variance if the amplitude
of the noise is small, namely if $2\alpha > \sigma^2$. In the case $2\alpha < \sigma^2$, the
distribution concentrates ultimately at the origin, so that extinction
of the species is expected to occur. For a thorough discussion of
interpretation and use of equations such as (8) to model population
growth process, We refer to May(1973) and Ricciardi(1977) and to
references therein.

The above two examples show that the stability property of the
equilibrium of a deterministic equation can be altered in the
presence of external noise. The following is a particularly
illuminating example. Consider a conservative mechanical system

$$\begin{cases} \dfrac{\partial x}{\partial t} = \dfrac{\partial H}{\partial p} \\[2mm] \dfrac{\partial p}{\partial t} = -\dfrac{\partial H}{\partial x} \end{cases} \tag{12}$$

where x denotes the position of a particle and p its momentum. Assume
that the Hamiltonian is given by

$$H(x,p) = \frac{p^2}{2} + V(x) \tag{13}$$

where V(x) is the potential energy attaining its minimum at the origin.

It can be shown that the expectation of H(x,p) will unlimitedly
increase with the increase of time in the presence of the random
external force σW̃(t), where W(t) is the white noise.  In fact this is
concluded by deriving the stochastic differential dH of the stochastic
Hamiltonian H by using the so-called Ito's lemma (McKean, 1969, p.32).
Thus noise appears to act, in general, in a way to weaken the stability
property of the dynamical system.  It is thus appropriate to talk
about destabilizing effects of noise*.  Another example has bee provided
by Arnold et al (1979).  They considered the stochastic Lotka-Volterra
equations and showed that on the average noise drives the system
toward orbits with larger and larger circumferences.

The fact that a dynamical system is generally destabilized by
external noise seems to indicate that noise can play an important
role in phase transition phenomena.  A simple mathematical model that
describes phase transitions consists of a system possising multiple
stable regimes each surrounded by an unstable boundary.  If the system
is autonomous, only external noise is allowed to trigger the transitions
between the different regimes.  We thus refer to this situation as to
noise induced phase transitions (see, for example, Arnold and Lefever,
1981).

## 3. Use of noise

So far we have discussed the average behaviours of non-linear
dynamical systems driven by external white noise and we have used the
transition probability density approach.  However, it is more
advantageous to represent the solution process X(t) in terms of the
individual paths of the external noise.  As is well known, if an
autonomous dynamical system is linear this is always possible.
However, such possibility also arises in non-linear cases.  We have
indeed already shown one such example in the foregoing when we have
pointed out a solution process that is explicitly expressed in terms
of the external noise, B(t) (cf. eq. (5)).  From such representation,
one can compute various average quantities of the solution process.

This is of course a very rare example of a non-linear system for
which an explicit relation exists between solution process and external
noise.  This provides us with a system-theoretic approach of the
stochastic differential equations which has become over the last
decade one of the most attractive subjects in engineering control
theory.  In this framework, the external noise is identified with the

---

* Arnold (9) showed that noise can sometimes stabilize a system.

input to a system whose dynamics is described by a set of differential equations. Thus a sample of the generally vector solution process coinciding with the output of the system is representable in the form of a functional of a sample of the white noise constituting the system's input. The output process may thus sometimes be referred to as the non-linear noise.

The possibility of representing the output of a time invariant physical system as a functional of the input white noise was originally proposed by Nobert Wiener (1958). The mathematical foundations were laid down by Wiener (1938). He was followed by many mathematicians and engineers (see for example, McKean, 1972). More recently his theory has been applied to a variety of fields including neuro-physiology. In this context the pioneering work has been carried out by Marmarelis and Naka (1973).

According to the theory of Wiener, noise can be viewed as an external stimulus to the system. Along these lines, it is possible to make prefitable use of the noise in order to obtain some information on the system itself. A comprehensive understanding of Wiener's theory requires a rather deep knowledge of the stochastic processes theory. The essential references may be Wiener (1958) and Hida(1980).

At some level of biological research, it is important to establish a quantitative relation between s muli and responses. Such relation has been exclusively investigated in biology from a static standpoint. The classical Weber-Fechner law is a good example. Michaelis-Menten type characteristics is another example concerned with the response sensitivity to light stimuli (Naka and Rushton, 1966) by a photo-receptive cell of fish.

We do not mean to deny the importance of finding out a static relation between stimuli and responses; nevertheless, it does not yield us the true input-output relation. Since a living organism receives stimuli of ever varying magnitude, it is more meaningful to watch the development of its response.

Quantitative description of stimulus-response dynamics is provided by systems theory and by modern control theory within the framework of dynamic system analysis. The study in this direction has nearly been brought to completion in the case of linear systems. The results obtained in such theories can also be applied to the analysis of various properties of biological and neurophysiological systems including a stimulus-response relation. However, most biological systems are essentially non-linear so that the mentioned

application is naturally limited to some extent.  On the contrary,
Wiener's theory of non-linear noise, about which we shall talk here-
after, can be applied to analyze a fairly wide class of time invariant
non-linear systems.  In system analysis, our concern is often twofold:
(i) To predict the response to a given stimulus provided that the
system structure or system dynamics is known; (ii) To determine the
transfer characteristics in terms of stimulus-response observations.
The latter is referred to as system identification.  What we have
studied in the previous section, in which the system dynamics was
described by a set of differential equations, can now be restated as
the estimation problem of the system's response statistics to the
noise input.

Wiener (1958) showed taht the output $X(t)$ of a stable and time
invariant system to the white noise input $W(t)$ ( $= dB(t)/dt$) can be
uniquely expanded in a series of orthogonal functionals of the input:

$$X(t) = \underset{N \to \infty}{l.i.m.} \sum_{n=0}^{\infty} I_n(k_n, B(\cdot))  \qquad (13)$$

where

$$I_n(k_n, B(\cdot)) = \int_{-\infty}^{\infty} \cdots \int k_n(t-t_1, \ldots, t-t_n) \times$$

$$\times h^{(n)}(dB(t_1), \ldots, dB(t_n))  \qquad (14)$$

The r.h.s. of (13) is called the Wiener-Hermite expansion of $X(t)$ (see
Ogura, 1972 for more information).  In the last equation, $k_n \in L^2(R^n)$
$n=1,2,\ldots$ are symmetric functions of the indicated arguments and $h^{(n)}$
$(x_1, \ldots, x_n)$ are polynomials associated with the multiple Hermite
polynomials, for which the orthogonal properties hold:

$$(2\pi)^{-\frac{N}{2}} \int_{-\infty}^{\infty} \cdots \int h^{(n)}(x_{i_1}, \ldots, x_{i_n}) h^{(m)}(x_{j_1}, \ldots, x_{j_m}) \times$$

$$e^{-\frac{1}{2} \sum_{i=1}^{N} x_i^2} dx_1 \ldots dx_N = \delta_{nm} \delta_{ij}^{(n)}  \qquad (15)$$

Here $\delta_{nm}$ is the Kronecker symbol; whereas $\delta_{ij}^{(n)}$ denotes the sum of all distinct products of n Kronecker deltas of the form $\delta_{i\nu j\mu}$, $i=(i_1,...,i_n)$, $j=(j_1,...,j_n)$, all $i\nu$ and $j\mu$ occurring once and only once in each product (hence $\delta_{ij}^{(n)}$ contains n! terms, Ogura, 1972).

If one could assign the sequence $k_n$ corresponding to a given system, one would solve the prediction problem to the white noise input. As any system is usually described by means of the standard formalism of systems theory, one needs to make the necessary correspondence between such formalism and the present one in order to solve the prediction problem in terms of Wiener's theory. Such correspondence is not so easy. There are a few examples of non-linear systems for which this task has been accomplished (see Yasui, 1979). Isobe and Sato (1981a) recently succeeded to obtain the Wiener-Hermite functional representation of X(t) which is the solution of the Ito type stochastic differential equation:

$$dX = f(X)\,dt + g(X)\,dB(t) \tag{16}$$

provided that f(x) and g(x) meet some regularity conditions similar to those mentioned in the previous section, the solution of the equation is a stationary Markov process, whose transition probability density function p(x,t|x',t') satisfies a Fokker-Planck equation. They showed that the kernels $k_n$ of X(t) are expressed by integrals of functions of x, p, and g.

Consider a scalar-valued Borel measurable function $h(x_1,...,x_n)$. Then for a solution process $X(t)=(X_1(t),...,X_n(t))$ of a vector stochastic equation, we may determine the Wiener-Hermite expansion of h(X(t)). Indeed Itô's lemma enables us to obtain the stochastic differential dh(X).

As far as the identification problem of a system is concerned, what kind of information on the system regarded as a black box is obtainable from the knowledge its response to a given input? One may establish a causal representation in terms of an impulse response function of an unknown system from the knowledge of the power spectrum of its output to the white noise input, if it satisfies the so-called Paley-Wiener's theorem (1934). I do not know of any identification method applicable to general non-linear systems except for the Wiener's method using the noise.

Wiener kernels of any stable and time invariant system can be determined from its output process to the white noise as a test input. Wiener (1958) indicated a method to determine the coefficients

$c_{m_1 m_2 \ldots m_n}$ of the Laguerre expansion of the kernel $k_n$ in $L^2(R^n)$.
While Lee and Schetzen (1965) have successively proposed an algorithm
to compute $k_n(t_1, \ldots, t_n)$ pointwise over the time domain $R^n$ by cross-
correlating the output value to the values of the test input at n
different instants of the past. French and Butz (1973) then gave a
method to obtain these kernels in the Fourier domain. These methods
are all based on the computation of the correlation between the test
input and the output.

The response characteristics of the retinal system of cat fish
to random fluctuating light stimuli has been investigated in terms of
Wiener kernels (Marmarelis and Naka, 1973). However, in the practical
use of Wiener's method, one finds it difficult to obtain kernels of
order higher than 3 because of the current computational limitations.
Isobe and Sato (1981b) gave an integro-differential formula on kernels
by varying the magnitude of the input noises. This formula appears
to be useful to obtain information on the function form of a kernel
of any order from the knowledge of that of lower order. They also
showed that a cascaded system of one linear, one non-linear without
memory and one linear subsystems, which is sometimes called a sandwich
system, is completely identified in the language of systems theory
by making use of the afore mentioned formula.

## 4. Conclusion

We showed examples of systems described by (non-) linear
differential equations that are destabilized if they are exposed to
noise. The destabilizing effect of the noise has been postulated to
play an essential role in the physical and biological world to make
a system switch from one to other stable regimes. We showed that one
is able to make a profitable use of the noise to analyze biological
systems and gave a short account of Wiener's analysis of non-linear
systems and of our related results.

## References

Arnold L., Horsthemke W. and Stucki J.: The influence of external
real and white noise on the Lotka-Volterra model, Biometrical J.
21, 459-471 (1979)

Arnold L. and Lefever R. (eds.): Stochastic Nonlinear Systems,
Springer-Verlag, 1981

Arnold L.: A new example of an unstable system being stabilized by
random parameter noise, Inform. Communication of Math. Chem.,
7, 133-140 (1979)

French, A.S. and Butz, E.G.: Measuring the Wiener kernels of a non-linear system using the fast Fourier transform algorithm, Int. J. Control, 3, 629-539 (1973)

Hida, T.: Borwnian Motion, Springer-Verlag, 1980

Isobe, E. and Sato, S.: On the Wiener Kernels of the Solution of the ITo type Stochastic Differential Equation (submitted for publication) (1981a)

Isobe, E. and Sato, S.: Integro-differential formula for Wiener kernels and its Application to Identification of Sandwich Systems (submitted for publication) (1981b)

Jazwinski, A.H.: Stochastic Processes and Filtering Theory, Academic Press, 1970

Kozin, F.: Stability of the linear stochastic system, in Stability of Stochastic Dynamical Systems(eds. Dold, A. and Eckmann, B.), Lecture Notes in Mathematics 294, Springer-Verlag, 1972

Ladde, G.S. and Lakshmikantham, V.: Random Differential Inequalities, Academic Press, 1980

Lee, Y.W. and Schetzen, M.: Measurement of the Wiener Kernels of a Nonlinear System by Cross Correlation, Int. J. Control, 2, 237-254 (1965)

Marmarelis, P.Z. and Naka, K.-I.: Nonlinear Analysis of Receptive Field Responses in the Catfish Retina I, II, III, J. Neurophysiol., 36, 605-648 (1973)

May, R.M.: Stability and Complexity in Model Ecosystems, Princeton, 1973

McKean, H.P.: Stochastic Integrals, Academic Press, 1969

McKean, H.P.: Wiener's Theory of Random Noise, in Stochastic Differential Equations(eds., Keller, J.B., McKean, H.P.) Courant Institute of Math. Sciences, 1972

Naka, K.-I. and Rushton, W.A.H.: S-potential from color units in the retina of fish(Cyprinidae), J. Physiol., 185, 536-555 (1966)

Ogura, H.: Orthogonal Functionals of the Poisson Process, IEEE Trans. on Inform. Theory, IT-18, 473-481 (1972)

Paley, R.E.A.C. and Wiener, N.: Fourier Transform in the Complex Domain, Amer. Math. Soc., 1934

Ricciardi, L.M.: Diffusion Processes and Related Problems in Biology, Lecture Notes in Biomathematics, 14, Springer-Verlag, 1977

Wiener, N.: Nonlinear Problems in Random Theory, The MIT Press, 1958

Wiener, N.: The Homogeneous Chaos, Amer. J. Math. 60, 897-936 (1938)

Yasui, S.: The Use of Nonlinearity and Wiener Kernel for Structural System Analysis, Proc. of the U.S.-Japan Joint Seminar on Advanced Analytical Techniques Applied to the Vertebrate Visual System, Oct. 28-31, 1979, Tokyo, Japan

# CHAPTER 8

## STOCHASTIC, QUANTAL MEMBRANE CONDUCTANCES

## AND NEURONAL FUNCTION

A.V. Holden, Department of Physiology,
University of Leeds, LEEDS LS2 9JT.

## Abstract

The electrical activity of a neurone is determined by both its geometry
and membrane conductance properties.  The excitable and synaptic
conductances are due to discrete, quantal channels that are controlled
by stochastic gating processes.  Coupling between gating processes, and
perhaps even between channels, can contribute to the rich macroscopic
dynamic behaviour of excitable membranes.

## Introduction

The current flowing across an excitable membrane is the sum of currents
produced by charge displacement and currents produced by movements of
charged particles across the membrane.  The displacement currents are
the linear displacement current associated with the membrane capacitance,
and the non-linear asymmetric displacement current associated with
dipole rotation or the displacement of charged particles bound to the
membrane.  The currents produced by charge movement through the membrane
are the ionic currents.

The capacitative current $I_c$ is proportional to the rate of change of
potential across the membrane V : $I_c = C_m dV/dt$, where the membrane
capacitance is constant, and so any fluctuations in $I_c$ reflect fluctuations in V.  Asymmetric displacement currents have been detected:
they are extremely small and can only be detected by averaging, after
compensation for the linear capacitative current, when ionic currents
are suppressed.  Thus the major cause of fluctuations in membrane
current is any fluctuation in ionic current.  Fluctuations in ionic
current may be measured by clamping the membrane potential at a constant
value, so $I_c = 0$.

Since Verveen and Derksen's (1965) pioneering measurements of the
spectral density of voltage fluctuations at the node of Ranvier of
frog myelinated fibres, the characteristics of microscopic voltage and

current fluctuations have been measured in a wide range of membranes: these experimental results are reviewed in Verveen and DeFelice (1974) and Conti and Wanke (1975). The major aim of these experiments has been to obtain a detailed description of the microscopic kinetics of ionic currents. Such a kinetic description cannot be used to infer the mechanisms producing the ionic currents: however, the kinetics can be inconsistent with proposed mechanisms. In this paper I will consider the stochastic kinetics of a number of models of how ions cross membranes: the spectral densities of fluctuations in current generated by these models are consistent with some components of membrane current fluctuation. However, a major component of membrane noise, flicker noise, is inconsistent with these simple models. Possible mechanisms which generate flicker noise will be considered and a synthesis of membrane noise models presented. This will show that mathematical problems in membrane noise modelling are complicated rather than difficult, and may best be approached by numerical computation.

The effects of mesoscopic fluctuations in current on the behaviour of a neurone are more complicated, and a variety of mathematical approaches may be used: see Arnold and Lefever (1981). A major problem is that, given the channel densities (Holden and Yoda, 1981a) and the small surface areas, the behaviour may be dominated by the activity of a restricted population of $10^2$ - $10^4$ channels (Holden, 1981). However, it is possible to relate the macroscopic dynamics of a neurone, through mesoscopic fluctuations and oscillations, to the stochastic, microscopic activity of the membrane channels, as experimental measurements are available at all three levels. In other areas of theoretical neuro- biology the macroscopic dynamics of a population are often explained by nonlinear interactions among microscopic components, when the activities of the population or its component elements are not observable.

## Carrier Models

Ions are hydrophilic, and so are unlikely to enter the lipid matrix of the membrane. One class of models for ion transport through the membrane assumes that ions penetrate the membrane bound to a lipid- soluble, mobile carrier. A low density of carriers in the membrane would account for the entropy barrier to diffusion offered by the membrane, and the voltage-dependence of ion transport could be achieved by a charged carrier. There are a number of detailed mechanisms which have been proposed for carriers, ranging from the carrier-ion complex diffusing through the membrane to rotational or conformational changes of a large structure. These different mechanisms may be reduced to the

same kinetic model, shown in Figure 1. This model is symmetric if the
rate coefficients and equilibrium constants of each carrier species are
the same on both sides of the membrane: $k'_S, k'_{MS}, k'_R, k'_D = k_S, k_{MS}, k_R, k_D$.
For such a symmetric system the carrier is equally distributed on both
sides of the membrane. The kinetics of symmetrical and asymmetrical
carrier systems are reviewed in Heinz (1978).

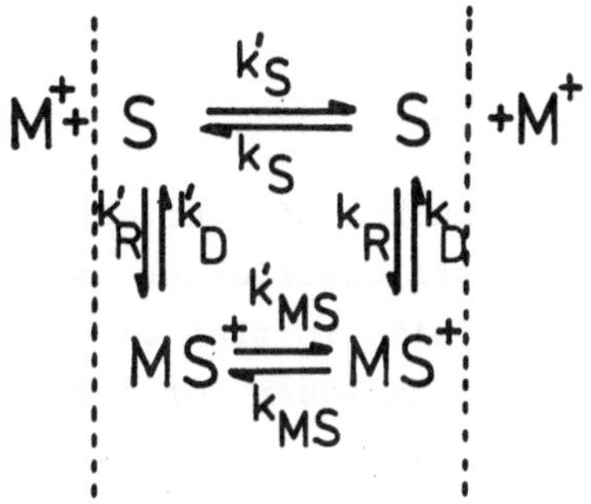

$$M^+ + S \underset{k_S}{\overset{k'_S}{\rightleftharpoons}} S + M^+$$

$$k'_R \updownarrow k'_D \qquad k_R \updownarrow k_D$$

$$MS^+ \underset{k_{MS}}{\overset{k'_{MS}}{\rightleftharpoons}} MS^+$$

Figure 1.   *Transport model for a carrier system for a univalent
cation $M^+$.*

An example of a carrier system which has been studied by both relaxation and fluctuation methods is provided by the ionophore valinomyosin
in a black lipid bilayer membrane. The kinetics obtained from relaxation experiments are as shown in Figure 1 with $k_{MS}$ and $k'_{MS}$ voltage-
dependent, and $k'_R = k_R, k'_D = k_D$ (Lauger and Stark, 1970). Kolb and
Lauger (1978) derived the spectral density for this system from the
complex admittance by the generalized Nyquist theorem. The spectrum
is flat at high and low frequencies. Between these two frequency-
independent regions there are two dispersions, with frequencies
determined by the relaxation times, and the spectral density increases
with frequency. This is in marked contrast to the spectral densities
of gated channels which are Lorentzians.

Fluctuations in current will be seen even at equilibrium when there is
no net current, and so this spectrum characterizes the transport
process, not its control by voltage. Fluctuations in current across
the membrane will be the same as fluctuations in current through the
membrane, and so a similar spectral density should be obtained if the
carrier S is charged and $M^+$ is absent. This is the situation of a lipid
membrane doped with a hydrophobic ion, such as the anion dipicrylamine
(Kolb and Lauger, 1977; Bruner and Hall, 1979). Such a hydrophobic
ion is soluble in the membrane matrix, and its movement through the
membrane can be considered in 3 steps: adsorption, migration by
electrodiffusion, and desorption. The adsorption and desorption are

extremely slow as the partition coefficient is high, and so on the fast time scale of fluctuation analysis the ions are trapped in the membrane, and fluctuation in current is due to random displacement of ions across the energy barrier in the centre of the membrane. If an ion moves from left to right, it can only move back again, and so the current is a sequence of correlated positive and negative going-pulses. A symmetrical energy profile is sketched in Figure 2 and the spectral density is obtained as

$$S(\omega) = 2 k_i (\alpha z e_o)^2 \frac{\omega^2 \tau_i^2}{1 + \omega^2 \tau_i^2} \qquad (1)$$

where $\tau_i = 1/2 k_i$, n = average number of ions adsorbed/unit area of membrane, $k_i$ is translocation rate coefficient and $\alpha \sim 2 s/d$ locates the energy minima. The spectral density is flat at high frequencies, and at low frequencies increases as the square of frequency with a single dispersion. Kolb and Lauger derived (1) from the Langevin equation for the number of ions one side of the barrier.

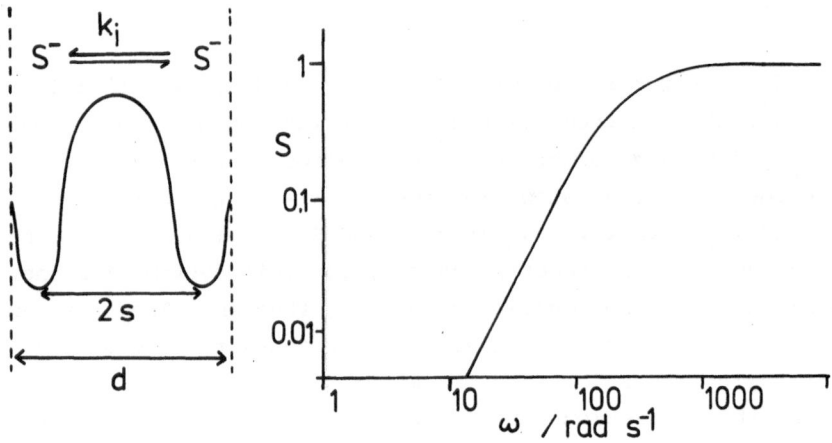

Figure 2. *Symmetric energy barrier for a membrane-bound, hydrophobic anion, and the spectral density calculated by equation (5), with $\tau_i$ = 5 ms.*

The current fluctuations at equilibrium for dipicrylamine noise are fluctuations in displacement currents, and so Eqn. (1) can be related to the spectral density of asymmetric displacement currents. Such asymmetric displacement currents are produced by the displacement of membrane-bound charge between two or more equilibrium positions in the membrane, and so the form of the spectral density will be similar.

If the measured asymmetric gating currents (e.g. Keynes and Rojas, 1974) are 'gating currents', then their kinetics control the kinetics of the gated ionic currents. The spectral densities of the ionic currents are Lorentzian, while the gating currents would have spectral densities given by Eqn. (1). However, asymmetric displacement currents may be epiphenomena, the large signal responses of a nonlinearity.

Accurate measurements of the kinetic parameters of carrier processes are best obtained from relaxation analysis - see Hladky (1979). Such a kinetic description is consistent with both carrier or pore models: the difference between the behaviour of these two classes of model is in the intensity, not the kinetics, of the currents. Fluctuation analysis can distinguish between carrier models where translation is the rate-limiting step and pore models, where the nonequilibrium spectral density is Lorentzian. This difference in behaviour is the difference between an energy storage, series RC network and a dissipative, parallel RC network. Thus the spectral density of carrier noise is similar to that of dielectric noise, derived by Lauger (1979) for conformational or rotational changes in dipole movements.

## Channel Transport Noise

The simplest model of a channel is that of an aqueous pore or hole through which ions can freely move by electrodiffusion. Although the macroscopic Nernst-Planck electrodiffusion equation describes the net flux of ions and is derived by treating the ions as independent particles in a gas-like state, microscopically ions are not freely mobile. Water has a structure with a high degree of short-range order. Around an ion the water molecules are immobilised and orientated. Thus electrodiffusion of an ion is closer to the hydrated ion jumping from hole to hole in a closely structured lattice than to free thermal movement. Microscopically the movement of an ion through water can be described in terms of Eyring's rate theory. Since the mean jump length << channel length (which is of the order of nm for a 10 nm thick membrane) the macroscopic description provides a good approximation for fluxes. However, for fluctuations in current produced by electrodiffusion through an open, conducting pore the details of the microscopic description need to be retained. The simplest case is where the relative concentrations are sufficiently low for there to be a negligible probability of there being more than one ion in the channel at any instant. The series of 'holes' which hydrated ions can jump into, or binding sites, may be represented as a series of potential energy wells separated by barriers. Lauger (1973, 1975, 1978, 1979) derived the permeability, conductance and spectral density of fluctuations at equilibrium for such a barrier model of the channel. He obtained the spectral density for fluctuations about equilibrium from the admittance by Nyquist's equation.

$$S(\omega) = 4 z^2 e_o^2 \{ \sum_{j=1}^{n} \xi_j / (1+\omega^2 \tau^2_i) + \sum_{i=1}^{n+1} \alpha^2_i F_i \} \tag{2}$$

where $F_i$ is the flux over the i-th barrier, $\alpha_i$ the fraction of the total voltage which drops over the i-th barrier, and $\xi_j$ are functions of the $\alpha_i$, $F_i$, $\tau_i$ and rate coefficients for jumps between energy wells. Thus for n binding sites there are n dispersion regions and the spectral density is flat at high and low frequencies. For a regular potential profile, with all the n barriers identical, $\xi_j = 0$ and so white noise results,

$$S(\omega) = 4kT/R_m.$$

Thus a large number of similar barriers give a white spectrum and the shape of the spectral density is dominated by the largest energy barriers: these are the barriers which determine the conductance of the channel.

Lauger (1973) considered 'gating' current noise, with a spectral density given by Eqn. (1), as a special case of Eqn. (2), where the rate coefficients over the outer barriers are zero. Thus the 'membrane bound carrier' may be considered to behave kinetically as a closed pore. Frehland (1978) generalized these results to the steady-state, and emphasized that the difference between pores and carriers is the closed-loop structure of the carrier mechanism.

Hille and Schwarz (1978) have reviewed the evidence that $K^+$ channels which show either H-H type delayed rectification or inward rectification are both multi-ion pores with at least 3 binding sites and often at least 2 ions/channel. Gramicidin A in lipid bilayers behaves as a multi-ion pore (Levitt, 1978). Thus there are a number of systems where relaxation kinetics and tracer fluxes suggest single-file diffusion, and so there is a need to generalize Eqn. (2) to permit more than 1 ion in a channel at any instant. As might be expected the resulting algebraic equations are extremely complicated even for steady-state fluxes. The kinetics of single-file diffusion are discussed in Aityan et al., 1977; Heckman, 1972; Hladky, 1965; Hladky and Harris, 1967; Kohler, 1977.

Frehland (1979) and Frehland and Stephan (1979) have discussed transport noise in a single-file diffusion system around equilibrium and non-equilibrium steady-states. With more than 1 ion/channel, and interactions between ions, the autocovariance function can show damped oscillations and the spectral density peaking. Chen (1978) has also obtained peaking when there are closed loops and the system is not at equilibrium, and so in principle peaking could be obtained in a carrier system at non-equilibrium.

## Gated Channel Noise

If a channel has two conductance states, with $\alpha\Delta t$ and $\beta\Delta t$ the transition probabilities of opening and closing, the durations of the open and closed states have exponential densities. Neher and Sakmann (1976) have recorded single channel currents from denervated frog muscle fibres, and from such records $\alpha$, $\beta$ and the channel conductance $\gamma$ can be estimated directly. However in most cases the membrane noise generated by a number of channels is observed - this is the superposition of the 'random telegraph waves' generated by each channel. The covariance for the current through a single channel of conductance $\gamma$ is

$$\rho(\tau) \quad = \quad \theta\gamma^2(\Delta V)^2 \ (1/\alpha+1/\beta).\exp(-\tau/\theta)$$

where $\theta = 1/(\alpha+\beta)$ and $\Delta V$ is the driving potential difference. By the Wiener-Khintchine theorem, for a membrane with N channels

$$S(\omega) \quad = \quad 4N\theta^2\gamma^2 \ (\Delta V)^2/(1/\alpha+1/\beta) \ (1+\omega^2\theta^2), \tag{3}$$

which is a Lorentzian. The single channel conductance $\gamma$ can be estimated from the ratio of the variance to the mean current - this equals $\Delta V\gamma(1-P)$, when P is the probability that a channel is open. Single channel conductances estimated from noise measurements are tabulated in Neher and Stevens (1977).

When there are N identical channels, each one of which can exist in more than two conductance states $\gamma_i$, i=1,2, .. r, the spectral density is a linear combination of (r-1) Lorentzian spectra. Equations for the autocovariance and spectral density for such a general multistate model are derived in Chen and Hill (1973). A more relevant multistate model is one in which there is only one conducting state, but a number of states of all zero conductance: this would be achieved by a gated channel, with more than one gate/channel. Figure 3 shows the transition diagram for a H-H $K^+$ channel, which has 16 states, only one of which is conducting. When the gates are independent, with rate coefficients $\alpha$ and $\beta$, this 16-state kinetic system is equivalent to the reduced scheme with 5 states. Hill and Chen (1972) and Stevens (1972) have derived the spectral density as

$$S_K(\omega) \quad = \quad 16Nn_\infty^4\tau_n \ (1-n_\infty)^4 \sum_{i=1}^{4} \frac{3!n_\infty^{4-i}(1-n_\infty)^{i-1}}{(i-1)!(4-i)!\{i^2+\omega^2\tau_n^2\}} \gamma_K^2 \ (V-V_K)^2. \tag{4}$$

The H-H $K^+$ system is a multi-state system with a single conducting state. Chen and Hill (1973) have shown that it is possible to reproduce the kinetics with a multiconductance model, where the effect of a gate 'closing' is not to block the channel, but to add a barrier which reduces the conductance by a factor $\kappa$. The closing of the second gate gives the reduction $\kappa^2$ or $\kappa/2$ in the $\kappa$ and $\kappa'$ models. For a multi-conductance model $\gamma$ will change with voltage. This method was used by Begenisich and Stevens (1975) to show that $K^+$ channels have a single conducting state.

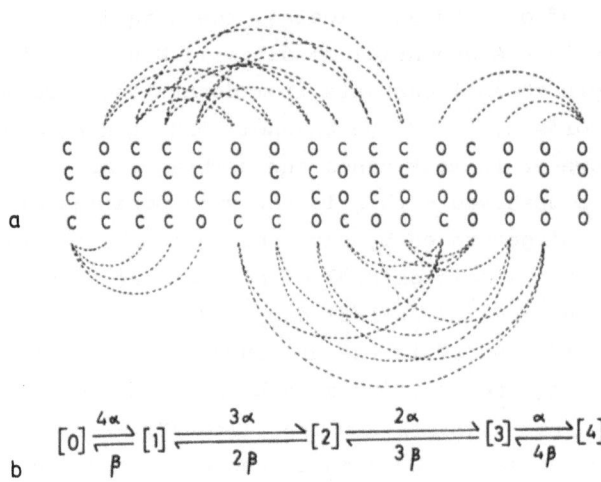

Figure 3. *Transition diagram for a H-H $K^+$ channel. When the gating reactions are independent, the 16-state kinetic system may be reduced to an equivalent scheme with 5 states.*

Fluctuation analysis can also distinguish between different single-conductance models: Grisell (1977) has compared the spectral densities of a number of different models for the kinetics of the $K^+$ channel. The motive for his calculations was to account for the peaking in the difference spectra (spectral density – spectral density when $K^+$ channels were blocked) observed by Fishman, Moore and Poussart (1975). Chen (1978) has suggested that the observed deviations from a Lorentzian spectral density – the peaking at the corner frequency and the greater negative curvature than that of a Lorentzian near the corner frequency – can be explained by a nonequilibrium model. The Nyquist theorem relates the admittance to the spectral density, so the admittance can be calculated from the observed non-Lorentzian spectral densities. Such calculations predict a low frequency feature in the complex admittance and impedance, which is not found in the admittance and impedance of the linearized H-H equations. This low frequency feature is found between 1 – 30 Hz and is a negative phase angle and dip in the impedance at low frequencies. This feature is linear as it is mirrored in the admittance. Fishman et al. (1977) have measured the low frequency admittance and impedance, and have found this feature. Grisell (1979) and Grisell and

Fishman (1979) have computed the admittance, impedance and spectra for a number of models and have concluded that the low frequency feature could be due to extracellular $K^+$ accumulation or a slow, voltage-dependent $K^+$ inactivation.

The spectral densities of gated channel noise are determined by the properties of the gating mechanism, and so provide a means of testing the behaviour of different kinetic schemes. Clay (1978) has calculated the spectral densities of a number of kinetic schemes for the $Na^+$-conductance, in particular models where activation and inactivation are coupled. Most of the coupled models yield spectra inconsistent with the observed Lorentzians.

In Eqns. 3 and 4, N is treated as a constant, given by the product of channel density and membrane area. For ionophores in lipid membranes, N may fluctuate, and an increase in N might be due to entry of ionophores into the membrane or formation of ionophores from inactive precursors. Such precursors would act as additional non-conductive states, and so would contribute further Lorentzians to the spectrum. Chen (1973) has discussed fluctuations and noise in such an open ensemble and has obtained a general expression for the spectral density: it can be written as a sum of Lorentzians.

## Flicker Noise

The spectral densities for carrier, gated channel and channel transport noise discussed above were all calculated for clearly defined kinetic systems. These kinetic systems are meant to represent the mechanisms of ion transport through excitable membranes. Verveen and Derksen (1965) measured voltage fluctuations in myelinated axons, and found that these fluctuations had a spectral density which was inversely related to frequency. This flicker or $\omega^{-\alpha}$, with $\alpha$ close to one, noise is found in a wide range of situations: Neumcke (1978) has reviewed $\omega^{-\alpha}$ noise in biological and synthetic membranes. The source or sources of flicker noise are unknown.

The spectral density of flicker noise varies as $\omega^{-\alpha}$ over a frequency range of at least several decades: the shape of the spectrum at low and high frequencies is not known and is not necessarily the same for different flicker noises; however, the spectrum must be integrable for $0 < \omega < \infty$. The exponent $\alpha$ is close to 1: Poussart (1971) measured flicker noise in lobster axons and found $\alpha$ ranging from 0.8 to 1.3, with a mean 1.002 and a standard deviation 0.076. No flattening at low frequencies has been observed, but this is usually assumed in order to maintain integrability. At high frequencies $\omega^{-\alpha}$ noise vanishes in thermal noise.

Flicker membrane noise is associated with the conductance systems for sodium, potassium and the constant leakage conductance. In the resting membrane, where $P_K \gg P_{Na}$, the $\omega^{-\alpha}$ noise is predominantly associated with the potassium system, as it is minimal at the Nernst potential for $K^+$ and is reduced by two orders of magnitude by tetraethylammonium and $Cs^+$, which block the $K^+$ conductance. When the $K^+$ conductance is blocked, the remaining membrane noise still has an $\omega^{-\alpha}$ component, part of which is removed by the sodium channel blocker tetrodotoxin. Thus membrane flicker noise is the sum of flicker noises associated with the separate membrane conductances.

If membrane ionic current is treated as a process, $I(t)$ can be represented as a sequence of pulses, where the amplitude h, duration $\tau$ and interval $\phi$ between adjacent pulses are random variables. Thus the ionic current flowing through an area of membrane will be

$$I(t) = h_1 Y_1(t, \tau_1) + h_2 Y_2(t - \phi_1, \tau_2) + \ldots$$

$$\ldots + h_N Y_N(t - \sum_{i=1}^{i=N-1} \phi_i, \tau_N), \tag{5}$$

where there are N pulses in an interval $(0, \theta)$. The intervals between adjacent current pulses have a density $\rho(\phi)$. Carson's theorem requires that the pulse functions $y(t)$ must be integrable for $0 < t < \infty$. Each pulse could correspond to the flow of a single ion by electrodiffusion, or bound to a carrier, or displacement of membrane-bound charge. Alternatively, it could be the flow of a large number of ions as through a channel: this description of the process of ion movement as a sequence of pulses does not require any assumptions about the mechanism of ion movement.

Stochastic pulse trains, as described by Eqn. (5), occur in a variety of contexts and the power spectral density of such a process is readily obtained by standard means and is well known as, for $N \to \infty$, $\theta \to \infty$

$$S(\omega) = r\langle h^2\rangle\langle|Y(\omega,\tau)|\rangle^2 \left\{1 + \frac{2\langle h\rangle^2 |\langle Y(\omega,\tau)\rangle|^2}{\langle h^2\rangle\langle|Y(\omega,\tau)|\rangle^2} \right.$$
$$\left. Re[\Phi(\omega)/1 - \Phi(\omega)]\right\}, \tag{6}$$

where $r = N/\theta$, the expected number of pulses/unit time, $Y(\omega,\tau) = \int_{-\infty}^{\infty} y(t,\tau)$ exp $(-i\omega t)dt$, the Fourier Transform of the pulse shape, and $\Phi(\omega) = \int_{-\infty}^{\infty} \rho(\phi)$ exp $(-i\omega t)dt$, the characteristic function. This expression for the spectral density can be considered as the product of two terms: the term outside the curly bracket, which is simply the average spectral density of the individual current pulses, and the term inside the curly bracket, which depends on the interval density.

For a Poisson pulse train, $\rho(\phi) = r \exp(-\phi r)$ and hence $\mathrm{Re}[\Phi(\omega)/(1 - \Phi(\omega))] = 0$. Thus the spectrum is simply

$$S(\omega) = r\langle h^2\rangle\langle|Y(\omega,\tau)|\rangle^2. \tag{7}$$

A flicker noise component in the spectral density can be produced by Eqn. (6) either by the average spectral density of the current pulses (the term outside the curly brackets) or by the term inside the curly brackets. If $\omega^{-\alpha}$ behaviour is to be produced by the term outside the curly brackets, then for $\omega^{-\alpha}$ behaviour to extend from <1 to $10^3$ Hz the time constants $\tau$ must include long time constants, of the order of a second or longer.

Although ionic currents with such slow time courses have been observed in some excitable membranes, such as molluscan somatic membrane, they are not found in axonal membranes, which have an $\omega^{-\alpha}$ spectral component. It is unlikely that such a slow current would account for $\omega^{-\alpha}$ noise, as a slow process would imply a large activation energy barrier, and hence a high $Q_{10}$. Such a high temperature-dependence for flicker noise has not been observed. Thus Holden (1976) suggested that the source of the $\omega^{-\alpha}$ component would be found inside the curly bracket of Eqn. (6), where the only terms which vary with frequency are $Y(\omega)$ and $\Phi(\omega)$, and $Y(\omega)$ has been eliminated as a possible source of flicker noise. If the $\omega^{-\alpha}$ component is generated by the term inside the curly brackets, then at the Nernst equilibrium potential where $\langle h\rangle = 0$, the shape of the spectral density will be given by Eqn. (7). Thus there would not be a zero current $\omega^{-\alpha}$ noise: the only noise at equilibrium would be thermal. The positive vlaue of the parameter A found for lobster axons need not be an equilibrium $\omega^{-\alpha}$ noise: it could be due to other ions flowing through $K^+$-selective channels, as the $K^+$ specificity is not absolute, or it could be due to $K^+$ movement through $K^+$ channels where the local field is not the $K^+$ equilibrium potential.

$\Phi(\omega)$ is the characteristic function of the intervals between adjacent current pulses. These intervals have a density $P(\phi)$ and the problem is to account for a $P(\phi)$ which will contribute an $\omega^{-\alpha}$ component to the spectral density. $S(\omega)$ is the spectral density of the current fluctuations through an area of membrane, within which there are a (large) number of channels. Thus the process with the interval density $P(\phi)$ has been generated by the superposition of the current pulses of a large number of current pulse trains, each of which has some interval density. Whatever the interval densities of the separate channels, the super-position of a large number of independent, uniformly sparse point processes generates a point process with an approximately Poisson distribûtion - this well-known superposition theory is discussed in Cinlar (1972). For the superposition of only a reasonably large, finite number of independent point processes the Poisson approximation is remarkably good, and so if the channels are independent, $P(\phi) = r \exp(-r\phi)$.

However, if $P(\phi)$ is exponential,

$$\text{Re}\{\phi/(1 - \phi)\} = 0,$$

and so if the channels are independent, $P(\phi)$ cannot côntribute an $\omega^{-\alpha}$ component to the spectral density. Thus an $\omega^{-\alpha}$ component in the spectral density requires that the activity of membrane channels is not independent. Possible mechanisms for non-independence are discussed in Holden (1976) and Holden and Rubio (1976), where a one-dimensional model for membrane flicker noise is developed. The development of this model is shown in Figure 4, where (4a) denotes a one-dimensional array of N interacting channels. The interaction is by a spatial analog of a Markov process, where a channel is directly influenced only by its nearest neighbours. Since the membrane is a closed surface there are no edge effects, so the system is in a spatial steady-state. Considering 3 channels, A, B, and C, as in (4b), the activity of channel B is influenced by the super-position $\Omega$ of the activities of A and C, as in (4c).

For an interval $(t_o, t_o + t]$, divided into n subintervals $I_k$, k = 1, ..., n, as in (4d), if there is a B event at $t_o$, the survivor function Q, with mean m, for the activity of any channel can be shown to satisfy the nonlinear integral equation

$$Q(t) = \exp(-\mu t) \{1-m^{-1}\int_o^t Q(\tau) \, d\tau\}^2$$
$$+ 2m^{-2}\int_o^t \exp(-\mu\tau)Q_1(t-\tau)\{1-m^{-1}\int_o^\tau Q(\lambda)d\lambda\}^2 Q(t-\tau)\int_{t-\tau}^\infty Q(u)du \, d\tau,$$

where $\mu$ is the Poisson parameter which describes the activity of the channels in the absence of interaction, and $Q_1(t-\tau)$ = Prob (No B event occurs on $(t_o+\tau, t_o + t]$ | an $\Omega$ event occurs at $t_o + \tau$). The auto-correlation of the process obtained by superposing the activities of all N channels is

$$R(\tau) = N R_{ii}(\tau) + N(R_{i,i+1}(\tau) + R_{i+1,i}(\tau)).$$

By reference to (4e), the autocorrelation is equal to the product of the probabilities that an event happens on $I_V$ and not before, an event does not happen on $(V,\tau]$, and the conditional probability of an event happening on $I_\tau$ under these circumstances. Thus $R_{ii}(\tau) = m^{-2}Q(\tau)F(\tau)$ and $R_{i,i+1}(\tau) = R_{i+1,i}(\tau) = m^{-2}Q(\tau)F_1(\tau)$, where F and $F_1$ are the distributions corresponding to Q and $Q_1$. By considering the Laplace transforms of these equations for $|s|$ near to the origin, $S(s)$ will vary as $1/s^{-\nu}$,

-1 < ν < 0, or the spectral density will vary as $\omega^{-\alpha}$, 0 < α < 1, if $Q_1(t)$ behaves asymptotically as t → ∞ as $t^\nu$ for -1 < ν < 0.  Thus the density corresponding to $Q_1$ decays slower than an exponential, as in (4f).

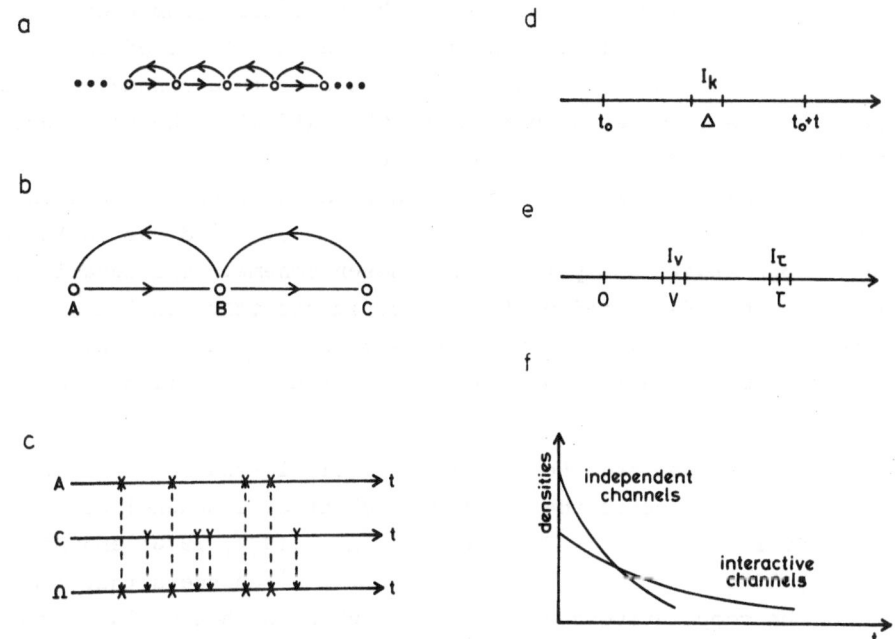

*Figure 4.* *Development of a one-dimensional model for flicker noise, with nearest-neighbour interactions between channels.*

Thus interaction between channels can generate a $\omega^{-\alpha}$ density:  the question whether this could be the mechanism responsible for $\omega^{-\alpha}$ noise in nerve membranes will be solved by experimental tests of the predictions of this model (e.g. see Holden, 1979) not by further analysis. The relaxation behaviour of this model is explored in Holden and Rubio (1978).

## Intrachannel cooperativity and coupling

The macroscopic $K^+$-current observed in the squid axon under voltage-clamp is delayed, and has a sigmoid approach to its steady-state value. Hodgkin and Huxley (1952) proposed a sequence of 4 independent, first order reactions:  $g_K = n^4 \bar{g}_K$.  Cole and Moore (1961) parodied this simple approach, with $g_K = n^{25} \bar{g}_K$ to describe $K^+$-currents after hyper-polarizing prepulses.  If the exponent is to be identified with the number of gating processes, one would expect some conductances to have

only one/channel and an exponential time course (as in cardiac muscle)
and a high exponent is an embarrassment. A different approach is to
propose some kind of cooperativity.

Blumenthal et al. (1970) coupled the analogy between membrane excitation
and allosteric enzymes with the non-equilibrium thermodynamic of dissipative structures, and described the membrane as a lattice of cooperative
protomers, where each protomer loosely corresponds to a channel. Related
Ising lattices are considered in Hill and Chen (1970). The introduction
of cooperative lattice schemes raises two classes of problem: the
details of the model and interaction, as a wide range of alternative
schemes can produce qualitatively similar behaviours, and the choice of
a method of approximate analysis of the chosen scheme. The details of
the model are often falsified by later experiments that still leave the
possibility of cooperative interactions, and the results obtained by the
Bragg-Williams approximation require confirmation by detailed numerical
simulations.

Membrane noise analyses and the detection of single-channel current
events has lent credence to the physical reality of a channel as a
structure: Starzak (1972, 1973), Bretag et al. (1974, 1978) and Bauman
et al. (1974, 1980) consider various models for $K^+$-channels that are
produced by the cooperative aggregation of subunits that could correspond
to gating mechanisms. Microscopic cooperativity was invoked to account
for the kinetics of macroscopic currents: measurements of asymmetric
displacement currents, that might correspond to the behaviour of the
processes controlling the conductances, provide a direct experimental
route to the intrachannel control processes. Thus it seems reasonable
to develop models with intrachannel cooperativity only when they are
required, not just suggested, by experiments. One macroscopic result
that strongly implies cooperativity is the Cole-Moore delay: although
in squid membrane it is possible to superimpose the delayed and normal
currents by time translation, Palti et al. (1976) and Begenisich (1979)
demonstrated that the $K^+$-currents of myelinated axons do not superimpose.
Hill and Chen (1971) showed that independent models gave delayed
currents that could be made to superimpose, and only models with intra-
or inter-channel cooperativity gave currents that did not superimpose.
Other possibilities include two populations of different $K^+$-channels,
or multi-state channels. Since two types of $K^+$-channels have been
demonstrated in myelinated axons intrachannel cooperativity in $K^+$-
channels is an attractive, but unnecessary conjecture.

Horsthemke and Lefever (1981) have shown that the qualitative macro-

scopic behaviour of the H-H $n^4(V)$ and $m^3(V)$ systems can be modified by
an external Markov voltage noise, to produce behaviours and transitions
that are not seen in a deterministic environment. Models with cooperativity produce a richer variety of noise-induced transitions than independent models, even when the cooperative and independent models have
the same steady-state behaviours in a deterministic environment. Thus
the nature of noise-induced transitions can expose cooperative nonlinearities.

Many membrane conductance systems are controlled by both activation and
inactivation processes: the responses to a maintained change in
potential is a transient change in conductance. In the H-H equations
the $Na^+$-conductance is controlled by independent activation m and
inactivation h variables. An alternative model developed by Hoyt
(1963) is that activation and inactivation are coupled and described
by a single second order variable, as in Goldman (1975). Measurements
of currents measured under double step depolarizing clamps support
coupled models.

Further support for coupling comes from measurements of asymmetric
displacement currents, if the inactivation of the gating current is
identified with the h process (see Meves, 1978; Khodorov, 1981).
Thus it seems likely that there is some intrachannel coupling between
gating systems and that inactivation of a channel is usually preceded
by activation.

Interchannel coupling

The interaction between channels that was postulated to account for
flicker noise is speculative; however, electrically mediated interactions between channels are well established.

The major source of apparent cooperativity is that the rate coefficients
for the gating processes are voltage-dependent. Within an isopotential
area of membrane, a homogenous population of channels will act in near
concert, as the gating mechanisms of the individual channels will
experience a common field. Further, the opening of channels will cause
changes in membrane potential that can favour further opening of channels:
this is the basis of the regenerative nature of the $Na^+$-conductance.
Thus although microscopically the channels could be opening and closing
independently, the macroscopic currents will suggest positive cooperativity. This tendency for synchronization within populations of channels
will make it difficult to detect any direct cooperativity within a
population.

One sensitive test of the independence of the individual channels would be to observe the opening and closing of a small number of adjacent channels: this is possible using the patch-clamp method introduced by Neher and Sakmann (1976). Colquhoun and Hawkes (1981) have begun to consider the interpretation of the bursts of pulses recorded as the summed activity of a small but unknown number of channels. If the activities of individual, separate channels can be separated any deviations from exponential densities would provide evidence for cooperative interactions.

In voltage-dependent ionic channels, movement of a bound sensor (the gating particle) leads to channel opening. Some $K^+$-selective channels are potential-independent, and are activated by intracellular $Ca^{2+}$ that could have entered through voltage-dependent, $Ca^{2+}$-selective channels (Meech, 1978). Thus the activity of $Ca^{2+}$-activated, $K^+$-selective channels is conditioned by the current flow through voltage-dependent $Ca^{2+}$-selective channels.

## Effects of channel density

The effects of channel density on neuronal function are reviewed in Holden and Yoda (1981a) - one important aspect is that the channel densities can act as bifurcation parameters. The standard Hodgkin-Huxley membrane equations lose their stable stationary solutions at a subcritical Hopf bifurcation as $\bar{g}_K$ is reduced below 19.762 mS cm$^{-2}$; there is a further subcritical bifurcation as $\bar{g}_K$ is reduced below 3.843 mS cm$^{-2}$ (Holden and Yoda, 1981b). Numerical solutions of the Hodgkin-Huxley equations jump to large amplitude, periodic solutions just above and below these critical values. Thus the probability that a $K^+$-channel is open not only varies, via its dependence on the stationary potential, with the $K^+$-channel density, but varies periodically with time for the range of $K^+$-channel densities that supports auto-rhythmic activity.

A bifurcation theoretic analysis examines the behaviour of the excitation differential system as one or more bifurcation parameters are continuous-ly varied. However, excitation is the result of quantal, not continuous, processes: single membrane channels open or close stochastically, and are either fully- or non-conducting. Thus membrane conductance or current is a jump Markov process that may be approximated by a differ-ential equation only when deviations from the macroscopic average are small - see Kurtz (1970). Internal fluctuations will be important when the number of channels is small, or close to a bifurcation point of the differential system. Membrane channel densities range from a few to several thousand/$\mu m^2$, and given the extensive surface areas of excitable cells a differential equation representation might provide a reasonable approximation for the cell properties. However, excitable cells are often not isopotential and the behaviour of a cell is often dominated by the activity of a small part of its membrane, such as the initial segment of myelinated axons. Within such a restricted membrane surface area there can be only a limited number of channels, so that the behaviour of the cell is dominated by the activity of a limited number of stochastic, quantal processes. These ideas are pursued in Holden (1981).

Erdi et al. (1979) consider the relation between the probability distributions of continuous time, discrete state stochastic models and the qualitative behaviour of the corresponding deterministic model, and prove there is no strong connection between the modality of the distribution and the number of equilibria of the deterministic systems. Thus the continuous time, discrete state stochastic systems that represent small areas of membrane have to be investigated directly. Gillespie (1977) gives a simple algorithm for simulating continuous time, discrete state stochastic systems.

## Conclusions

The rich macroscopic electrical activity of neurones is reminiscent of the effects of cooperative interactions, but closer examination shows little direct evidence for cooperativity. The behaviour of the neurone results from nonlinear membrane properties that are produced by systems of stochastic, quantal channels.

## References

Aityan, S.K., Kalandadze, I.L. and Chizmadjev, Y.A. (1977): Ion transport through the potassium channels of biological membranes. Bioelectrochem. Bioenerg. 4, 30-44.

Arnold, L., Lefever, R. (ed) (1981): Stochastic Nonlinear Systems. Berlin: Springer.

Baumann, G., Mueller, P. (1974): A molecular model of membrane excitability. J. Supramolec. Struct. 2, 538-557.

Baumann, G., Easton, G.S. (1980): Micro-Macrokinetic behaviour of the subunit gating channel. J. Membrane Biol. 52, 237-243.

Begenisch, E. (1979): Conditioning hyperpolarization-induced delays in the potassium channels of myelinated nerves. Biophys. J. 27, 257-265.

Begenisch, T., Stevens, D.F. (1975): How many conductance states do potassium channels have? Biophys. J. 15, 843-846.

Blumenthal, R., Chaugeux, J.P. and Lefever, R. (1970):  Membrane excitability and dissipative instabilities.  J. Membrane Biol.  2, 351-374.

Bretag, A.H., Davis, B.R. and Kerr, D.I.B. (1974).  Potassium conductance models related to the interactive subunit membrane.  J. Membrane Biol.  16, 363-380.

Bretag, A.H., Hurst, C.A. and Kerr, D.I.B. (1978):  Potassium conductance models related to the interactive subunit membrane II.  J. theor. Biol.  73, 367-382.

Chen, Y. (1978):  Noise analysis in kinetic systems and its application to membrane channels.  Adv. Chem. Phys.  37, 67-97.

Chen, Y. (1978):  Differentiation between equilibrium and non-equilibrium kinetic systems by noise analysis.  Biophysical J.  21, 279-285.

Chen, Y. (1978):  Further studies on noise and fluctuations in kinetic systems.  J. Chem. Phys.  68, 1871-1875.

Chen, Y., Hill, T.L. (1973):  Fluctuations and noise in kinetic systems. Application to $K^+$ channels in the squid axon.  Biophys. J.  13, 1276-1295.

Chen, Y., Hill, T.L. (1973):  On the theory of ion transport across the nerve membrane.  VII.  Cooperativity between channels of a large square lattice.  Proc. Nat. Acad. Sci.  USA  70, 62-65.

Cinlar, E. (1972):  Superposition of point processes.  In:  Stochastic Point Processes:  Statistical analysis, theory and applications. P.A.W. Lewis (ed.) New York, Wiley.

Clay, J.R. (1979):  Comparison of ion current noise predicted from different models of the sodium channel gating mechanism in nerve membrane.  J. Membrane Biol.  42 (3), 215-229.

Cole, K.S., Moore, J.W. (1960):  Potassium ion current in the squid giant axon.  Biophys. J.  1, 1-14.

Colquhoun, D., Hawkes, A.W.  (1981):  On the stochastic properties of single ion channels.  Proc. Roy. Soc. B  211, 205-235.

Conti, R., Wanke, E. (1975):  Channel noise in nerve membranes and lipid bilayers.  Quart. Rev. Biophysics  8, 451-506.

Erdi, P., Toth, J., and Hars, V. (1979):  Some kinds of exotic phenomena in chemical systems.  Colloquia Mathematica Societatis Janos Bolyai 30, 205-229.

Fishman, H.M., Dorset, D.L. (1973):  Comments on electrical fluctuations associated with active transport.  Biophys. J.  13, 1339-1342.

Fishman, H.M., Moore, L.E. and Poussart, D.J.M. (1975):  Potassium ion conduction noise in squid axon membrane.  J. Membrane Biol.  24, 305-328.

Fishman, H.M., Moore, L.E. and Poussart, D.J.M. (1977):  Ion movements and kinetics in squid axon.  II.  Spontaneous electrical fluctuations.  Ann. N.Y. Acad. Sci.  303, 399-423.

Fishman, H.M., Poussart, D.J.M., Moore, L.E. and Siebenga, E. (1977): $K^+$ conduction description from the low frequency imepdance and admittance of squid axon.  J. Membrane Biol.  32, 255.

Frehland, E. (1978):  Current noise around steady-states in discrete transport systems.  Biophysical Chemistry  8, 255-265.

Frehland, E. (1979):  Theory of transport noise in membrane channels with open-closed kinetics.  Biophys. Struct. Mechanism  5, 91-106.

Gillespie, G.T. (1977): Exact stochastic simulation of coupled chemical reactions. J. Physical Chem. 81, 2340-2361.

Goldman, L. (1975): Quantitative description of the sodium conductance of the giant axon of Myxicola in terms of a generalized second order variable. Biophys. J. 15, 119-126.

Grisell, R.D. (1977): Models of potassium conduction in excitable membrane examined by spectral analysis. J. theor. Biol. 66, 399-436.

Grisell, R.D.S. (1979): Toward a multimembrane model for potassium conduction in squid giant axon. J. theor. Biol. 76 (3), 233-266.

Grisell, R.D., Fishman, H.M. (1979): $K^+$ conduction phenomena applicable to the low frequency impedance of squid axon. J. Membrane Biol. 46, 1-25.

Heckmann, K. (1972): Single-file diffusion. Biomembranes 3, 127-153. In: Passive permeability of cell membranes. ed. F. Kreuzer and J.F.G. Slegers. Plenum Press: NY.

Heinz, E. (1978): Mechanics and energetics of biological transport. Springer: Berlin, Heidelberg and New York.

Hill, T.L., Chen, Y. (1970): Cooperative effects in models of steady-state transport in membranes I, II, III. Proc. Nat. Acad. USA 65, 1069-1076; 66, 189-196; 66, 607-614.

Hill, T.L., Chen, Y. (1972): On the theory of ion transport across the nerve membrane. IV, V. Biophys. J. 12, 948-59, 960-77.

Hille, B., Schwartz, W. (1978): Potassium channels as multi-ion single-file pores. J. Gen. Physiol. 72, 409-442.

Hladky, S.B. (1965): The single file model for the diffusion of ions through a membrane. Bull. Math. Biophys. 27, 79-86.

Hladky, S.B. (1979): Ion transport and displacement currents with membrane-bound carriers. The theory for voltage-clamp currents, charge-pulse transients and admittance for symmetrical systems. J. Membrane Biol. 46, 213-237.

Hladky, S.B., Harris, J.D. (1967): An ion displacement membrane model. Biophys. J. 7, 535-543.

Hodgkin, A.L., Huxley, A.F. (1952): A quantitative description of membrane current and its application to conduction and excitation in nerve. J. Physiol. 117, 500-544.

Holden, A.V. (1976): Flicker noise and structural changes in nerve membrane. J. theor. Biol. 57, 343-346.

Holden, A.V. (1979): Cybernetic approach to global and local mode of action of general anaesthetics. J. Cybernetics 9, 143-150.

Holden, A.V. (1981): Membrane current fluctuations and neuronal information processing. Adv. Physiol. Sci. 30, 23-41.

Holden, A.V., Rubio, J.E. (1976): A model for flicker noise in nerve membranes. Biol. Cybernetics 24, 227-236.

Holden, A.V., Rubio, J.E. (1978): Retardation currents in excitable membranes and models of flicker noise. Biol. Cybernetics 30, 45-54.

Holden, A.V., Yoda, M. (1981a): The effects of channel density on neuronal function. J. theoret. Neurobiol. 1 (in press).

Holden, A.V., Yoda, M. (1981b): Ionic channel density of excitable membranes can act as a bifurcation parameter. Biol. Cybernetics (in press).

Horsthemke, W., Lefever, R. (1981): Voltage-noise-induced transitions in electrically excitable membranes. Biophys. J. 35, 415-432.

Keynes, R.D., Rojas, E. (1974): Kinetics and steady-state properties of the charged system controlling sodium conductance in the squid giant axon. J. Physiol. (Lond.) 239, 393-434.

Khodorov, B.I. (1981): Sodium inactivation and drug-induced immobilization of the gating change in nerve membrane. Prog. Biophys. Molec. Biol. 37, 49-89.

Kohler, H.H. (1977): A single file model for potassium transport in squid giant axon. Stimulation of potassium currents at normal ionic concentrations. Biophys. J. 19, 125-140.

Kolb, H.A., Lauger, P. (1977): Electrical noise from lipid bilayer membranes in the presence of hydrophobic ions. J. Membrane Biol. 37, 321-345.

Kolb, H.A., Lauger, P. (1978): Spectral analysis of current noise generated by carrier-mediated ion transport. J. Membrane Biol. 41, 167-187.

Kubo, R. (1957): Statistical mechanical theory of irreversible processes. I. General theory and simple application to magnetic and conduction problems. J. Phys. Soc. Japan 12, 570-586.

Kurtz, T.G. (1970): Solutions of ordinary differential equations as limits of pure jump Markov processes. J. Applied Prob. 7, 49-58.

Lauger, P. (1973): Ion transport through pores: a rate theory analysis. Biochimica Biophysica Acta 311, 423-441.

Lauger, P. (1975): Shot noise in ion channels. Biochimica et Biophysica Acta 413, 1-10.

Lauger, P. (1976): Diffusion-limited ion flow through pores. Biochim. Biophys. Acta 455, 453-509.

Lauger, P. (1978): Transport noise in membranes. Current and voltage-fluctuations at equilibrium. Biochim. Biophys. Acta 507, 337-349.

Lauger, P. (1979): Transport of noninteracting ions through channels. In: Membrane Transport Processes, Volume 3, ed. C.F. Stevens and R.W. Tsien. Raven Press, N.Y., 17-26.

Lauger, P. (1979): Dielectric noise in membranes. Biochim. Biophys. Acta 557, 283-294.

Lauger, P., Stark, G. (1970): Kinetics of carrier-mediated ion transport across lipid bilayer membranes. Biochim. Biophys. Acta 211, 458.

Levitt, D.G. (1975): General continuum analysis of transport through pores. I. Proof of Onsager's reciprocity postulate for uniform pores. Biophys. J. 15, 533-552.

Levitt, D.G. (1975): General continuum analysis of transport through pores. II. Nonuniform pores. Biophys. J. 15, 553-564.

Levitt, D.G. (1978): Electrostatic calculations for an ion channel. I. Energy and potential profiles and interactions between ions. Biophys. J. 22, 209-220.

Levitt, D.G. (1978): Electrostatic calculations for an ion channel. II. Kinetic behaviour of the gramicidin A channel. Biophys. J. 22, 221-248.

Meech, R.W. (1978): Calcium-dependent potassium activation in nervous tissue. Ann. Rev. Biophys. Bioeng. 7, 1-18.

Meves, H. (1978): Inactivation of the sodium permeability in the squid giant nerve fibres. Progr. Biophys. Molec. Biol. 33, 207-230.

Neher, E., Sakmann, B. (1976): Single-channel currents recorded from membrane of denervated frog muscle fibres. Nature 260, 779-802.

Neher, E., Stevens, C.F. (1977): Conductance fluctuations and ionic pores in membranes. Ann. Rev. Biophys. Bioeng. 6, 345-381.

Neumcke, B. (1978): 1/f noise in membranes. Biophys. Struc. Mechanisms 4, 179-199.

Palti, Y., Ganot, G., Stampfli, R. (1976): Effects of conditioning potential on potassium current kinetics in the frog muscle. Biophys. J. 16, 261-273.

Poussart, D.J.M. (1971): Membrane current noise in lobster axon under voltage clamp. Biophys. J. 11, 211-234.

Starzak, M.E. (1972): A model for conductance changes in the squid giant axon I, II. J. theor. Biol. 39, 487-504, 505-522.

Starzak, M.E. (1973): Properties of membrane stationary states. J. Membrane Biol. 15, 37-60.

Stevens, C.F. (1972): Inferences about membrane properties from electrical noise measurements. Biophys. J. 12, 1028-1047.

Verveen, A.A. and DeFelice, L.J. (1974): Membrane noise. Progr. Biophys. Molec. Biol. 28, 189-265.

Verveen, A.A. and Derksen, H.E. (1965): Fluctuations in membrane potential and the problem of coding. Kybernetik, 2, 152-160.

Weissman, M.B. (1976): Models for 1/f noise in nerve membranes. Biophys. J. 16, 1105-1108.

# Chapter 9

DIFFUSION APPROXIMATIONS AND COMPUTATIONAL PROBLEMS
FOR SINGLE NEURONS' ACTIVITY

Luigi M. Ricciardi[*]

Department of Biophysical Engineering, Faculty of Engineering Science,
Osaka University, Toyonaka, Osaka (Japan)

(*) On leave of absence from Istituto di Matematica dell'Universita',
Napoli (Italy)

1.  Introduction

    Stochastic processes, and in particular stochastic differential
equations, are being increasingly exploited in the realm of Biology
mainly due to the frequent practical impossibility of taking into
account the details of the interaction of the system under study with
its surroundings.  Yet, such interaction can be at times extremely
effective to the point of becoming responsible alone to determine the
ultimate faith of the system itself.  In the absence of a complete
knowledge of the system's input and of all the environmental effects,
one is then often forced to come up with some kind of "heat bath"
assumption and thus limit the attention to the description of only a
part of the original system.  The remaining effects are then represented by some "random force" term to be inserted in the equations
describing the system's dynamics.  The stochastic properties of the
random force are usually conjectured on the grounds of intuition and
common sense.  Eamples of use of such procedure are widespread in the
biological literature.  Some are discussed in Ricciardi (1977) with
a view to point out just their essential features.

    The stochastic process approach has also been exploited to arrive
at some statistical description of the output of spontaneously active
neurons belonging to complex networks.  Indeed, the large degree of
variability and randomness exhibited by the responses of most neurons
in the absence of specific external stimuli has lead many investigators
to give up deterministic models for the description of neuronal outputs.  The literature on this subject is too vast to be referenced
here.  We only wish to point out that the observed variability in the
neuron's interspike intervals could be due to fluctuations in the
synaptic inputs, in the mechanism transforming the chemicals released
by the synaptic knobs into postsynaptic potentials or else in the
neuron's threshold.  The lack of arguments to select specific types
of synaptic inputs when dealing with neurons that are part of complex
neuronal networks induces one to resort to asymptotic theorems on the
superposition of uncorrelated random variables or stochastic point
processes.

    In the sequel we shall exclusively refer to diffusion approxi-
mations to the single neuron problem that can be justified by making
use of the mentioned asymptotic theorems.  Although many people have
been interested in deriving and exploiting some kind of diffusion
approximations in the neurobiological context, most of the related
mathematical problems are as yet either unsolved or only partly solved.
A brief account of the origin of the diffusion approximations and of

some of the related questions will be given in the sequel. A more comprehensive report will apprar elsewhere ( Ricciardi, Sacerdote and Sato, in preparation ).

It should be mentioned that the interest of diffusion models for the spontaneous activity of single neurons belonging to complex networks has very frequently emerged in the neurophysiological literature. Particularly important from an historical view point is the contribution by Gerstein and Mandelbrot (1964) in which the neuron's membrane potential is depicted as a Wiener process with drift in order to interprete a number of experimentally recorded spike trains. The firing time distribution thus identifies with the first passage time distribution for the Wiener precess through a boundary representing the neuron's threshlod. Besides some problems of interpretation, this model is oversimplified in that an infinitely large membrane time constant is considered in order to overcome the difficulty of spontaneous exponenentially varying membrane potentials. While referring to Ricciardi and Sacerdote (1979) and to the references therein for some complementary remarks, in Section 2 we shall sketch the procedure leading to a sensible diffusion approximation for the neuron's membrane potential under the influence of very many uncorrelated excitatory and inhibitory inputs expressing the result of the interaction of the considered neuron with the rest of the neural network and with the external world. In Section 3 we shall then outline two numerical procedures to evaluate the firing p.d.f. for the case of arbitrarily varying neuronal thresholds.

## 2. The Diffusion Model

Let x denote the variation of the potential difference across the membrane of the neuron. In the absence of inputs x exponentially decays toward the (zero) resting potential with a time constant $\theta$. We shall now assume that the neuron's input consists of p+q sequences of approximately zero-width pulses Poisson distributed in time with rates $\alpha_1, \alpha_2, \ldots, \alpha_p$ and $\beta_1, \beta_2, \ldots, \beta_q$, respectively. The pulses characterized by rates $\alpha_k$ (k=1,2,...,p) are taken as excitatory while the others are inhibitory, Denoting by $e_k > 0$ (k=1,2,...,p) and by $i_k < 0$ (k=1,2,...,q) the corresponding EPSP's and IPSP's , the instantaneous transition $x \rightarrow x + e_k$ (k=1,2,...,p) represents the effect of an excitatory input pulse belonging to the sequence characterized by the rate $\alpha_k$ whereas $x \rightarrow x + i_k$ (k=1,2,...,q) is the instantaneous transition occurring when the neuron is hit by a pulse belonging to the sequence that has

rate $\beta_k$ . The underpinning of this schematization is clearly origi-
nating from the superposition properties of infinitely many uncorrelated
point processes. Setting:

$$f(x,t|y,\tau) \equiv \frac{\partial}{\partial x} \Pr\{X(t) < x | X(\tau) = y\} \qquad (2.1)$$

for the membrane potential transition p.d.f., from our assumptions it
follows that the membrane potential $X(t)$ is a time homogeneous Markov
process. Hence, for any $t+\Delta t > t > 0$ we have:

$$f(x,t+\Delta t|x_0) = \int_{-\infty}^{\infty} dz\, f(x,\Delta t|z)\, f(z,t|x_0), \qquad (2.2)$$

where we have set $X(0) = x_0$ and have omitted the specification of the
initial time. We now note that, apart from infinitesimal quantities
$o(\Delta t)$, there results:

$$f(x,\Delta t|z) = \{1-\Delta t[\sum_{k=1}^{b} \alpha_k + \sum_{k=1}^{q} \beta_k]\}\delta\left[x-(z-z\tfrac{\Delta t}{\theta})\right]$$
$$+\Delta t \sum_{k=1}^{b} \alpha_k \delta\left[x-(z-z\tfrac{\Delta t}{\theta} + e_k)\right]$$
$$+\Delta t \sum_{k=1,..,q} \beta_k \delta\left[x-(z-z\tfrac{\Delta t}{\theta} + i_k)\right] , \qquad (2.3)$$

where $\delta(\cdot)$ is the Dirac delta-function. Substituting (2.3) in (2.2),
after some straightforward calculations and in the limit when $\Delta t \to 0$ we
obtain:

$$\frac{\partial f}{\partial t} = \frac{\partial}{\partial x}\left(\frac{x}{\theta}f\right) + \sum_{k=1}^{b} \alpha_k [f(x-e_k,t|x_0) - f(x,t|x_0)] \qquad (2.4)$$
$$+ \sum_{k=1,..,q} \beta_k [f(x-i_k,t|x_0) - f(x,t|x_0)].$$

In order to smooth down the process' sample paths, we first expand the
functions on the r.h.s. of (2.4) as Taylor series about $x$ and then apply
a suitable limit precedure to the magnitude of PSP's and of input
arrival rates. After the mentioned expansion, Eq. (2.4) yields:

$$\frac{\partial f}{\partial t} = -\frac{\partial}{\partial x}\left[(-\tfrac{x}{\theta} + \mu_1) f\right] + \sum_{j=2}^{\infty} \frac{(-1)^j}{j!} \mu_j \frac{\partial^j f}{\partial x^j} \qquad (2.5)$$

where we have set:

$$\mu_j = \sum_{k=1}^{b} \alpha_k e_k^j + \sum_{k=1}^{q} \beta_k i_k^j \qquad (j=1,2,\ldots,n). \qquad (2.6)$$

We then proceed with the smoothing and take the parameters appearing in (2.5) and (2.6) as indicated in Eq. (2.7) below. Note that, without loss of generality we have assumed $p > q$.

$$\alpha_k = \lim_{y \to 0} \frac{a_k}{y^2}, \qquad a_k > 0$$

$$\beta_k = \lim_{y \to 0} \frac{b_k}{y^2}, \qquad b_k > 0$$

$$i_k = \lim_{y \to 0} d_k y, \qquad d_k < 0 \qquad\qquad (k=1,2,\ldots,q)$$

$$e_k = \lim_{y \to 0} c_k y, \qquad c_k = \frac{|d_k| b_k}{a_k}$$

$$\alpha_r = \lim_{y \to 0} \frac{a_r}{y}, \qquad a_r > 0$$

$$\beta_r = \lim_{y \to 0} \frac{b_r}{y}, \qquad b_r > 0$$

$$i_r = \lim_{y \to 0} d_r y, \qquad d_r < 0 \qquad\qquad (r=q+1,q+2,\ldots,p)$$

$$e_r = \lim_{y \to 0} c_r y, \qquad 0 < c_r \neq \frac{|d_r| b_r}{a_r} \qquad\qquad (2.7)$$

Here $e_k$'s, $b_k$'s, $c_k$'s and $d_k$'s are otherways arbitrary constants. From (2.6) and (2.7) we are finally lead to the following result:

$$\mu_1 \to \lim_{y \to 0} \left[ y^{-1} \sum_{k=1}^{q} (a_k c_k - b_k |d_k|) + \sum_{r=q+1}^{p} (a_r c_r - b_r |d_r|) \right] \equiv \delta$$

$$\mu_2 \to \sum_{k=1}^{q} b_k d_k^2 \left( 1 + \frac{b_k}{a_k} \right) + \lim_{y \to 0} \left[ y \sum_{r=q+1}^{p} (a_r c_r^2 + b_r d_r^2) \right] \equiv \mu > 0$$

$$\mu_j \to \lim_{y \to 0} \left\{ y^{n-2} \sum_{k=1}^{q} b_k d_k \left[ 1 + (-1)^n \left( \frac{b_k}{a_k} \right)^{n-1} \right] \right.$$

$$\left. + y^{n-1} \sum_{r=q+1}^{p} \left[ a_r c_r^n + (-1)^n b_r |d_r|^n \right] \right\} = 0 \quad (j=3,4,\ldots). \qquad (2.8)$$

We have thus proved that Eq. (2.5) tends to the Fokker-Planck equation

$$\frac{\partial f}{\partial t} = - \frac{\partial}{\partial x} \left[ \left( - \frac{x}{\theta} + \delta \right) f \right] + \frac{\mu}{2} \frac{\partial^2 f}{\partial x^2}, \qquad (2.9)$$

where $\mu$ denotes the infinitesimal variance. The quantity $\delta$ in the drift is the net rate of excitation. The case $\delta=0$ corresponds to the situation described in Capocelli and Ricciardi (1971).

The limit procedure carried out in the foregoing models the neuron's membrane potential as a diffusion process of the Ornstein-Uhlenbeck type. Hence, the firing p.d.f. coincides with the first passage time p.d.f. for this process through a boundary representing the neuron's threshold. Up to this day, the information available on the firing

probability and on the firing time statistics can be summarized as
follows.

a. Constant threshold. Let S denote such threshold, and let
$g(S,t|x_0)$ be the firing p.d.f., i.e. the first passage time p.d.f.
through the threshold S for the diffusion process described by Eq.
(2.9). It is possible to obtain a closed form expression for $g(S,t|x_0)$
only if $S=\delta\theta$. If $S\neq\delta\theta$, the Laplace transform $g_\lambda(S|x_0)$ of $g(S,t|x_0)$
can be obtained (cf. Ricciardi and Sacerdote, 1979). Hence, it is
possible to evaluate the moments of the firing time (that turn out
to be all finite) in terms of the derivatives of the function $g_\lambda(S|x_0)$.
It should be pointed out that such evaluation is very cumbersome
because the expression of $g_\lambda(S|x_0)$ includes a ratio of two parabolic
cylinder functions. While a compact expression for the firing time
moments is due to Sato (1978), a set of numerical tables including
mean, variance and skewness of the firing time has been prepared by
Cerbone et al. (1981) for a variety of parameters values of neurophysiological interest. This information can be profitably complemented
by calculating the mode of the firing distribution. This can indeed
be proven to be solution of the equation

$$\frac{\partial}{\partial t}\left\{A(S)\ f(S,t|x_0) - \frac{1}{2}\frac{\partial}{\partial x}\ [\ B(x)f(x,t|x_0)\ ]\ \right\}_{x=S} = 0 \quad (2.10)$$

for any arbitrary time homogeneous diffusion process having drift $A(x)$,
infinitesimal variance $B(x)$ and transition p.d.f. $f(x,t|x_0)$.

b. Time varying threshold. Here the neuron's threshold is
realistically viewed as a function $S(t)$ of time, typically attaining
a sharp maximum at time zero to reset subsequently to its stationary
value with a time constant somewhat shorter than $\theta$. As far as the
neuron's membrane potential is described by Eq. (2.9), it is fair to
claim that the most general case in which a closed form expression
for the firing p.d.f. is known is when $S(t)=Ae^{-t/\theta} + Be^{t/\theta}$, A and B
denoting arbitrary constants (Ricciardi, 1977). This is not useful
per se; however, it indirectly helps one to set upper and lower bounds
to the firing p.d.f. and to its mode when more realistic threshold
functions are taken into account (Ricciardi, 1977; Ricciardi et al.,
in preparation). Alternatively, one can resort to numerical arguments
to obtain an evaluation of the firing p.d.f. This can be done in
essentially two ways: either by approximating the threshold by some
more convenient function (Durbin, 1971; see also Sato, 1976) and by
using Doob's transformation to change the membrane potential process

into the zero-drift Wiener process; or by numerically solving an
integral equation in which the unknown function is the firing p.d.f.
(Anderssen et al., 1973). In Section 3, we shall briefly discuss
alternative numerical methods that have been proven useful to handle
threshold functions of the type to be considered in neruophysiological
instances. A more complete discussion will appear elsewhere (Favella
et al., 1981).

3. Computational Methods

The starting point is the integral equation

$$f[S(t), t|x_0] = \int_0^t f[S(t),t|S(\tau), \tau]g[S(\tau),\tau|x_0]d\tau \qquad (3.1)$$

by Fortet (1943). Here $S(t)$ is the threshold function, which is assumed
to be continuous together with its derivative; $f$ denotes the transition
p.d.f. of the membrane potential while $g[S(t),t|x_0]$ is the unknown
firing p.d.f. It is convenient to re-write Eq. (3.1) as:

$$\int_0^t \frac{K(t,\tau)}{\sqrt{t-\tau}} g(\tau)d\tau = x(t), \qquad (3.2)$$

where we have set

$$K(t,\tau) = \sqrt{t-\tau} \ f[S(t),t|S(\tau),\tau] \qquad (3.3)$$

and where the specification of threshold and initial state has been
omitted. We now partition the interval $(0,t)$ into adjacent intervals
of width $T$:

$$t = nT+\theta, \quad 0 \leqslant \theta \leqslant T. \qquad (3.4)$$

Eq. (3.2) can then be written as follows:

$$\int_0^\theta \frac{K(nT+\theta,nT+u)}{\sqrt{\theta-u}} g(nT+u) \ du = x(nT+\theta) - I_n(\theta) \equiv \Phi_n(\theta), \qquad (3.5)$$

$$(n=0,1,2,...)$$

having set:

$$I_n(\theta) = \int_0^{nT} \frac{K(nT+\theta,\tau)}{\sqrt{nT+\theta-\tau}}\, g(\tau)\, d\tau \qquad (n=0,1,2,\ldots) \qquad (3.6)$$

$$\Phi_n(\theta) = x(nT+\theta) - I_n(\theta).$$

Eq. (3.5) can be viewed as an equation in the unknown function $g(t)$, with $nT \leqslant t \leqslant nT+\theta$ if $g(t)$ is assumed to be known for $0 \leqslant t \leqslant nT$. We now express $g(nT+u)$, with $0 \leqslant u \leqslant T$, in terms of Lagrange polynomials with $p$ equally spaced nodes:

$$g(nT+u) = \sum_{k=1}^{p} L_k(u)\, g(nT+u_k), \qquad (n=0,1,2,\ldots) \qquad (3.7)$$

where

$$L_k(u) = \frac{\prod\limits_{r \neq k} (u-u_r)}{\prod\limits_{r \neq k} (u_k-u_r)} \qquad (k=1,2,\ldots,p). \qquad (3.8)$$

Making use of (3.7), Eq. (3.5) becomes:

$$\sum_{k=1}^{p} g(nT+u_k) \int_0^{\theta} du\, \frac{L_k(u)\, K(nT+\theta,\ nT+u)}{\sqrt{\theta-u}} = \Phi_n(\theta). \qquad (3.9)$$

Further, setting in (3.9) $\theta=\theta_r$ $(r=1,2,\ldots,p)$, with $\theta_r$'s coinciding with the $p$ nodes of Lagrange polynomials, we obtain:

$$\sum_{k=1}^{p} c_{rk}^{(n)}\, g(nT+u_k) = \Phi_n(\theta_r), \qquad \begin{array}{l} (n=0,1,2,\ldots) \\[4pt] (r=1,2,\ldots,p) \end{array} \qquad (3.10)$$

where coefficients $c_{rk}^{(n)}$ have been defined as follows:

$$c_{rk}^{(n)} = \int_0^{\theta_r} du\, [(\theta_r-u)^{-1/2}\, L_k(u)\, K(nT+\theta_r,\ nT+u)] \qquad \begin{array}{l} (r,k=1,2,\ldots,p) \\[4pt] (n=0,1,2,\ldots). \end{array} \qquad (3.11)$$

With the choice $\theta_1 = 0$ equations (3.10) take the simpler form:

$$\sum_{k=2}^{p} C_{rk}^{(n)} \, g(nT+u_k) = \Phi_n(\theta_r) - C_{r1}^{(n)} \, g(nT) \qquad \begin{array}{l} (n=0,1,2,\ldots) \\ (r=1,2,\ldots,p) \end{array} \qquad (3.12)$$

due to the vanishing of $C_{1k}^{(n)}$ for all k. Solving Eqs. (3.12) with the initial condition $g(0)=0$ finally leads us to a numerical evaluation of the firing p.d.f. $g(t) \equiv g[S(t), t \mid x_0]$. More explicitly, Eqs. (3.12) read as follows:

$$\sum_{k=2}^{p} C_{rk}^{(n)} \, g(nT+u_k) = x(nT+\theta_r)$$

$$-\sum_{m=0}^{n-1} \int_{mT}^{(m+1)T} \frac{K(nT+\theta_r, \tau)}{\sqrt{nT+\theta_r - \tau}} \, g(\tau) \, d\tau - C_{r1}^{(n)} \, g(nT) \qquad (3.13)$$

$$(r=2,3,\ldots,p)$$
$$(n=0,1,2,\ldots).$$

The integrals on the r.h.s. of (3.13) are conveniently evaluated by means of four and six points Newton-Cotes quadrature formulas; coefficients $C_{rk}^{(n)}$ can instead be computed by means of four and six points Gauss-Legendre quadrature formulas (cf., for instance, Abramowitz and Stegun, 1964). It should be pointed out that this method can be used for arbitrary diffusion processes for which the singular term of the square root type can be factored out. Note that, differently from Anderssen et al. (loc. cit.), in the above outlined method the firing p.d.f. alone is approximated by means of the Lagrange polynomials.

An alternative method to determine the firing p.d.f. as solution of Eq. (3.2) consists of partitioning the interval (0,t) into subintervals of width T and of searching an approximate solution $\tilde{g}(t)$ in the form of polynomials of degree r (De Griffi and Favella, 1978; Favella et al., 1981):

$$\tilde{g}(t) = \sum_{k=0}^{r} a_k^{(n)} \, t^k, \qquad \begin{array}{l} nT \leqslant t \leqslant (n+1)T \\ (n=0,1,2,\ldots) \end{array} \qquad (3.14)$$

where the coefficients $a_k^{(n)}$ are to be determined by minimizing the "integrated square error":

$$\varepsilon_n^2 = \int\limits_{nT}^{(n+1)T} dt \left[ \int\limits_{nT}^{t} \frac{K(t,\tau)}{\sqrt{t-\tau}} \, \tilde{g}(\tau) \, d\tau - \Phi_n(\tau) \right]^2 \qquad (n=0,1,2,\ldots). \quad (3.15)$$

We explicitly note that the functions $\Phi_n(\tau)$ differ from those defined in (3.5) in that now the approximate solution $\tilde{g}(\tau)$ is to be inserted in definition (3.6). By the foregoing procedure the following system of equations in the unknown $a_k^{(n)}$ is obtained:

$$\sum_{k=0}^{r} A_{ik}^{(n)} a_k^{(n)} = B_i^{(n)}, \qquad \begin{matrix} (i=0,1,\ldots,r) \\ \\ (n=0,1,2,\ldots) \end{matrix} \qquad (3.16)$$

where

$$A_{ik}^{(n)} = \int\limits_{nT}^{(n+1)T} \phi_i^{(n)}(t) \, \phi_k^{(n)}(t) \, dt$$

$$B_i^{(n)} = \int\limits_{nT}^{(n+1)T} \phi_n(t) \, \phi_i^{(n)}(t) \, dt \qquad \left\{ \begin{matrix} n=0,1,2,\ldots \\ k=1,2,\ldots,r \\ nT \leqslant t \leqslant (n+1)T \end{matrix} \right. \qquad (3.17)$$

$$\phi_i^{(n)}(t) = \int\limits_{nT}^{t} \frac{K(t,\tau)}{\sqrt{t-\tau}} \, \tau^i d\tau \, .$$

The functions $\phi_i^{(n)}(t)$ can be evaluated by means of modified Gauss-Legendre quadrature formulas while the functions $i_n(\theta)$ in the expression of $\phi_n(\theta)$ and the quantities $A_{ik}^{(n)}$ and $B_i^{(n)}$ can be computed by means of ordinary Gauss-Legendre quadrature formulas. The procedure just described is clearly a variant of that proposed by Bubnov and Galearkin (cf. for instance, Mikhlin and Smolitshi, 1967).

The system of equations (3.16) has been solved in a variety of cases to construct the approximate firing p.d.f. (3.14). While referring to the quoted papers by De Griffi and Favella and by Favella et al. for a more exhaustive discussion, we limit ourselves to pointing out that it is possible to prove the oscillatory character about the origin of the difference $\tilde{g}-g$ between approximate and actual solutions. Furthermore, upper bounds to the absolute value of such difference can be explicitly computed.

## 4. Concluding Remarks

In the foregoing we tried to provide a bird's eye view at some techniques for modeling single neuron's activity as a first crossing time problem through some suitable threshold function for a diffusion process of the Ornstein-Uhlenbeck type (linear drift and constant infinitesimal variance). Our aim has been to point out very briefly facts, results and references little known, or unknown at all, to most of the non-specialists of the field and to display alternative computational methods to evaluate the firing p.d.f. both for the case of constant as well as time-varying threshold functions. While for the case of uncorrelated neuronal inputs the effect of the time dependence of the firing threshold can be expected to be small at law firing rates, it certainly plays an essential role when other than delta-type correlated Gaussian processes are taken into account to model the neuron's input. However, in this case the computational complexity increases in an astonishing fashion, as it will be shown elsewhere (Ricciardi, Sacerdote and Sato, in preparation).

We would also like to point out that alternative diffusion models can be constructed to account for the effect of "burst like" neuron's stimulation. This requires a "modulation" of the constants appearing in the limit expressions of PSP's and input arrival rates. The final diffusion equation can for instance be proved to become.

$$\frac{\partial f}{\partial t} = - \frac{\partial}{\partial x} [(ax+b) f] + \frac{1}{2} D(t) \frac{\partial^2 f}{\partial x^2}$$

where a<0 and b are arbitrary constants and where the diffusion term D(t) can be any non-negative function. Clearly, such an equation appears to be particularly suitable to account for non-stationary effects. However, all this is the object of a future report.

In conclusion, we would like to remark that any contribution to the neuronal firing problem within the framework outlined in the foregoing also proves to be valuable in a variety of fields other than Biology (cf., for instance, Durbin, 1971 and references therein). As far as neurons are concerned, the use of diffusion approximations and the conputation of firing distributions provides a tool to look into parameters' identification problems and to achieve a better understanding of purely "noisy" versus "informative" spike trains.

At different levels of investigation - for instance when one is concerned with transmission properties of neural tissues or when the

issue is a global analysis of the functions performed by certain
cortical areas - one may easily be lead to believe that a detailed
study of the input-output behavior of single neurons bears no specific
relevance.  However, this would be a dangerously perfunctory conclusion.
The problem of "microscopic" versus "global" approaches to the study
of the nervous systems of complex organisms is indeed, in our view,
still up to this day one of the key issues in theoretical neurobiology
that surely deserves the highest consideration.

Acknowledgements.  I am grateful to the Faculty of Engineering Science
of Osaka University and to the Japanese Ministry of Education for
providing facilities, intellectual stimulation and financial support.
I also wish to express my sincere thanks to Professors S. Sato and
R. Suzuki for their warm hospitality and for many interesting
discussions.

### REFERENCES

Abramowitz, M. and Stegun, I.A. 1964. Handbook of Mathematics.
    National Bureau of Standards, N.Y.

Anderssen, R.S., De Hoog, F.R. and Weiss, R. 1973.  On the numerical
    solution of Brownian motion processes. J. Appl. Prob. 10, 409.

Capocelli, R.M. and Ricciardi, L.M. 1971.  Diffusion Approximation
    and first passage time problem for a model neuron.  Kybernetik
    8, 214.

Cerbone, G., Ricciardi, L.M. and Sacerdote, L. 1981.  Mean, variance
    and skeweness of the first passage time for the Ornstein-Uhlenbeck
    process.  Cybernetics and Systems 12:395

De Griffi, E.M. and Favella, L.F. 1978.  On a weakly singular Volterra
    integral equation.  Pubblicazione IAC. Serie III N. 167, 1978,
    Roma.

Durbin, J. 1971.  Boundary crossing probabilities for Brownian motion
    and Poisson processes and techniques for computing the power of
    the Kolmogorov-Smirnov test.  J. Appl. Prob. 8, 431.

Favella, L., Reineri, M.T., Ricciardi, L.M. and Sacerdote, L.  1981.
    First passage time problems and some related computational methods.
    Preprint.

Fortet, R. 1943.  Les functiones aléatoires du type de Markov associées
    a certaines èquationes linèaires aux derivées partielles du type
    parabolique.  J. Math. Pures Appl. 22, 177.

Gerstein, G.L. and Mandelbrot, B. 1964.  Random walk models for the
    spike activity for a single neuron.  Biophy. J. 4, 41.

Mikhlin, S.G. and Smolitshi, K.I. 1967. Approximate Methods for
     Solution of Differential and Integral Equations. Elsevier. N.Y.
Ricciardi, L.M. 1977. Diffusion Processes and Related Topics in
     Biology. Lecture Notes in Biomathematics. Springer-Verlag. N.Y.
Ricciardi, L.M. and Sacerdote, L. 1979. The Ornstein-Uhlenbeck
     process as a model for neuronal activity. Biol. Cyb. 35, 1.
Ricciardi, L.M., Sacerdote, L. and Sato, S. Diffusion approximations
     in neurobiology. In Preparation.
Sato, S. 1976. (In Japanese) Numerical solution of a first passage
     time problem for the Wiener process by Durbin's method. 1976.
     Report CST. 76-101 JIEE.
Sato, S. 1978. On the moments of the firing interval of the diffusion
     approximated model neuron. Math. Biosciences 39, 53.

# CHAPTER 10

PERIODIC PULSE SEQUENCES GENERATED

BY AN ANALOG NEURON MODEL

S. Yoshizawa
Department of Mathematical Engineering
and Instrumentation Physics
University of Tokyo
Tokyo 113, Japan

## 1. Introduction

Neuron models have been investigated from various points of view. One of the most fundamental viewpoints is to study the response characteristic of the neuron model to the periodic pulse input.

Based on the experiments on an analog neuron model using transistors, Harmon [1] first demonstrated that the relation between the amplitude of the input pulse and the firing rate of the neuron model has a very complicated form as shown in Fig. 1.

In order to explain the result, Nagumo and Sato [2] studied the following Caianiello nerve equation:

$$x_{n+1} = 1[\beta - \alpha \sum_{r=0}^{n} c^r x_{n-r} - \theta], \qquad (1.1)$$

where $1[z] = 1$ for $z > 0$, and $= 0$ for $z \leq 0$. They showed that the relation between the amplitude of the input pulse and the firing rate is an extended Cantor function. By the use of new variable

$$y_n = \sum_{r=0}^{n} c^r x_{n-r} , \qquad (1.2)$$

(1.1) is rewritten into the following difference equation:

$$y_{n+1} = Cy_n + \lambda \qquad \text{for } y_n \leq 0,$$

$$y_{n+1} = Cy_n - (1 - \lambda) \qquad \text{for } y_n > 0, \qquad (1.3)$$

$$0 < \lambda < 1,$$

where $\lambda = (\beta - \theta)/\alpha$ represents the input to the neuron model.

Hata [3] studied (1.3) mathematically and showed that (1.3) has non-periodic output sequence if and only if λ belongs to the exceptional set whose measure is zero.

With the intention of showing the similar result for more realistic neuron models, we investigate the BVP model which was proposed by FitzHugh [4] and Nagumo et al. [5] as a simplified realization of the Hodgkin-Huxley equation. First, we consider the BVP model with a simple nonlinear characteristic and then we treat a more general case.

The results obtained in this paper will be applicable to the study of the entrainment problem such as arrythmias.

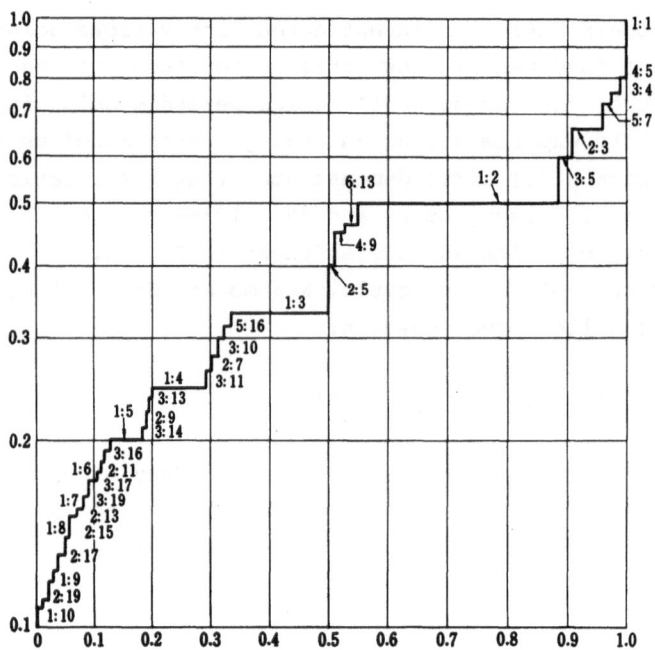

Fig.1    Response chracteristic of Harmon's electronic neuron
    model.   The relation between the pulse amplitude of the
    stimulating pulse sequence with a fixed frequency (abscissa)
    and the firing rate of the stimulated neuron model (ordinate).

## 2. Neuron Model

Figure 2(a) shows an electronic realization of the BVP model, TD is a tunnel diode whose voltage-current characteristic (V vs. $f_1(V)$) is shown in Fig. 2(b), and e(s) represents the voltage output of a pulse generator.

The behavior of the circuit is described by the following differential equation:

$$\frac{dV}{ds} = \frac{1}{C}(I - f_1(V)),$$

$$\frac{dI}{ds} = \frac{1}{L}(E + e(s) - V - RI). \tag{2.1}$$

We assume that the circuit has only one equilibrium point $(V_0, I_0)$ when e(s) = 0 as shown in Fig. 2(b). Let $r = 1/(df_1(V_0)/dV)$ and $I_a = f_1(V_p) - f_1(V_v)$, where $V_p$ and $V_v$ represent peak voltage and valley voltage of the tunnel diode, respectively. If we introduce the following normalized variables:

$$t = sr/L, \quad v = (V - V_0)/(rI_a), \quad w = (I_0 - I)/I_a,$$

$$\varepsilon = r^2 C/L, \quad \gamma = R/r, \quad f(v) = (f_1(V_0 + rI_a v) - I_0)/I_a,$$

$$g(t) = e(Lt/r)/(rI_a),$$

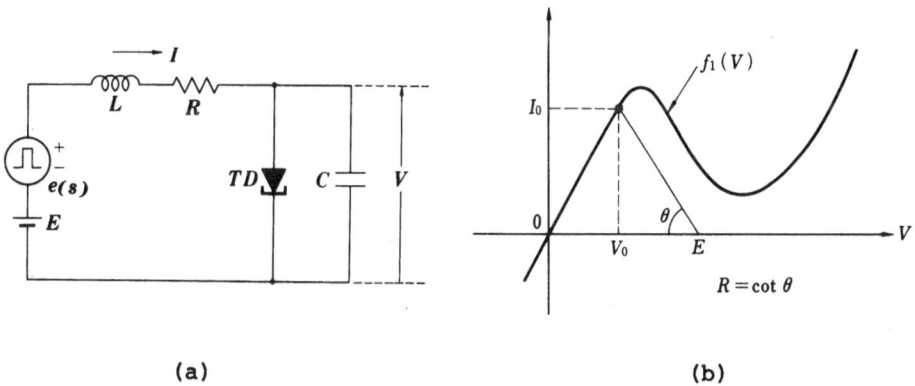

(a)                          (b)

Fig.2     (a) Neuron model using a tunnel diode (TD) proposed by Nagumo et al., (b) Voltge-current characteristic (V vs. $f_1(V)$) of the tunnel diode and the equilibrium point ( $V_0$, $I_0$ ) in the ( V, I )-plane when e(s) = 0.

(2.1) is reduced to

$$\varepsilon \frac{dv}{dt} = -f(v) - w,$$

$$\frac{dw}{dt} = v - \gamma w - g(t), \qquad (2.2)$$

where $f(v)$ is the function shown in Fig. 3(a).

For simplicity, let us assume $f(v)$ as follows (Fig. 3(b)):

$$f(v) = \begin{cases} v & \text{for } v \le a, \\ -(v - b)/(b - a) + a - 1 & \text{for } a < v < b, \\ v - 1 - (b - a) & \text{for } v \ge b, \end{cases} \qquad (2.3)$$

$$0 < a < b < 1.$$

Also, let $g(t)$ be a periodic function of period $T$ given by

$$g(t) = \begin{cases} g^* & \text{for } 0 < t \le \tau, \\ 0 & \text{for } \tau < t \le T. \end{cases} \qquad (2.4)$$

To proceed futher, we restrict our consideration to the "degenerate" case $\varepsilon = 0$ as the limit of small $\varepsilon$. Namely, we assume that the representative point of (2.2) is confined on the following two half lines:

$$B_r = \{(v,w) \mid w = -v, \ v \le a\},$$

$$B_e = \{(v,w) \mid w = -v + 1 + b - a, \ v \ge b\},$$

and that $v$ does not take value in $a < v < b$.

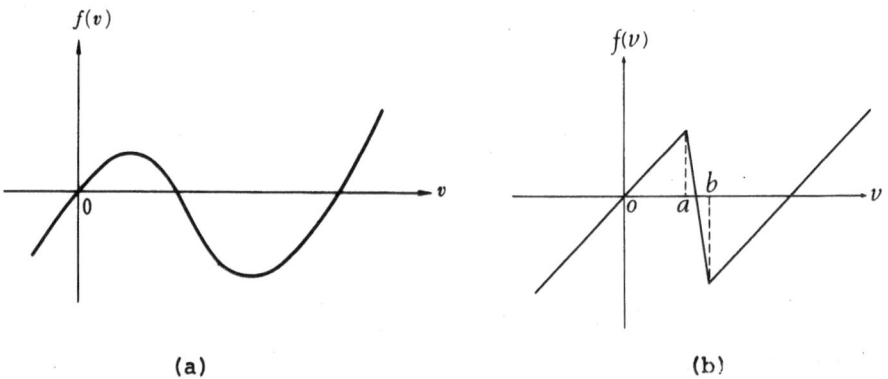

(a)                                              (b)

Fig.3    (a) General form of function $f(v)$, (b) a simplified form of function $f(v)$.

On these half lines, (2.2) reduces to

$$\frac{dv}{dt} = -((1 + \gamma)v - g(t)) \qquad\qquad \text{for } v \leq a,$$

$$\frac{dv}{dt} = -((1 + \gamma)v - \gamma(1 + (b - a)) - g(t)) \quad \text{for } v \geq b.$$

(2.5)

Transition between $B_r$ and $B_e$ is gorverned by the following instantaneous jumps:

$$v = b \longrightarrow v = a - 1 \qquad \text{for } B_e \longrightarrow B_r,$$

$$v = a \longrightarrow v = b + 1 \qquad \text{for } B_r \longrightarrow B_e,$$

(2.6)

where w is kept constant. This condition on w comes from the physical requirement that current through an inductance must change continuously with time.

## 3. Mapping P

Let us assume the following condition:

$$\gamma < a/(1 - a), \qquad g* > \max\{(1 + \gamma)a, \ b + \gamma(a - 1)\}, \qquad (3.1)$$

which guarantees that (2.2) has only one equilibrium point $(v,w) = (0, 0)$ for $g(t) = 0$ and also only one equilibrium point $(e_1, 1 + b - a - e_1)$ for $g(t) = g*$, where

$$e_1 = (\gamma(1 + b - a) + g*)/(1 + \gamma). \qquad (3.2)$$

Henceforth, we consider the case where the solution of (2.5) with initial value $v(0) \leq a$ satisfies $v(T) \leq a$. The condition for this being given later (see (3.6)), here, let us define the state of the neuron model. If the solution of (2.5) with the given initial value $v(0) \leq a$ becomes $v(t) > a$ for some $t$ ($0 < t < T$), then the neuron model is said to be in "excited state" (Ex.) and if $v(t) \leq a$ for all $t$ ($0 < t < T$), the neuron model is said to be in "resting state" (Re.). In this context, we call two half lines $B_e$ and $B_r$ "exciting branch" and "resting branch", respectively.

Further, the following notations are introduced.

$$A = \exp(-(1 + \gamma)\tau), \quad B = \exp(-(1 + \gamma)(T - \tau)),$$

$$r_1 = g*/(1 + \gamma), \quad e_0 = \gamma(1 + b - a)/(1 + \gamma). \tag{3.3}$$

In (3.2) and (3.3), $v = 0$ and $e_0$ represent the attractor on the resting branch $B_r$ and the virtual attractor on the produced line of $B_e$ when $g(t) = 0$, respectively, and $r_1$ and $e_1$ represent the virtual attractor on the produced line of the resting branch $B_r$ and the attractor on $B_e$ when $g(t) = g*$, respectively. We assume

$$g* < a\gamma + b + 1 \tag{3.4}$$

so that these quantities satisfy the followig relation. See Fig. 4.

$$0 < e_0 < b, \quad a < r_1 < e_1 < b + 1 \tag{3.5}$$

The transition of the solution of (2.5) from $v_0 = v(0)$ to $v_T = v(T)$ during one period of the stimulating periodic pulse sequence is is regarded as a mapping

$$P: v_0 \longrightarrow v_T.$$

The mapping P is calculated on four intervals of the initial value $v_0$.

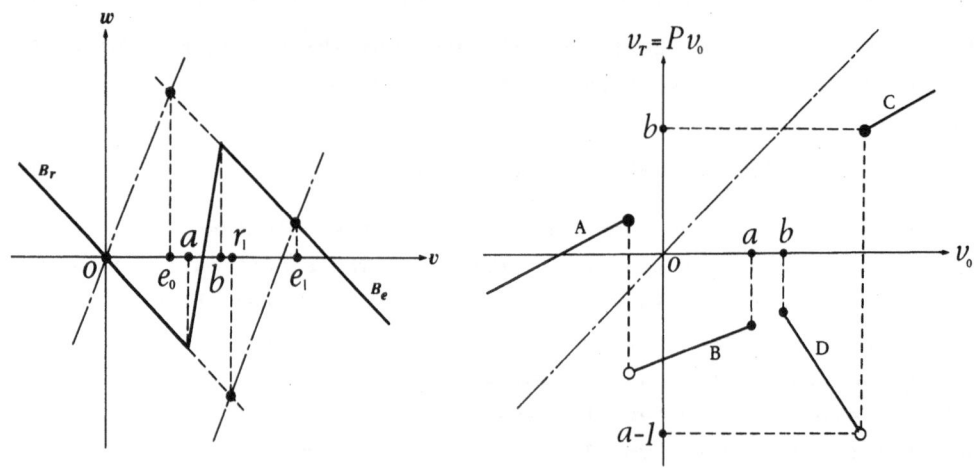

Fig.4    Difinitions of $e_0$, $e_1$ and $r_1$.

Fig.5    Mapping P.

**Interval A**  If $v_0 \leq r_1 - (r_1 - a)/A$ , then there occurs no jump during $0 \leq t \leq T$ and we have

$$v_T = (r_1 - (r_1 - v_0)A)B.$$

**Interval B**  If $r_1 - (r_1 - a)/A < v_0 \leq a$, then jump from $v = a$ to $v = b + 1$ occurs at some $t_1 < \tau$. In order to guarantee the second jump from $v = b$ to $v = a - 1$ to take place at some $t_2 < T$, we assume the following condition.

$$B < (b - e_0)/(1 + b - e_0) \tag{3.6}$$

Then, $v_T$ is given by

$$v_T = (a - 1)\{r_1 + (r_1 - v_0)A(1 + b - e_1)/(r_1 - a)\}B / (b - e_0) < a.$$

**Interval C**  If $v_0 \geq ((b - e_0)/B - r_1)/A + e_1$, then no jump occurs during $0 \leq t \leq T$ and we have

$$v_T = ((v_0 - e_1)A + r_1)B + e_0.$$

**Interval D**  If $b \leq v_0 < ((b - e_0)/B - r_1)/A + e_1$, then the jump from $v = b$ to $v = a - 1$ occurs and we have

$$v_T = (a - 1)((v_0 - e_1)A + r_1)B/(b - e_0).$$

Summing up these results, we obtain Fig. 5. From Fig. 5, it is clear that $Pv_0 = v_T \leq a$ if $v_0 \leq a$ and there exists a finite $n$ such that $P^n v_0 \leq a$ even if $v_0 > a$. Thus, we can assume $v_0 \leq a$ so far as the asymptotic behavior of the solution is dealt with. Therefore, we can consider that the mapping $P$ is described by the following piecewise linear difference equation:

$$v_T = (r_1 - (r_1 - v_0)A)B \qquad \text{for } v_0 \leq r_1 - (r_1 - a)/A$$

$$v_T = (a - 1)\{r_1 + (r_1 - v_0)A(1 + b - e_1)/(r_1 - a)\}B/(b - e_0)$$
$$\text{for } v_0 > r_1 - (r_1 - a)/A \tag{3.7}$$

In order to see the relation between the pulse width of the stimulating pulse and the characteristic of the solution of (3.7), let us regard the pulse width $\tau$ as a variable parameter.

Put

$$C = \exp(-(1 + \gamma)T) = AB = \text{constant},$$

$$B_0 = r_1/((r_1 - a)/C + a),$$

$$B_1 = r_1/\{(r_1 - a)/C - (1 - a)(1 + b - e_0)/(b - e_0)\}.$$

Introducing new veriable X and new variable parameter $\lambda$ defined by

$$X = (v - r_1 + (r_1 - a)/A)/\{(B_1 - B_0)(a - (a - r_1)/C)\},$$

$$\lambda = (B - B_0)/(B_1 - B_0),$$

we rewrite (3.7) into

$$X_T = CX_0 + \lambda \qquad \text{for } X_0 \leq 0,$$

$$X_T = CDX_0 - E(1 - \lambda) \qquad \text{for } X_0 > 0,$$

(3.8)

where

$$D = (1 - a)(1 + b - e_1)/\{(b - e_0)(r_1 - a)\},$$

$$E = B_0/B_1.$$

Moreover, if C satisfies the following relation

$$0 < C < (b - e_0)(r_1 - a)/\{(1 - a + r_1)(1 + b - e_0)\}, \qquad (3.9)$$

it is shown that

$$0 < \lambda < 1, \quad 0 < C < 1, \quad 0 < CD < 1,$$

$$0 < B_0 < B_1 < (b - e_0)/(1 + b - e_0),$$

(3.10)

and $\lambda$ changes from 0 to 1 as B changes from $B_0$ to $B_1$.

## 4. Periodic Output Sequences and Firing Rate

Let us investigate the time sequences generated by recursive operations of the mapping P. It is convenient to rewrite (3.7) into the form of the following difference equation:

$$y_{n+1} = Cy_n + \lambda \qquad \text{for } y_n \leq 0,$$

$$y_{n+1} = CDy_n - E(1 - \lambda) \qquad \text{for } y_n > 0,$$

(4.1)

$$0 < \lambda < 1, \quad 0 < C < 1, \quad 0 < CD < 1, \quad E > 0,$$

where $y_n = X_{nT}$. See Fig. 6. Note that (4.1) includes the term D in the second equation in contrast with (1.3).

If we represent the resting state and the excited state as 0 and 1, respectively, then we have a sequence of 0 and 1, which we call the

output sequence of (4.1), corresponding to each time sequence $\{y_n\}$.
Namely, from the definition of the excited and the resting states of
the neuron model, the output sequence of (4.1) can be written as the
sequence of $x_n$ defined by

$$x_n = 1[y_{n-1}].$$

Remark that, by the use of $x_n$, (4.1) is reduced to

$$y_{n+1} = CD^{x_{n+1}} y_n + LQ(1 - CD)^{-1}(1 - CD^{x_{n+1}}) - Lx_{n+1}, \qquad (4.2)$$

where

$$L = \lambda(1 - CD)/(1 - C) + E(1 - \lambda),$$
$$Q = Q(\lambda) = \lambda(1 - CD)/(L(1 - C)). \qquad (4.3)$$

Here, let us introduce several notations concerning with the
periodic sequences. A periodic sequence consisting of 0 and 1
$\ldots; x_m x_{m-1} \cdots x_1; x_m x_{m-1} \cdots x_1; \ldots$ is denoted $\underline{x}$ or $(x_m x_{m-1} \cdots x_1)$.
Finite sequence $x_m \cdots x_1$ is also written as $\underline{x}$, but this will not cause
any confusion. The concatenation of two periodic sequences $\underline{x} = (x_m \cdots$
$x_1)$ and $\underline{y} = (y_n \cdots y_1)$ is defined by the periodic sequence $(x_m \cdots x_1$
$y_n \cdots y_1)$ and is written as $\underline{xy}$. The cyclic shift $(x_i x_{i-1} \cdots x_{i+1})$
of $(x_m x_{m-1} \cdots x_1)$ is called the rotation of $\underline{x}$ and denoted as $\underline{x}^*$.

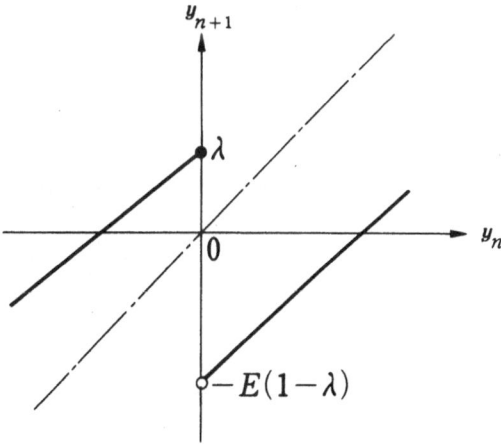

Fig.6    Graphic representation of difference equation (4.1).

For each periodic sequence $\underline{x}$, the firing rate is defined by

$$\Phi(\underline{x}) = (\text{number of 1's in } \underline{x})/(\text{length of } \underline{x}). \tag{4.4}$$

Also, the following functions are used.

$$W(\underline{x}) = \sum_{r=1}^{m} x_r \, C^{m-r} \, D^{\sum_{i=r+1}^{m} x_i}, \qquad \tilde{x} = C^m \, D^{\sum_{r=1}^{m} x_r}, \tag{4.5}$$

$$H(\underline{x}) = (1 - CD)W(\underline{x})/(1 - \tilde{x}),$$

where $\underline{x} = (x_m \, x_{m-1} \, \cdots \, x_1)$.

Now, we define the periodic output sequence of (4.1) as follows: A periodic sequence $\underline{x}$ is said to be the periodic output sequence of (4.1) if there exist input $\lambda$ and initial value $y_0(\underline{x}^*)$ for any rotation $\underline{x}^*$ such that the output sequence of (4.1) is given by the periodic sequence $\underline{x}^*$.

The periodic output sequence of (4.1) is characterized as follows [6].

(P1) A periodic sequence $\underline{x}$ is the periodic output sequence of (4.1) if and only if

$$H_L(\underline{x}) \equiv \max_{x^*_1 = 1} H(\underline{x}^*) < H_U(\underline{x}) \equiv \min_{x^*_1 = 0} H(\underline{x}^*), \tag{4.6}$$

where the maximum and the minimum are taken with respect to the rotations of $\underline{x}$ with $x^*_1 = 1$ and $x^*_1 = 0$, respectively. When (4.6) is satisfied, the periodic sequence $\underline{x}$ is generated if the input parameter $\lambda$ satisfies

$$H_L(\underline{x}) \leq Q(\lambda) \leq H_U(\underline{x}), \tag{4.7}$$

where $Q(\lambda)$ is the function defined in (4.5).

The following uniqueness property is shown in [7].

(P2) If (5.2) has a periodic output sequence $\underline{x}$, any output sequence of (4.1) approaches to $\underline{x}$.

## 5. Farey Series F and Set S of Periodic Sequences

In order to study the detailed structure of the output sequences of (4.1), we shall construct a set of periodic sequences corresponding to the Farey Series.

The Farey series of order n, denoted $F_n$, is the ascending series of irreducible fractions between 0 and 1 whose denominators do not exceed n. Numbers 0 and 1 are included in $F_n$ in the forms of 0/1 and 1/1, respectively. $F_{n+1}$ is constructed from $F_n$ by interpolating the mediants whose denominators are equal to n+1, where the mediant of two successive fractions h/k and h'/k' of $F_n$ is the irreducible fraction (h+h')/(k+k'). Clearly, $F_1 \subset F_2 \subset F_3 \subset \ldots \subset F_n$ and $F = \bigcup_n F_n = \lim_{n \to \infty} F_n$ contains every rational number included in the closed interval [0, 1].

Now, let us construct the set $S_n$ of periodic sequences as follows. First, let $S_1 = \{(0), (1)\}$ corresponding to $F_1 = \{0/1, 1/1\}$. Then, $S_{n+1}$ is constructed from $S_n$ by adding each 'median periodic sequence' corresponding to each newly interpolated mediant of $F_{n+1}$, where the median periodic sequence means the concatenation xy of two periodic sequences x and y of $S_n$ which correspond to two irreducible fractions p and q (p < q) of $F_n$ which are adjacent to the newly interpolated mediant of $F_{n+1}$. Put $S = \bigcup_n S_n = \lim_{n \to \infty} S_n$. Examples of $F_n$ and $S_n$ are shown in Table 1.

Table 1   Examples of $F_n$ and $S_n$.

| | | | | | | | | |
|---|---|---|---|---|---|---|---|---|
| $F_1$: $\frac{0}{1}$ | | | | | | | | $\frac{1}{1}$ |
| $S_1$: (0) | | | | | | | | (1) |
| $F_2$: $\frac{0}{1}$ | | | | $\frac{1}{2}$ | | | | $\frac{1}{1}$ |
| $S_2$: (0) | | | | (01) | | | | (1) |
| $F_3$: $\frac{0}{1}$ | | $\frac{1}{3}$ | | $\frac{1}{2}$ | | $\frac{2}{3}$ | | $\frac{1}{1}$ |
| $S_3$: (0) | | $(0^2 1)$ | | (01) | | $(01^2)$ | | (1) |
| $F_4$: $\frac{0}{1}$ | $\frac{1}{4}$ | $\frac{1}{3}$ | | $\frac{1}{2}$ | | $\frac{2}{3}$ | $\frac{3}{4}$ | $\frac{1}{1}$ |
| $S_4$: (0) | $(0^3 1)$ | $(0^2 1)$ | | (01) | | $(01^2)$ | $(01^3)$ | (1) |
| $F_5$: $\frac{0}{1}$ $\frac{1}{5}$ | $\frac{1}{4}$ | $\frac{1}{3}$ | $\frac{2}{5}$ | $\frac{1}{2}$ | $\frac{3}{5}$ | $\frac{2}{3}$ | $\frac{3}{4}$ $\frac{4}{5}$ | $\frac{1}{1}$ |
| $S_5$: (0) $(0^4 1)$ | $(0^3 1)$ | $(0^2 1)$ | $(0^2 101)$ | (01) | $(0101^2)$ | $(01^2)$ | $(01^3)$ $(01^4)$ | (1) |

From the construction of S, it is clear that each periodic sequence of S has the form $(0 \; x_{m-1} \; \ldots \; x_2 \; 1)$ except (0) and (1). Periodic sequence $\underline{x}$ of S corresponding to fraction h/k of F has a period k and includes h 1's and (k-h) 0's, accordingly, the firing rate $\Phi(\underline{x}) = h/k$. See [6] as for the details of discussions in this section.

The periodic sequences of S have the following property.

(P3)  For each periodic sequence $\underline{x} = (0 \; x_{m-1} \; \ldots \; x_2 \; 1)$ of S, the rotations $\underline{x}_U$ and $\underline{x}_L$ of $\underline{x}$ which give $H_U(\underline{x})$ and $H_L(\underline{x})$, respectively, are represented as follows:

$$\underline{x}_U = \underline{u} \; 10, \qquad \underline{x}_L = \underline{u} \; 01, \tag{5.1}$$

where $\underline{u} = (x_{m-1} \; \ldots \; x_2)$.

Depending on (P3), we see that the difference $\Delta(\underline{x})$ of $H_U(\underline{x})$ and $H_L(\underline{x})$ for any periodic sequence $\underline{x}$ in S is given by

$$\Delta(x) \equiv H_U(\underline{x}) - H_L(\underline{x}) = H(\underline{x}_U) - H(\underline{x}_L)$$

$$= (1 - CD)\{(W(\underline{u}) + \tilde{u}) - (W(\underline{u}) + \tilde{u}C)\}/(1 - \tilde{x})$$

$$= (1 - CD)(1 - C)\tilde{u}/(1 - \tilde{x}) > 0.$$

This and (P1) imply that any periodic sequence of S is the periodic output sequence of (4.1) and the interval of $Q(\lambda)$ corresponding to the output sequence $\underline{x}$ is given by (4.7). Also, for any adjacent pair of periodic sequences $\underline{x}$ and $\underline{y}$ in $S_n$ with $\Phi(\underline{x}) < \Phi(\underline{y})$, we can show, by the mathematical induction, that

$$H_U(\underline{x}) < H_L(\underline{y}).$$

Moreover, it is shown that the sum $\sum \Delta(\underline{x})$ of the lengths of the intervals of $Q(\lambda)$ corresponding to all $\underline{x}$ in S is equal to 1.

Thus, noting (P2) and the fact that $Q(\lambda)$ is a monotonically increasing smooth function varying from 0 to 1, we have the following conclusion.

(P4)  The periodic sequences of S are the periodic output sequences of (4.1). The corresponding intervals of input $\lambda$ are separated each other and are odered in accordance with the values of the firing rate of the periodic output sequences. Moreover, the sum of the lengths of the intervals is equal to 1.

Conversely, the periodic output sequences of (4.1) are only those periodic sequences belonging to S.

In the above investigation, we regarded the pulse width as the input to the neuron model for simplicity. The result is applicable to get the similar conclusion for the case where the pulse height is variable, although the fundamental difference equation corresponding to (4.1) changes to the following:

$$y_{n+1} = Cy_n + \lambda \qquad \text{for } y_n \leq 0,$$

$$y_{n+1} = CD(\lambda)y_n - E(1 - \lambda) \qquad \text{for } y_n > 0,$$

where, under suitable additional assumptions,

$$0 < \lambda < 1, \quad 0 < C < 1, \quad 0 < CD(\lambda) < 1, \quad dD(\lambda)/d\lambda < 0.$$

## 6. BVP Model with More General Nonlinear Characteristic

If the nonlinear characteristic of the BVP model has a more complicated form than that given by (2.3), the mapping P fails to be piecewise linear as (4.1). To include more general cases, let us write the mapping P as

$$y_{n+1} = P(y_n, \lambda), \quad x_{n+1} = 1[y_n]. \tag{6.1}$$

In many cases, we can assume that the mapping P satisfies the following condition. See Fig. 7.

$$0 < \kappa \leq P_y(y, \lambda) \leq \mu \quad \text{for } y \neq 0,$$

$$P_\lambda(y, \lambda) > 0 \quad \text{for } y \neq 0, \tag{6.2}$$

$$P^2(0-, \lambda) < P^2(0+, \lambda),$$

$$P(0-, \lambda) = P(0+, \lambda) + 1.$$

For example, this is true for the nonlinear characteristic in Fig. 3(b) with different positive coefficients and also true for that in Fig. 3(a) under appropriate conditions.

Keener [7] investigated this type of difference equation and showed that relation between the firing rate and the parameter has the Cantor property.

The following discussion is concerned with the pattern of the periodic pulse sequences generated by (6.1) with (6.2).

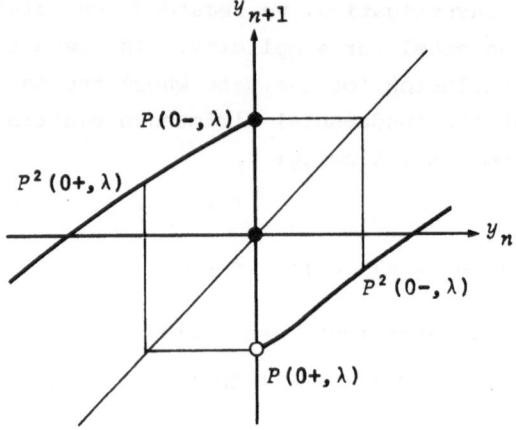

Fig.7    Graphic representation of difference equation (6.1).

First, let us cite from Yoshida [8] the following result which characterizes the periodic sequences not belonging to S.

(P5)  If a periodic sequence $\underline{x}$ does not belong to S, there are two rotations $\underline{x}^*$, $\underline{x}^{**}$ such that

$$\underline{x}^* = (1 \ \underline{a} \ 1 \ \underline{b}),$$

$$\underline{x}^{**} = (0 \ \underline{a} \ 0 \ \underline{c}).$$

(6.3)

The following property is easily obtained.

(P6)  It never occurs that the periodic output sequence of (6.1) with (6.2) contains $0^m$ and $1^n$ (m, n > 1), simultaneuosly.

By (P6), it is sufficient to consider the case where 0 appears on both sides of 1 such as ...010....

Now, take a periodic sequence $\underline{x}$ which does not belong to S and suppose that $\underline{x}$ is the output sequence of (6.1). Then, by (P5), there exist two rotations $\underline{x}^*$ and $\underline{x}^{**}$ and corresponding sequences of $\underline{y}_n : \underline{y}^*$ and $\underline{y}^{**}$.  Let us express them as follows:

$$\underline{x}^* = 1 \ 0 \ \ a_{n-1} \cdots a_2 \ 0 \ 1 \ \underline{b}$$

$$\underline{y}^* = \ \ v_2 \ z_n \ z_{n-1} \cdots \ \ z_1 \ v_1 \ \underline{s},$$

$$\underline{x}^{**} = 0 \ 0 \ \ a_{n-1} \cdots a_2 \ 0 \ 0 \ \underline{c},$$  (6.4)

$$\underline{y}^{**} = \ \ u_2 \ w_n \ w_{n-1} \cdots \ \ w_1 \ u_1 \ \underline{r},$$

where we put $a_n = a_1 = 0$ depending on (P6). The following relation clearly holds.

$$v_2, \ v_1 > 0, \ \ z_n, \ z_1 \leq 0,$$  (6.5)

$$u_2, \ u_1 \leq 0, \ \ w_n, \ w_1 \leq 0,$$

Now, suppose that

$$z_1 = P(v_1, \lambda) \geq w_1 = P(u_1, \lambda),$$  (6.6)

then, by (6.5), we have

$$P^2(0-, \lambda) \geq P(v_1, \lambda) = z_1 \geq w_1 = P(u_1, \lambda) > P^2(0+, \lambda),$$

which contradicts (6.2).

Contrary, if we suppose

$$z_1 < w_1,$$

then inequalities $z_i < w_i$ (i = 2, ..., n) hold, since $a_2, ..., a_n$ are common in $\underline{x}^*$ and $\underline{x}^{**}$ as shown in the expression (6.4). Therefore, we have

$$z_n < w_n \leq 0,$$

which implies

$$P(z_n, \lambda) = v_2 > 0 \geq u_2 = P(w_n, \lambda).$$

The last inequality contradicts (6.2).

Thus we obtain the following property.

(P7) Any periodic sequence which does not belong to S can not be the periodic output sequence of (6.1) with (6.2).

If P satisfies

$$\mu < 1, \ \ P(0-, 0) = 0, \ \ P(0+, 1) = 0,$$

in addition to (6.2), then, from the continuity of $P(y, \lambda)$ with respect to $\lambda$, it follows that every periodic sequence of S is generated as the periodic output sequence of (6.1) by a suitable choice of $\lambda$ and $y_0$.

7. Conclusion

We investigated the response characteristic of the BVP model to periodic pulse stimulations.  It was shown that the periodic output sequences of the degenerated BVP model belong to a special class of periodic sequences generated by a simple algorithm.  In a special case, the relation between the firing rate and the strength of the input (pulse width or pulse height) becomes an extended Cantor function.

References

[1] Harmon, L.D.: Studies with artificial neurons, I: Properties and functions of an artificial neuron,  Kybernetik 1, 89-101 (1961).
[2] Nagumo, J., Sato, S.: On a response characteristic of a mathematical neuron model,  Kybernetik 10, 155-164 (1972).
[3] Hata, M.: Dynamics of Caianiello's equation, preprint (1980).
[4] FitzHugh, R.: Impulses and physiological states in theoretical models of nerve membrane, Biophysical J. 1, 445-466 (1961).
[5] Nagumo, J., Arimoto, S., Yoshizawa, S.: An active pulse transmission line simulating nerve axon,  Proc. Inst. Radio Engineers 50, 2061-2070 (1962).
[6] Yoshizawa, S., Nagumo, J., Osada, H.: A Cantor function-like response characteristic of a neuron model, J. Faculty Eng., Univ. Tokyo (B) XXXVI, 59-90 (1981).
[7] Keener, J.P.: Chaotic behavior in piecewise continuos diffrence equations, Trans. American Math. Soc. 261, 589-604 (1980).
[8] Yoshida, T.: Response chracteristics of a mathematical neuron model, Dr. Eng. thesis, Univ. Tokyo (1977).

# Chapter 11

## ON A MATHEMATICAL NEURON MODEL

Masaya YAMAGUTI and Masayoshi HATA


Department of Mathematics
Faculty of Sciences
Kyoto University, Japan

Abstract.    J. Nagumo and S. Sato proposed a nonlinear difference
equation as mathematical neuron model and explained the unusual and
unsuspected phenomenon which was found by L.D. Harmon in experimental
studies with his transistor neuron model. From the mathematical stand
point, we investigate rigorously the complete dynamics of this model.
We can define a topological number called the average firing rate, which
is analogous to the rotation number of a homeomorphism of the circle,
and show the existence of Cantor attractor if and only if the average
firing rate is irrational.

## 1. Introduction.

In 1961, L.D. Harmon
found an unusual and unsuspected
phenomenon between the amplitude
of the input pulses and the
firing frequency in experimental
studies with his transistor
neuron model.   The relationship
between them is shown in Fig. 1,
which was reproduced from Harmon
with a minor modification.   Using
a Caianiello's equation, J. Nagumo
and S. Sato [2]   explained the

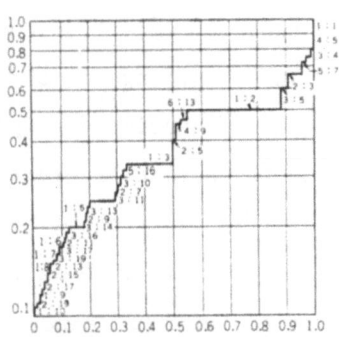

Fig. 1.  Relationship between the
amplitude of driving pulses $a$  and
the pulse frequency ratio of the
driven unit to that of the driving
unit  $F$.

above phenomenon. Indeed, they regarded a neuron as a threshold element with a refractory period and assumed that the inhibitory influence of a past firing upon the excitability of the neuron at the present decreases exponentially with time. Also they assumed that the magnitude of the input stimulus is constant. Under these assumptions, Caianiello's equation takes the form as follows.

(1.1)
$$x_{n+1} = \mathbb{1}[A - \alpha \sum_{r=0}^{n} \frac{x_{n-r}}{b^r} - \theta] \ , \alpha > 0, \ b > 1.$$

where $\mathbb{1}[x] = 1 \ (x \geq 0), \ = 0 \ (x < 0)$. The value $x_n$ represents the state of the neuron at the instant $n$. The resting state is represented by $0$ and the exciting state by $1$. Constant $A$ is the magnitude of the input stimulus and $\theta$ is the threshold value.

Let $y_n = 1 + \dfrac{A - \theta}{\alpha b} - \displaystyle\sum_{r=0}^{n} \frac{x_{n-r}}{b^r}$ . (1.1) implies

(1.2)
$$y_{n+1} = f(y_n)$$

where $f(x) = \beta(x - c) + 1 \ (x < c), \ = \beta(x - c) \ (x \geq c), \ \beta = \dfrac{1}{b}$ ,
$c = 1 - \dfrac{A - \theta}{\alpha}(1 - \dfrac{1}{b})$ , and $x_{n+1} = \mathbb{1}[y_n - c]$.
We assume that $0 < c < 1$, because the neuron always excites or rests for large $n$ according to $c < 0$ or $c > 1$ respectively. So we investigate the dynamics of a piecewise-linear map $f(x) \equiv f(x,\beta,c)$ in (1.2) on the parameter plane $(\beta,c) \in (0,1) \times (0,1)$. It is sufficient to examine $f(x)$ only on $I = [0,1)$ because the iterated point of the initial value out of $I$ must fall into $I$ after some iterations by $f$.

Recently, James P. Keener [3] started to study piecewise continuous difference equations. Our model is included in his model. But we can give a finer results than Keener's one because of the particularity of our model. Note that the map $f$ in (1.2) is a nonoverlapping piecewise continuous one. All proofs of theorems in the sequel sections are omitted. See [1] for the proofs.

## 2. Periodic attractor and Cantor attractor.

Let $I = [0,1)$. We define Case A and Case B as follows.

Case A : there exists some integer $N > 1$ such that $c \in \text{Int } f^i(I)$
for $i = 0,1,\ldots,N-1$ and $c \notin \text{Int } f^N(I)$.

Case B : for any integer $n \geq 1$, $c \in \text{Int } f^n(I)$.

We call Case A a periodic case and Case B a singular case. These
definitions are justified by the following theorems.

Theorem 2.1. In Case A, for any $x \in I$, the orbit of $x$ is
asymptotically periodic with period $N+1$, where $N$ is given in the
definition of Case A. Moreover, the map $f$ is regarded as a
permutation $\pi = \sigma^Q$ on the subintervals, $I_1, I_2, \ldots, I_{N+1}$ where $\bigcup_{j=1}^{N+1} I_j$
$= I$, $I_i \cap I_j = \phi(i \neq j)$, $\sigma = \begin{pmatrix} 12 \cdots \cdot N+1 \\ 23 \cdots \cdot 1 \end{pmatrix}$, and $Q$ is the number of $1 \leq j$
$\leq N+1$ such that $I_j \subset [c,1)$ and $(Q, N+1) = 1$.

Theorem 2.2. In Case B, $f(x)$ has neither periodic orbits nor
asymptotically periodic orbits. Moreover, $\Lambda = \bigcap_{n=1}^{\infty} f^n(I)$ is an
f-invariant Cantor set with zero Lebesgue measure and also is the $\omega$-limit
set of $x$ for any $x \in \Lambda$. We call $\Lambda$ a Cantor attractor.

## 3. Average firing rate.

Definition 3.1. We call the following value $\rho(x,\beta,c)$ the average
firing rate if the limit exists.

$$\rho(x,\beta,c) = \lim_{n \to \infty} \frac{1}{n} \sum_{j=1}^{n} \mathbf{1}[f^{j-1}(x,\beta,c) - c]$$

This number is analogous to the rotation number of a homeomorphism of
the circle, and we will see later the close relation between $f(x,\beta,c)$
and the rigid rotation on the unit interval.

Theorem 3.2.    For any $(x,\beta,c) \in [0,1] \times (0,1) \times (0,1)$, the average firing rate exists.  Moreover, we have  $\rho(x,\beta,c) = \rho(0,\beta,c)$  for any $x \in [0,1]$, that is, the average firing rate is independent of the initial value  x.


Theorem 3.3.    The singular case is valid if and only if the average firing rate is an irrational number.


So the average firing rate characterizes the dynamics of  $f(x,\beta,c)$.


Theorem 3.4.    For fixed  $\beta \in (0,1)$, $\rho(0,\beta,c)$  is a continuous monotone decreasing function of  c.


4. Distribution of periodic cases and singular cases on the parameter plane.

Fix the value of  $\beta \in (0,1)$.  We will investigate the two orbits which start from  0  and  1, and these orbits may be regarded as the sequences of functions of variable  c.  This idea is analogous to the investigation of the orbit of a critical point of one dimensional endomorphism.  (See [4])

Let
$$f(x,c) = \begin{cases} \beta(x - c) + 1 & (x < c) \\ \beta(x - c) & (x \geq c) \end{cases}$$

And define  $F_n(c)$  and  $G_n(c)$  inductively as follows.

$$F_1(c) = 1 - \beta c, \quad F_{n+1}(c) = f(F_n(c),c),$$

$$G_1(c) = \beta(1 - c), \quad G_{n+1}(c) = f(G_n(c),c).$$

(See Fig. II).  A set of discontinuity points of  $F_n(c)$  and  $G_n(c)$  plays an important role in this section.

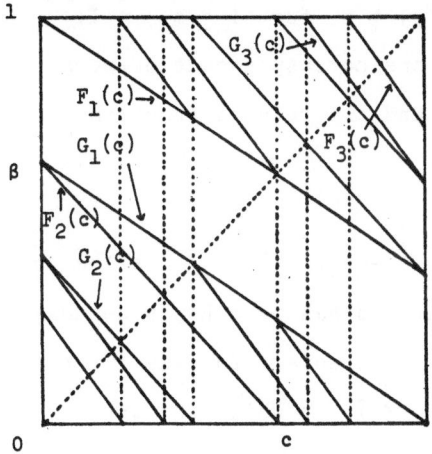

Fig.2.  Graphs of $F_i(c)$ and $G_i(c)$ for $i=1,2,3.$

## Lemma 4.1.

(a) $p$ is a discontinuity point of $F_n(c)$ if and only if there exists $1 \leq m \leq n-1$ such that $p$ is a fixed point of $F_m(c)$ for $n \geq 2$.

(b) Any discontinuity point of $F_n(c)$ is also that of $F_{n+1}(c)$.

(c) $F_n(c)$ is continuous at any discontinuity point of $G_n(c)$ for any $m \geq 1$.

Also we obtain the similar results for $G_n(c)$.

## Theorem 4.2.

For fixed $\beta$ and for any reduced fraction $\frac{q}{p} \in (0,1)$, there exists uniquely a closed interval $\Delta(\frac{q}{p})$ on the $c$ axis, such that for any $c \in \Delta(\frac{q}{p})$, we get always a periodic case with period $p$ and with the average firing rate $\frac{q}{p}$. Moreover, let $\Delta(\frac{q}{p}) = [a,b]$, then

$$a = 1 - q \frac{\beta^{p-1}(1 - \beta)}{1 - \beta p} - \frac{1}{1 - \beta p}(\frac{1 - \beta}{\beta})^2 \sum_{j=1}^{p} [\frac{q}{p} j] \beta^j$$

$$b = a + (\frac{1 - \beta}{\beta})^2 \frac{\beta^p}{1 - \beta p} .$$

And if $\frac{q}{p} < \frac{s}{r}$, then $\Delta(\frac{q}{p}) > \Delta(\frac{s}{r})$.

Indeed, we define inductively $\Delta(\frac{m}{n})$ by using primitive fixed points of $F_{n-1}(c)$ and $G_{n-1}(c)$. We say that $p$ is a primitive fixed point of $F_n(c)$, if $F_n(p) = p$ and $F_m(p) \neq p$ for $1 \leq m \leq n-1$.

The above theorem gives an one to one correspondence between disjoint subintervals in the unit interval and all reduced fractions in the unit interval.

**Theorem 4.3.** Let $\sum = (0,1) - \cup \Delta(\frac{q}{p})$. Then $\sum$ is the set removed a countable set from the Cantor set and has zero Lebesgue measure. We get a singular case for any $c \in \sum$. Moreover, for any irrational number $\gamma \in (0,1)$, there exists uniquely $c \in \sum$ at which the average firing rate is $\gamma$ and satisfies

$$c = 1 - (\frac{1 - \beta}{\beta})^2 \sum_{m=2}^{\infty} [\gamma n] \beta^n.$$

The above theorem gives an one to one correspondence between $\sum$ and all irrational numbers in the unit interval.

5. Conjugacy problem.

We say that $f$ is topologically conjugate to $g$ if $g = h \bullet f \bullet h^{-1}$ for some homeomorphism $h$.

**Theorem 5.1.** If $\rho(0,\beta,c) = \rho(0,\xi,\lambda) \notin \mathbb{Q}$, then $f(x,\beta,c)$ is topologically conjugate to $f(x,\xi,\lambda)$.

We say that $f$ is topologically semi-conjugate to $g$ if $h \bullet g = f \bullet h$ for some continuous monotone onto map $h$.

**Theorem 5.2.** If $\rho(0,\beta,c) \notin \mathbb{Q}$, then $f(x,\beta,c)$ is topologically semi-conjugate to $R_\gamma : x \rightarrow x + \gamma \pmod 1$ where $R_\gamma$ is the rigid rotation on the unit interval and $\gamma$ is the average firing rate of $f(x,\beta,c)$.

REFERENCE

[1] M. Hata, Dynamics of Caianiello's equation, to appear in J. Math. Kyoto Univ., 1982.

[2] J. Nagumo and S. Sato, On a responce characterisitc of a mathematical neuron model, Kybernetic 10 (1972) p.155-164.

[3] James P. Keener, Chaos behavior in piecewise continuous difference equations, Trans. Amer. Math. Soc., 261 (1980), p.589-604.

[4] J. Milnor and W. Thurston, On iterated maps of the interval I and II, Princeton University and the Institute for Advanced Study, Princeton, 1977.

# Chapter 12

# CONTROL OF DISTRIBUTED NEURAL OSCILLATORS

Ryoji SUZUKI, Sumiko MAJIMA, Hitoshi TATUMI
Department of Biophysical Engineering
Faculty of Engineering Science
Osaka University

Toyonaka, Osaka, Japan

## 1. Introduction

Recently, decentralized control system has received considerable attention in the field of control engineering for the control of large-scale and complex systems. The decentralized control system has a multiple number of control centers. Each of them controls its own subsystem with specific goal and need to coordinate its activity to satisfy the overall goal.

On the other hand, animal has a high degree of ability to organize autonomous activities of individual organs into one ordered behavior of corrective motion. For example, when starfish is placed upside down, it turns itself over by coordinating the motions of five arms. It should be noted that coordination is preceded by individual random motion of each arm during which the leading arm for coordination is selected from the five arms. Thereafter the individual random motions are inhibited and a systematic turn-over motion starts. (1)

In the author's opinion, animal system can be conceived as a typical example of the decentralized control system.

What mechanism controls this organized process? Answer to this question might be interesting for system engineer as well as for neurophysiologist.

We have taken rhythmic behaviros such as locomotion, swimming as the first step to be studied.

It was found that these rhythmic behaviors are controlled by mutually coupled endogeneous neural oscillators. For example, the coordinated movement of swimmerets in the crayfish is controlled by the distributed neural oscillators in abdomen interacting each other. (2)

The problems are now reduced to the investigation of the behaviors of coupled oscillator neurons and to find how to control them.

of electrotonically coupled neurons. (7) Studying our model according
to the theory of Hopf bifurcation, we found the conditions over the diffusion constants of the electrical junctions which give two kinds of
periodic solutions. One is the solution where two neurons oscillate in
phase synchrony. The other is the solution of 180 degree out of phase
oscillation.

Furthermore, the general two-oscillator system was used to explain the splitting phenomena which were first found by Pittendrigh in
circadian rhythms. By the theory of Hopf bifurcation, we could show two
stable periodic solutions. One is the in-phase and the other is the anti
phase solution. The latter corresponds to a splitting pattern. (8)


## 3. Control of coordinated rhythmical behavior of crayfish swimmeret System

It has been suggested that the rhythmical behavior of invertebrate such as locomotion, swimming or flying are controlled by the neural
networks which receive control signals from the command fibers. The crayfish swimmeret system is a good example to find the control mechanism
of such rhythmical behaviors. Wiersma and Ikeda (9) showed that rhythmical motor activity can be released in crayfish swimmeret system by
stimulating command fibers with constant frequency pulse trains. They
found that the period of the rhythm and the latency depend on the stimulus frequency of command fiber. Stein (10) studied intersegmental
coordination in swimmeret and proposed the existence of coordinating
neuron of interappendage phases.

In this paper, we will show the relationship between the stimulus
frequency of command fiber and the burst period more quantitatively
and the interappendage phase constancy. The phase is kept constant even
when the burst period changes more than twofold. These results are analyzed in terms of a neural model consisting of Wilson Cowan type oscillator.

### 3.1 Physiological Experiment

The preparation for recording the burst periods was as follows:
The abdomen detached from thorax was dissected in the way described by
Kennedy and Takeda (11), except that the sternal ribs of the segments
were cut along their long axis in the saline solution ( Van Harreveld's

## 2. Phase entrainment of two coupled neural oscillators

The entrainment behavior of stable oscillators interacting each other can be investigated by phase response curves (PRCs). This method was developed by Perkel and coworkers and applied by Pittendrigh et al. extensively to circadian rhythms. After Winfree's excellent works, we studied topological properties of PRCs by the theory of a dynamical system and the homotopy theory. (3)

We applied PRC method to the study of human finger tapping.

Assuming that human finger tapping is controlled by an oscillatory neural network, we studied the functional interaction between the finger tapping neural network and neural networks which control some psychological tasks imposed on the subject as perturbations of the phase resetting experiments. (4)

We also investigated the phase entrainment between finger tapping of left hand and that of right hand. The subjects were instructed to coordinate the finger tapping by both hands so as to keep the phase difference between two hands constant. The performance was evaluated by a systematic error and a standard deviation of phase differences. It was shown that the performance is better at the phase difference 0.0 and 0.5 than at other phase difference, which means that we can achieve synchronous or alternate rhythm by both hands more easily than other rhythms. This result was analyzed by using of the same method of Daan and Berde. (5)

PRCs were measured for left and right hand finger tappings and steady phase differences were obtained from the intersections of the following two graphs.

$$\phi_2 = 1 - (\phi_1 + f(\phi_1))$$
$$\phi_1 = 1 - (\phi_2 + g(\phi_2))$$

where $f(\phi)$, $g(\phi)$ are PRCs for left and right hand finger tapping respectively. This analysis showed a good agreement with the results of experiments. (6)

Oscillator neuron can be described mathematically using nonlinear oscillator model. Thus the theory of coupled nonlinear oscillators, in particular, the theory of Hopf bifurcation is useful to study the phase entrainment between two interacting oscillators and the stability of solutions.

We studied two BVP model neurons coupled by diffusion as a model

solution ) kept in the range of 14°C- 20°C. This made the first root of
each segment free for recording. The rami of the swimmerets were removed
so that their movements did not affect the electrodes. The recording
was performed with silver-wire suction electrodes and the position of
each suction electrode was ajusted along the posterior side of the main
first root so as to record the discharges of powerstroke motorneurons.
Male animals were used mainly because of their relatively few connective tissue.  Recordings were made only from the roots of the third and
fifth ganglia and the signals were amplified by conventional means and
recorded on FM tape for later analysis.

     Command fibers located along the lateral edges of the intergan-
glionic connectives between the second and the third ganglion were stimulated to induce rhythmic activity. This was accomplished with two
platinum wire electrodes on which the small nerve bundles from the lateral edges of the connective were lifted in a bath of paraffin oil,
reference electrode is placed in the bath. Square wave pulses with the
duration of 100 μsec were used for stimulation. The frequency of the
pulse trains was varied between 2 and 100 pulses per second.

     Increasing the stimulus frequency to the command fibers were
usually accompanied by monotonic decreasing of the burst period. The
stable interappendage coordinated rhythms were obtained for up to 50
pulses per second of the stimulation. At the same time, the interappendage phases during locomotion were measured. They were almost constant even when the frequency of the motor activity changes more than
twofold . (Fig.1)

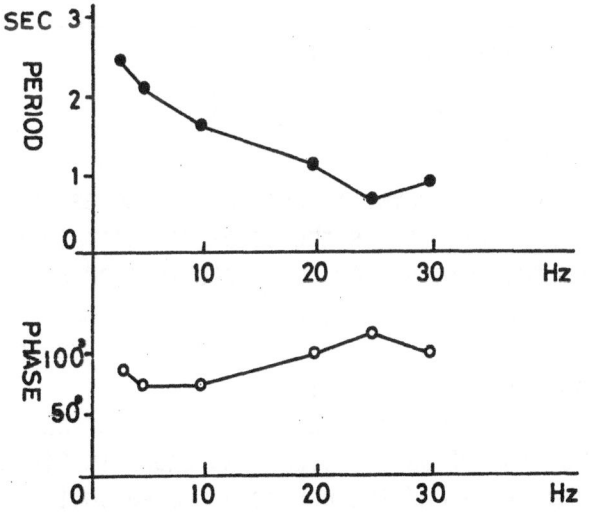

Fig.1

Relation between
burst period and
stimulus frequency

Phase delay between
the fifth and fourth
swimmeret motor
discharge

We also could change the oscillating period of swimmeret by increasing stimulating voltage while stimulus frequency fixed. The monotonic decreasing of the period was also obtained. This result can be interpreted as follows. There exist several command fibers in the stimulated portion and each fiber has different threshold to be activated. The increase of stimulating voltage recruits them. In this case the interappendage phase also is kept constant, even though the burst period changes more than twofold.

## 3.2 Model

In order to analyze such a rhythmical behavior as mentioned above, a neural oscillator model whose oscillating period can be controlled externally must be used. Wilson-Cowan type model (12) has such a property. The model shows the autonomous oscillation within a range of parameter values and the period decreases with increasing the excitatory input to the excitatory population.

We adopted the simplified version of their model as follows:

$$\tau_e \frac{dE}{dt} = -E + (k_e - rE) S_e (c_1 E - c_2 I + P)$$

$$\tau_i \frac{dI}{dt} = -I + (k_i - rI) S_i (c_3 E - c_4 I + Q)$$

E(t) and I(t) denote respectively the time coarse-grained activities of excitatory and inhibitory neural population at time t. $S_e(x)$ and $S_i(x)$ are called the response function because they give the expected proportion of cells in each subpopulation which can respond to a given level of the input. The form we chose for S(x) is

$$S(x) = \frac{1}{2} \left\{ 1 + \tanh (ax - a\theta) \right\}$$

here a gives average sensitivity and $\theta$ gives threshold. $\tau$ denotes the neural membrane time constant. r represents the absolute refractory period. $c_1$, $c_2$, $c_3$ and $c_4$ are the interaction coefficients amongst the excitatory and inhibitory populations . $k_e$ and $k_i$ are the maximum value of the response function. P and Q are the external input to the excitatory and inhibitory population respectively.

Fig.2 (A) shows typical limit cycles in E-I plane for various values of P and (C) shows waveforms of oscillations. As shown in Fig.2 (B), with increasing P, the amplitude and the period decrease until oscillation stops.

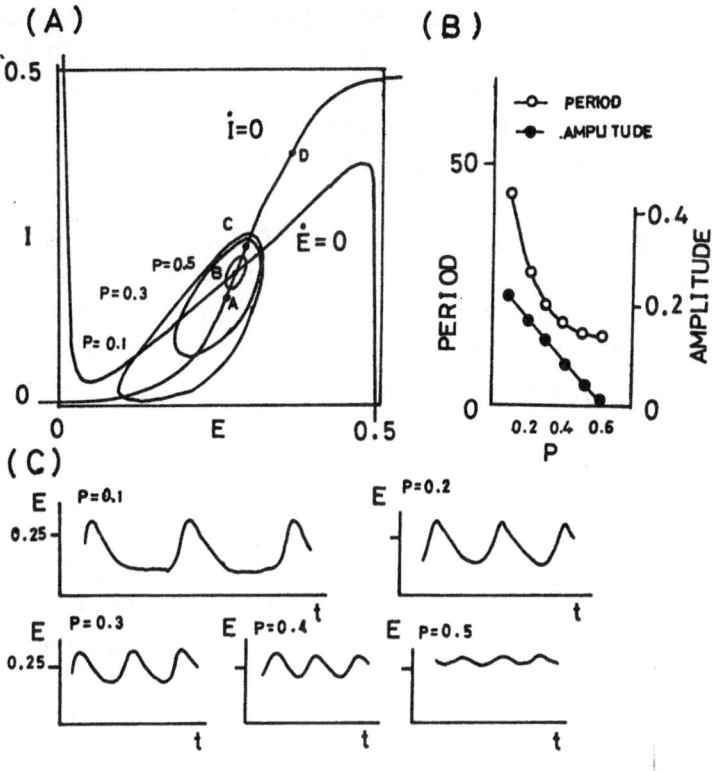

Fig.2

(A) Phase plane showing limit cycles in
response to constant stimulation.  P=0.1,
0.3, o.5.      Points A,B,C,D are equilibrium
points for P=0.1, 0.5, 1.0 and 2.0
(B) Dependencies of the period and the
amplitude on the stimulus strength P
(C) Waveforms of oscillation of E(t)
For numeriacl calculation,  r=1, $a_e$=0.75,
$a_i$=1.25, $\theta_e$ =2.8,  $\theta_i$=4.0 are used.

## 3.3  Phase response curve of Wilson-Cowan model

Imposing short time increment of P as disturbance, PRCs of Wilson
Cowan model were calculated. Fig.3 shows PRCs for different values of
disturbance $\Delta P$. All of the PRCs intersect with negative slope at the
same point of phase axis. And that the slope is less than two in the

case of small perturbation, which means that the oscillator can be entrained at this phase, if it receives pulse type input with the same
period of the oscillator. PRCs were also obtained by changing the period of oscillation while the magnitude of disturbance fixed. As shown
in Fig.4, the entrainmental phase varies only 0.15, even when the period
changes about twofold.

PRCs described above are the steady PRCs. In experiments with
real system, it is difficult to obtain the steay PRC. The PRC with
swimmeret of crayfish was reported only by Stein which is the first
transient PRC. He plotted the first phase delay of the powerstroke motor
neuron discharge in the third abdominal ganglion correlated with the
discharge of medial ascending coordinating neurons in the 4-3 connective.

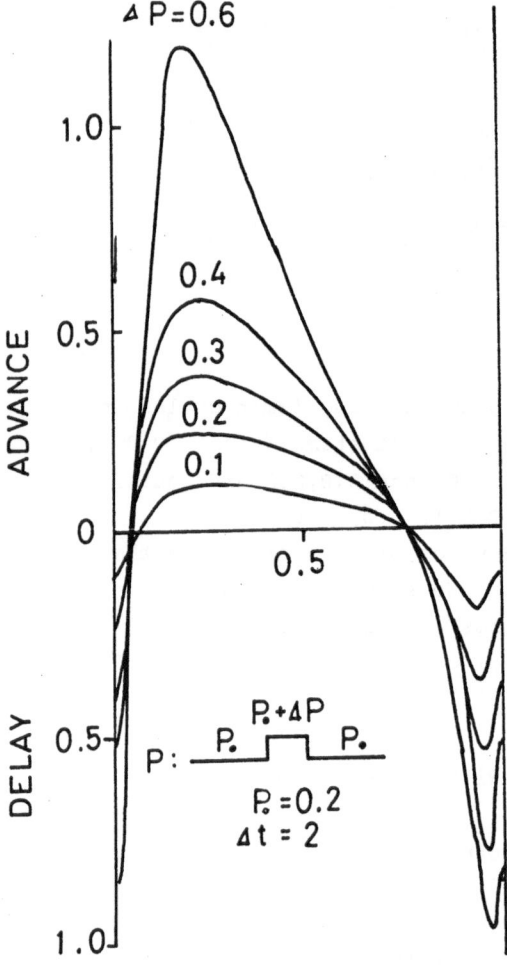

Fig.3
PRCs for various

magnitude of pert-
urbation $\Delta P$

Duration of pertur-
bation: 2 time unit
Chronic stimulation

$P = 0.2$

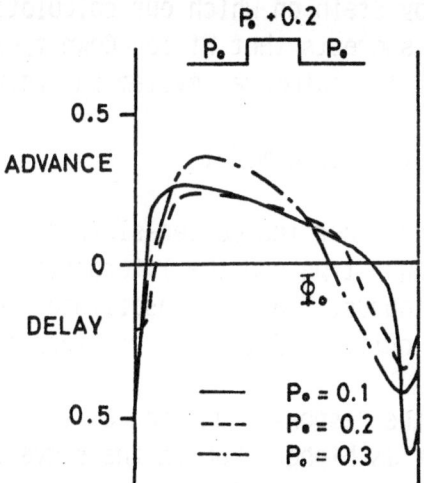

Fig.4
PRCs for different input P with same
magnitude of disturbance ΔP
The period of oscillator is 44msec at P=0.1
                            28msec at P=0.2
                            21msec at P=0.3

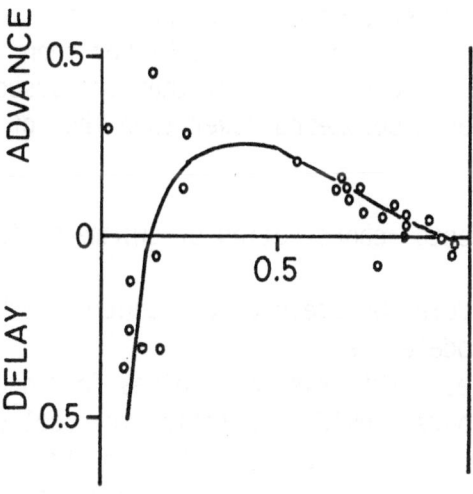

Fig.5  The first transient PRC by Stein ( o )
superimposed the result of calculation (—)
P=0.2 and  ΔP=0.6

Fig.5 is the transient PRC by Stein on which our calculation is super-imposed. The good agreement suggests that Wilson-Cown type oscillator is suitable for the model of the swimmeret system in crayfish.

## 3.4 Phase entrainment of master-slave model

In this section, we consider the master-slave oscillator model. Each oscillator is Wilson-Cowan type and the interaction between them is uni-directional and continuous. The slave oscillator receives input proportional to the average activity $E(t)$ of excitatory subset of master oscillator.

If the threshold of the response function of the excitatory sub-set in the master oscillator is higher than in the slave oscillator, which means the period of the slave oscillator is shorter than that of the master, the two oscillators synchronize each other with an almost constant phase difference over a wide range of the oscillating period as shown in Fig.6.

In real system, there might exist threshold in interaction. We assumed that control signal from master to slave is active only when the average activity $E(t)$ of the master exceeds some threshold $E_t$. Fig.7 shows the dependence of phase difference on the period of oscillation. When two oscillators are identical, phase difference decreases monotonically as their periods become longer. If the threshold of the response function in master oscillator is higher than in slave, then the phase difference decreases at first then increases. So the range of the phase difference becomes narrower than that of coupling of identical oscillator.

## 3.5  A model of intersegmental coordination of swimmeret in crayfish

Considering the studies by Stein et al, control mechanism of swimmeret system can be modeled as follows.

As suggested by Stein, this system involves four neural process-es. First, the command fibers provide excitation to local oscillator by sending impulses continuously. Secondly, the oscillator neurons translate the unpatterned command input into patterned output. In mathe-matical sense, it corresponds to the bifurcation phenomenon. Third, the coordinating neurons send to each oscillator the temporal information from the other neurons. Fourth, the motorneurons receive oscillator in-put and drive the swimmeret movements.   Stein is saying that the goal

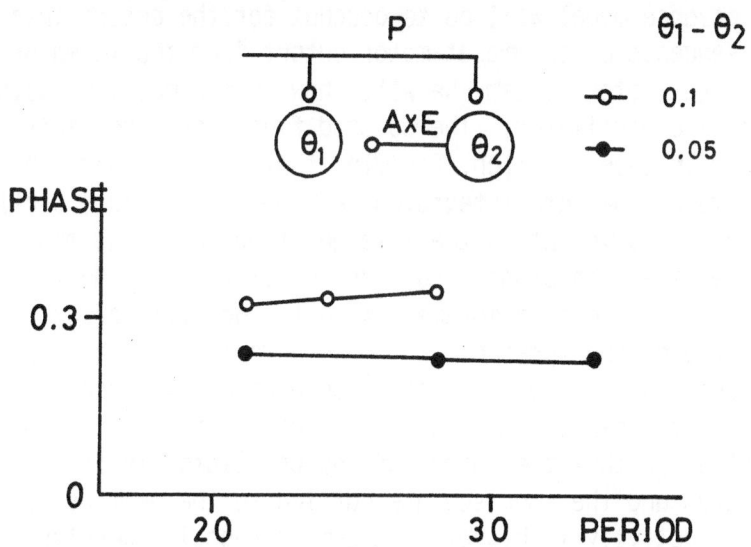

Fig.6
The relationship between the phase difference
and their period.  Continuous interaction without
threshold.  A=0.5

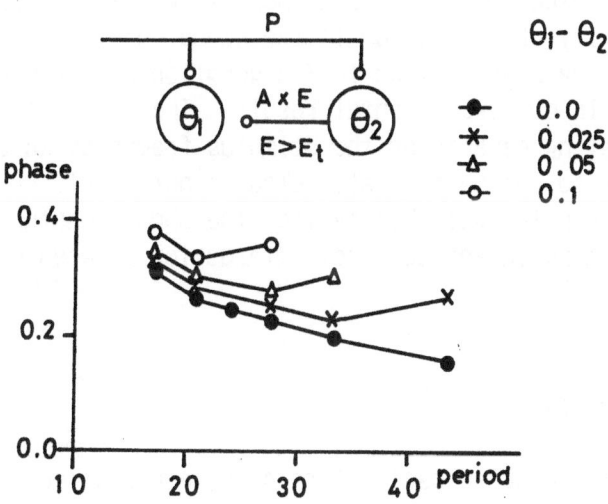

Fig.7
The relationship between the phase difference
and their period. Continuous interaction with
threshold. A=0.5, $E_t$=0.25

of the more advance model will be to account for the entire intersegmental timming sequence of swimmeret motor output from the known properties of a coupled oscillator system. We will show such a model by using of Wilson-Cowan type oscillator. The command fibers provide excitation to the oscillator in each ganglion. Assuming that the impulses running along the command fiber are integrated before entering each oscillator, the frequency is equivalent to the external input P of the model. The motorneuron receives the signal from the oscillator in abdominal ganglion and drives the swimmeret movement when the activity of excitatory neuron E(t) exceeds the threshold $E_t$. We have two possibilities of the type of coordination. One is that each excitatory neuron in caudal oscillator directly sends impulses to the oscillator in the next ganglion. The other is that the output of the oscillator is integrated in some interneuron and the impulses are carried to the next oscillator only when this activity is higher than some threshold. Results in the previous section show that the variation of phase difference with increasing of oscillating period is small in the former type of coordination. So we adopted the former type of coordination in our modeling.

Fig.8 shows the periods of the oscillator and the phase difference between two oscillators in 3rd ganglion and 4th ganglion, and also between 4th and 5th ganglion with changing external input P. The result resembles to the experimental result shown in Fig.1. We also assume that the sensitivity of the neuron in the caudal ganglion is higher so that the activity begins first in 5th ganglion when the command fiber is stimulated as in our experimental result. The fact that the delay of the first bursting depends on the stimulus frequency which was found by Wiersma and Ikeda can be also simulated in our model.

We have shown the possibility that the phase constancy can be explained by assuming one-directional interaction between oscillators in each ganglion.


4. Concluding remarks

The mutually coupled oscillators have been applied as a mathematical model to explain several biological phenomena such as circadian rhythms. However there have been few attempts to explain rhythmical behavior whose frequency is changed. We showed that the one directionally coupled oscillator are entrained at the frequency of the caudal oscillator and the phase difference is almost independent of the frequency by using Wilson-Cowan type model.

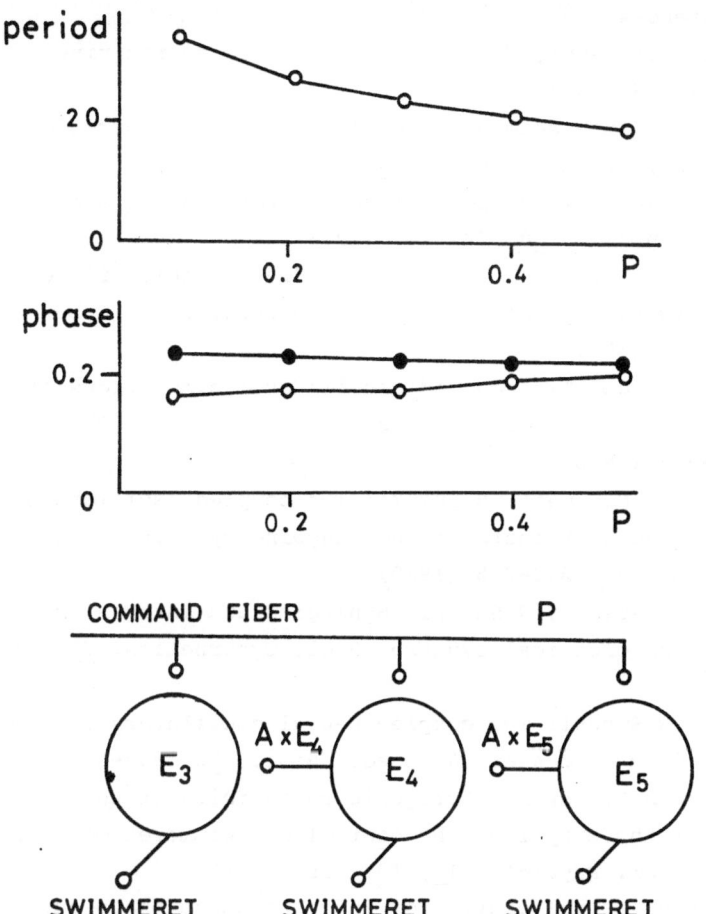

Fig.8
The model of the crayfish swimmeret
upper: the relationship between the period and
       stimulation P
middle: phase difference and P
            - ● -  3rd and 4th
            - ○ -  4th and 5th
threshold of response function  θ
            3rd ganglion: 2.83, 4th : 2.825
            5th: 2.8

References

(1) R.Suzuki, I.Katsuno, K.Matano: Dynamics of neuron ring, Kybernetik, 8, 39-45 (1971)

(2) P.S.G.Stein: Intersegmental coordination of swimmeret motorneuron in crayfish, J.Neurophysiol.,34, 310-318 (1971)

(3) M.Kawato, R.Suzuki: Biological oscillators can be stopped, Biol. Cybernetics, 30, 241-248 (1978)

(4) J.Yamanishi, M.Kawato, R.Suzuki: Studies on human finger tapping neural networks by phase transition curves, Biol. Cybernetics, 33, 199-208 (1979)

(5) S.Daan, C.Berde: Two coupled oscillators: simulations of the circadian pacemaker in mammalian activity rhythms, J.Theor. Biology, 70, 297-313 (1978)

(6) J.Yamanishi, M.Kawato, R.Suzuki: Two coupled oscillators as a model for the coordinated finger tapping by both hands, Biol. Cybernetics, 37, 219-225 (1980)

(7) M.Kawato, M.Sokabe, R.Suzuki: Synergism and antagonism of neurons caused by an elctrical synapse, Biol. Cybernetics, 34, 81-89 (1979)

(8) M.Kawato, R.Suzuki: Two coupled neural oscillators as a model of the circadian pacemaker, J. Theor. Biol., 86, 547-575 (1930)

(9) K.Ikeda, C.A.G.Wiersma: Autogenic rhythmicity in the abdominal ganglia of the crayfish: the control of swimmeret movements. Comp. Biochem. Physiol., 12, 107-115 (1964)

(10) P.S.G.Stein: Neural control of interappendage phase during locomotion, Amer. Zool., 14, 1003-1036 (1974)

(11) D.Kennedy, K.Takeda: Reflex control of abdominal flexor muscles in the crayfish, J.Exp. Biol., 43, 211-227 (1965)

(12) H.R.Wilson, J.D.Cowan: Excitatory and inhibitory interactions in localized populations of model neurons. Biophysical J., 12, 1-24 (1972)

(13) Phase-locked phenomena of swimmeret were first reported in H.Tatsumi, S.Majima, R.Suzuki: Analysis of phase-locked phenomena in coordinated movements of crayfish swimmeret, Neuroscience Letters, Supplement, 6, S48 (1981)

(14) Detail on model work is in preparation as S.Majima, H.Tatsumi, R.Suzuki: A model of coordinated rhythmical behavior of crayfish swimmeret.

Acknowledgements

This work is supported by grants from the Japanese Ministry of Education, Science and Culture.

# CHAPTER 13

## CHARACTERISTICS OF NEURAL NETWORK WITH UNIFORM STRUCTURE

Shoichi Noguchi          *Tetsuo Araki
Research Institute of Electrical Communication
Tohoku University, SENDAI JAPAN
*Research Institute of Electrical Communication
N.T.T., YOKOSUKA JAPAN

[Introduction]

The aim of this paper is to make clear the organizing capability of the neural network under the general environment. But as the first step, we discuss the characteristics of the neural network with uniform structure. Main problems of this paper are to consider the dynamic behavior, and the mechanism of the generation of the cycle mode of this network.

As the result, number and kind of cycle modes of this network are analysed, and the information processing capability of the neural networks under the uniform structure environment can be estimated.

[Definition of Uniform Neuronic Network]

Neuronic network $\mathcal{N}$ is defined by the following:

$$\mathcal{N} = (C, N, Q, f)$$

where

(1) C is the ordered set of neurons,

(2) N is the neighbourhood function such that $C \to C^k$ and represents the connection among each neurons,

(3) Q is the finite set of the state of the neuron,

(4) $f$ is the local function such that $Q^k \to Q$ and defines the next state of each neuron.

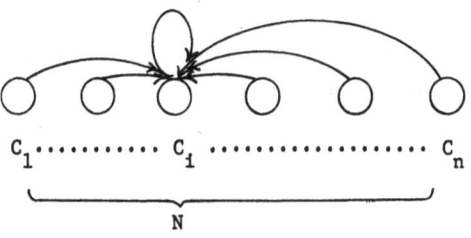

Fig.1

[Definition of Configuration of Neuronic Network]

Let $\mathcal{N}$ be the neuronic network, then the configuration of $\mathcal{N}, \mathcal{C}(\mathcal{N})$ is defined by the mapping $\mathcal{C}$ such that

$$\mathcal{C} : C \to Q.$$

Let C be given by an order set such that $C = (C_1, \cdots\cdots, C_n)$ and $\mathcal{C}(C_i)$

represent a state of neuron corresponding to $C_i$.

Then $\mathcal{C}$ is given by $(\mathcal{C}(C_1), \cdots, \mathcal{C}(C_n))$.

The set of $\mathcal{C}$ on $\mathcal{N}$ is denoted by $\mathcal{C}(\mathcal{N})$.

The neighbourhood function N is given by the ordered set:

$$N(C_i) = (C_{i_1}, \cdots, C_{i_k})$$

which represents the connection to $C_i$ from $C_{i_j}$ (i = 1~k).

The neighbourhood configuration of $C_i$, $\tilde{N}(C_i)$, is also represented by

$$\tilde{N}(\mathcal{C}(C_i)) = (\mathcal{C}(C_{i_1}), \cdots \mathcal{C}(C_{i_k}))$$

For simplicity, we may use $\tilde{N}(C_i)$ instead of $\tilde{N}(\mathcal{C}(C_i))$ in the following.

Let $\mathcal{C}'$ be the next state configuration of $\mathcal{C}$, then $\mathcal{C}'(C_i)$ is given by

$$\mathcal{C}'(C_i) = f(\tilde{N}(C_i)) \qquad\qquad\text{(1)}$$

The fig.1 represents the basic model.


[Definition of Global Function F on $\mathcal{N}$]

It is clear that a local function decides a next state of the corresponding cell, and in the synchronous system, appling the same local function to all cell, we have the next step configuration $\mathcal{C}'$ from the present configuration $\mathcal{C}$.

This mapping is defined as global function F , and

$$F: \mathcal{C}(\mathcal{N}) \to \mathcal{C}(\mathcal{N}) \text{ is defined}$$

as follows:

$$F(\mathcal{C})(C_i) = \mathcal{C}'(C_i) \Leftrightarrow \mathcal{C}'(C_i) = f(\tilde{N}(C_i)) \qquad\qquad\text{(2)}$$

It should be mentioned that $\mathcal{N}$ becomes autonomous network with the transition function F in the synchronous system.

By the above definition, $\mathcal{N}$ is characterized by the neighbourhood function $\mathcal{N}$ and the local function $f$ .

N plays the very important role in $\mathcal{N}$ and there are many kinds of connections generally.

In this paper, we consider the case that each neuron is fully connected and its connection is represented by the group graph.


[Definition of Group Graph]

Let G be a group such that $G = (g_1, g_2, \cdots, g_n)$ and H be its subset $H = (h_1, \cdots, h_k)$, then group graph $G_H$ is given by

$$G_H = (G, E_H) \qquad\qquad\text{(3)}$$

where G is a set of vertices and $E_H$ a set of edges such that

$$(g_i, g_j) \in E_H \Leftrightarrow g_i = g_j h, \; g_i, g_j \in G, \; h \in H.$$

Let $\mathcal{N}$ be represented by a group graph $G_H$, then the position of neuron $C_i$ is given by $g_i$, and the neighbourhood connection $N_H(g_i)$ is given by

$$N_H(g_i) = (g_ih_1, g_ih_2, \cdots\cdots, g_ih_k)$$ ————————(4)

[Isomorphism on Graph]

Let $\alpha$ be a mapping from graph $\Gamma = (V, E)$ to $\Gamma' = (V', E')$, such that

    (1) $\alpha : V \to V'$ injective and surjective

    (2) $(v_i, v_j) \in E \Leftrightarrow (\alpha(v_i), \alpha(v_j)) \in E'$,

then $\alpha$ is defined as a graph isomorphism.

[Isomorphism on Group Graph]

Let $\alpha$ be a graph isomorphism such that $\forall g \in G$, $\alpha N_H(g) = N_H(\alpha(g))$,

then $\alpha$ is defined as a group graph isomorphism.

[Theorem 1]

An automorphism $\alpha$ on a connected group graph $G_H$ is given as follows:

    $\alpha = \rho \, \alpha^*$

where $\alpha^*$ is an automorphism of G such that $\alpha^*(H) = H$, and $\rho$ is a left translation

on G such that $\rho_g g' = gg'$.

By utilizing the concept of the isomorphism, the local functions $f$ on $\mathcal{N}$ are characterized and classified into the standard form, but we neglect about this discussion.

[Characteristics of Global Function on Uniform Neuronic Network]

In the following, we consider the neuronic network represented by group graph $G_G$.
This means that each neuron is connected with all neuron in the network.

Then the network $\mathcal{N}$ is given as follows:

    $\mathcal{N} = (G, N_G, Q, f)$

where   $G = (g_1, g_2, \cdots\cdots, g_n)$,     and     $N_G(g) = (gg_1, gg_2, \cdots\cdots; gg_n)$

The following relationship is clear,

    $\mathcal{C}' = F(\mathcal{C}) = (f(\tilde{N}_G(\mathcal{C}(g_1))), \cdots\cdots, f(\tilde{N}_G(\mathcal{C}(g_n))))$

[Definition 1]

Let $\mathcal{C}_1$ an $\mathcal{C}_2$ be the configurations on $\mathcal{N}$ such that there exist the following

relation

    $\tilde{N}_G(\mathcal{C}_1(g)) = \mathcal{C}_2$,     $g \in G$

then $\mathcal{C}_1$ and $\mathcal{C}_2$ is defined as G conjucate.

By the definition, the following relationship is clear;

    $N_G(\mathcal{C}_1(g_i)) = \mathcal{C}_2 \Rightarrow N_G(\mathcal{C}_1(g_ig_j)) = N_G(\mathcal{C}_2(g_j))$

[Proposition 2]

On $\mathcal{C}(\mathcal{N})$, G conjucate relation is an equivalent relation.

By this proposition $\mathcal{C}(\mathcal{N})$ is classified by G conjugate relation and an equivalent class $[\mathcal{C}]$ corresponding to $\mathcal{C}$ is given as follows;

$$[\mathcal{C}] = \{\mathcal{C}' \in \mathcal{C}(\mathcal{N}) \mid g \in G, \ \tilde{N}_G(\mathcal{C}(g)) = \mathcal{C}'\}$$

[Proposition 3]

Let $H_{\mathcal{C}} = \{ g \mid \tilde{N}_G(\mathcal{C}(g)) = \mathcal{C}\}$,

then $H_{\mathcal{C}}$ is a subgroup of G and defined as an invariant group to $\mathcal{C}$.

[Definition 2]

Let H be a subgroup of G, then the number of left coset on G by H is defined as an index of H on G and denoted by [G:H].

[Proposition 4]

Let $H_{\mathcal{C}}$ be the invariant group to $\mathcal{C}$,

then $|[\mathcal{C}]| = [G:H_{\mathcal{C}}]$

where $||$ mean the cardinality of the set, and $|H_{\mathcal{C}}| \mid |G|$.

[Theorem 5]

$\quad F(\mathcal{C}) = \mathcal{C}', \ \mathcal{C} \in \mathcal{C}(\mathcal{N})$ implies $f(\tilde{N}_G \mathcal{C}(g)) = \mathcal{C}'(g)$,

that is $F(\tilde{N}_G(\mathcal{C}(g_1)) = \tilde{N}_G(\mathcal{C}'(g))$, and the following relations hold:

$\quad F(\mathcal{C}) = \mathcal{C}' \Rightarrow H_{\mathcal{C}} \subseteq H_{\mathcal{C}'}$

$\quad H_{\mathcal{C}} \subseteq H_{\mathcal{C}'} \Rightarrow \exists F', \ F'(\mathcal{C}) = \mathcal{C}'$.

[Corollary 6]

Let $[\mathcal{C}]$, $[\mathcal{C}']$ be the equivalent classes such that

$\quad F(\mathcal{C}) = \mathcal{C}'$,

then $\forall \mathcal{C}_i \in [\mathcal{C}], \ \exists \mathcal{C}'_i \in [\mathcal{C}'], \ H_{\mathcal{C}_i} \leq H_{\mathcal{C}'_i}$

[Proposition 7]

Let $\mathcal{C}_j \in [\mathcal{C}], \ \mathcal{C}_j = N_G(\mathcal{C}(g_j))$ and $H_{\mathcal{C}}$ be the invariant subgroup of $\mathcal{C}$, then

$\quad H_{\mathcal{C}_j} = g_j \ H_{\mathcal{C}} \ g_j^{-1}$.

By this proposition, we have

[Corollary 8]

Let G be commutative, then there is an unique invariant group to $[\mathcal{C}]$.

Still, we have

[Corollary 9]

Let G be cyclic group, then we have the following:

(1) $|[\mathcal{C}]| \mid |[\mathcal{C}']| \Leftrightarrow H_{\mathcal{C}} \in H_{\mathcal{C}'}$

(2) $|[\mathcal{C}]| = |[\mathcal{C}']| \Leftrightarrow H_{\mathcal{C}} = H_{\mathcal{C}'}$

Let F be the global function on $\mathcal{C}(\mathcal{N})$, then, we have the following proposition:

[Proposition 10]

Let $F(\mathcal{C}) = \mathcal{C}'$, $\mathcal{C} \in \mathcal{C}(\mathcal{N})$

then F is injective and surjective $\Leftrightarrow H_{\mathcal{C}} = H_{\mathcal{C}'}$ for all $\mathcal{C}, \mathcal{C}' \in \mathcal{C}(\mathcal{N})$

[Generation of Surjective and Injective Global Function F by $f$ ]

In the following, we discuss the algorithm to obtain the surjective and injective global function considering the properties of local function $f$ and show that this algorithm is the only way to generate the desired F.

First, we define the followings:

Let $G = (g_1, g_2, \cdots\cdots, g_n)$

and $G = Hg_{i_1} + Hg_{i_2} + \cdots\cdots + Hg_{i_r}$. $(g_{i_1} = e)$

where H is a subgroup of G.

We define, $[\mathcal{C}]_H : \{\mathcal{C}' \mid \mathcal{C}' \in [\mathcal{C}], \mathcal{C} \text{ is invariant to } H \}$.

As defined before,

$$[\mathcal{C}]_H = \{\mathcal{C}, \tilde{N}_G \mathcal{C}(g_{i_2}), \cdots\cdots, \tilde{N}_G \mathcal{C}(g_{i_r}) \}.$$

We define also

$$x^H[\mathcal{C}] = (\mathcal{C}(g_{i_1}), \mathcal{C}(g_{i_2}), \cdots\cdots, \mathcal{C}(g_{i_r}))$$

where $\mathcal{C}$ is invariant to H and

$$x^H[\mathcal{C}](g) = (\mathcal{C}(gg_{i_1}), \mathcal{C}(gg_{i_2}), \cdots\cdots, \mathcal{C}(gg_{i_r})),$$

$$*x^H[\mathcal{C}] = \bigcup_{g \in G} x^H[\mathcal{C}](g).$$

It is clear that $| [\mathcal{C}]_H | = [G:H] = r$

It is sure that there are many equivalent classes $[\mathcal{C}]$ which is invariant to H in general. These classes are denoted by

$$[\mathcal{C}_1^H], [\mathcal{C}_2^H], \cdots\cdots, [\mathcal{C}_i^H].$$

Then, let

$$Y(H) = \{ [\mathcal{C}_1^H], [\mathcal{C}_2^H], \cdots\cdots, [\mathcal{C}_i^H] \}.$$

Y(H) is defined for all subgroup of G.

Let M(H) be the number of equivalent classes in $\mathcal{C}(\mathcal{N})$ corresponding to H, and P(G) the number of all equivalent classes in $\mathcal{C}(\mathcal{N})$, then we have

$$P(G) = M(H_1) + M(H_2) \cdots\cdots + M(H_k) \qquad\qquad\text{————(5)}$$

where $\{ H_i \}$ is set of all subgroup of G.

Let each neuron have P states and $Z_G(x_1, x_2, \cdots\cdots, x_n)$ be the cycle index polynominal of G, then we have following relation.

$$P(G) = Z_G(p, p, \cdots\cdots, p).\tag{6}$$

In the following, we consider some special case.

Let G be a cyclic group with degree $n$ , then, there is one subgroup with degree r such that $r \mid n$ .

Next, let $\theta(r)$ be a number of circular permutation of length r, then we have the following equation:

$$\sum_{r \mid n} r \cdot \theta(r) = p^n.\tag{7}$$

By using Möbius formula,

$$n\theta(n) \big| = \sum_{r \mid n} \mu(r) p^{n/r}\tag{8}$$

where $\mu(x)$ is möbius function.

It is clear that

$$\theta(r) = \mid Y(H_r) \mid\tag{9}$$

where $H_r$ is a subgroup of G with degree r.

We have the following:

If n is prime, then

$$n \cdot \theta(n) = p^n - p\tag{10}$$

Next, we define several functions in the following.

The first is permutation function $\mathcal{P}$ on $Y(H)$ such that

$$\mathcal{P}_H \{ (i_1^{(1)} \cdots i_{t_1}^{(1)}), (i_1^{(2)} \cdots i_{t_2}^{(2)}) \cdots \cdots , (i_1^{(s)} \cdots i_{t_s}^{(s)})$$

where $\qquad (i_1^{(1)} \cdots i_{t_1}^{(1)}) \cdots (i_1^{(s)} \cdots i_{t_s}^{(s)}) \qquad$ means the permutation

on $\qquad \{ 1, 2, \cdots\cdots, i \}$

The second is the selecting function $\mathcal{S}_H$

on $\quad [\mathcal{C}_k^H] \qquad (k = 1, 2, \cdots\cdots, i) \qquad$ such that

$$\mathcal{S}_H ([\mathcal{C}_k^H]) = \tilde{\mathcal{C}}, \ \tilde{\mathcal{C}} \in [\mathcal{C}_k^H]$$

Each $[\mathcal{C}_k]$ contain just r elements, so $\mathcal{S}_H$ means to select one representative among r elements of $[\mathcal{C}]$.

For selected $\tilde{\mathcal{C}}$, we have the following set such that

$$(\tilde{\mathcal{C}}(g_1), \mathcal{C}(g_2) \cdots\cdots \tilde{\mathcal{C}}(g_n), (\tilde{\mathcal{C}}(g_{i_2} g_1), \cdots\cdots \tilde{\mathcal{C}}(g_{i_2} g_n))$$
$$\cdots\cdots (\tilde{\mathcal{C}}(g_{i_r} g_1), \cdots\cdots \tilde{\mathcal{C}}(g_{i_r} g_n)).$$

This set is denoted by $S_H^*([\mathcal{C}_k])$.

We also use $S_H^*$ as the operation on $[\mathcal{C}_k]$ to obtain $S_H^*([\mathcal{C}_k])$ from $\tilde{\mathcal{C}}$.

We defined two functions, $\mathcal{P}_H$ and $\mathcal{S}_H$ and also $S_H^*$.

$\mathcal{P}_H$ gives the transition among the equivalent classes by the permutation, and $\mathcal{S}_H$ gives to select one representative from each equivqlent class.

It is proved that by appling $\mathcal{P}_H$ and $S_H$ sequentially for all $H \in G$, we can decide the complete transition on $C(N)$.

We call this operation as $\mathscr{A}_H \mathcal{P}_H$ operation.


[Decision of Local Function]

Let $\mathscr{S} = (s_1, s_2, \cdots\cdots, s_n) \to \mathscr{S}' = (s_1', s_2', \cdots\cdots, s_n')$

be the transition induceed by $\mathscr{A}_H \mathcal{P}_H$ operation.

Then, let $S_H^*([\mathscr{S}])$ be $(\mathscr{S}, \mathscr{S}_{i_2} \cdots\cdots, \mathscr{S}_{i_r})$

and set $(\mathscr{S}'(g_{i_1}), \mathscr{S}'(g_{i_2}), \cdots\cdots, \mathscr{S}'(g_{i_r}))$.

We define the local function $f$ by the following way.

$$\left.\begin{aligned}
f(\mathscr{S}) &= f(s_1, s_2, \cdots\cdots, s_n) = \mathscr{S}'(g_{i_1}) \\
f(\mathscr{S}_{i_2}) & \qquad\qquad\qquad\quad = \mathscr{S}'(g_{i_2}) \\
f(\mathscr{S}_{i_r}) & \qquad\qquad\qquad\quad = \mathscr{S}'(g_{i_r}) \,.
\end{aligned}\right\}$$

By the above procedure, the local function $f$ is decided on the domain $Y(H)$.
Appling $\mathscr{A}_H \mathcal{P}_H$ operation for all $H \in G$, $f$ is completly decided on $C(N)$.

Then we have the following theorem.


[Theorem 11]

Local function $f$ decided by the above algorithm is well defined.
Still we have the following theorem.


[Theorem 12]

The global function F induceed by $f$ is surjective and injective iff $f$ is decided by
the above procedure.


[Number of Surjective and Injective Function]

We have defined two function $\mathcal{P}_H$ and $\mathscr{A}_H$ for the fixed subgroup H of G.
Two functions $\mathcal{P}_H$ and $\mathscr{A}_H$ are independent and those are also independent for the
different H.

So, the number of surjective and injective function is obtained straight foward by
calculating the number of permutations on $Y(H)$ and of selections on $[C]$.

Let $| Y(H) | = i$, then the permutation is obtained by deciding the cycles of length
$\ell_k$ such that $\sum_k n_k \cdot \ell_k = i$ where $n_k$ is the number of cycles of length $\ell_k$.

Let $\pi(n_1, \ell_1 ; n_2, \ell_2 : \cdots\cdots, n_p, \ell_p)$ be number of cycles with these parameters
on $\{i\}$ then, we have easily the following proposition.


[Proposition 13]

The number of permutation $\Pi(n_1, \ell_1, : n_2, \ell_2 : \cdots\cdots : n_{p_1}, \ell_R)$ on $\{i\}$ with $n_i$ cycles of length $\ell_i$ is obtained as follows :

$$\Pi(n_1, \ell_1 : n_2, \ell_2 : \cdots\cdots : n_p, \ell_p) = {}_iC_{n_1\ell_1} \cdots\cdots {}_{n_p\ell_p}C_{n_p\ell_p} \cdot$$

$$_{n_1\ell_1}C_{\ell_1} \cdots\cdots {}_{\ell_1}C_{\ell_1} \cdots\cdots {}_{n_p\ell_p}C_{\ell_p} \cdots\cdots {}_{\ell_p}C_{\ell_p} \cdot (\ell_1 - 1)! \cdots (\ell_p - 1)!$$

[The number of Selecting Function]

Let $|Y(H)| = i$, and $|[\mathcal{C}]| = r$, then the number of selecting function $S(H)$ is easily found to be

$$S(H) = |r|^i.$$

Then, by the above procedure, we obtained the following theorem ;

[Theorem 14]

The number of surjective and injective function $n(\mathcal{N})$ on $\mathcal{C}(\mathcal{N})$ is given as follows:

$$n(\mathcal{N}) = \sum_i \sum_H S(H) \cdot \Pi_H(n_1, \ell_1 : \cdots\cdots : n_p, \ell_p)$$

where the first $\Sigma$ means to take all permutation on $|Y(H)|$ , and the second $\Sigma$ to take all subgroup of G.

[The Cycle Mode Generated on $\mathcal{N}$ by F]

In this section , we show the behaviour of uniformly structured neuronic network under the fixed connection environment.

The discussions we have explained before gives us the mechanism of the dynamical behaviour of the neuronic network.

The cycle mode of the network is completly decided by giving $\mathscr{A}_H \mathcal{P}_H$ operation and the initial configuration $\mathcal{C}$.

If $\mathscr{A}_H \mathcal{P}_H$ operation gives $(n_1^{(H)}, \ell_1^{(H)}, : \cdots\cdots :, n_p^{(H)}, \ell_p^{(H)})$ permutations, the network has $n_i^{(H)}$ cycles of length $\ell_i^{(H)}$ and behaviour of network is decided by the initial configuration.

Next, we consider about the cycle length.

As the summary, we conclude that the main factors to decide the behaviour of the network are two operations, those are $\mathcal{P}_H$ and $\mathscr{A}_H$.

$\mathcal{P}_H$ decides the transition among the equivalent classes and let its cycle length be equal to $\ell$ such that $[\mathcal{C}_1] \rightarrow [\mathcal{C}_2] \rightarrow \cdots\cdots \rightarrow [\mathcal{C}_\ell] \rightarrow [\mathcal{C}_1]$. $\mathscr{A}_H$ decides the start point $\mathcal{C}_{11}$ in $[\mathcal{C}_1]$ and the next transition point $\mathcal{C}_{21}$ in $[\mathcal{C}_2]$ and so on in the above transition.

After one cycle of transition on equivalent class $[\mathcal{C}_1] \rightarrow \cdots\cdots \rightarrow [\mathcal{C}_1], \mathcal{C}_{11} \in [\mathcal{C}_1]$

changes to $\mathcal{C}_{12} \in [\mathcal{C}_1]$ and after $r$ time cycles, it returns back again to $\mathcal{C}_{11}$.

Then, the total length of the transition cycle becomes $r \cdot \ell$.

The mechanism of transition is well explained in fig. 2.

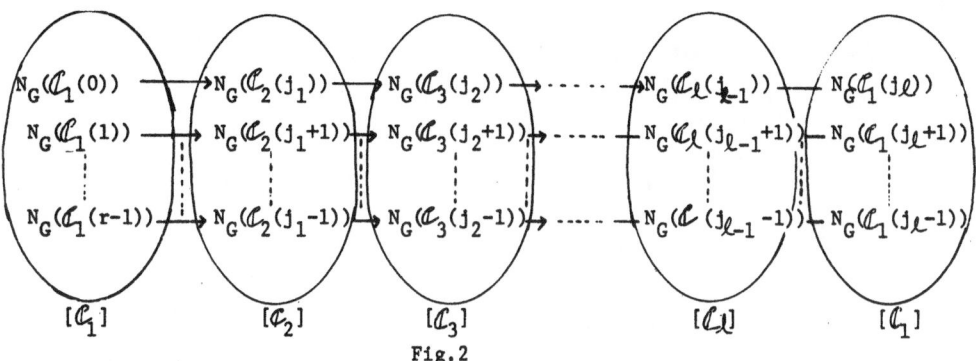

Fig.2

In order to consider in more detail, we analyse the case when G is a cyclic group. Then, we have the following theorem.

[Theorem 15]

Let G be a cyclic group, and its subset $H_1$, $\cdots\cdots$, $H_k$ with the order of $r_1$, $r_2$, $\cdots$ $\cdots$, $r_k$ respectively such that $1 = r_1 < r_2$, $\cdots\cdots < r_k = n$.

Then realizable cycle lengthes on $\mathcal{L}(\mathcal{N})$ are 1, 2, $\cdots\cdots$, $r_1|Y(H_1)|$, $r_2$, $2r_2$, $\cdots\cdots$, $r_2|Y(H_2)|$, $\cdots\cdots$, $r_k$, $2r_k$, $\cdots\cdots$, $r_k|Y(H_k)| = n|Y(G)|$.

By using this theorem and by eq.(10) we have,

[Theorem 16]

Let n, order of G be prime, then the maximul cycle length of this network is $p^n - p$. The results of a cyclic group can be extended to the general finite group without any difficulty.

As an example, we give one cycle mode generated by permutation group G of order such as

$$G = \left\{ (a_1, a_2)\ a_1^4 = a_2^2 = e \right\}.$$

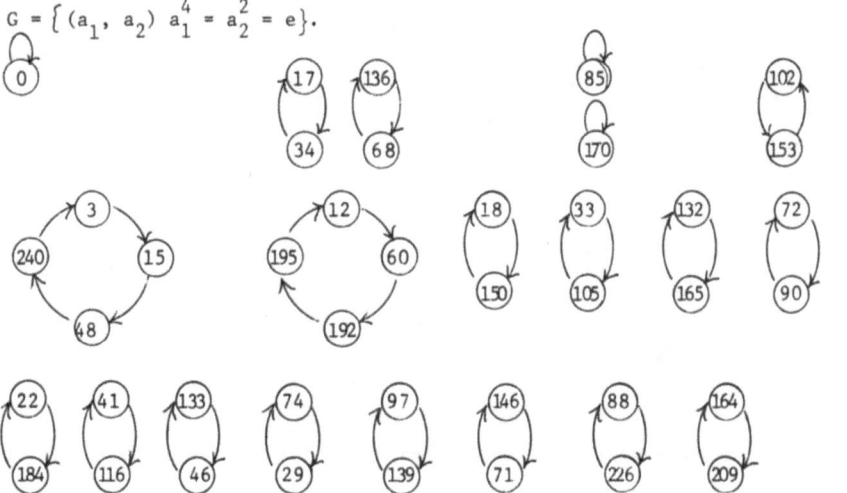

200

Fig.3

In this figure, configuration is represented by binary decimal number (half of all cycle modes are given).

[Conclusion]

We give the fundamental consideration about the mechanism of the generation of the cycle mode of the uniformly structured network, and show the two fundamental operation $\mathcal{P}_H$ and $\mathcal{S}_H$.

The organization of the network is given by those two operations, however, we don't discuss what is the fundamental keys for organizing of $\mathcal{P}_H$ and $\mathcal{S}_H$ in the neural network. These are very important problems in the future, but there might be several approaches from neurophysical view point.

[Reference]

(1) S.Amoroso and Y.Wpatt "Decision procedure for surjective and injectivity of parallel maps for tesellation structure" J.C.S.S., 6 (1972)

(2) H.Yamada and S.Amoroso "Structure and behavioral equivalences of tessellation automata" Information and control, 18 (1971)

(3) M.A.Harrison "Introduction to switching automata theory" Mc-Graw=Hill (1965)

(4) T.Araki, M.Harao, S.Noguchi "Algebraic properties of cellular automata" Proc. of six Hawaii International Conf. on System Science.

(5) T.Araki, M.Harao, S.Noguchi "On the structure preserving isomorphism of the cellular automaton and its classification" Trans. IECE'75/7 vol.58-d,No.7

# CHAPTER 14

SYSTEMS MATCHING AND TOPOGRAPHIC MAPS:
THE BRANCH-ARROW MODEL (BAM)[1]

K.J. OVERTON and M.A. ARBIB
Center for Systems Neuroscience and
Computer and Information Science Department
University of Massachusetts
Amherst, Massachusetts, USA

Abstract

This paper presents BAM (the Branch-Arrow Model), a new model of the development of the retino-tectal topographic mapping to test the limits of system-matching principles in forming such maps. The updating process employed by BAM is distributed in nature and depends upon interactions between branches of retinal fibres, and between the branches and the boundaries of the tectum and of grafts. Results of computer simulation of the model are related to experimental data obtained from tectal and retinal graft and lesion studies and comparisons are also made with other models. Although the model accords well with most of the data, we shall present evidence that indicates the need for an additional mechanism, interaction between the branches and the tectal surface. A model of this kind, the Extended Branch Arrow Model (XBAM), will be described in a sequel.

I.   Introduction

Many experiments study the development of the topographic mapping between the retina and tectum of various lower vertebrates. Goldfish, frog, and toad visual systems have generally been the targets of these studies. In these animals the fibres from each retina project to the contralateral tectum. Early behavioral studies (Sperry 1943, 1944, 1945) (Maturana et al., 1959) showed that the visual fields of these animals would regrow to map in an orderly way after surgical interruption. With the development of electrophysiological recording techniques, investigators have been able to better understand the details of the mapping (Gaze et al. 1963, 1965, 1970, 1974) (Jacobson 1965). Stimuli in the superior section of the visual field project to the medial section of the contralateral tectum while those in the inferior field project to the lateral side. Similarly, stimuli in the

1  The research reported in this paper was supported in part by grant no. NS14971-03 from the National Institutes of Health.

nasal portion of the visual field project to the rostral end and temporal stimuli to the caudal end of the contralateral tectum (see Figure 1). As the body of experimental data has grown, numerous models of the process by which the mapping is formed have been proposed. The models can be divided into two general classes: those subscribing to the idea of a point-to-point chemoaffinity between the retinal and tectal cells and those using the idea of systems matching.

Sperry (Sperry 1944, 1945, 1963) first proposed the idea of chemoaffinity between the layers of cells. Under this hypothesis, each retinal cell is uniquely labeled according to its position on the retina. The tectal surface is considered to be labeled in a similar manner and organization of the map is the result of each retinal cell axon seeking the point on the tectum which matches its own retinal label. The Marker Induction model of Willshaw and von der Malsburg (Willshaw 1979) may be viewed as a sophisticated development of this idea in that the tectal "addresses" are not prespecified, but rather develop as a result of the interactions between the tectum and retinal fibres.

In system matching models (Gaze 1972), the information available to the retinal fibres is considerably less specific. Retinal fibres do not seek a particular point on the tectum, but rather seek a neighbourhood where the interactions with the surrounding fibres match the activity on the retina. The Arrow Model proposed by Hope et al. (1976) may be placed in this class.

While the Marker Induction and Arrow models employ different underlying assumptions as to the amount and type of information required by the organization process, they both explain many aspects of the experimental data. The model presented in this paper incorporates new modelling ideas to explore the limits of systems-matching mechanisms. The sequel (Overton and Arbib, to appear) extends the present model by use of tectal markers to produce a hybrid model XBAM (the Extended Branch-Arrow Model) which explains an ever wider body of experiments.


II.  System Matching

In this section, we discuss the system matching idea as represented by the Arrow Model in more detail. The Arrow Model of Hope, Hammond, and Gaze (1976), uses the relative spatial positions of the points of origin of the retinal fibres and their relative termination positions on the tectum to determine the "sorting out" of retinal fibres at the tectal surface. The use of this information in a distributed, iterative process is sufficient to account for a majority of the physiological data without recourse to any "absolute addressing" of the tectum in terms of retinal coordinates.

In the Arrow Model, the tectum is modelled as a discrete grid with retinal fibres allowed to terminate only at the intersections of the grid lines, called

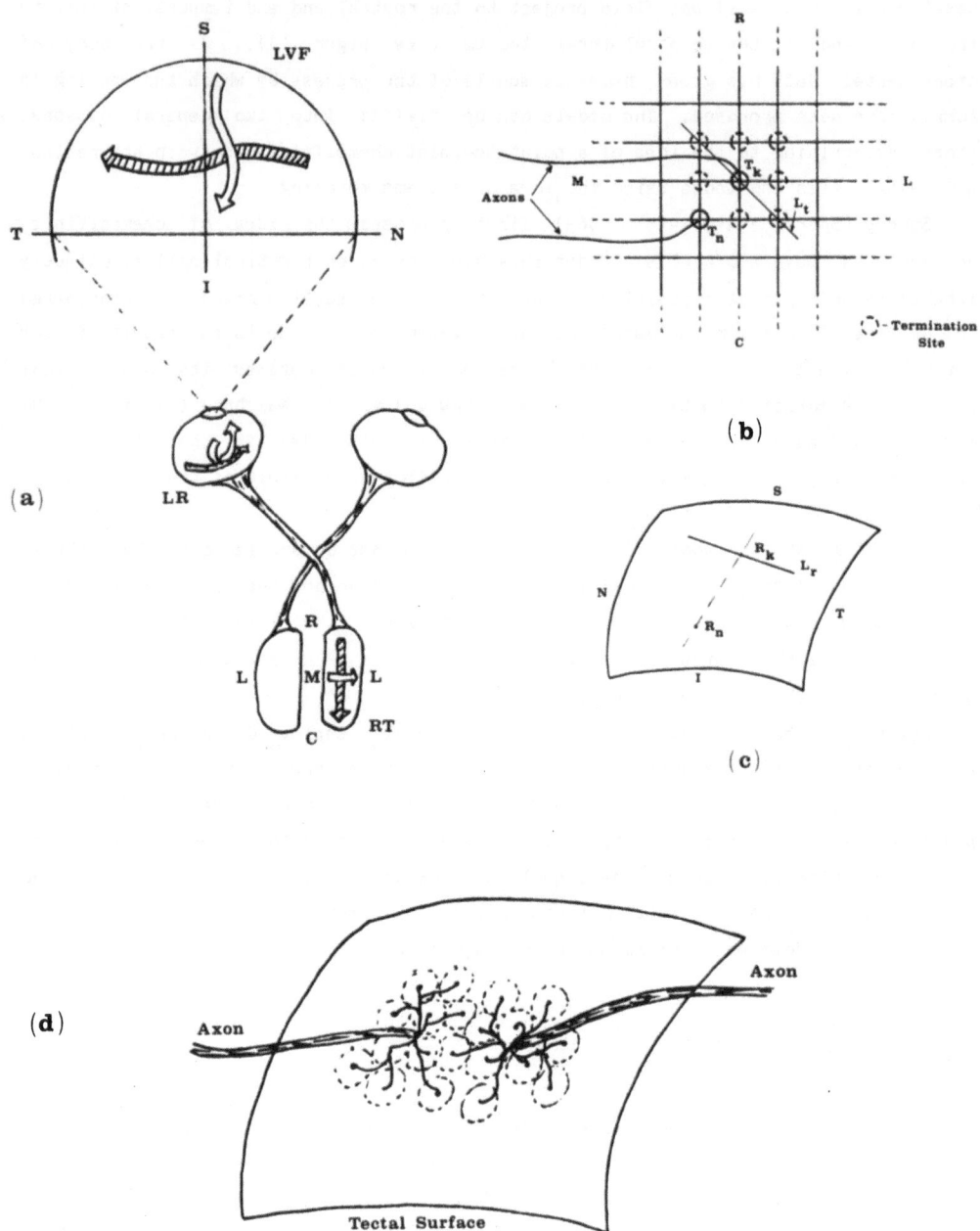

Figure 1. (a) Schematic of the frog visual system. Fibres from the retina project
onto the contralateral tectum. Tectum: R-Rostral, C-Caudal, M-Medial, L-Lateral.
Visual Field: N-Nasal, T-Temporal, S-Superior, I-Inferior. (b) The grid configuration used in the Arrow Model. Each retinal fibre is constrained to terminate at a
single vertex of the discrete tectal grid. (c) Retinal locations $R_n$ and $R_k$ of the
somas of fibres terminating at tectal sites $T_n$ and $T_k$. (d) In BAM, each retinal
fibre terminates in several branches which may terminate at different loci on the
tectum, which is now modelled as a continuous surface.

tectal sites. Each iteration of the sorting process may assume one of the two forms: switching interaction or random walking.

One iteration of switching interaction applies the following interchange rule to each retinal fibre, $T_k$. One of the eight sites immediately adjacent to the present tectal site of $T_k$ (see Figure 1b). is chosen at random until one containing another fibre, $T_n$, is found. It should be noted that the boundaries of the tectum are only implicitly considered in that a termination site on a boundary has only five neighbours. The retinal locations, $R_k$ and $R_n$, of the somas from which fibres $T_k$ and $T_n$, respectively, emanate are compared and the following process is used to determine the new grid location of the termination of the fibres (Figure 1c).

* Construct a line, $L_t$, on the tectum which passes through the site of termination of fibre $T_k$ and is orthogonal to the line connecting the termination sites of both fibres, $T_k$ and $T_n$.

* Construct a line, $L_r$, on the retina which passes through the retinal position, $R_k$, of the soma of fibre $T_k$ and is orthogonal to the line passing through the retinal locations of both fibres, $R_k$ and $R_n$ (Figure 1c).

* With the superior-inferior retinal axis mapped onto the medial-lateral tectal axis and the nasal-temporal retinal axis mapped onto the rostral-caudal tectal axis, switching is determined as follows: if $T_n$ resides on the same side of line $L_t$ as $R_n$ does with respect to line $L_r$, then $T_k$ and $T_n$ retain their original locations, otherwise they switch positions.

This process is applied to all fibres in such a way that no fibre interacts with more than one other fibre during any iteration. Repeated application of this process produces an ordered mapping from one in which the initial positions were randomly assigned.

Using this mechanism alone, there is no way for fibres to move into previously unoccupied termination sites, an ability required to explain the expansion data for a hemiretina projecting onto a complete tectum (see Section IV). In order to circumvent this problem, the fibres are periodically allowed to take random steps. During an iteration of random walking, a site adjacent to each fibre is chosen at random. If the site is empty, the fibre moves to occupy that location. If the site is occupied or, one or more other fibres are trying to move into it, the fibre retains its original location. One iteration of random walking consists of applying this process to each fibre on the tectum. The updated positions form the initial state for the succeeding iteration.

The overall sorting method involves the use of switching interaction and random walking. The majority of the iterations will employ the switching interaction while a few, spaced at predetermined intervals, will employ the random walk. The use of the random walk allows the fibres to disperse to all parts of the tectum while the switching interaction provides a degree of ordering in the mapping.


III.  The Branch-Arrow Model


Our Branch-Arrow Model (BAM) redefines the Arrow Model in several ways.  Recall that in the Arrow Model, retinal fibres must terminate at discrete points on a grid. In BAM, retinal fibres form several branches as they reach the tectal surface which is now modelled as a continuum rather than a grid.  The termination of each branch is surrounded by a circle which represents the area of interaction with other branches (see Figure 1d).  Further, each branch explicitly interacts with the tectal and graft boundaries.  These changes also dictate that the neighbourhood interaction rules be modified.  In our model the neighbourhood interaction process is applied to each branch so that the actual position of a fibre as a whole is determined implicitly by the locations of its branches.  The resultant model, BAM, seems to more closely resemble the physiology of fibre movement.  However, there is a small number of experiments which cannot be accounted for by either the Arrow Model or BAM.  The Extended Branch-Arrow Model, XBAM, adds fibre-surface interaction to the Branch-Arrow Model.  This extended version appears to account for an even larger body of experimental data and will be discussed in (Overton and Arbib, to appear).

The BAM updating process is obtained by averaging three components: the interaction influence, $\vec{I_b}$, the boundary effect, $\vec{E_b}$, and the average influence, $\vec{A_b}$. The interaction influence component, $\vec{I_b}$, is a continuous analogue of the Arrow-Model interaction process, employed at the level of the branches of a fibre combined with a term describing the local interactions between the branches and the boundaries of the tectum and the various grafts.  The average of the physical influence, $\vec{A_b}$, felt by all of the branches of a given fibre is calculated.  The ultimate movement of a particular branch is then determined as the weighted sum of these influences.

$$\vec{M_b} = a_1 \vec{I_b} + a_2 \vec{E_b} + a_3 \vec{A_b} \tag{1}$$

where $a_1$, $a_2$, and $a_3$ are weighting constants, $\vec{I_b}$ and $\vec{E_b}$ are described in equations (3) and (4) below, and the average influence is given by

$$\vec{A_b} = \frac{1}{m} \sum_{k \in F_b} (a_4 \vec{I_k} + a_5 \vec{E_k}) \tag{2}$$

where the sumation ranges over the set $F_b$ of all branches k from the same retinal fibre as b, m is the number of branches in $F_b$, and $a_4$, $a_5$ are weighting constants.

The first term in the physical influence component provides the extension of the Arrow Model.  In the Arrow Model, only one of a fibre's eight immediate neighbours is involved during each iteration of its updating.  As a result, the fibre in question receives no influence from any of the other neighbouring fibres, nor from any of the fibres which do not occupy immediately adjacent sites.  In BAM, the continuous nature of the tectal surface, the fact that each fibre has a set of branches, and the circle of interaction for each branch eliminate these restrictions. Due to the continuous nature of the tectal surface, the updating process is truly a neighbourhood interaction rather than an interchange. During each iteration, a branch interacts with all other branches whose circles of interaction have a non-empty intersection with the circle of the branch in question. to produce the interaction component, $\vec{I_b}$.

To see the shape of this interaction, let $B_b$ be the set of fibre-branches whose interaction circles intersect that of branch b. Let $\vec{U}(b,k)$ be the unit vector in the "interchange direction" for the current position of b and that of k. Then the movement of b induced by its interaction with k is

$$W_d(b,k) \ W_g(b,k) \ \vec{U}(b,k)$$

where the weights $W_d$(due to the distance of separation) and $W_g$(due to interaction across a boundary) are described below in equations (4) and (5). Thus the total interaction component is given by

$$\vec{I_b} = \sum_{k \in B_b} W_d(b,k) \ W_g(b,k) \ \vec{U}(b,k) \tag{3}$$

The direction determination, $\vec{U_b}$, is very similar to the interchange rule used in the Arrow Model. The main difference lies in the fact that if the somas are separated by a distance greater than a specified value, the direction of the influence is chosen at random. We feel that retinal cells may communicate in a meaningful way only if their somas are within a certain distance, e.g. if their receptive fields overlap. Thus if the cells are separated by a great distance, no communication is possible. When the soma of two interacting branches are within the distance allowing meaningful communication, the direction is chosen as in the Arrow Model. If the branch terminations are oriented on the tectum the same relative to some axis system as are the somas on the retina relative to the corresponding axis system, the influence tends to force the branches apart. If, however, the branch positions are reversed compared to the relative soma locations of the retina, the influence tends to force the branches to move past one another, i.e. interchange positions. In the case where both of the interacting branches in question belong to the same fibre, the influence felt by one from the other always tends to force the branches apart thereby attempting the maximize the area of the tectum covered by a fibre. The second component of the physical influence involves the interaction between the branches and the tectal and graft boundaries.

In the Arrow Model, the interchange of two fibres occurs in discrete steps. When it has been determined that two branches are oriented in the reverse of their retinal locations, they simply exchange positions. In BAM, the influence between two branches, $W_d$, is graduated depending upon separation. The weight is linear in nature with a value of 1 when the two branches terminate at the same point and 0

when the branches are separated by a distance of twice the radius of their circles
of interaction:

$$W_d(b,k) = \begin{cases} 1 - \dfrac{d(b,k)}{2r} & \text{if } d(b,k) < 2r \\ 0 & \text{otherwise} \end{cases}$$

(4)

where r is the radius of interaction on the tectum and d(b,k) denotes the distance
between the tectal terminations of fibres b and k.

The weight due to intervening graft boundaries, $W_g$, is intended to model the
discontinuous nature of such edges. Since the edges of grafts are actual surgical
disruptions of the surface, we feel that communication across a graft edge should
be attenuated. This is expressed mathematically by including a mulplicative
constant for each boundary between the two branches, so that two branches separated
by a boundary exhibit less influence on one another than do two branches separated
by a similar distance but with no intervening boundaries.

$$W_g(b,k) = a_g^n$$

(5)

where $a_g$ is the mulplicative constant determining cross-boundary communication
effectiveness, $0 \le a \le 1$ and n is the number of graft boundaries intersecting the
line segment connecting the terminations of branches b and k.

The simple Arrow Model does not include the boundary of the tectum nor the edges
of the grafts as influencing factors. The second term of equation (1) includes this
factor explicitly as $\vec{E}_b$.

$$E_b = \sum_{q \in Q} W_d(b,q)\, W_g(b,q)\, \vec{U}(b,q)$$

(6)

where Q is the index set of all tectal and graft boundaries; $\vec{U}(b,q)$ is the unit
vector along the line perpendicular to boundary q and passing through the
termination of branch b; $W_d$ is the weight due to the distance of separation; and
$W_g$ is the weight due to graft boundaries.

Tectal and graft edges are physical discontinuities in the surface of the
tectum. It should, therefore, be more difficult for an axon to migrate across such
a boundary than to move across an unobstructed surface. The boundaries thus have
influence by restricting the movements of the branches. As in the case of
interacting branches, the magnitude of the influence, $W_d$, is proportional to the
distance from the center of the branch circle to the boundary along a line
perpendicular to the boundary.

$$W_d(b,q) = \begin{cases} 1 - \dfrac{d(b,q)}{r} & \text{if } d(b,q) < r \\ 0 & \text{otherwise} \end{cases}$$

(7)

where r is the radius of interaction on the tectum and d(b,q) denotes the distance
between the termination of branch b on the tectum and boundary q. In addition, due
to the physically discontinuous nature of a boundary, the influence of one branch on

another across a boundary is decreased, via $W_g$. Mathematically, the influence due to boundary interaction felt by branch k is given in (5). The direction of the influence, $\vec{U}$, is always away from the boundary along a line perpendicular to the boundary through the point of termination of the branch in question.

The actual influence, $\vec{M_b}$, used to update the position of a branch b during an iteration is determined as the weighted sum of the physical influences $\vec{I_b}$ and $\vec{E_b}$ felt by the branch and the average $\vec{A_b}$ of the physical influences of the branches from the same fibre, as we saw in equation (1). Since, by definition, the branches of a fibre are connected to one another, we feel that this form of information transfer can take place.

The Branch-Arrow Model incorporates the above defined neighbourhood and boundary interaction mechanisms to produce behavior accounting for essentially all of the experimental data. However, this model cannot account for the translocation experiment results since the branches are supplied with only directional information. The Extended Branch-Arrow Model builds upon the Branch-Arrow Model by the addition of global information. As described in (Overton and Arbib, to appear) this will be incorporated in XBAM as a fourth factor in the physical influence component which describes the interaction of the branches and the tectal surface.


IV. Experimental data compared to simulation behavior for BAM


This section presents some of the typical experimental results obtained during electrophysiological recording after various tectal and retinal lesioning. The experiments performed to date have include studies of the mapping after complete and partial ablations of the tectum as well as studies of mappings to tecta which have had sections surgically excised, rotated, inverted, or translocated and then reimplanted. Experiments involving hemiretina and compound eyes have also been performed as have studies of the initial development of the retinotectal projection. The following is a discussion of the simulation behavior of the BAM in light of these physiological experiments. Results of computer simulation of the BAM are compared with the results of the experiments. The results presented below were obtained through computer simulation of a one-dimensional retina/tectum pair containing 40 fibres, each with 4 branches. A one-dimensional simulation was utilized to reduce the amount of computation required.

Experiment I.

Visual fields of lower vertebrates such as frogs and goldfish map in an ordered and predictable way onto the tectum (Yoon 1973) (Gaze 1974). In the simplest of the physiological experiments, the optic nerve is severed and both the retina and tectum are left intact. Figure 2 illustrates the typical findings from a normal animal with an intact retina projecting to a normal tectum. Once the tectum has been

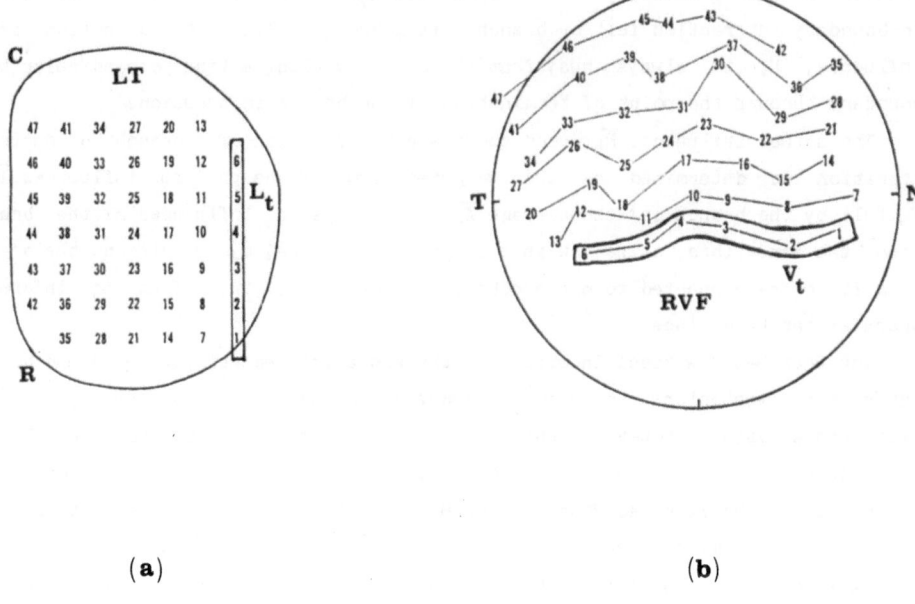

(a)                                                    (b)

Figure 2. The projection of right visual field onto the left tectum in a normal animal (Gaze 1974). (a) Lt: a single line of electrode positions which is analogous to the one-dimensional tectum in BAM (cf. Figure 3 below). (b) Vt: the correspon-ding single row of points in the visual field which excite the tectal position -- analogous to the vertical axis of the BAM display.

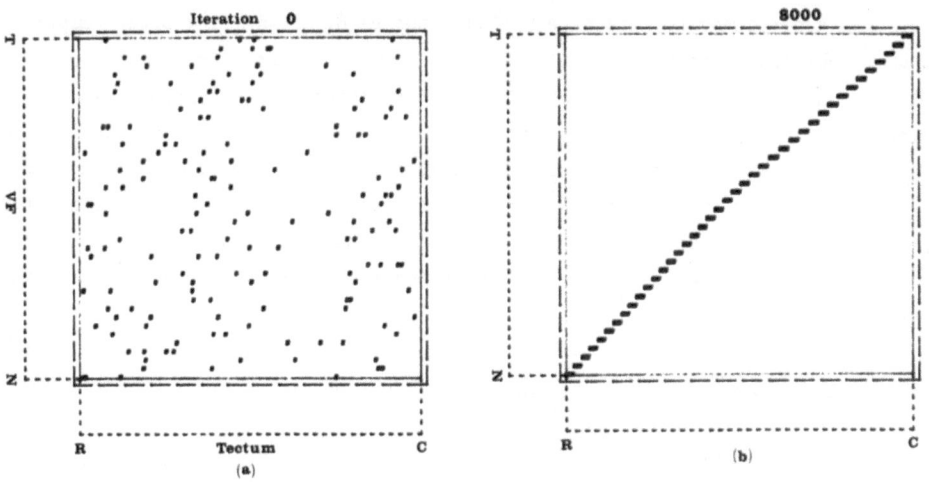

Figure 3. (a) The initial distribution of branch terminations for the BAM simula-tion. 40 fibres with 4 branches/fibre. (b) Organization of the mapping after 8000 iterations of BAM using a retinal information exchange distance equal to the entire extent of the retina.

reinnervated, the mapping between the visual field and the tectal surface is studied. The data are typically obtained by electrophysiologically recording from a point on the tectum while presenting the animal with controlled visual stimuli. The location of the stimulus eliciting the greatest response from the area being recorded is said to indicate the point of emanation on the retina of the fibres being recorded.

The simulation results appear in Figure 3. The one-dimensional tectal surface is represented along the horizontal axis of the display. Similarly, the visual field appears along the vertical axis. Each horizontal row of symbols represents one fibre with each symbol in the row marking the position of the termination of an individual branch of the fibre. Thus for each branch of each fibre, the position in the visual field of the stimulus exciting the fibre, aNd thus the retinal location of the branch's soma is indicated by the height of the row on which the symbol appears. The position of the branch on the tectum is indicated by the horizontal position along the row. Further, the lower end of the visual field display maps to the leftmost end of the tectum and the upper end of the visual field display maps to the rightmost end of the tectal representation. Thus a normal mapping is depicted by a diagonal line of symbols from the lower left corner of the display to the upper right.

The physiological data are inherently two dimensional. To compare the simulation results with these data, equate the vertical axis of the simulation display with V in Figure 2b and the horizontal axis with L in 2a. The left end of the tectal representation corresponds to the lower, rostral, end of the bar in Figure 2a while the lower end of the retinal representation corresponds to the right, nasal, end of the bar in Figure 2b.

Figure 3a shows the initial configuration of the fibres and branches. The branches are randomly distributed across the tectal surface. Figure 3b depicts the simulation results after 8000 iterations of the Branch-Arrow Model. Global organization is apparent at iteration 8000 as indicated by the diagonal nature of the display. During the course of the simulation, another interesting feature is apparent. If the size of the projection field of a fibre is determined by the extent of its branches, one sees that the projection field is initially quite large. As the mapping organizes, the branches of each fibre tend to move together resulting in progressively smaller projection fields. Similar results have recently been observed by Humphery and Beazley (1981) in the frog visual system after optic nerve section.

For this simulation, the radius of effective communication among the somas on the retina was set such that all cells, regardless of their separation, could exchange information. The effect of varying this parameter is the subject of the next experiment. This situation most closely resembles the Arrow Model. The weighting constants $a_1$, $a_2$, $a_3$, $a_4$ and $a_5$ in equations 1 and 2 were assigned the values of 0.25, 0.25, 0.5, 0.35, and 0.65 respectively. These simulation results demonstrate that at least in the simplest case, the Branch-Arrow Model can produce an ordered mapping.

Experiment II.

The purpose of this experiment was to study the effects of varying the maximum distance allowing effective communication between cells on the retina. The initial state and weighting constants remained as in Experiment I. Figure 4a shows the organization resulting after 8000 iterations with the radius set to allow effective communication over roughly two thirds of the retina. That is, if two cells are separated by a distance greater than two thirds of the total retinal expanse, then they cannot meaningfully communicate. Thus branches at opposite ends of the retina interact at random. The resulting configuration shows two organized maps on the tectum. The simulation results depicted in Figure 4b show the state after 8000 iterations with the radius set to roughly one half of the total retinal size. In this case, three maps are produced. Figure 4c shows the map resulting when the distance is reduced to one fifth of the retinal size. Again, several organized pieces are seen. With the radius reduced to one tenth, the configuration in Figure 4d results. The map contains many small pockets of organization yet lacks global organization. The final subfigure, Figure 4e shows the resulting map when the retinal interaction distance is reduced to approximately one twentieth of the total retinal expanse. Though some areas of organization can be seen, no global organization is apparent.

This experiment demonstrates that as the distance on the retina within which an effective exchange of information can take place is reduced, the amount of global organization is also reduced. The Arrow Model is essentially the discrete analog of the Branch-Arrow Model when the retinal interaction distance in the latter is equal to the width of the retina. It seems to us that the physiology of the visual systems being studied would indicate that an assumption of effective communication between any two retinal cells regardless of their separation is questionable. Alternately, the distance should be reduced to some fraction of the total width. The exact amount is unknown. The lack of global organization which results when the distance is reduced is evidence that a local neighbourhood interaction mechanism alone is insufficient to account for the organizational behavior observed in the physiological experiments. This point will be addressed in greater detail in the sequel dealing with XBAM. In the remainder of the experiments with BAM, the distance is assumed to be equal to the width of the retina.

Experiment III.

Work has also been conducted in regard to the compression of the projection onto tectum of which one half has been completely ablated. Udin (1977) and others (Sharma 1977)(Yoon 1976) have studied the form of the retina-tectal projection in

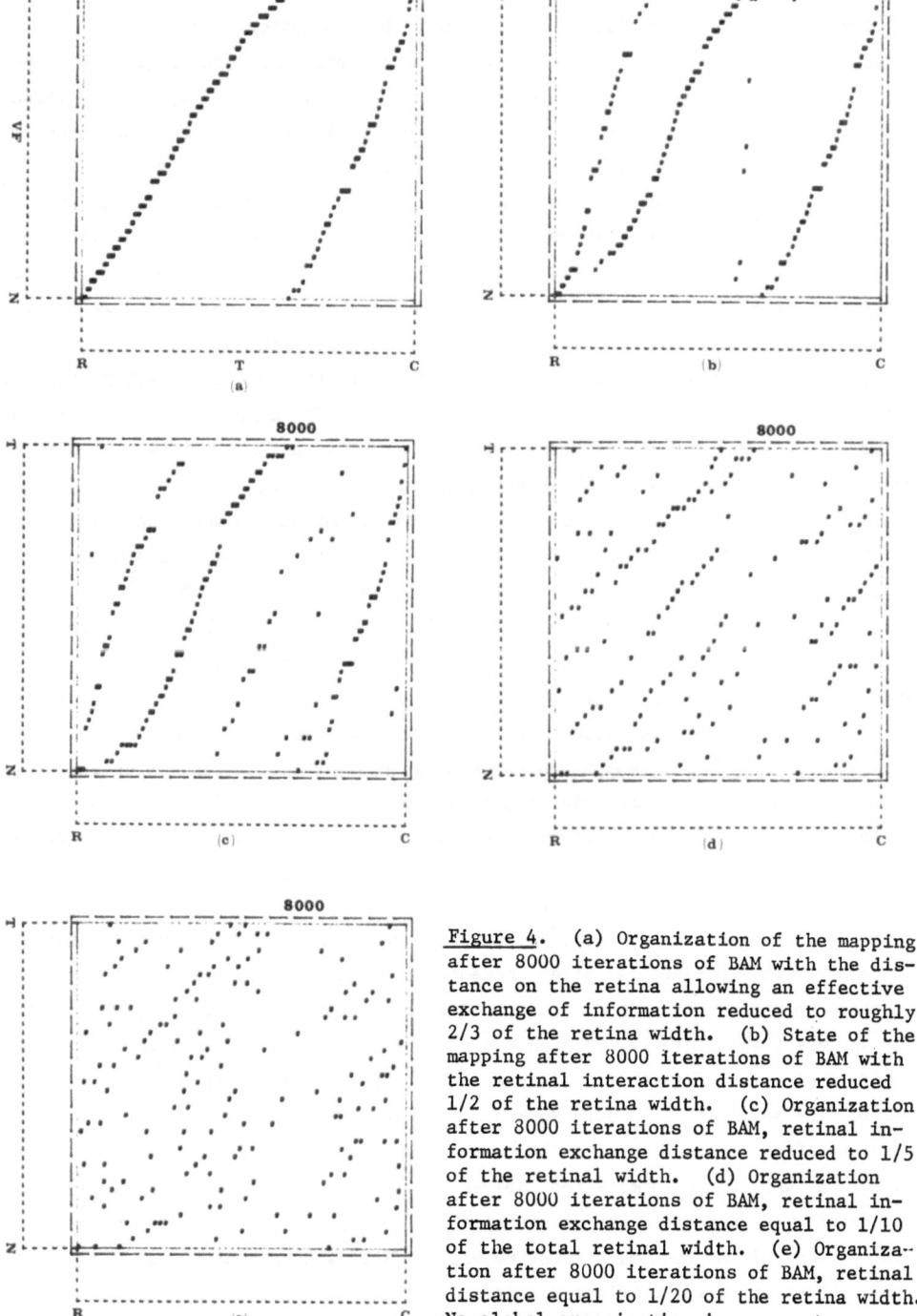

Figure 4. (a) Organization of the mapping after 8000 iterations of BAM with the distance on the retina allowing an effective exchange of information reduced to roughly 2/3 of the retina width. (b) State of the mapping after 8000 iterations of BAM with the retinal interaction distance reduced 1/2 of the retina width. (c) Organization after 8000 iterations of BAM, retinal information exchange distance reduced to 1/5 of the retinal width. (d) Organization after 8000 iterations of BAM, retinal information exchange distance equal to 1/10 of the total retinal width. (e) Organization after 8000 iterations of BAM, retinal distance equal to 1/20 of the retina width. No global organization is apparent.

such a paradigm in the frog visual system. The results obtained from these experiments are generally consistent. An ordered mapping of the entire visual field is found on the intact portion of the tectum. The projection is compressed to fill the available space. The physiological results are displayed in Figure 5, from (Udin 1977). The key features to note here are the facts that the entire visual field is represented along the dimension where only half of the original surface remains and that along the other dimension, all of the space is utilized.

Figure 6 contains the results from the computer simulation. The initial termination locations were randomly distributed over the tectal surface. For this experiment, the surface was reduced to half of its original size. All simulation other parameters were defined as in the previous two experiments. The organization of the projection after 8000 iterations is shown in the figure. This map shows excellent global organization. The branches in the extreme lower left portion of the field have been forced off of the tectal surface. The magnitude of the effect is determined in part by the relative magnitudes of the unit movement vector and the boundary strength parameter in the particular simulation run.

Some investigators have noticed that a mapping identical to the original mapping with the normal tectum appears first, followed by a trend toward a complete, compressed projection (Sharma 1977) (Cook 1974). Horder (1977) found duplicate maps initially which later appeared compressed. He further found that if one third or less of the surface was ablated, the projection moved immediately to a compressed state. It has been posited that this initial mapping is due, in part, to the debris left on the tectum when the optic nerve is sectioned and then degenerates. The fibres which originally mapped to the rostral half of the tectum may be guided by the debris remaining from the prior mapping. Sharma (Sharma 1977) performed an experiment to test this hypothesis and found that when the fibres are forced to reinnervate a tectum previously devoid of fibres, the compressed mapping appeared with no initially uncompressed projection.

Another tectal ablation experiment involves removing 1/4 of the tectal surface and mapping the projection after regeneration. Schmidt and Easter (1978c) have performed experiments in which the medial-caudal quarter of the left tectum of the goldfish has been surgically ablated. A similar experiment designed to investigate the effect of removing part of the tectal surface was performed by Sharma (1972) and involved the ablation of a rostroventral strip on the tectal surface. An organized mapping was found compressed onto the remaining surface. The purpose of these experiments was to determine the degree to which the axes of the tectum are independent with respect to the compression of the mapping. They found that, after reinnervation of the tectum, the entire visual field was represented on the tectum. Further, the mapping was completely ordered and was compressed with respect to both axes. The compression appears to have been uniform across the tectum, that is, the fibre arbors appeared to organize in a fashion which resulted not only in an ordered representation of the visual field but also in a uniform distribution across the

215

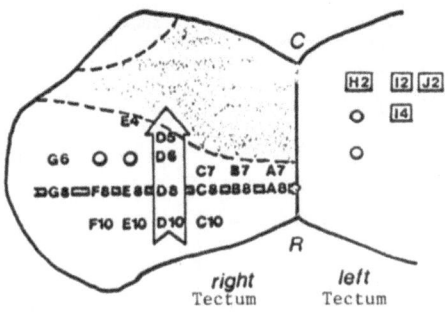

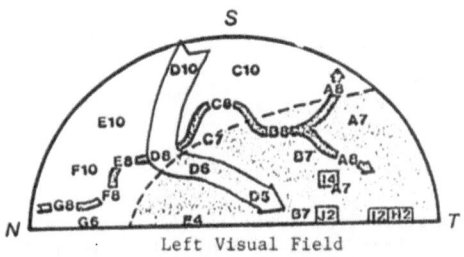

Left Visual Field

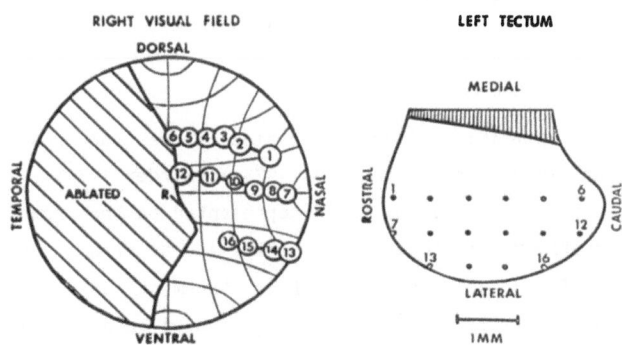

Figure 5. Physiological data
from caudal 1/2 tectal ablation
experiment (Udin 1977). The mapping from the entire visual field
is compressed onto the remaining
tectal surface.

Figure 6. Results after 20000
iterations of the BAM simulation in which the entire
retina is forced to map onto
a tectum which is only 1/2
the size of that in a normal
projection.

Figure 7. Physiological data from an animal in which the nasal half
of the retina has been ablated. The remaining visual field is represented by an ordered projection covering all of the tectal surface.
(Schmidt 1978b)

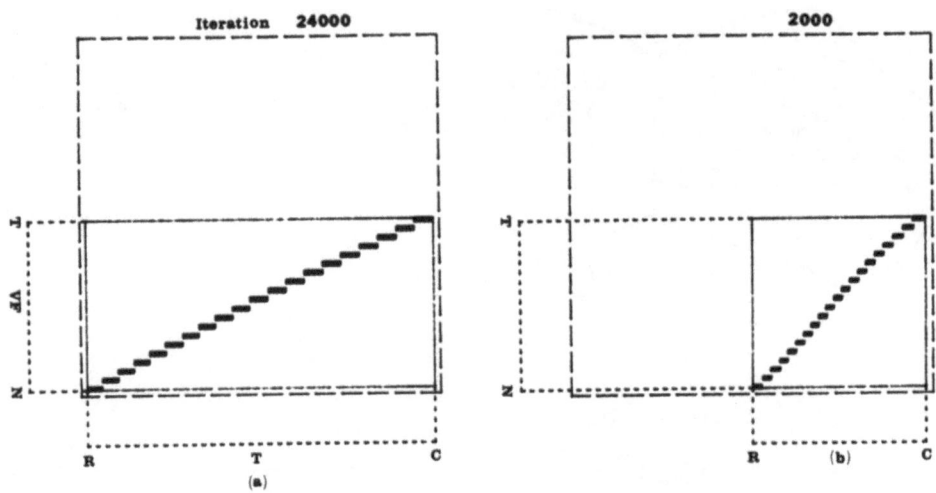

Figure 8. (a) Configuration of the projection from a retina 1/2 of the
normal size onto a full tectum after 24000 iterations of BAM. (b) Simulation results in the case where a hemiretina projects onto the "wrong" half
tectum. That is, this half of the retina normally projects onto the other
half of the tectum. Results after 2000 iterations of BAM.

available tectal surface. While we cannot duplicate this paradigm due to the one
dimensional nature of our simulation, we can predict the behavior of the two
dimension version of the model in such a case.

Since our model tends to minimize the overlap of adjacent projection fields, as
illustrated in the compression results above, and produces a continuous mapping, we
would predict that the projection resulting when 1/4 of the tectal surface is
removed would be uniformly compressed in all directions.

Experiment IV.

Another class of experiments involves studies of the map resulting between a
hemiretina and an intact tectum (Schmidt 1978a) (Horder 1971). In this case, the
projection of the half of the visual field represented on the remaining hemiretina
expands in an orderly manner to completely fill the available space on the tectum.
Typical results are found in Figure 7, from (Schmidt 1978b). Feldman (1975)
conducted an experiment in which one eye was removed before it differentiated and
the fibres from the other eye were directed to the ipsilateral tectum. He found a
normal projection of the entire visual field on the ipsilateral tectum. However,
some of the retinal fibres had managed to innervate the contralateral tectum to
produce an expanded projection of the represented area of the visual field.

The simulation results for this situation are given in Figure 8. The initial
locations for the branches were randomly assigned. The state after 24000 iterations
shows complete organization with half of the normal number of fibres mapping onto

the complete tectal surface.

Figure 8b illustrates the results of a related experiment. Again half of the retinal surface has been ablated. In addition, the half of the tectal surface to which the remaining retinal surface should project has been ablated. Thus the projection from the retina is to the foreign half of the tectum. The preojection shows complete organization. This result is expected since the BAM does not contain information about the specific labelling of the tectal surface, only information describing relative orientation.

Experiment V.

A direct extension to the hemiretina, full tectum experiments involves studying the projection resulting when hemiretinae from different eyes are fused to form a single eye. Once the nerve connections have regenerated, the mapping is determined electrophysiologically. Gaze et al (1963, 1965) and Hunt (1973) have conducted such experiments. Their results are shown in Figure 9.

In the case of Figure 9a, (Hunt 1973), the compound eye was composed of the nasal hemiretinae of two eyes with the division along the superior-inferior axis. The interesting point to note is that the projection from the two halves of the compound eye are superimposed. That is, the projection from each hemiretina expanded in an organized way to cover the entire tectal surface. The projection from one of the hemiretina is rotated 180 degrees. This is the case since one of the hemiretinae had to rotated through 180 degrees in order to create the complete eye. Figure 9b contains a schematic of the results from a compound eye composed of two temporal hemiretinae. The projections are ordered the same since the neither piece required rotation when the eye was constructed.

Figure 10 contains the results from computer simulations of the BAM under similar circumstances. Figure 10a was produced with the orientation of one piece reversed. Figure 10b results from hemiretinae with their orientations retained so that the maps are oriented the same. Recall that the simulation proceeds from a known retina and visual field, and the results are depicted in terms of the organization on the tectum. Also, the simulation is one dimensional.

Notice that in both cases the global organization of the field represented on each hemiretinae is maintained. However, the projection has expanded to cover the entire available tectal surface. This result is expected in the systems matching paradigm if it is assumed that the exchange of information between retinal cells separated by the junction of the two hemiretinae is negligible or random at best.

Experiment VI.

To this point, the experiments have focused on the mappings when there is a mismatch in the size or type (e.g. the compound eye experiments) of retinal versus tectal tissue. Another class of experiments involves the excision of a section of the tectum and its subsequent reimplantation after some form of inversion or

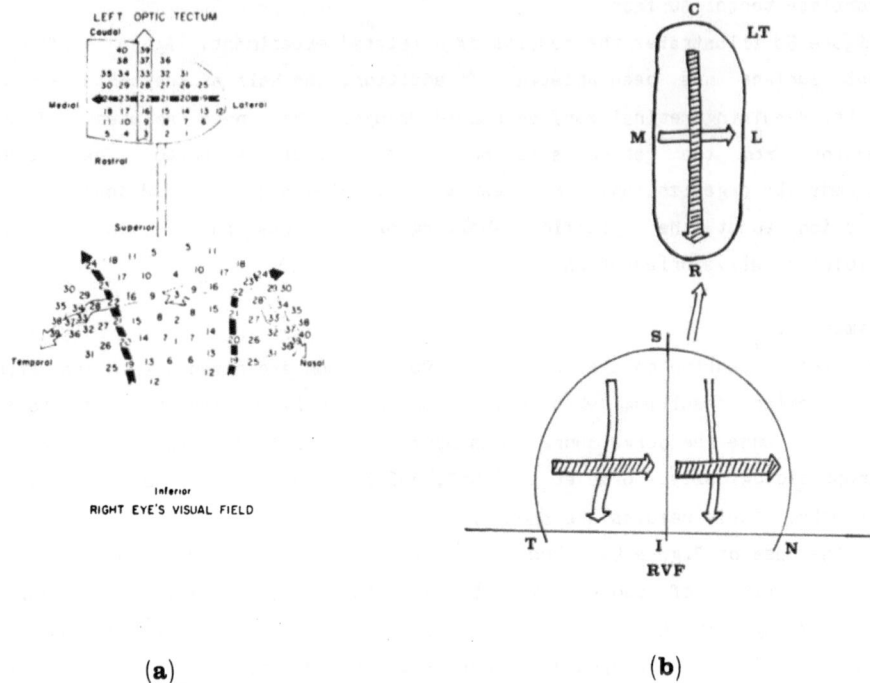

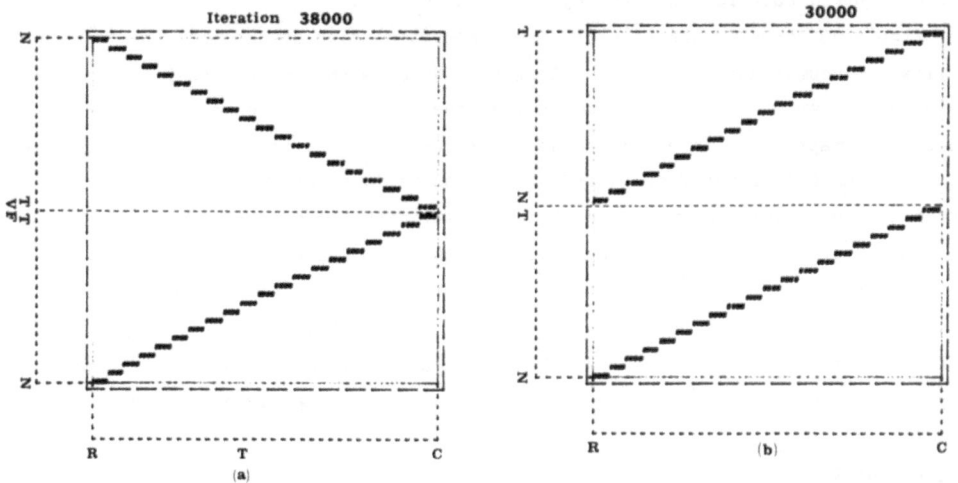

**Figure 9.** (a) Projection of the visual field from a double-nasal compound eye onto a normal tectum (Hunt 1973). Notice that there are two overlapping projections, one from each half of the eye. (b) Schematic of the projection resulting from a compound eye composed of the nasal hemiretina from one eye and the temporal hemiretina from another.

**Figure 10.** (a) Simulation results of the double-nasal compound eye experiment. 38000 iterations. (b) Simulation of the nasal-temporal compound eye. 30000 iterations.

rotation. Yoon (1975) and Sharma and Gaze (1971) have studied the projection following a 90 degree rotation of the graft tissue about the dorsoventral axis. Completely ordered projections are found both within and surrounding the graft. The projection within the graft, however, is found to be rotated in the same manner relative the the tectum as the graft. Figure 11, taken from (Yoon 1975) illustrates these results.

Several neurophysiologists (Yoon 1973, 1975) (Levine 1974) have experimented with animals in which a single tectal graft has been rotated 180 degrees, again about the dorsoventral axis. Once the tectum has been reinnervated the mapping is found to be completely ordered both within and around the graft. The orientation of the map within the graft is again found to be rotated in a manner identical to the graft, see Figure 12, again from (Yoon 1975).

Yoon (1975) has also performed experiments in which a graft has been excised, rotated 180 degrees about the rostocaudal axis, and reimplanted. This rotation causes the graft to retain its normal orientation along the rostocaudal axis while having an inverted orientation along the superior-inferior axis. The resulting projection map is illustrated in Figure 13, (Yoon 1975), and shows that the map retains its normal rostocaudal organization while the mediolateral organization within the graft is reversed. Again, the local orientation of the map follows that of the underlying tectal surface.

Since the Simulation of the DAM is one dimensional, the 90 degree rotation experiments cannot be performed and the two forms of the 180 degree rotations are equivalent. Figure 14 illustrates the state of organization of the branch termination locations after 8000 iterations. The initial configuration was again random.

The areas both within and surrounding the graft are organized. However, it is interesting to note that the entire visual field is represented in each of the three sections. Ideally, only the central portion of the visual field should be represented within the graft. This behavior is due to the interaction mechanism employed by the model and is an inherent difficulty.

Consider a branch located at the left most edge of the graft. The influence felt from the branches to its right tends to force the branch to move to the left while the opposite is true for the branches located immediately to the left, off the graft. Thus the branch is held at the edge of the graft. With this interplay at the graft boundaries, it is impossible for a branch to move across the graft to its proper side.


Experiment VII.

The previous experiment illustrated that the simple BAM cannot produce behavior exactly as that found in graft rotation experiments. This experiment emphasizes this point more dramatically.

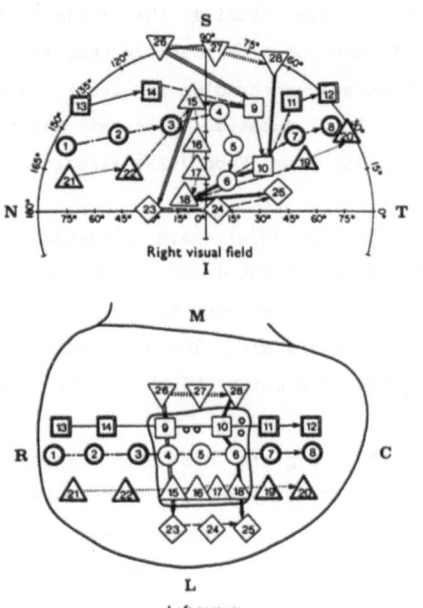

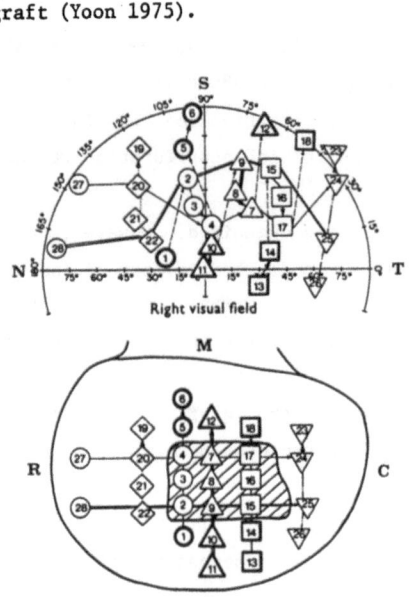

**Figure 11.** Mapping recorded after section of the optic nerve and 90 degree counterclockwise rotation of a graft (Yoon 1975).

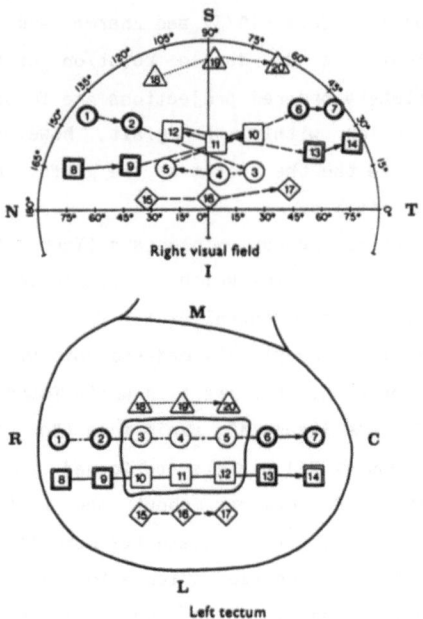

**Figure 12.** Visual field projection organization after optic nerve section and 180 degree rotation about the dorsoventral axis of a graft (Yoon 1975).

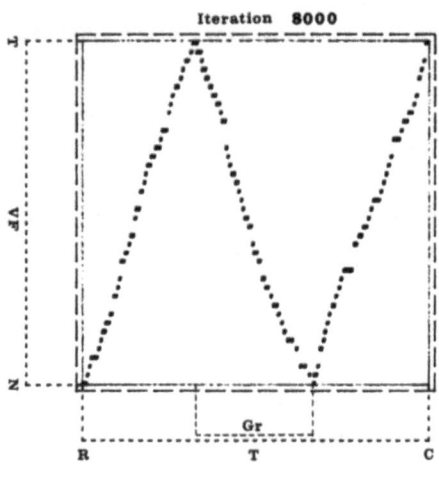

**Figure 13.** Projection organization after optic nerve section and graft inversion along the rostrocaudal axis (Yoon 1975).

**Figure 14.** Results of 8000 iterations of BAM in the case of a single inverted graft. Notice that the entire visual field is represented in each of the three sections and that the organization within each section is ordered.

An extension to the graft rotation experiments consists of excising two pieces of tissue from the surface of the tectum and reimplanting them without rotation yet with their positions switched, (Hope 1976). This is a difficult experiment to perform physiologically. In some cases, a completely normal projection is seen. This result would favor the "systems matching" theory of organization. However, different results have been obtained. In these cases the projection is ordered both within and surrounding the grafts but the grafts retain their original spatial mapping. Thus two sections of the mapping are interchanged. Since no information is available to differentiate between two tectal locations, these data support the "point-to-point chemospecificity" approach. Physiological results from such experiments are given in Figure 15, from (Hope 1976).

Simulation results from this situation appear in Figure 16. Note that the projection is organized as in the case of a normal tectal surface, compare with Figure 3b. This result is expected since the model contains information allowing only direction determination on the tectum, not absolute position of individual locations.

Experiment VIII.

This experiment was designed to emulate two particular points in the development of the retina and tectum. All of the experiments outlined above used an initial configuration which consisted of the initial termination locations being randomly distributed over the surface of the tectum. Since the tectum is innervated from the rostromedial area, this assumption is not accurate. Figure 17 illustrates the configuration apparent after 20000 iterations with the fibres initially randomly distributed over a very narrow band located toward the left end of the tectal surface.

The mapping displays excellent global organization with the "tails" slowly speading to fill all of the available space. If allowed to continue to iterate, a complete mapping identical in structure to that in Figure 3b would result.

V. The Problem

From Experiments I through VIII with the simulation of BAM, it is apparent that this model, using the principles of system matching, accounts for a great deal of the experimental data. Relative retinal and tectal orientation information is sufficient to produce behavior similar to most of the experimental results when utilized by neighbourhood and boundary interaction mechanisms. If the effect of retinal separation an the ability of two retinal cells to communicate is investigated, as in Experiment II, degraded behavior results. Restricting the distance allowing a meaningful exchange to a fraction of the retinal expanse results

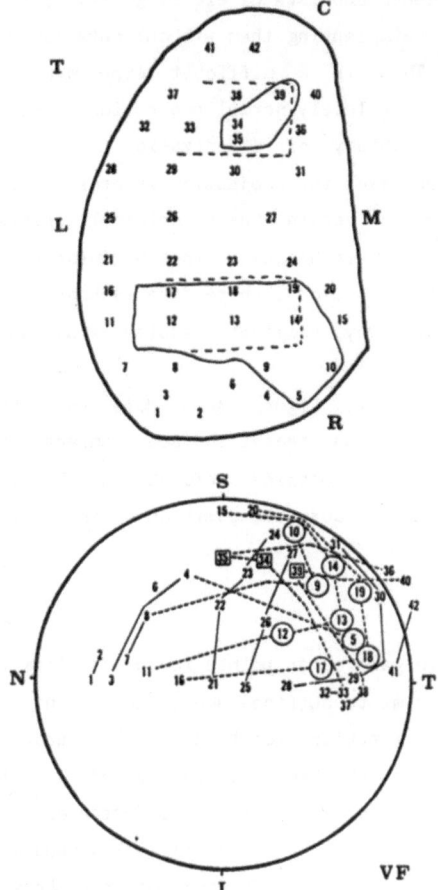

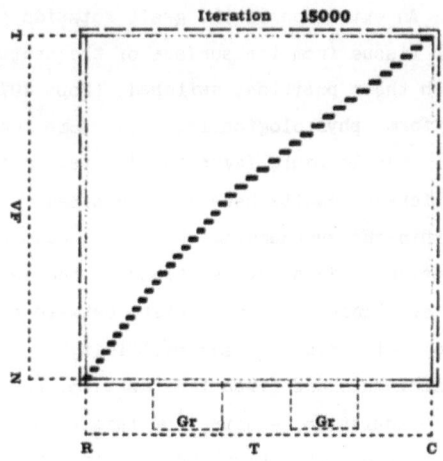

**Figure 16.** BAM simulation results. 32500 iterations, of the paradigm in which two grafts are translocated but their orientations are maintained.

**Figure 15.** Mapping data from an animal with the positions but not orientations changed for two grafts. (Hope 1976)

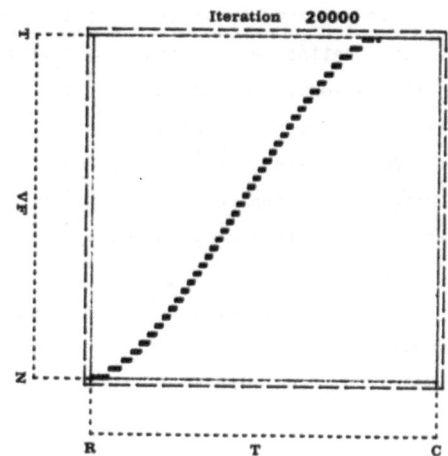

**Figure 17.** BAM simulation results when a normal retina innervates an intact tectum. In this case the branches were distributed initially over a very small section of the tectal surface.

in a locally continuous yet globally discontinuous mapping. Experiments VII and VIII illustrated that BAM cannot accurately produce the graft rotation and translocation behavior seen experimentally. The systems matching approach then appears inadequate to account for the physiological data. This is due to the lack of specific information differentiating individual tectal locations and the inability for information to be shared over distances which are large relative to the neighbourhood interaction. This amount of information is insufficient, however, to account for the class of experiments in which the locations but not the orientations are changed for a pair of grafts. Unlike the system matching ideas, point-to-point chemoaffinity provides specific information describing every point on the two surfaces.

Experiment II was designed to demonstrate the effect of varying the retinal distance allowing effective communication between cell somas. The results indicate that any two cells on the retinal surface must be capable of communicating regardless of their separation if the simple systems matching mechanism is to produce a globally ordered mapping. Experiments III through VIII were run under this assumption. As stated in the discussion of the BAM, we feel that this assumption is not necessarily true. Instead, the maximum separation allowing effective communication should be reduced to some relatively small value, perhaps the diameter of the receptive field. However, if the retinal distance allowing communication is reduced to a fraction of the total retinal expanse, the behavior degrades considerably. This leads to the conclusion that neighbourhood interaction mechanisms alone lack sufficient information to produce global organization.

The chemoaffinity theory as embodied in the Marker Induction Model of Willshaw and von der Malsburg (1979) can produce behavior similar to the experimental results but requires a degree of initial organization in order to produce a final map which is globally continuous. While this requirement may not be unreasonable, the more general question of the amount of information, or specificity, required by a model to explain the physiological results remains. As we describe in the sequel (Overton and Arbib, to appear), we have developed the Extended Branch-Arrow Model, XBAM, as a compromise between these two approaches. XBAM combines the local systems matching mechanisms apparent in BAM with a component describing a rough, inaccurate global positioning mechanism derived from the chemoaffinity theory. The behavior of this hybrid model was investigated through computer simulation and was shown to be in good agreement with the physiological data.

## Bibliography

Cook, J.E. & Horder, T.J. 1974 Interactions between optic fibers in their regeneration to specific sites in the goldfish tectum. J. Physiol. 241, 89-90P.

Feldman, J.D., Keating, M.J., & Gaze, R.M. 1975 Retino-tectal mismatch: a serendipitous experimental result. Nature, Lond., 253, 445-446.

Gaze, R.M., Jacobson, M., & Szekely, G. 1963 The retino-tectal projection in Xenopus with compound eyes. J. Physiol., 165, 484-499.

Gaze, R.M., Jacobson, M., & Szeleky, G. 1965 On the formation of connections by compound eyes in Xenopus. J. Physiol., 176, 409-417.

Gaze, R.M. & Sharma, S.C. 1970 Axial differences in the reinnervation of the goldfish optic tectum by regenerating optic fibers. Expl. Brain Res. 10, 171-181.

Gaze, R.M. & Keating, M.J. 1972 The visual system and "neuronal specificity". Nature 237, 375-378.

Gaze, R.M., Keating, M.J., & Chung, S.H. 1974 The evolution of the retinaotectal map during development in Xenopus. Proc. R.Soc. Lond. 185, 301-330.

Hope, R.A., Hammond, B.J., & Gaze, F.R.S. 1976 The arrow model: retinotectal specificity and map formation in the goldfish visual system. Proc. R. Soc. Lond. 194, 447-466.

Horder, T.J. 1971 Retention by fish optic nerve fibres regenerating to new terminal sites on the tectum, "chemospecific" affinity for their original sites. J. Physiol. Lond. 216, 53-55.

Horder, T.J. & Martin, K.A.C. 1977 Translocation of optic fibers in the tectum map be determined by their stability relative to surrounding fiber terminals. J. Physiol., Lond., 271, 23-24P.

Humphery, M.F. and Beazley, L.D. 1981 An electrophysiological study of early patterns of the retinotectal projection during optic nerve regeneration in Hyla Moorei. submitted to Brain Research.

Hunt, R.K. & Jacobson, M. 1973 Development of neuronal locus specificity in Xenopus retinal ganglion cells after surgical transection or after fusion of whole eyes. Dev. Biol., 40, 1-15.

Jacobson, M. & Gaze, R.M. 1965 Selection of appropriate tectal connections by regenerating optic nerve fibers in adult goldfish. Exp. Neurol., 13, 418-430.

Levine, R. & Jacobson, M. 1974 Development of optic nerve fibres is determined by positional markers in the frog tectum. Exp. Neurol. 43, 527-538.

Maturana, H.R., Lettvin, J.Y., McCulloch, W.S., & Pitts, W.H. 1959 Evidence that the cut optic nerve fibers in a frog regenerate to their proper places in the tectum. Science, 130, 1709-1710.

Overton, K.J. and Arbib, M.A. To Appear. The Extended Branch-Arrow Model (XBAM) of the Formation of Retino-Tectal Connections.

Schmidt, J.T. 1978 Expansion of the half retinal projection to the tectum of in goldfish: an electrophysiological and anatomical study. J. comp. Neurol. 177, 257-278.

Schmidt, J.T. 1978 Retinal fibers alter tectal positional markers during the expansion of the half retinal projection in goldfish. J. comp. Neurol. 177, 279-300.

Schmidt, J.T. & Easter, S.S. 1978 Independant biaxial reorganization of the retinotectal projection: a reassessment. Exp. Brain Res. 31, 155-162.

Sharma, S.C. 1972 Redistribution of visual projections in altered optic tecta of adult goldfish. Proc. Nat. Acad. Sci. U.S.A. 69, 2637-2639.

Sharma, S.C. & Romeskie, M.  1977 Immediate 'compression'of the retinal  projection to a tectum devoid of degenerating debris, Brain Res.  134, 367-370.

Sperry, R.W.  1943 Visuomotor coordination in the newt (Triturus viridescens)  after regeneration of the optic nerve.  J.  Comp.  Neurol., 79, 33-55.

Sperry, R.W.  1944 Optic nerve regeneration with return of vision  in  Anurans.  J. Neurophysiol.  7, 57-70.

Sperry, R.W.  1945 Restoration of vision after crossing of optic  nerves  and  after contralateral transplantaion of eye.  J.  Neurophysiol.  8, 15-28.

Sperry, R.M.  1963 Chemoaffinity in  the  orderly  growth  of  nerve  patterns  and connections.  Proc.  natn.  Acad.  Sci., U.S.A.  50, 701-709.

Willshaw, D.J. & von der Malsburg, C.  1979 A marker induction  mechanism  for  the establishment  of  Ordered neural mappings:  its application to the retinotectal problem.  Phil.  Trans.  Proc.  R.  Soc.  Lond.  B, 287, 203-243.

Udin, S.B.  1977 Rearrangements of the retinotectal  projection in Rana Pipiens after unilateral caudal half-tectum ablation.  J.  comp.  Neurol.  173, 561-582.

Yoon, M.G.  1973 Retention of the original toopgraphic polarity by the  180  degree rotated tectal reimplant in young goldfish.  J.  Physiol.  233, 575-588.

Yoon, M.G.  1975 Readjustment of retinotectal projection following reimplantation of a  rotated  or  inverted  tectal  tissue  in adult goldfish.  J.  Physiol.  252, 137-158.

Yoon, M.G.  1976 Progress of topographic regulation of the visual projection in  the halved optic tectum of adult goldfish.  J.  Physiol.  257, 621-643.

# CHAPTER 15

Differential localization of plastic synapses in the visual
cortex of the young kitten: Evidence for guided development
of the visual cortical networks

Keisuke Toyama and Yukio Komatsu

Department of Physiology, Kyoto Prefectural University of
Medicine and Department of Physiology, School of Medicine, Nagoya
University

Visual cortical plasticity has been studied extensively
during the last two decades (1,2,6,7,9,15,16). It has been
estabilished that neuronal responsiveness in the visual cortex is
highly modifiable by visual experience. The modifiability is
found only during the early infancy of the animal (2,8,15-17),
and the modified state persists throughout life of the
animal. It is postulated that neuronal networks in the visual
cortex are still immature in infancy and that their structure can
be modified by visual inputs within certain limitations
(2,8,15-17).

In contrast to its remarkable plasticity during infancy, the
visual cortex of the adult animal is characterized by an orderly,
fixed pattern of organization. As schematically illustrated in
Fig. 1, cells responding predominantly to the inputs through the
right eye are clustered into a slab perpendicular to the cortical
surface (R columns) and those responding to the left eye into
another slab (L columns). Orthogonal to the slabs of ocular

dominance, are smaller slabs which contain cells with the same preference for the orientation of light stimuli. The orientation slabs are arranged in a well-ordered sequence, from the one for the vertical orientation to that for the horizontal one and further back to the vertical one (9).

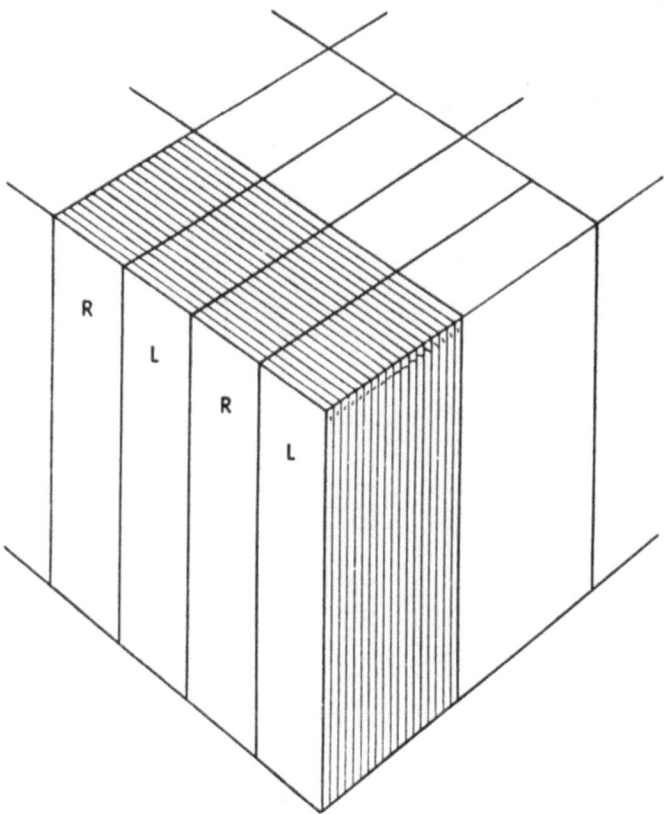

Fig. 1 Three-dimensional schema of columnar structures in the visual cortex

Two organizations, ocular dominance and orientation columns, are represented along the two coordinates orthogonal to each other. The columns denoted by L and R are those for the inputs through the left and right eyes; those denoted by vertical and horizontal bars represent the columns for the vertical and horizontal orientaion, respectively (9).

Finally there is a tangential pattern of organization (14).

Cells of the same efferent connectivity are arranged in a lamina;

1) cells whose axons project to the contralateral visual cortex

or to the ipsilateral visual association cortex are located in

layer III ($E_1$ in Fig. 2A), 2) those projecting to the superior

colliculus in layer V ($E_2$), and those to the lateral geniculate

nucleus (LGN) in layer VI ($E_3$). Tangential organization is also

found in afferent connectivity. Only cells in layers III-V

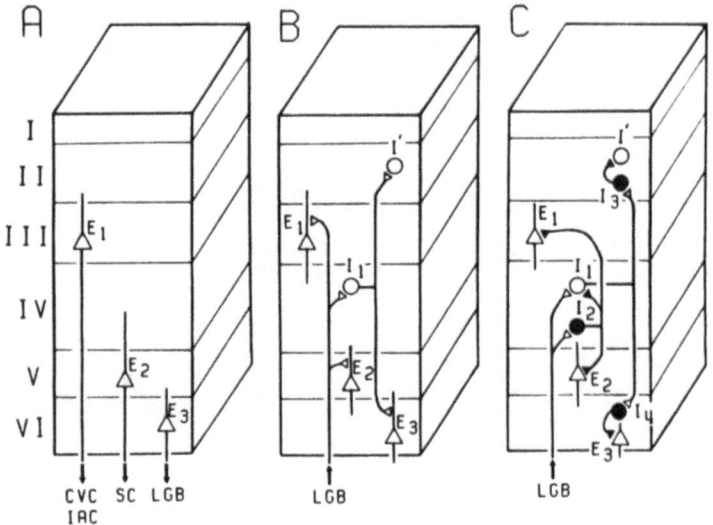

Fig. 2 Laminar stuctures in cat's visual cortex

A, efferent organization. $E_1$: efferent cell projecting to
the contralateral visual cortex (CVC) or the ipsilateral
association cortex (IAC), $E_2$: corticotectal cell to the superior
colliculus (SC), $E_3$: corticogeniculate cells to the lateral
geniculate nucleus (LGN). B and C, organizations of geniculate
inputs. B for excitation, C for inhibition. $I_1$: interneuron
mediating disynaptic excitation of cells in layers II (I') and VI
($E_3$), $I_2$: interneuron mediating disynaptic inhibition of cells
in layers III-V ($E_1$, $E_2$, $I_1$), $I_3$ and $I_4$: interneurons mediating
trisynaptic inhibition of cells in layers II and VI (I' and $E_3$)
(14).

$(E_1, E_2, I_1$ in Fig. 2B) receive excitatory inputs from LGN. Cells in layers II and VI $(I'$ and $E_3)$ recieve the inputs indirectly through an excitatory interneuron in layer IV $(I_1)$. Inhibition is exerted on cells in layers III-V $(E_1, E_2$ and $I_1)$ through an inhibitory interneuron in layer IV $(I_2$ in Fig. 2C), which receives excitation from LGN. On cells in layers II and VI $(I'$ and $E_3)$ inhibition is exerted through additional inhibitory interneurons $(I_3$ and $I_4)$. These perpendicular and tangential organizations in the visual cortex have been demonstrated either functionally as clusters of cells with similar responsiveness to light stimuli (9) or structurally as clusters of cells of similar input-output connections (14).

The reconciliation of this rigid pattern of organization with plasticity in photic responsiveness is a basic problem in developmental neurobiology. One approach has been to assume that only certain features of cortical organization can be modified by visual experience. This assumes that the modification of photic responsiveness by visual experience involves plastic change at a specific subset of synapses in the visual cortex. This approach is plausible because a population of cells in the newborn animal display mature photic responsiveness, while responsiveness is immature in most neurons (2).

We attempted to identify plastic and rigid synapses in the visual cortex, using a slice preparation of the visual cortex of the young kittens. In the slice preparation experimental manipulations such as placement of stimulating and recording electrodes are controlled under direct visualization. Our results indicate that plastic synapses are confined to layers II and IV of the visual cortex. This differential localization of plasticity is consistent with the hypothesis of the guided development of the visual cortical networks.

Kittens (7-49 days old) were decapitated, and transverse slices (0.5-0.7 mm thick) of striate cortex were dissected and incubated in Krebs-Ringer solution saturated with a mixture of $O_2$ and $CO_2$ (95:5) at $33^{\circ}$ C for one hour. After the incubation, a slice was placed in a recording chamber which was perfused with the same solution as that used for the incubation. The temperature of the perfusing solution was maintained at $33^{\circ}$ C. A glass microelectrode ($m_1$ in Fig. 3A) was inserted into the visual cortex (VC) for recording extracellular field potentials (FP) and impulses from individual cortical cells. Another microelectrode ($m_2$) was placed in the white matter (WM) to monitor afferent volleys produced by WM stimulation. Two pairs of bipolar stimulating electrodes ($s_1$ and $s_2$) each of which were composed of teflon-coated platinum wires (interpolar distance, 0.15 mm), were placed in WM with about 1 mm separation. Pulse stimuli (0.5-3 mA 0.1 ms) were supplied to these stimulating electrodes from constant current stimulators (internal resistance, 8 Mohm). $s_1$ was used for both conditioning and test stimulation of WM. Paired stimulation with $s_1$ and $s_2$ was used for studying neuronal connections of cortical cells with WM. The conduction velocities were determined according to latency difference of orthodromic impulses evoked by $s_1$ and $s_2$

stimulation and distance between $s_1$ and $s_2$. The central delays (time spent for orthodromic activation of cortical cells after arrival of impulses at the synapses) were determined by extrapolating the latency-distance relationship in orthodromic activation with $s_1$ and $s_2$ to zero distance. For conditioning, repetitive pulses at 2 Hz were given for one hour. Test stimulation was repeated throughout the entire period of recording session at 0.05 Hz. The 0.05 Hz stimulation was low enough to avoid cumulative effects such as depression or potentiation (13).

Figures 3B-E illustrate FPs in VC and afferent volleys in WM before and 2 hours after the onset of conditioing stimulation. The FPs were 2-6 times larger after conditioning stimulation (C) than those before conditioning stimulation (B), while the afferent volleys in WM remained essentially unchanged (D and E). The potentiation started 30 min. after the onset of conditioning stimulation, and continued after cessation of the conditioning stimulation as long as the slice preparation remained intact (observed up to 15 hours). The potentiation of FPs was dependent upon the age of the kittens used; it occurred in more than 80 % (30/36 ) of slices sampled from kittens at ages of 21-34 days, about 50 % (3/6 and 4/10) at 14-20 and 35-41 days, and none (0/2 and 0/2) of those at 7-13 and 42-49 days.

Events in individual cortical cells were compared between conditioned and unconditioned slices which had been dissected from the same hemisphere of the same kittens. Seven to sixteen cells were sampled extracellularly from each slice during a recording session which lasted for 3-7 hours. Ninety-three and 77 cells were collected from 8 pairs of unconditioned

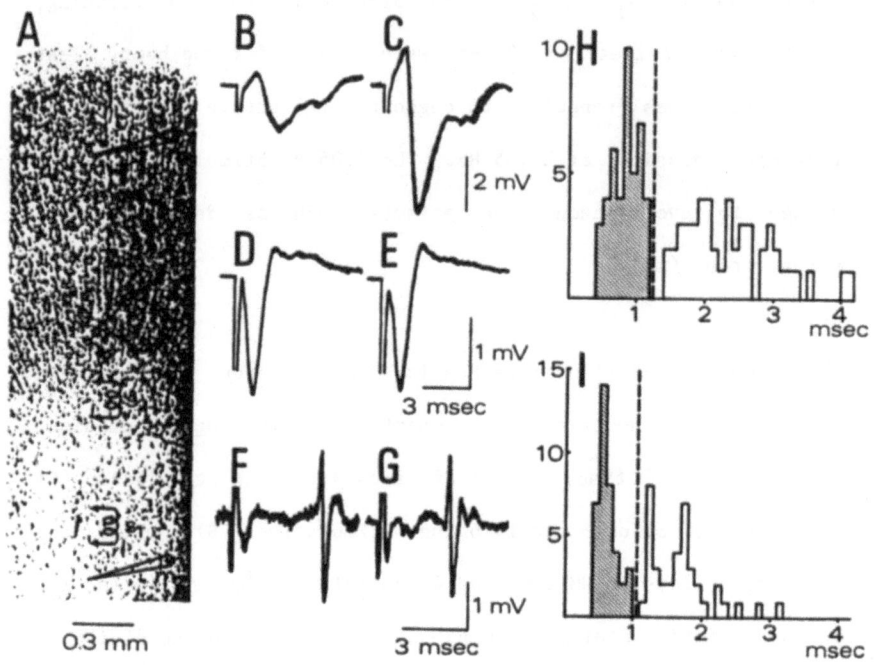

Fig. 3 Potentiation of cortical field potentials and shortening of latencies of cortical units

A illustrates arrangements of stimulating and recording electrodes on a histological section of a cortical slice (Nissl staining). B and C, cortical field potentials evoked by test stimulation. B, before and C, 2 hours after the onset of conditioning stimulation at 2 Hz. D and E, volleys in WM evoked by test stimulation before and after conditioning. F and G, impulses extracellularly recorded from cortical cells in response to test stimulation. F, sampled from an unconditioned slice. G, from a conditioned slice. H and I, histograms of central delays. H for unconditioned slices. I, for conditioned slices. Hatched columns represent monosynaptic, and blank columns polysynaptic values. Dotted lines indicate critical delay discriminating between mono- and polysynaptic cells.

and conditioned slices, respectively. All conditioned slices
exhibited FPs potentiated to a level more than twice the
control. Recordings in the conditioned slices started 2 hours
after the onset of conditioning stimulation. The sampled cells
were located in an area of 0.3-0.8 mm from the cortical surface,
where a large current source increase for FPs was revealed by
current source-density analysis (see below). In each cell the
conduction velocity and the central delays were determined for
orthodromic impulses evoked by $s_1$ and $s_2$ stimulation. Values of
central delays in the unconditioned slices comprised two peaks at
0.8 and 2 ms, having clear separation at 1.2 ms (dotted line in
Fig. 3 H). These two peaks probably represent monosynaptic
(hatched columns) and polysynaptic activations (blank columns).
Central delays in the conditioned slices were also distributed
bimodally (Fig. 3I). However the central delays in the
conditioned slices were significantly shorter than those in
control slices for both monosynaptic (mean delay, $0.61 \pm 0.15$ ms
in conditioned slices vs. $0.80 \pm 0.21$ ms in unconditioned
slices) and polysynaptic activation ($1.71 \pm 0.46$ vs. $2.35 \pm 0.67$
ms). Both of these differences are statistically significant (p
$< 0.001$). It is concluded that the FP potentiation may involve
both mono- and polysynaptic transmission from cortical afferents
to cortical cells. By contrast, there was no significant
difference in conduction velocities of afferent impulses between
cells from the conditioned ($0.74 \pm 0.31$ m/sec) and uncoditioned
($0.70 \pm 0.32$ m/sec) slices ( $p > 0.4$). These slow conduction
velocities of afferent impulses are consistent with the finding
that the visual afferents remain unmyelinated until 32 days after

the birth (5,12).

Location of the potentiated synapses was determined by current source-density analysis of FPs (11,12). Before conditioning stimulation, FPs evoked by test stimulation at $s_1$ were recorded at 100 μm increments along a track perpendicular to the cortical laminae (Fig. 4A). Current densities (CD),

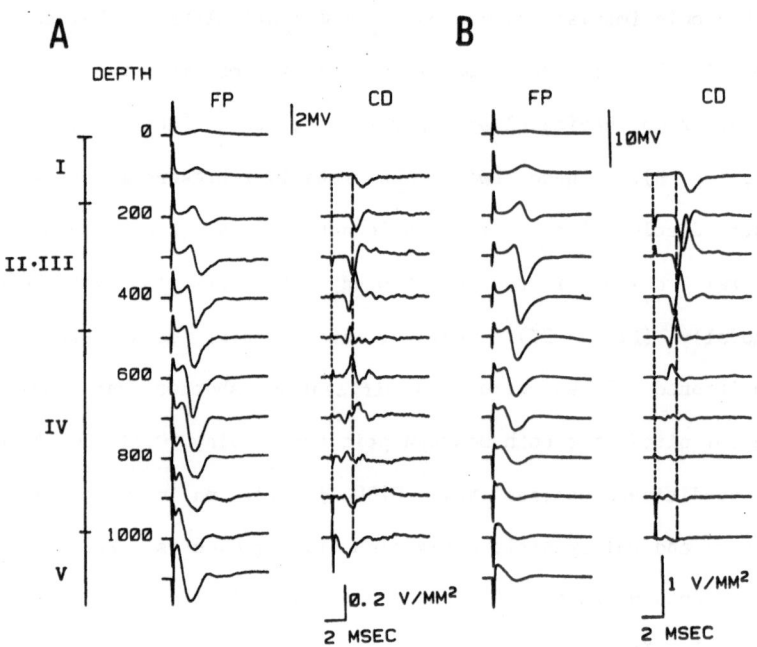

Fig. 4 Current source-density analysis of cortical field potentials

A and B, cortical field potentials (FP) and current source-densities (CD) evoked by test stimulation. A, before condition. B, after conditioning. Dotted lines in traces of CD represent the onsets of test stimulation and interrupted lines the border value to discriminate mono- and polysynaptic current-sinks.

determined as the second-order differential of the FPs, reveal current-sinks (upward deflection) representing excitatory postsynaptic currents and action currents generated in cortical cells in the local area. The current-sinks were divided into mono- and polysynaptic components by taking 2.2 ms as the border (interrupted line in Fig. 4A), which is the sum of 1.2 ms for central delay (see above) and 1 ms for afferent impulse conduction over a distance of 0.7 mm from WM to layer IV at the mean conduction velocity of 0.7 m/s (see below).

A monosynaptic current sink was most prominent in layer IV (traces at 500 and 600 mμ in Fig. 4A, CD), while a polysynaptic sink was found in layer II·III (traces at 300 and 400 mμ). Eight hours after conditioning stimulation, the FPs increased markedly; 5 times as large as control at 0.3 mm from cortical surface (cf. traces at 400 μm in Figs. 2A and B, FP). Correspondingly, there was an enhancement of CD in two restricted zones: a sixfold increase in the polysynaptic current-sink in layer II·III (cf. traces at 300 μm in Fig. 4B and at 400 μm in Fig. 4A) and a threefold increase in the monosynaptic current-sink in layer IV (traces at 600 um in A and B). The results suggest that the potentiation of FPs after conditioning stimulation is due to facilitation of transmission in layers II · III and IV. This localization of the current sinks evoked in the striate cortex by WM stimulation is similar to those produced by stimulation of the optic chiasm (11). It is also consistent with the anatomical findings that the geniculate afferents preferentially terminate in layer IV, while non-geniculate afferents terminate in other layers (3,4,10). The FPs evoked by WM stimulation in this study, therefore, might be related with those evoked by excitation of the geniculate afferents.

In summary, the present study indicated that conditioning activation of WM fibers causes a long-term enhancement of synaptic transmission in the visual cortex. The enhancement was greatest at intracortical transmission in layer II, moderate at geniculo-cortical transmission in layer IV, and negligible at intracortical transmission in layers V and VI. Therefore it is likely that not all synapses are plastic in the visual cortex of the infant animal, but plastic synapses are mostly confied to layer II and partly to layer IV. This differential location of plasticity in cortical networks may provide the basis for the guided organization of cortical networks.

## References

1. Blakemore, C. and Cooper, G. F., Development of the brain depends on the visual environment, Nature, 228 (1970) 477-478.

2. Blakemore, C. and Van Sluyters, R. C., Innate and environmental factors in the develpment of the kitten visual cortex, J. Physiol. 248 (1975) 663-716.

3. Ferster, D. and Levay, S., The axonal arborizations of lateral geniculate neurones in the striate cortex of the cat, J. comp. Neurol. 182 (1978) 923-944.

4. Gilbert, C. D. and Wiesel, T. N., Morphology and intracortical projections of functionally characterized neuroes in the cat visual cortex, Nature, 280 (1979) 120-125.

5. Grafstein, B., Postnatal development of the transcallosal evoked response in the cerebral cortex of the cat, J. Neurophysiol. 26 (1963) 79-99.

6. Hirsch, H. V. B. and Spinelli, S. N., Visual experience modifies distribution of horizonatally oriented receptive fields in cats, Science, 168 (1970) 869-871.

7. Hubel, D. H. and Wiesel, T. N., Receptive fields of cells in striate cortex of very young, visually inexperienced kittens. J. Neurophysiol. 26 (1963) 994-1002.

8. Hubel, D. H. and Wiesel, T. N., The period of susceptibility to the physiological effects of unilateral eye closure in kittens, 206 (1970) 419-436.

9. Hubel, D. H. and Wiesel, T. N., Sequential regularity and geometry of orientation columns in the monkey striate cortex. J. Comp. Neurol. 158 (1974) 267-293.

10. Levay, S. and Stryker, M. P., The development of ocular dominance columns in the cat. In Society for Neuroscience Symposia Vol. IV, Aspects of Developmental Neurobilogy, ed. by Ferrendell, T. A., The Society for Neuroscience, Bethesda, 1978.

11. Mitzdorf, V. and Singer, W., Prominent excitatory pathway in cat visual cortex (A17 and A18): A current source-density analysis of electrically evoked potentials, Exp. Brain Res., 33 (1978) 371-394.

12. Nicolson, C. and Freeman, J. A., Theory of current source-density analysis and determination of conductivity tensor for anuran cerenbellum, J. Neurophysiol., 38 (1975) 356-368.

13. Tsumoto, T. and Suda, K., Cross-depression: an electrophysiological manifestation of binocular competition in the developing visual cortex, Brain Res. 168 (1979) 190-194.

14. Toyama, K., Matsunami, K, Ohno and Tokashiki, A., An intracellular study of neuronal organization in the visual cortex. Brain Res. 21 (1974) 45-66.

15. Wiesel, T N. and Hubel, D. H., Single-cell responses in striate cortex of kittens deprived of vision in one eye, J. Neurophysiol. 26 (1963) 1003-1017.

16. Wiesel, T. N. and Hubel, D. H., Comparison of the effects of unilateral and bilateral eye closure on cortical unit responses in kittens, J. Neurophysiol., 28 (1965) 1029-1040.

17. Wiesel, T. N. and Hubel, D. H., Extent of recovery from the effects of visual deprivation in kittens, J. Neurophysiol. 28 (1965) 1060-1072.

SELF-ORGANIZATION OF NEURAL NETS WITH

COMPETITIVE AND COOPERATIVE INTERACTION

R.Sawada and N.Sugie
Department of Information Science
Faculty of Engineering, Nagoya University

Furo-cho, Chikusa-ku, Nagoya 464, Japan

## 1. INTRODUCTION

It is now well accepted that the response characteristics of the cells in the cat striate cortex are modified significantly depending on the early visual experience (Blakemore, 1978). The modification is known to occur in relation to orientation selectivity, directional selectivity, and binocular responsiveness.

Several models have been proposed regarding the above observations (von der Malsburg 1973, Nagano 1977). It can be said that they are concerned with the local feature extraction in the visual system. In order to account for the global nature of binocular stereopsis, a mechanism is required for competitive and cooperative interactions among local feature detectors, i.e., binocular disparity detectors (Sugie and Suwa, 1974;1977 : Marr and Poggio, 1976). It is known that cells exist in the cat for local depth extraction and that the sensitivity to binocular disparity improves during the period of several weeks after birth as shown in Fig.1 (Pettigrew, 1974). In view of the plasiticity of the local depth extraction, we might postulate that the mechanism for global depth extraction is also plastic.

This research intended to demonstrate by computer simulation that a model can acquire the binocular stereopsis through self-organization by presenting a sequence of visual stimuli.

## 2. STRUCTURE OF MODEL

For simplicity, it was assumed that the retina consisted of a horizontal array of receptors or feature detectors. As is shown in Fig.2, subsets of neurons are grouped in direction columns (Blakemore, 1970). Neurons belonging to a column respond if a visual stimulus with a certain feature is presented within a certain range of depth along a particular oculocentric visual direction. Each column consists of several units or compartments. Each unit contains six neurons. One excitatory

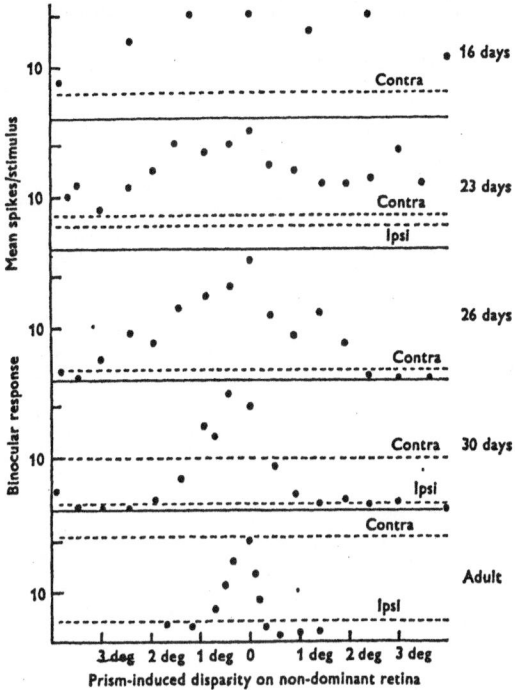

Fig.1    Development of binocular disparity detector (Pettigrew, 1974).


neuron in each unit receives excitatory inputs from both eyes.   Thus
the neuron, $Y_{ik}$, in the unit $i$ of the column $k$ receives inputs from the
neuron $L_k$ in the left eye with the synaptic weight $W_{Lk}$ and from the neurons $R_{k-2} \sim R_{k+2}$ in the right eye with the synaptic weights $W_{Rnki}$, where
$k-2 \le n \le k+2$.

The remaining five interneurons are connected with the excitatory
neuron via fixed synaptic weights.   It is evident that the neurons in
the direction column $k$ is excited if the visual stimulus is presented
at the site $k$ in the left eye and at the site between $(k-2)$ and $(k+2)$
in the right eye.   The synaptic weight, $W_{Lk}$, is assumed to be fixed for
simplicity, whereas the weight, $W_{Rnki}$, is nearly uniform initially with
regard to $n$ and is modified in accordance with the rules described herein.

One inhibitory interneuron (e.g. $Z_{ik}$) in each unit is for competitive intracolumnar interaction.   The neuron sends inputs to every unit

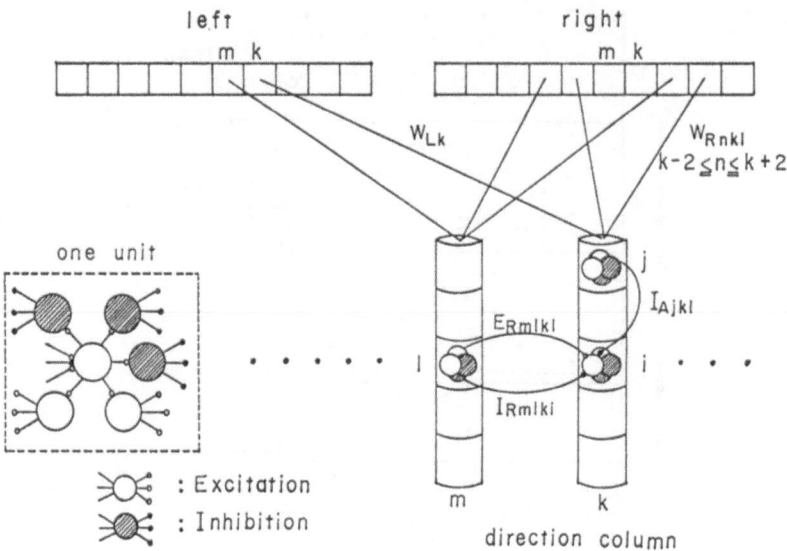

Fig.2.　Neural network model of binocular stereopsis.

in the same column. The synaptic weight $I_{Ajki}$ is modifiable, but uniform initially independent of $i$ and $j$ .

　　　Each column is connected with the left and right immediately neighboring columns for cooperative (or excitatory) and competitive (or inhibitory) intercolumnar interactions. Two excitatory and two inhibitory interneurons ($Y_{Dik}$ and $Z_{Dik}$, respectively) are allocated for this purpose. Each interneuron sends inputs to every unit in the neighboring column, the synaptic weight $E_{Rmlki}$ (or $I_{Rmlki}$) is uniform initially with respect to $i$ and is modified later on.

　　　Assuming that the name of each neuron represent its output and that each monosynaptic delay correspond to a unit time, then the post-synaptic potential $X_{ik}(t)$ for $Y_{ik}$ is represented as follows.

$$X_{ik}(t)= W_L L_k(t-1)+ \sum_{n=k-2}^{k+1} W_{Rnik}R_n(t-1)- \sum_{j=1}^{5} I_{Ajki}Z_{jk}(t-1)$$

$$+ \sum_{\substack{m=k-1 \\ m\neq k}}^{k+1} \sum_{l=1}^{5} E_{Rmlki}Y_{Dml}(t-1)- \sum_{\substack{m=k-1 \\ m\neq k}}^{k+1} \sum_{l=1}^{5} I_{Rmlki}Z_{Dml}(t-1)$$

Let $\theta$ be the threshold of the neuron $Y_{ik}$.   It is assumed that

$$Y_{ik}(t) = \{ X_{ik}(t) - \theta \}^2 \qquad \{\ \}^2 \leq Y_{max}$$

$$\phantom{Y_{ik}(t)} = Y_{max} \qquad \text{for } \{\ \}^2 > Y_{max}$$

Furthermore

$$Z_{ik}(t) = 0.1\ Y_{ik}(t-1)$$

$$Y_{Dik}(t) = Y_{ik}(t-1)$$

$$Z_{Dik}(t) = Y_{ik}(t-1)$$

## 3.  PROCEDURES FOR SYNAPTIC MODIFICATION

Basically, two rules are postulated for the modification of synaptic weight.

**Rule 1**    The excitatory interaction between two neurons tends to be strengthened if the chance of the simultaneous firing is high.  Otherwise the interaction tends to be weakened.

**Rule 2**    The inhibitory interaction between two neurons tends to be strengthened if only the presynaptic neuron is excited.  Otherwise the interaction tends to be weakened.

It is obvious that the rule 1 is responsible for the establishment of cooperative interaction and that the rule 2 for the establishment of competitive interaction.

The detailed procedure for synaptic modification is essentially due to Hirai's method (Hirai, 1980).  The outline of the procedure is summarized in Fig.3.  If the excitatory presynaptic neuron Y fires then the postsynaptic neuron $X_i$ sends a demand signal $e_i^*$ to Y.  $e_i^*$ is positive only if Y and $X_i$ five simultaneously and is zero otherwise.  This relation is expressed by the following formula.

$$e_i^* = f[\ k_1 YX_i - k_2\ (k_3 - Y) X_i]$$

where $k_i'$s are appropriate positive constants.   Y and $X_i$ are the output of the pre- and post-synaptic neurons.   Further

$$f(x) = x \qquad \text{for } x > 0$$

$$\phantom{f(x)} = 0 \qquad \text{otherwise.}$$

Then each synaptic weight $e_i$ is modified by $\Delta e_i$, which is determined by the following formula.  Namely,

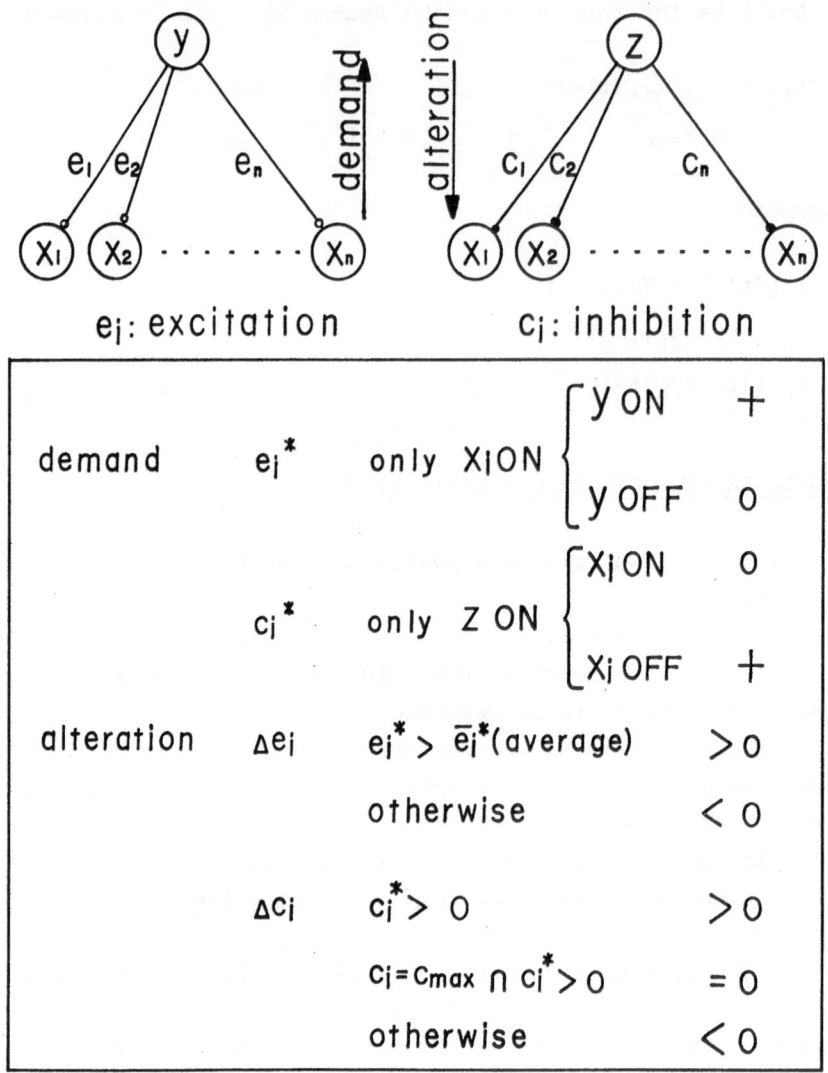

Fig.3. Conditions of modification of synaptic weight.

$$\Delta e_i = k_4 \left( e_i^* - \frac{1}{n^o - 1} \sum_{j \neq i} e_j^* \right)$$

where $n^o$ designates the number of postsynaptic neurons with $e_i^* > 0$ and the summation is carried out for positive $e_i^*$'s. Note that $\Sigma \Delta e_i = 0$. The range of $e_i$ is set to an interval $[0, e_{max}]$.

As for the inhibitory synapse, the demand signal $c_i^*$ is generated

$$c_i^* = f[\ Z(k_5 c_i Z - k_6 X_i)]$$

where $k_5$ and $k_6$ are appropriate positive constants. This formula forces the range of $c_i$ to be $[\ c_{min},\ c_{max}]$, where $c_{max} > c_{min} > 0$. That is, if $c_i$ is allowed to be equal to one, $c_i^*$ cannot be positive and therefore $c_i$ cannot be modified. If the synaptic weight $c_i$ is equal to $c_{max}$ and $c_i^*$ is positive, then $c_i$ is not modified. Otherwise $c_i$ is modified by $\Delta c_i$ using the following formula

$$\Delta c_i = k_7 (\ c_i - \frac{1}{n-1} \sum_{l \neq i} c_l^* )$$

where $k_7$ is an appropriate positive constant. Thus $\Delta c_i$ can be either positive or negative depending on the values of $c_i^*$ and $c_l^*$'s.

## 4. COMPUTER SIMULATION

### Initial synaptic weights

Since it is assumed that each column contains five units, there are 25 $W_{Rnki}$'s per column, of which five were assigned as 1.0 and the the remaining twenty were assigned random values ranging from 0 to 1.0
The other weights were set in the following way.

$$I_{Ajki} = 0.5$$
$$E_{Rmlki} = 0.01$$
$$I_{Rmlki} = 0.01$$

### Learning procedure

Each retina consisted of ten units. The stimuli consisting of five types of features (e.g. light edge, P, dark edge, N, etc.) were presented to the retinae. In each unit of the retina, there were five types of feature detectors. The binocular unit $Y_{ik}$ cannot be excited unless the feature at the site k of the left retina is the same as that at the site $(k-2) \sim (k+2)$ of the right retina.

In order to present stimuli of various disparities, the right stimuli were obtained by successively shifting the left stimulus horizontally. After each stimulation, several units of time were spent until the

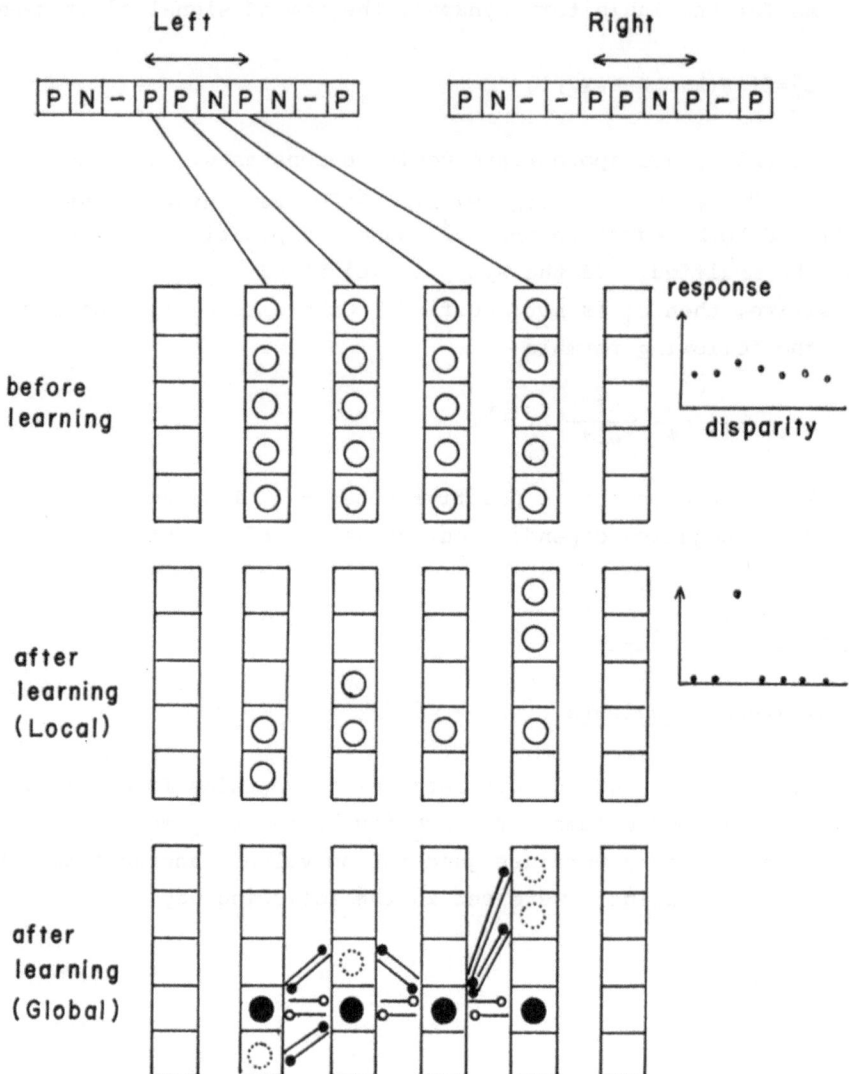

Fig.4.   Responses of model before and after self-organization.

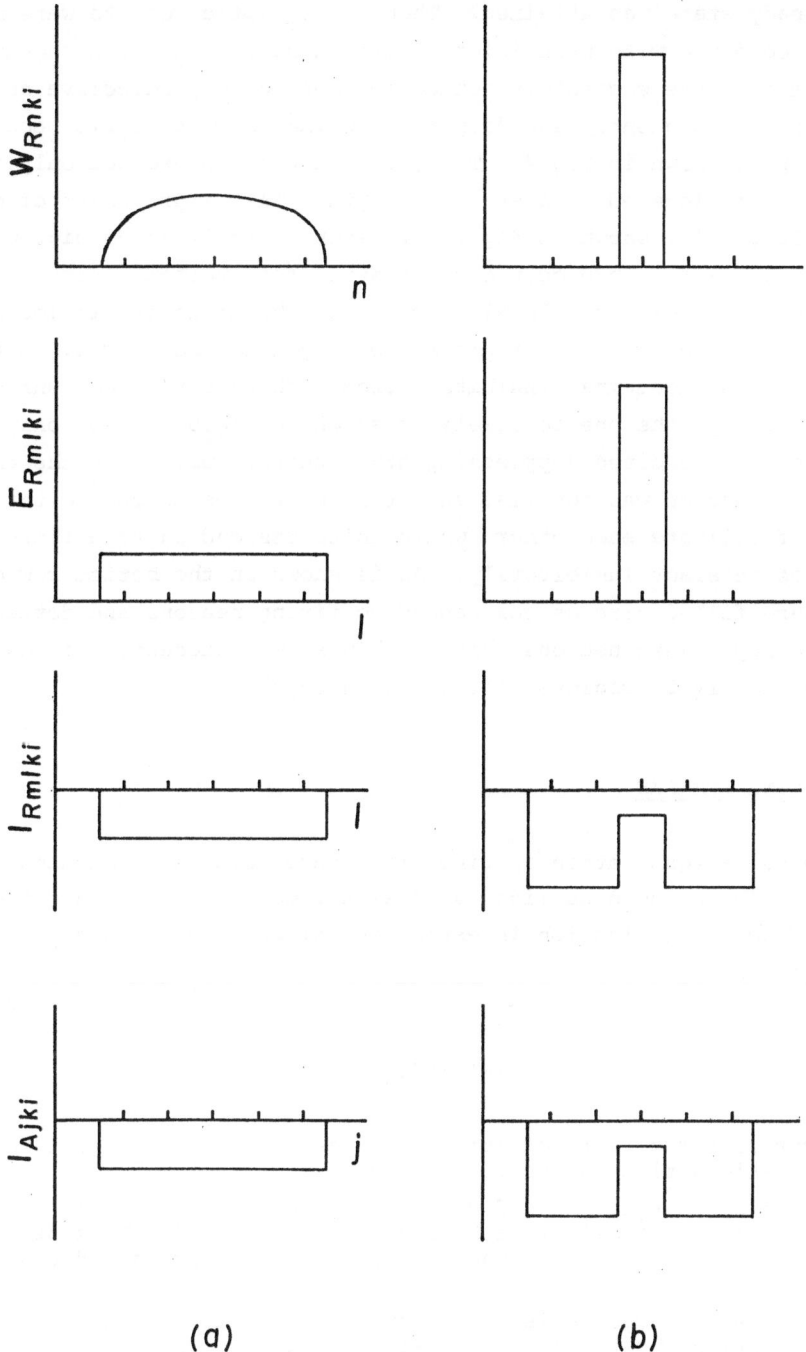

Fig.5. Modifications of synaptic weights before and after self-organization.

nearly steady state was attained. Then the synaptic weights were modified in accordance with principles described above. After a sequence
of stimulation, the excitatory connection became very selective as shown
in Fig.5 and consequently the disparity tuning characteristic became
quite sharp as shown in Fig.4. Thus, each neuron is excited only if
the stimulus is located at a specific depth. This improvement of disparity selectiveity shown in Fig.4 corresponds nicely to Pettigrew's
finding in kittens, where open circles stand for excited neurons.

In order to extract global 3D regions, the ambiguity in local correspondence as shown in the middle row of Fig.4 has be solved. After
learing, the intracolumnar inhibition from each unit to other units was
reinforced except the one to itself as shown in Fig.5. Thus, only one
unit tends to be excited suppressing other units. Obviously the intercolumnar connection was modified such that units responding to the same
disparity facilitate each other, while units responding to different
disparities interact inhibitorily. As is shown in the bottom row of
Fig.4, where filled circles correspond to firing neurons and dotted cir-
circles to suppressed neurons, this mechanism was successful in extracting unambiguously 3D regions continuous in depth.


## 5. CONCLUDING REMARKS

The self-organization in binocular stereopsis was simulated successfully. The principles might give significant insights into the mechanism of self-organization in extracting global regions such as in
texture perception.

## REFERENCES

1. Blakemore,C.: The representation of three dimensional visual space
   in the cat's striate cortex. J.Physiol. 193 (1970) 155-178.

2. Blakemore,C.: Maturation and modification in the developing visual
   system. In: R.Held,Leibowitz,H.,Teuber,H.-L.(eds) Handbook of Sensory
   Physiology. Vol.VIII (1978) Chapt.12.pp.337-436. Springer Verlag,
   Berlin.

3. Hirai,Y.: A new hypothesis for synaptic modification: an interactive
   process between postsynaptic competition and presynaptic regulation.
   Biol.Cybern. 36 (1980) 41-50.

4. Marr,D.,Poggio,T.: Cooperative computation of stereo disparity. Science, 194 (1976) 283-287.

5. Nagano,T.: A model of visual development. Biol.Cybern. 26 (1977) 45-
   52.

6. Pettigrew,J.D.: The effect of visual experience on the development of stimulus specificity by kitten cortical neurones. J.Physiol. <u>237</u> (1974) 49-74.

7. Sugie,N.,Suwa,M.: A neural network model of Binocular depth perception. Nat.Conv.Rec. of Inst. Electronics & Commun.Engrs. Japan (1974) No.59-3.

8. Sugie,N.,Suwa,M.: A scheme for binocular depth perception suggested by neurophysiological evidence. Biol.Cybern. <u>26</u> (1977) 1-15.

9. von der Malsburg,C.: Self-organization of orientation selective cells in the striate cortex. Kybernetik <u>14</u> (1973) 85-100.

CHAPTER 17

## A SIMPLE PARADIGM FOR THE SELF-ORGANIZED FORMATION
## OF STRUCTURED FEATURE MAPS

Teuvo Kohonen
Department of Technical Physics
Helsinki University of Technology
SF-02150  Espoo 15, Finland

## 1. Introduction

There exist many kinds of maps or images in the brain; the most familiar ones may be the retinotopic, somatotopic, and tonotopic projections in the primary sensory areas, as well as the somatotopic order of cells in the motor cortex. There is also some experimental evidence (cf., e.g. Lynch et al., 1978) for that topographic maps of the exterior environment are formed in the hippocampus. In this work it is claimed that the brains might also more generally produce ordered maps that are directly or indirectly related to sensory or somatic information. It does not seem impossible that formation of *feature maps* at various levels of abstraction is the main mode of information processing in the brain.

Without doubt the main structures of the brain network are determined genetically. However, there also exists plenty of experimental evidence for sensory projections to be affected by experience. For instance, after ablation of sensory organs or brain tissue, or sensory deprivation at young age, some projections are not developed at all, and the corresponding territory of the brain is occupied by the remaining projections (cf., e.g. Schneider, 1977; Sharma, 1972; Hunt et al., 1975). Recruitment of cells to different tasks depending on experience is well-known. These effects should then be explained by neural plasticity and they exemplify self-organization that is mainly controlled by sensory information.

The present discussion is related to an idealized *neural* structure. The purpose is by no means to claim that self-organization is mainly of neural origin; genetic control, mediated by chemical labels, might produce similar effects. The present discussion is mainly aimed at pointing out what kind of a simple mechanism *alone* is able to implement self-organization.

A few related works on self-organizing phenomena should be mentioned
(Malsburg, 1973; Willshaw and Malsburg, 1976, 1979; Malsburg and
Willshaw, 1977; Amari, 1980; Swindale, 1980). The content of the present work might be characterized as an attempt to extract the basic self-
organizing function in the simplest and purest form in which it lends
itself to strict quantitative analysis. The model thereby developed
has proved very robust with respect to its parameters, and produced many
intriguing new results some of which are reported below. These results
are now subjected to consideration as an idealized and generalized self-
organizing paradigm. More detailed discussions of this model can be
found elsewhere (Kohonen, 1981a-e).

This work first describes empirically found results as such; after that,
in Sec. 7, a simple theoretical explanation is delineated.

## 2. Localized responses and ordered mappings

Assume a physical system that receives a set of concomitant *input signals*; these are transformed into *output responses* (Fig. 1). Characteristic of the systems discussed in this work is that there operates some
mechanism by which the active response is concentrated on some *location*
in the output plane; for different sets of input signals this location
is in general different.

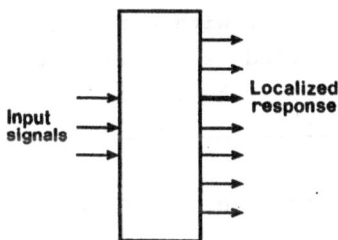

Fig. 1. On the definition of
localized responses

Every set of stationary input signals can be termed *pattern*. In matrix
notation, a pattern is a column vector $x \in R^n$ which is typographically
expressed as $x = [\xi_1, \dots, \xi_n]^T$, where the $\xi_j$, j=1...n are the real-
valued scalar inputs written as a row, and T means transposition of
a row into a column.

One of the simplest systems which is able to produce localized responses
is a linear array (row) of functional units, each of which receives the
same set of input signals in parallel. Assume that such a unit i then
produces a different response to the different input patterns:

$\eta_i(x_1)$, $\eta_i(x_2)$,... . Next we define a transformation that is characteristic of the present self-organizing systems, and obviously of many
brain areas, too. Assume first for simplicity that a set of input patterns $\{x_i : i = 1,2,...\}$ can be *ordered* in some metric or topologic
way such that $x_1 \, R \, x_2 \, R \, x_3$ ... where $R$ stands for a simple ordering
relation with respect to a single *feature* that is implicit in the
representations.

<u>Definition</u>: The system is said to implement a *one-dimensional ordered
mapping* if for $i_1 > i_2 > i_3 > \ldots$

$$\eta_{i_1}(x_1) = \max_i\{\eta_i(x_1) : i=1,2,\ldots,n\}$$

$$\eta_{i_2}(x_2) = \max_i\{\eta_i(x_2) : i=1,2,\ldots,n\}$$

$$\eta_{i_3}(x_3) = \max_i\{\eta_i(x_3) : i=1,2,\ldots,n\}$$

etc.

The above definition is readily generalizable to two- and higher-
dimensional arrays of processing units. In this case the *topology* of
the array is simply defined by the definition of *neighbours* to each
unit. As for the input signals some *metric or topological order* may
then be definable for the patterns $x_k$ , induced by more than one ordering relations with respect to different features or attributes. If the
units form, say, a two-dimensional array, and the unit with the maximum response to a particular input pattern is regarded as the *image* of
the latter, then *the mapping is said to be ordered if the topological
relations of the images and the patterns are similar.*

A rectangular two-dimensional array with an identical set of input
signals to each unit is exemplified in Fig. 2.

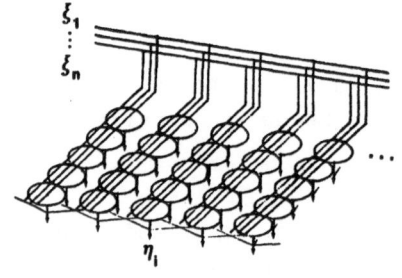

Fig. 2.   A two-dimensional array
of processing units

## 3. A simple two-dimensional self-organizing system

If the transfer characteristics of every unit are assumed time-dependent
and a function of the occurring signals, the system is said to be
*adaptive*. Such processing units (abbreviated PU in the sequel) may be
identified with concrete physical adaptive components much of the same
type as the Perceptrons (Rosenblatt, 1961), or even with simpler ones,
as shown below.

Let each PU receive the same scalar signals $\xi_1, \xi_2, \ldots, \xi_n \in R$. Unit i,
in the simplest case, may then form a linear functional of these:

$$\eta_i = \sum_{j=1}^{n} \mu_{ij} \xi_j , \tag{1}$$

where the $\mu_{ij} \in R$ are (variable) parameters. In fact, every ordered set
$(\mu_{i1}, \mu_{i2}, \ldots, \mu_{in})$ may be regarded as a kind of *referential system* or
*image* that shall be matched or compared against a corresponding ordered
set $(\xi_1, \xi_2, \ldots, \xi_n)$; our aim is to devise adaptive processes in which
the parameters of all units converge to such values that every unit
becomes specifically matched or sensitive to a particular domain of
input signals in a regular order.

Before proceeding further, it may be useful to emphasize that many other
functionals, even simple ones, could be chosen instead of that expressed
in Eq. (1). Moreover, if the output mapping were not signified by the
maximum but rather by the minimum response, then another simple functional which indicates a match between input signals and system parameters
might be

$$\eta_i = \sum_{j=1}^{n} (\xi_j - \mu_{ij})^2 . \tag{2}$$

Both (1) and (2) have been applied successfully in simulation experiments, without major difference.

The system represented by the array becomes adaptive if the $\mu_{ij}$ are
made variable in time and they change with the signals. One of the simplest laws of adaptation that can be imagined to lead to a matched filter is a corrective process that updates the parameters $\mu_{ij}$ according
to

$$d\mu_{ij}/dt = \alpha_i(t) \cdot (\xi_j - \mu_{ij}) \tag{3}$$

where t is the time parameter. It can be shown that if $\alpha_i(t)$ is further chosen in a particular way, then (3) indeed leads to self-ordering. Let $\alpha_i(t)$ be determined in a *decision process* in the following fashion.

We shall first assume that *the units can communicate mutually* to decide, e.g., on the basis of Eq. (1) or (2) which one of the units $\eta_i$ indicates the best match:

Relating to Eq. (1):     Relating to Eq. (2):

$$\eta_m = \max_i \{\eta_i\} \quad\quad (4) \quad\quad \eta_m = \min_i \{\eta_i\} \quad\quad (5)$$

In the first place it is not necessary to specify how (4) or (5) is implemented physically; let it suffice to mention that they are readily realizable by simple physical elements.

Next we shall define $\alpha_i(t)$ analytically. There again exist many alternative forms for this coefficient that lead to self-organization. Let Eq. (4) or (5) specify unit m, and let $S_m$ be the set of units which consists of unit m and its *immediate neighbours*. Then a possible choice for $\alpha_i(t)$ might be

$$\alpha_i(t) = \begin{cases} \alpha(t) & \text{for } i \in S_m , \\ 0 & \text{otherwise} . \end{cases} \quad\quad (6)$$

Of course, $\alpha_i(t)$ could be made to depend on the distance of units i and m. *Notice that the updating process (3) increases the similarity of the parameter sets of the selected units.*

The time function $\alpha(t)$ can similarly take on many different forms. In the simplest case $\alpha(t)$ could be constant, preferably a small one. However, if we want to achieve fast ordering and convergence to stationary parameter values, then we have to require that $\lim_{t \to \infty} \alpha(t) = 0$. This condition is similar to that encountered in the theory of stochastic approximation and mathematical learning systems where $\alpha(t)$ is often chosen to be of the form $\alpha_1/t$ , $\alpha_1 = \text{const}$.

A slightly modified updating law is obtained if the second term in (3) ("forgetting") is multiplied by $\beta_i(t) \cdot \eta_i$ where $\beta_i(t)$ is another freely selectable, time-dependent coefficient:

$$d\mu_{ij}/dt = \alpha_i(t) \cdot (\xi_j - \beta_i(t) \, \eta_i \cdot \mu_{ij}) . \quad\quad (7)$$

It has recently been pointed out (Oja, 1981) that if this law is expressed in a discrete-time domain where the input variables $\xi_j$ are held
stationary ($=\xi_j(t)$) during sufficiently long intervals, and especially
if $\beta_i(t)$ is constant, then an expression almost equivalent to (7) is

$$\mu_{ij}(t+1) = \frac{\mu_{ij}(t) + \alpha_i(t) \cdot \xi_j(t)}{\left(\sum_{j=1}^{n} [\mu_{ij}(t) + \alpha_i(t) \cdot \xi_j(t)]^2\right)^{1/2}} \tag{8}$$

where we have chosen the sampling instants of t as integers,
$t = 0,1,2,\ldots$ . In fact, since the simulations are anyway carried out
as discrete-time processes, (8) could directly be taken to define the
updating process. It only superficially looks more complicated than,
say, (3); in practice it leads to very similar system behaviour.

It might further be noted that (8) always *normalizes* the parameter
sets: if $(\mu_{i1}, \mu_{i2}, \ldots, \mu_{in}) \in R^n$ is regarded as a Euclidean vector,
then after every recursive step defined in (8), the vectors are normalized to unit length. To put it in another way, (8) only *rotates* the
parameter vector in a proper direction.

## 4. Self-organized formation of the image of a two-dimensional
## distribution of vectors

To elucidate the behaviour of the above system, we first demonstrate
that *if the input vector x is a random variable with a stationary
distribution p(x), then an ordered image of p(x) will be formed onto
the input weights* $\mu_{ij}$ *of the processing units.* In other words, every
PU becomes maximally sensitized to a particular x, but for different
PU this sensitization occurs in an orderly fashion, corresponding to
the distribution p(x).

In the first simulation we shall now restrict ourselves to three-
dimensional input vectors $x \in R^3$ because they can be easily visualized
by computer graphics. Accordingly, every PU has three input weights
$\mu_{i1}, \mu_{i2}, \mu_{i3}$ which constitute the *weight vector* $m_i \in R^3$ of unit i.

The distribution p(x) is first assumed to have a simple *structured* form;
the vectors $x \in R^3$ are normalized to unit length, and thus they lie on
the surface of a unit sphere in $R^3$. Since the $m_i$ are also normalized to
unit length according to Eq. (8), they shall lie on the surface of a
unit sphere, too. Fig. 3 represents a projection of this sphere; the

"training vectors" are shown as distinct *dots*. The $m_i$ vectors are shown
as points, too, but in order to indicate to which unit each $m_i$ belongs,
the end points of the $m_i$ are connected by a lattice of lines which
conforms with the topology of the processing unit array. A line connecting two weight vectors $m_i$ and $m_j$ is thus only used to indicate that the
two corresponding units i and j are adjacent in the array.

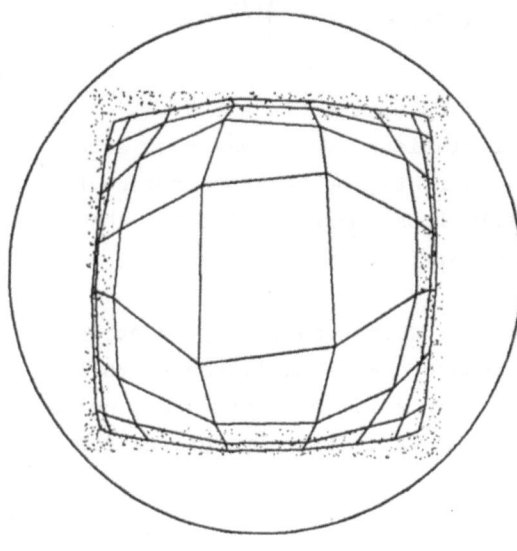

Fig. 3. Projection image of a unit
sphere in $R^3$ showing the
distribution of training
vectors (dots) and the
weight vectors (lattice)

Fig. 3 shows a typical distribution of the vectors $m_i(t)$ after 2 000
steps; hereby $\alpha(t)$ = 200/t. The "training vectors" x(t) were drawn from
the distribution p(x) in a completely random order.

The distribution of the $m_i$ is slightly "contracted" with respect to
p(x), and there are also values of $m_i$ inside the distribution where
p(x) = 0. Such *boundary effects*, however, can be shown to disappear
when the size of the lattice is increased (cf. Kohonen 1981e).

## 5. Self-organized formation of the image of a hierarchical distribution
## of vectors

Obviously the distribution of the weight vectors always tends to *imi-
tate* that of the training vectors x(t). To demonstrate that this is
due for a higher dimensionality of the x(t), too, we shall introduce the
following example that performs a kind of self-organized *taxonomy* of a
hierarchically structured pattern distribution.

Hereupon we shall employ a further innovation which has yielded a superior speed of convergence in the ordering process as compared with earlier experiments. Instead of restricting the updating operation to the unit with the maximum response and its immediate neighbours, it seems more effective to impose the corrections first over a larger set $S_m$ of selected units around and including the maximally responding one, and then to let the size of this set *shrink* to improve the final resolution. This strategy sounds intuitively reasonable and it has also proved to be very effective in practice; no mathematical justification is given here.

The general setting described above is directly applicable to many kinds of data. The input signals can alternatively be continuous- or discrete-valued, even binary ones, as long as it is possible to define a metric between such items.

There is neither anything in the present scheme which would prevent it from being discussed on a more abstract level. The physical system can then be replaced by a mathematical algorithm, whereby the input signals assume the role of *attributes* or *characteristics*, each one eventually with a well-defined meaning (such like an observed feature, e.g., number of legs of an animal). In this way one pattern would correspond to a set of observations of an *item*.

We shall retain the two-dimensional ordering of the PU's; for a change, the topology of the array is now selected *hexagonal*; every unit has thereby only six neighbours with which it interacts in the updating process.

Consider Table 1 which represents a given input data matrix. The columns, labeled "A" through "6", are hypothetical recordings of observable items, and each row position indicates an attribute value. To be quite exact, only $\xi_1$ through $\xi_5$ are given attribute values whereas the sixth component $\xi_0$ has been introduced to *normalize* the pattern vectors of the items, i.e. to make their lengths equal to 10. (The length is equivalent with the Euclidean norm as seen from the definition of $\xi_0$ below Table 1.) An alternative possibility to normalize the items is to use the signals $\xi_1$ through $\xi_5$ only but to multiply each of them by a suitable constant. It has to be emphasized that normalization is in no way necessary for self-organization but it usually makes the process "smoother" and faster. Moreover normalization aids in the final evaluation of the map as will be seen. (The small nonzero value s = 0.01 has been taken to represent zero in floating-point computations.)

Table 1. Data matrix used in the self-ordering experiments

| Item<br>Attr. | A | B | C | D | E | F | G | H | I | J | K | L | M | N | O | P | Q | R | S | T | U | V | W | X | Y | Z | 1 | 2 | 3 | 4 | 5 | 6 |
|---|---|---|---|---|---|---|---|---|---|---|---|---|---|---|---|---|---|---|---|---|---|---|---|---|---|---|---|---|---|---|---|---|
| $\xi_0$ | * | * | * | * | * | * | * | * | * | * | * | * | * | * | * | * | * | * | * | * | * | * | * | * | * | * | * | * | * | * | * | * |
| $\xi_1$ | 1 | 2 | 3 | 4 | 5 | 3 | 3 | 3 | 3 | 3 | 3 | 3 | 3 | 3 | 3 | 3 | 3 | 3 | 3 | 3 | 3 | 3 | 3 | 3 | 3 | 3 | 3 | 3 | 3 | 3 | 3 | 3 |
| $\xi_2$ | s | s | s | s | s | 1 | 2 | 3 | 4 | 5 | 3 | 3 | 3 | 3 | 3 | 3 | 3 | 3 | 3 | 3 | 3 | 3 | 3 | 3 | 3 | 3 | 3 | 3 | 3 | 3 | 3 | 3 |
| $\xi_3$ | s | s | s | s | s | s | s | s | s | s | 1 | 2 | 3 | 4 | 5 | 6 | 7 | 8 | 3 | 3 | 3 | 3 | 6 | 6 | 6 | 6 | 6 | 6 | 6 | 6 | 6 | 6 |
| $\xi_4$ | s | s | s | s | s | s | s | s | s | s | s | s | s | s | s | s | s | s | 1 | 2 | 3 | 4 | 1 | 2 | 3 | 4 | 2 | 2 | 2 | 2 | 2 | 2 |
| $\xi_5$ | s | s | s | s | s | s | s | s | s | s | s | s | s | s | s | s | s | s | s | s | s | s | s | s | s | s | 1 | 2 | 3 | 4 | 5 | 6 |

$$*) \quad \xi_0 = \left(100 - \sum_{i=1}^{5} \xi_i^{2}\right)^{1/2} , \quad s = 0.01$$

The data given in Table 1 have actually been chosen according to a particular rule which defines a *data structure*. If the items were chained together in a similar way as the so-called *dendrograms* or *cladograms* are formed in taxonomy, i.e., by linking those pairs of items which have the smallest mutual distance, and neglecting couplings between more dissimilar items, then it is possible to see that Table 1, at least approximatively, corresponds to the diagram shown in Fig. 4. This is actually the best binary tree which spans the data, although here it has been constructed á priori.

A B C D E
  F
  G
  H K L M N O P Q R
  I     S    W
  J     T    X 1 2 3 4 5 6
       U    Y
       V    Z

Fig. 4. Data structure corresponding to Table 1

Next a self-organizing system of the type depicted in Fig. 2 (with six inputs), and defined by Eqs. (1), (4), (6), and (8) is applied to this data matrix. The initial values of the input parameters $\mu_{ij}$ can be chosen as random numbers. Using a random number generator, samples of

the items of Table 1 are then picked up in a completely haphazard
order, and each of the samples is let to update the "memory array". So,
in a sufficiently long sequence of "training steps", the parameters
$\mu_{ij}$ finally assume values which represent some *image* of the data structure given in Table 1 and Fig. 4.

Six-dimensional vectors, however, cannot be visualized easily. Therefore the ordering result has been displayed in more natural ways.

The first "map" of the results, shown in Fig. 5, is directly based on
the definition of ordered mapping as done in Sec. 2. For every item of
Table 1, the various responses of the PU's are compared, and the position in the array corresponding to the maximal response is labelled by
the symbol of ·this item. Intact positions are labelled by a star (*).

```
E D C B A * * 6 * 5
 * * * * * Z * * * *
* F * * V * Y * 3 *
 G * * U * * 1 2 * 4
H * * T * X * * * *
 I K * S * * W * * R
J * L M N O P * Q *
```

Fig. 5.  Ordered mapping of the
hierarchical data structure
of Table 1

The second "map" of the results is formed in a similar way as single
cell responses are represented in neurophysiology. For every position
in the array, a test over all items is performed, and this position is
labelled according to that item for which a maximum response is obtained. These results are given in Fig. 6.

```
D D C B B U 3 6 5 5
 D C C F U Z 2 5 4 5
G G G U V Y Y 1 3 4
 G G U U U X 1 2 3 4
H H S T T X X 1 2 Q
 I K L S O W W W R R
I K L M N O P P Q R
```

Fig. 6.  Ordered mapping of the
hierarchical data structure
of Table 1

A remark concerning the results is due. As the array is completely
symmetric with respect to the lattice directions and no auxiliary reference information ("seeds") is used, there are many possible orientations for topologically correct organization. If it were necessary
to define a particular orientation for the structure, this could easily
be implemented by the introduction of some anisotropy into the system,
or forcing the image of some items into a fixed position.

It may also be necessary to emphasize that the hierarchical ordering
is defined to be topologically correct although the "legs" of the
graph are bent in different ways. It is the set of neighbourhood relations and not the true form of the graph which is important.

## 6. Construction of relational maps for phonemes by the self-organizing process

In order to demonstrate the self-organizing power of the above system
on natural data, the process described in this paper was applied to a
geometric visualization of relationships between *phonemes*, in particular those of the Finnish language. The system model was similar to that
used above; the number of inputs to each unit was 15. The input data
were derived from utterances of Finnish words.

The Finnish phoneme paradigm consists of eight vowels/u o a æ ø y e i/
and thirteen consonants /s m n ŋ l r j v h k p t d/. It seems that most
of these phonemes can be characterized by stationary frequency spectra
except the voiceless stop consonants /k p t/ which lack energy during
occlusion; they can be distinguished mutually only by context effects,
e.g., transition of formants of the neighbouring voiced phonemes. For
this purpose we have included only 18 phonemes, not regarding the voiceless stop consonants, to this experiment.

For the following demonstration, we have used experimental speech data
that were collected earlier (Kohonen et al., 1980a). The speech waveform
samples were spoken by one male speaker (S.H.). Spectral decomposition
of 4 kHz lowpass-filtered waveforms was made by a programmed DFT algorithm, using a 25.6 ms Hamming window and 10 kHz sampling rate. Of a
128-point logarithmic power spectrum, 15 sample points with 12 equally
spaced points were selected from the frequency range [200 Hz, 3 kHz ] ,
and three points from the range [ 3 kHz, 5 kHz ] . From these spectra,
samples were chosen at those instants of time which corresponded to
stationary phonemes. These phonemic samples were picked up manually

from continuous speech, and in total, 540 phonemes were collected for
the present experiment. These samples thus constituted the 15-component
input pattern vectors $x(t) \in R^{15}$.

The array of processing units was two-dimensional, 7 by 10, and the units
were arranged in a hexagonal topology. As in earlier experiments, the
initial values $m_i(0)$ were chosen as random numbers. The "gain coefficient"
$\alpha(t)$ was taken to be of the form $\alpha(t) = 1\ 000/t$ for $1 \leq t \leq 10\ 000$; for
$t > 10\ 000$, $\alpha(t)$ was made to decrease to zero linearly. The size of the
neighbourhood set $S_m$ decreased linearly in t (in discrete steps) from a
maximum (say, 37) to 7 units for $1 \leq t \leq 10\ 000$ and then stayed constant.

As mentioned above, 540 samples of phonemic spectra were used for "training". The vectors $x(t)$ were picked up from this sample set at random. In
this way, every sample occurred on the average some tens of times during
"training". *No information concerning the classification of phonemes was
given during this phase.*

After a certain number of steps, say, 30 000 in total, the "training"
process was stopped. The map formed in the $m_i$ was tested by the application of input vectors $x$ with known classification. Two alternative ways
to test the map can be used:
A. One could compare to which phoneme each unit was most sensitive, giving the maximum response;  the unit is labelled by the symbol of this
phoneme.
B. One could look at which unit a given phoneme caused the maximum response;  this unit is labelled accordingly.


```
e   e   ø   ø   æ   æ   æ   æ   a   a

  e   ø   y   y   æ   m   m   o   a   a        Fig. 7.  Self-organized relational
i   i   y   y   v   m   m   m   o   o                   map of the Finnish phonemes

  i   i   y   n   n   m   m   ŋ   o   o

j   i   l   v   n   n   ŋ   ŋ   l   u

  h   h   v   n   n   n   d   v   u   u

s   s   r   r   r   r   d   v   v   u
```

Fig. 7 represents one organizing result which was tested by method A. Notice that in the resulting map, the dimensions do not represent any particular features; e.g., this is not a map where the coordinates would correspond to any formant frequencies. One might characterize the map as having the same topology as the set of phonemes has in the frequency space; the form of the map is determined by the density function of empirically recorded phonemes, whereas its orientation has several symmetrically equivalent possibilities.

In the evaluation of experimental results, one has to realize that the phonemic samples used in training *are only believed to be correct;* in principle, there is a possibility that the speaker did not really pronounce them in the same way. It is plausible that different utterances of, say, /v/ or /ℓ/ represent different phonemes if they occur in different context.

## 7. On the theoretical explanation of the self-organizing effect

The reasons for the above self-ordering phenomena are actually very subtle and have strictly been proved only in the simplest cases. In this presentation we shall only delineate a qualitative explanation that should help to understand the basic nature of the process. Details of the proof can be found in (Kohonen, 1981e).

We shall restrict ourselves to a one-dimensional array of functional units to each of which the same scalar-valued input signal $\xi$ is connected. Let the units be numbered $1,2,\ldots,\ell$. Each unit i has only a single input weight $\mu_i$ whereby the similarity between $\xi$ and $\mu_i$ is deduced according to their difference. In fact we can take

$$\eta_i = |\xi - \mu_i| \quad , \tag{9}$$

and the best match is indicated by

$$\eta_m = \min_i |\xi - \mu_i| \quad . \tag{10}$$

We shall define the selected set of units $S_m$ for updating as follows:

$$S_m = \{\max (1,m-1), m, \min (\ell,m+1)\} \quad . \tag{11}$$

In other words, every unit i has the neighbours i-1 and i+1, except at the borders of the array whereby the neighbour of unit 1 is 2, and the neighbour of unit $\ell$ is $\ell$-1, respectively. Then $S_m$ is simply the set of units consisting of unit m and its neighbours.

The updating of the $\mu_i$ is effected in the following way:

$$d\mu_i/dt = \alpha(t)(\xi - \mu_i) \quad \text{for } i \in S_m \quad ,$$

$$d\mu_i/dt = 0 \quad \text{otherwise} \quad . \tag{12}$$

Equations (9-12), almost directly, correspond to those of the higher-dimensional process, i.e., (2), (3), (5), and (6).

Let us now define the *index of disorder* D of the $\mu_i$ in the following way:

$$D = \sum_{i=2}^{\ell} |\mu_i - \mu_{i-1}| - |\mu_\ell - \mu_1| \quad . \tag{13}$$

Obviously $D \geq 0$ whereby the equality holds only if the values $\mu_1$, $\mu_2$, ... , $\mu_\ell$ are ordered in an ascending or descending sequence.

The self-organizing phenomenon now directly follows from the fact that if $\xi$ is a random variable with a stationary, dense distribution $p(\xi)$, then D more often decreases than increases in updating (Kohonen, 1981e). Computation of dD/dt on the basis of Eqs. (9-12) can be performed in a straightforward way. For instance, if $3 \leq m \leq \ell-2$, then only $\mu_{m-1}$, $\mu_m$, and $\mu_{m+1}$ are changed in updating, and there are only five terms in D which can be affected. For any particular sign combination of the $\mu_i - \mu_{i-1}$ corresponding to these terms, the derivative dD/dt can then be expressed in closed form and simplified, after which its sign can be deduced. The discussion, even in this simple case, is a bit too lengthy to be reviewed here. Let it be mentioned that if the $\mu_i$ were initially chosen at random, then D will decrease in 13 cases, stay constant in two cases, and increase in one case out of 16. With time, when partial ordering of the values starts to build up, the cases in which D remains constant at an updating step become more and more frequent. Once D becomes zero, it stays so; in other words, an ordered system cannot become disordered.

After the values $\mu_i$ have become ordered, their final convergence to the asymptotic values is of particular interest since the latter represent the image of the input distribution $p(\xi)$. Assume first for simplicity that $p(\xi)$ is constant over a support $[a,b]$ and zero otherwise. Since a selected unit can affect only its immediate neighbours, it is easily deducible (cf. Kohonen, 1981e) that $\mu_i$ can be affected only if $\xi$ hits an interval $S_i$ which is defined in the following way. Referring to Fig. 8, with $\ell \geq 5$, we have:

Support of $p(\xi)$

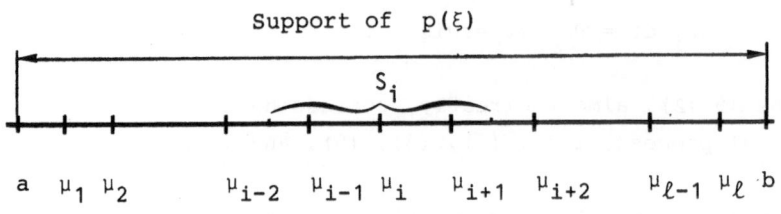

$$S_i$$

$a$ $\mu_1$ $\mu_2$ $\quad$ $\mu_{i-2}$ $\mu_{i-1}$ $\mu_i$ $\quad$ $\mu_{i+1}$ $\mu_{i+2}$ $\quad$ $\mu_{\ell-1}$ $\mu_\ell$ $b$

Fig.8.

for $3 \leq i \leq \ell - 2$: $\quad S_i = [\frac{1}{2}(\mu_{i-2} + \mu_{i-1}), \frac{1}{2}(\mu_{i+1} + \mu_{i+2})]$ ,

for $\quad i = 1 \quad : \quad S_i = [a, \frac{1}{2}(\mu_2 + \mu_3)]$ ,

for $\quad i = 2 \quad : \quad S_i = [a, \frac{1}{2}(\mu_3 + \mu_4)]$ ,

for $\quad i = \ell - 1: \quad S_i = [\frac{1}{2}(\mu_{\ell-3} + \mu_{\ell-2}), b]$ ,

for $\quad i = \ell \quad : \quad S_i = [\frac{1}{2}(\mu_{\ell-2} + \mu_{\ell-1}), b]$ . $\hfill (14)$

The conditional expectation values with respect to $\mu_1, \ldots, \mu_\ell$ of the $d\mu_i/dt = \dot{\mu}_i$ read

$$< \dot{\mu}_i > \overset{d.}{=} E\{\dot{\mu}_i\} = \alpha(E\{\xi \mid \xi \in S_i\} - \mu_i) . \hfill (15)$$

Now $E\{\xi \mid \xi \in S_i\}$ is the center of gravity of $S_i$ (cf. (14)) which is a function of the $\mu_k$ when $p(\xi)$ has been defined. In order to first solve the problem in simplified closed form, it is assumed that $p(\xi) = $ const., whereby one first obtains:

for $3 \leq i \leq \ell - 2$: $\quad < \dot{\mu}_i > = \frac{\alpha}{4} (\mu_{i-2} + \mu_{i-1} + \mu_{i+1} + \mu_{i+2} - 4\mu_i)$ ,

$< \dot{\mu}_1 > = \frac{\alpha}{4} (2a + \mu_2 + \mu_3 - 4\mu_1)$ ,

$< \dot{\mu}_2 > = \frac{\alpha}{4} (2a + \mu_3 + \mu_4 - 4\mu_2)$ ,

$< \dot{\mu}_{\ell-1} > = \frac{\alpha}{4} (\mu_{\ell-3} + \mu_{\ell-2} + 2b - 4\mu_{\ell-1})$ ,

$< \dot{\mu}_\ell > = \frac{\alpha}{4} (\mu_{\ell-2} + \mu_{\ell-1} + 2b - 4\mu_\ell)$ . $\hfill (16)$

Starting with arbitrary initial conditions $\mu_i(0)$ , the most probable, "averaged" trajectories $\mu_i(t)$ are obtained as solutions of an equivalent differential equation corresponding to (16), namely,

$$d\underline{z}/dt = F\underline{z} + \underline{h} \tag{17}$$

where

$$\underline{z} = [z_1, z_2, \ldots, z_\ell]^T ,$$

$$F = \frac{\alpha}{4} \begin{bmatrix} -4 & 1 & 1 & 0 & 0 & 0 & 0 & \cdots & & & & & & \\ 0 & -4 & 1 & 1 & 0 & 0 & 0 & & & & & & & \\ 1 & 1 & -4 & 1 & 1 & 0 & 0 & & & & & & & \\ 0 & 1 & 1 & -4 & 1 & 1 & 0 & & & & & & & \\ & \vdots & & & & & & & & & & & & \vdots \\ & & & & & & & 0 & 1 & 1 & -4 & 1 & 1 & 0 \\ & & & & & & & 0 & 0 & 1 & 1 & -4 & 1 & 1 \\ & & & & & & & 0 & 0 & 0 & 1 & 1 & -4 & 0 \\ & & & & & & \cdots & 0 & 0 & 0 & 0 & 1 & 1 & -4 \end{bmatrix}$$

and

$$\underline{h} = \frac{\alpha}{2} [a, a, 0, 0, \ldots, 0, b, b]^T ,$$

with the initial condition

$$\underline{z}(0) = [\mu_1(0), \mu_2(0), \ldots, \mu_\ell(0)]^T .$$

The averaging, producing (17) from (16), could be made rigorous along the lines given by (Geman, 1979). Equation (17) is a first-order differential equation with constant coefficients, and its solutions are well-established in system theory. It has a fixed-point solution, a particular solution with $d\underline{z}/dt = 0$ which is

$$\underline{z}_o = -F^{-1}\underline{h} \tag{18}$$

provided that $F^{-1}$ exists (which can be shown). The general solution of (17) then reads formally

$$\underline{z} = \underline{z}_o + e^{Ft} \cdot \underline{z}(0) \tag{19}$$

where the exponential function is a square matrix operator. It is a known fact that $\underline{z}$ will converge to $\underline{z}_o$ if all eigenvalues of $F$ have negative real parts (which can be shown to be true).

The asymptotic values of the $\mu_i$ have been calculated for a few lengths of the array and presented in Fig. 9.

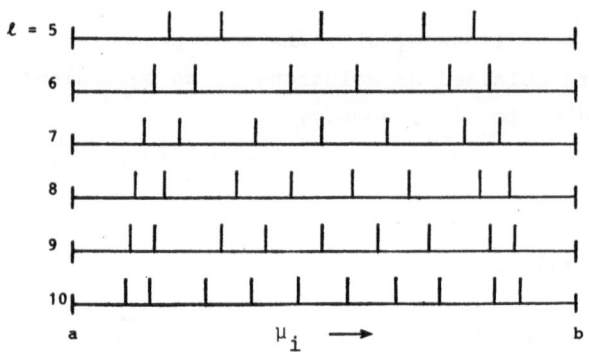

Fig. 9. Asymptotic values of $\mu_i$ on a line of real numbers, for various values of $\ell$

It can be concluded that

- the outermost values $\mu_1$ and $\mu_\ell$ are shifted inwards by an amount which is approximately $1/\ell$; consequently this effect vanishes with increasing $\ell$.
- the values $\mu_3$ through $\mu_{\ell-2}$ seem to be distributed almost evenly.

## 8. The magnification factor

Let us express the result obtained above (which can also be proved easily more rigorously) by stating that, let alone the values near the edge, the $\mu_i$ *are distributed evenly* on the line of real numbers for $p(\xi) =$ const.

Reverting now to Eq. (15) with arbitrary $p(\xi)$, it still seems obvious (although a dynamic analysis would then become much more cumbersome) that *the asymptotic values of the* $\mu_i$ *coincide with the centers of gravity of the* $S_i$ , *with respect to the weighting function* $p(\xi)$. In simple terms this means that *the densities of the* $\mu_i$ *values on the line of real numbers become approximately proportional to their weights* $p(\mu_i)$, *i.e., the point density function of the* $\mu_i$ *approximates* $p(\xi)$.

If, on the other hand, we consider the set of the original units and their associated values $\mu_i$ as a *map* of the input distribution $p(\xi)$, then a result equivalent to the above one is that *the number of units which have their* $\mu_i$ *values confined to a particular interval* $[\mu',\mu'']$ *is approximately proportional to the average of* $p(\xi)$ *over* $\xi \in [\mu',\mu'']$. This number may be termed the *magnification factor* of the map in the due location, and it seems to be directly related with a similar factor which is met in neural fields.

# 9. Conclusions

The main purpose of this work has been to show that very general self-organizing phenomena follow from certain simple structural and functional assumptions in a collective system model. This model was aimed at the demonstration of a possible information processing paradigm; whether it has any explanatory power for biological functions depends finally on the fact how accurate the basic assumptions are. It is not necessary, however, that *exactly* similar structures and functions exist in the nature; the functions may be realized in varying degrees of perfection.

It should be pointed out that this kind of models may have predictive power, too. As the self-organizing principle seems to be of a general nature and thus amenable to implementation by neural signals of many modalities, even the conventional single-cell recording techniques ought to be sufficient for the discovery of new topological feature maps from the brain.

Another implication of this model is that it may be able to conciliate two opposing views of the biological memory, namely, about the memory traces being *localized* vs. *distributed*. It seems that the *responses* can be localized with a high spatial accuracy (at specific cells) while *selection* of such cells is made by means of a very great number of adaptive synaptic-connections which thus constitute the collective and distributed memory traces (Kohonen, 1977; Kohonen et al., 1977, 1980b, 1981).

# References


Amari, S.-I.: Bull Math. Biol. 42, 339-364 (1980)
Geman, S.: SIAM J. Appl. Math. 36, 86-105 (1979)
Hunt, R.K., Berman, N.: J. Comp. Neurol. 162, 43-70 (1975)
Kohonen, T.: Associative Memory - A System-Theoretical Approach. Berlin-He
    delberg-New York: Springer-Verlag (1977)
Kohonen, T. In: Proc. 2nd Scand. Conf. on Image Analysis, 214-220(1981a)
Kohonen, T.: Biological Cybernetics 43, 59-69 (1982)

Kohonen, T.: Helsinki University of Technology Report TKK-F-A461 (1981c)
Kohonen, T.: Helsinki University of Technology Report TKK-F-A462 (1981d)
Kohonen, T.: Helsinki University of Technology Report TKK-F-A463 (1981e)
Kohonen, T., Lehtiö, P., Rovamo, J., Hyvärinen, J., Bry, K., Vainio, L.:
    Neuroscience (IBRO) 2, 1065-1076 (1977)
Kohonen, T., Riittinen, H., Jalanko, M., Reuhkala, E. and Haltsonen, S.:
    Proc. 5th Int. Conf. on Pattern Recognition, 158-165 (1980a)

Kohonen, T., Lehtiö, P., Oja, E.: To appear in a book based on the workshop "Synaptic modification, neuron selectivity, and nervous system organization", Brown University (1980b)

Kohonen, T., Lehtiö, P., Oja, E. In: Hinton, G., Anderson, J.A. (Eds.): Distributed Models of Associative Memory, pp. 105-143. New Jersey: Lawrence Erlbaum Associates (1981)

Lynch, G.S., Rose, G., Gall, C.M. In: Functions of the Septo-Hippocampal System, 5-19, Amsterdam: Ciba Foundation, Elsevier (1978)

Malsburg, Ch. von der: Kybernetik 14, 85-100 (1973)

Malsburg, Ch. von der, Willshaw, D.J.: Proc. Natl. Acad. Sci. USA 74, 5176-5178 (1977)

Oja, E.: A Simplified Neuron Model as a Principal Component Analyzer; to be published (1981)

Rosenblatt, F.: Principles of Neurodynamics: Perceptrons and the Theory of Brain Mechanisms. Washington, D.C.: Spartan Books (1961)

Schneider, G.E. In: Sweet, W.H., Abrador, S., Martin-Rodriquez, J.G. (Eds.): Neurosurgical Treatment in Psychiatry, Pain and Epilepsy. Baltimore, London, Tokyo: University Park Press (1977)

Sharma, S.C.: Exp. Neurol. 34, 171-182 (1972)

Swindale, N.V.: Proc. R. Soc. B 208, 243-264 (1980)

Willshaw, D.J., Malsburg, Ch. von der: Proc. R. Soc. B 194, 431-445 (1976)

Willshaw, D.J., Malsburg, Ch. von der: Phil. Trans. R. Soc. Lond. B 287, 203-243 (1979)

NEOCOGNITRON:   A SELF-ORGANIZING NEURAL NETWORK MODEL

FOR A MECHANISM OF VISUAL PATTERN RECOGNITION

Kunihiko Fukushima   and   Sei Miyake
Auditory and Visual Science Research Division
NHK Broadcasting Science Research Laboratories

1-10-11, Kinuta, Setagaya, Tokyo 157 / Japan

ABSTRACT    A neural network model, called a "neocognitron", is proposed for a mechanism of visual pattern recognition.  It is demonstrated by computer simulation that the neocognitron has characteristics similar to those of visual systems of vertebrates.

The neocognitron is a multilayered network consisting of a cascade connection of many layers of cells, and the efficiencies of the synaptic connections between cells are modifiable.  Self-organization of the network progresses by means of "learning-without-a-teacher" process:  Only repetitive presentation of a set of stimulus patterns is necessary for the self-organization of the network, and no information about the categories to which these patterns should be classified is needed.  The neocognitron by itself acquires the ability to classify and correctly recognize these patterns according to the differences in their shapes:  Any patterns which we human beings judge to be alike are also judged to be of the same category by the neocognitron.  The neocognitron recognizes stimulus patterns correctly without being affected by shifts in position or even by considerable distortions in shape of the stimulus patterns.  If a stimulus pattern is presented at a different position or if the shape of the pattern is distorted, the responses of the cells in the intermediate layers, especially the ones near the input layer, vary with the shift in position or the distortion in shape of the pattern.  However, the deeper the layer is, the smaller become the variations in cellular responses.  Thus, the cells of the deepest layer of the network are not affected by the shift in position or the distortion in shape of the stimulus pattern.

## 1.   INTRODUCTION

The neural mechanism of visual pattern recognition in the brain is little known, and it seems to be almost impossible to reveal it only by conventional physiological experiments.  So, we take a slightly different approach to this problem.  If we could make a neural network model which has the same capability for pattern recognition as a human being, it would give us a powerful clue to the understanding of the neural mechanism in the brain.  In this paper, we discuss how to synthesize a neural network model in order to endow it an ability of pattern recognition like a human being.

Several models were proposed with this intention (Rosenblatt 1962; Block, et al.

1962; Fukushima 1975). The response of most of these models, however, was severely affected by the shift in position and by the distortion in shape of the stimulus patterns. In other words, these conventional models did not have an ability to recognize position-shifted or deformed patterns.

In this paper, we discuss an improved neural network model, which is capable of recognizing stimulus patterns correctly without being affected by shifts in position or even by considerable distortions in shape of the stimulus patterns (Fukushima 1979, 1980).

The new model, which is called a "neocognitron", is a multilayered network consisting of a cascade connection of many layers of cells, and the efficiencies of the synaptic connections between cells are modifiable. It has a function of self-organization, and the self-organization of the network progresses by means of "learning-without-a-teacher" process: Only repetitive presentation of a set of stimulus patterns is necessary for the self-organization of the network, and no information about the categories to which these patterns should be classified is needed. The neocognitron by itself acquires the ability to classify and correctly recognize these patterns according to the differences in their shapes: Any pattern which we human beings judge to be alike are also judged to be of the same category by the neocognitron. The neocognitron recognizes stimulus patterns correctly without being affected by shifts in position or even by considerable distortions in shape of the stimulus patterns.

The "neocognitron" was named after the "cognitron" which was proposed by one of the authors before (Fukushima 1975, 1981). The conventional cognitron was also a self-organizing multilayered neural network and was capable of recognizing patterns. The difference between the neocognitron and the conventional cognitron is that the neocognitron can recognize correctly even position-shifted or shape-distorted patterns while the conventional cognitron could not.

The neocognitron has a hierarchical structure. The information of the stimulus pattern given to the input layer of the neocognitron is processed step by step in each stage of the multilayered network: A cell in a deeper layer generally has a tendency to respond selectively to a more complicated feature of the stimulus patterns, and at the same time, has a larger receptive field, and is less sensitive to shifts in position of the stimulus patterns. Thus, in the deepest layer of the network, each cell responds only to a specific stimulus pattern without being affected by the position or the size of the stimulus patterns.

## 2. STRUCTURE OF THE NEOCOGNITRON

### 2.1. Physiological Background

The structure of the neural network of the neocognitron somewhat resembles the hierarchical model of the visual nervous system proposed by Hubel and Wiesel (1962, 1965).

According to the hierarchical model by Hubel and Wiesel, the neural network in the visual cortex has a hierarchical structure: LGB (lateral geniculate body) → simple cells → complex cells → lower order hypercomplex cells → higher order hypercomplex cells. It is also suggested that the neural network between lower order hypercomplex cells and higher order hypercomplex cells has a structure similar to the network between simple cells and complex cells. In this hierarchy, a cell in a higher stage generally has a tendency to respond selectively to a more complicated feature of the stimulus pattern, and, at the same time, has a larger receptive field, and is more insensitive to the shift in position of the stimulus pattern.

It is true that the hierarchical model by Hubel and Wiesel does not hold in its original form. In fact, there are several experimental data contradictory to the hierarchical model, such as monosynaptic connections from LGB to complex cells. This would not, however, completely deny the hierarchical model, if we consider that the hierarchical model represents only the main stream of information flow in the visual system.

Hubel and Wiesel do not tell what kind of cells exist in the stages higher than hypercomplex cells. Some cells in the inferotemporal cortex of the monkey, however, are reported to respond selectively to more specific and more complicated features than hypercomplex cells (for example, triangles, squares, silhouettes of a monkey's hand, etc.), and their responses are scarcely affected by the position or the size of the stimuli (Gross et al., 1972; Sato et al., 1980). These cells might correspond to so-called "grandmother cells".

Suggested by these physiological data, we extend the hierarchical model of Hubel and Wiesel, and hypothesize the existence of a similar hierarchical structure even in the stages higher than hypercomplex cells. In the extended hierarchical model, the cells in the highest stage are supposed to respond only to specific stimulus patterns without being affected by the position or the size of the stimuli.

The neocognitron proposed here has such an extended hierarchical structure. Figure 1 shows the relation between the hierarchical model by Hubel and Wiesel, and the structure of the neural network of the neocognitron.

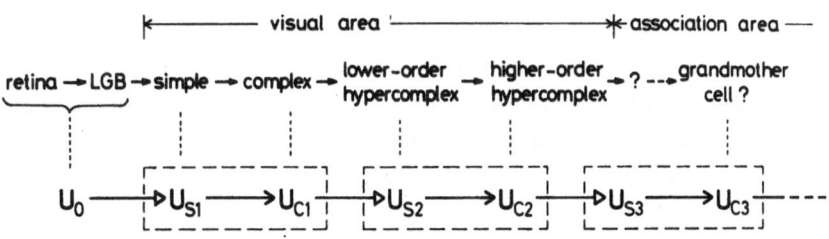

Fig. 1. Comparison between the hierarchical model by Hubel and Wiesel, and the structure of the neural network of the neocognitron.

## 2.2. Structure of the Network

As shown in Fig. 1, the neocognitron consists of a cascade connection of a number of modular structures preceded by an input layer $U_0$. Each of the modular structures is composed of two layers of cells, namely a layer $U_S$ consisting of S-cells, and a layer $U_C$ consisting of C-cells. The layers $U_S$ and $U_C$ in the $\ell$-th module are denoted by $U_{S\ell}$ and $U_{C\ell}$, respectively. An S-cell has a response characteristic similar to a simple cell or a lower order hypercomplex cell according to the classification by Hubel and Wiesel, while a C-cell resembles a complex cell or a higher order hypercomplex cells. In the neocognitron, only the afferent synapses to S-cells are modifiable, and the other synapses are fixed and unmodifiable.

The input layer $U_0$ consists of a photoreceptor array. The output of a photo-receptor is denoted by $u_0(\mathbf{n})$ where $\mathbf{n}$ is the two-dimensional co-ordinates indicating the location of the cell.

S-cells or C-cells in any single layer are sorted into subgroups according to the optimum stimulus features of their receptive fields. Since the cells in each subgroup are set in a two-dimensional array, we call the subgroup as a "cell-plane". We will also use a terminology, S-plane and C-plane to represent the cell-planes consisting of S-cells and C-cells, respectively.

It is hypothesized that all the cells in a single cell-plane have afferent synapses of the same spatial distribution, and only the positions of the presynaptic cells are shifted in parallel depending on the position of the postsynaptic cells. This situation is illustrated in Fig. 2. Even in the process of learning, in which the efficiencies of the afferent synapses to S-cells are modified, the modification is performed always under this restriction.

We will use notations $U_{S\ell}(k_\ell,\mathbf{n})$ to represent the output of an S-cell in the $k_\ell$-th S-plane in the $\ell$-th module, and $u_{C\ell}(k_\ell,\mathbf{n})$ to represent the output of a C-cell in the $k_\ell$-th C-plane in that module, where $\mathbf{n}$ is the two-dimensional co-ordinates representing the position of these cells' receptive fields on the input layer.

Figure 3 is a schematic diagram illustrating the synaptic connections between layers. Each tetragon drawn with heavy lines represents an S-plane or a C-plane, and each vertical tetragon drawn with thin lines, in which S-planes or C-planes are enclosed, represents an S-layer or a C-layer.

In Fig. 3, for the sake of simplicity, only one cell is shown in each cell-plane. Each of these cells receives afferent synapses from the cells within the area enclosed by the ellipse in its pre-ceding layer. All the other cells in the same cell-plane have afferent synapses of the same spatial distribution, and only the positions of the pre-synaptic cells are shifted in parallel from cell

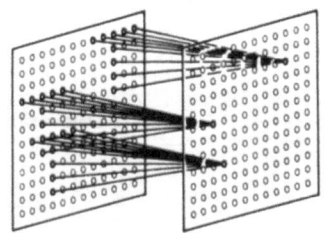

Fig. 2. Illustration showing the afferent synapses to the cells of an arbitrary cell-plane.

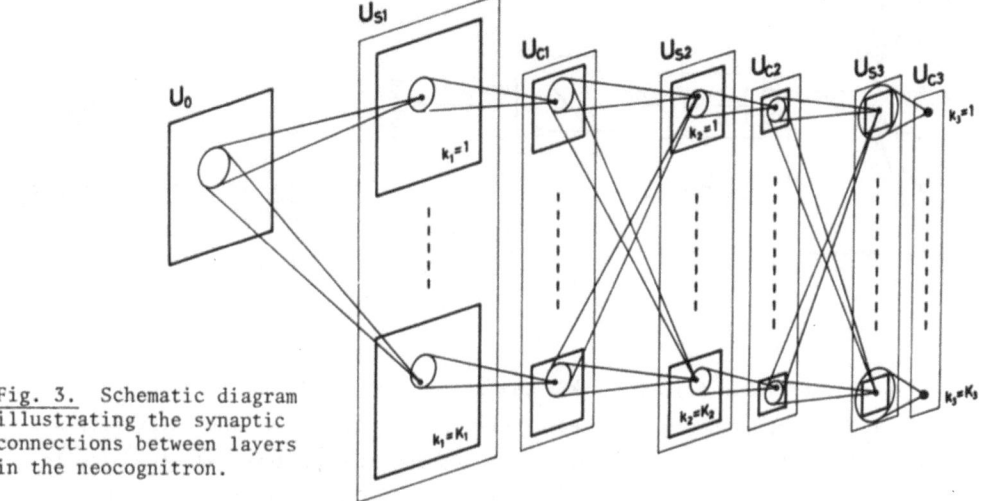

Fig. 3. Schematic diagram
illustrating the synaptic
connections between layers
in the neocognitron.

to cell. Hence, all the cells in a single cell-plane have receptive fields of the
same function but at different positions.

Since the cells in the network are interconnected in a cascade as shown in Fig.
3, the deeper the layer is, the larger becomes the receptive field of each cell of
that layer. The density of the cells in each cell-plane is so determined as to decrease in accordance with the increase of the size of the receptive fields. Hence,
the total number of the cells in each cell-plane decreases with the depth of the cell-
plane in the network. In the deepest module, the receptive field of each C-cell becomes so large as to cover the whole input layer, and each C-plane is so determined
as to have only one C-cell.

S-cells and C-cells are excitatory cells. Although it is not shown in Fig. 3,
we also have inhibitory cells, namely, $V_S$-cells in S-layers and $V_C$-cells in C-layers.
All the cells employed in the neocognitron are of analog type: That is, the input and
output signals of the cells take non-negative analog values proportional to the instantaneous firing frequencies of actual biological neurons.

Here, we will describe the outputs of these cells with numerical expressions.

As shown in Fig. 4, S-cells have inhibitory inputs with shunting mechanism.
Incidentally, S-cells have the same characteristics as the excitatory cells employed
in the conventional cognitron (Fukushima 1975, 1981). The output of an S-cell of the
$k_\ell$-th S-plane in the $\ell$-th module is given by

$$
u_{S\ell}(k_\ell,\mathbf{n}) = r_\ell \cdot \phi \left[ \frac{1 + \sum_{k_{\ell-1}=1}^{K_{\ell-1}} \sum_{\mathbf{v} \in S_\ell} a_\ell(k_{\ell-1},\mathbf{v},k_\ell) \cdot u_{C\ell-1}(k_{\ell-1},\mathbf{n}+\mathbf{v})}{1 + \frac{r_\ell}{1+r_\ell} \cdot b_\ell(k_\ell) \cdot v_{C\ell-1}(\mathbf{n})} - 1 \right], \quad (1)
$$

where $\phi[x] = \max(x,0)$ .

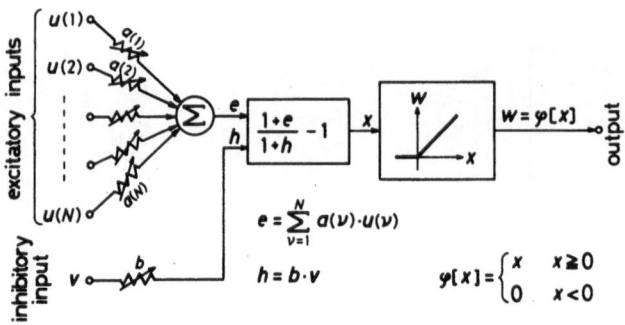

**Fig. 4.** Input-to-output characteristics of an S-cell: A typical example of the cells employed in the neocognitron.

In the case of $\ell=1$ in Eq. (1), $u_{C\ell-1}(k_{\ell-1},\mathbf{n})$ stands for $u_0(\mathbf{n})$, and we have $K_{\ell-1}=1$.

Here, $a_\ell(k_{\ell-1},\mathbf{v},k_\ell)$ and $b_\ell(k_\ell)$ represent the efficiencies of the excitatory and inhibitory modifiable synapses, respectively. As described before, all the S-cells in the same S-plane are assumed to have identical set of afferent synapses. Hence, $a_\ell(k_{\ell-1},\mathbf{v},k_\ell)$ and $b_\ell(k_\ell)$ do not contain any argument representing the position $\mathbf{n}$ of the receptive field of cell $u_{S\ell}(k_\ell,\mathbf{n})$.

Parameter $r_\ell$ in Eq. (1) controls the intensity of the inhibition. The larger the value of $r_\ell$ is, the more selective becomes the cell's response to its specific feature. A more detailed discussion on the response of S-cells will be given in Section 3.3.

The inhibitory cell $v_{C\ell-1}(\mathbf{n})$, which is sending an inhibitory signal to cell $u_{S\ell}(k_\ell,\mathbf{n})$, receives afferent synapses from the same group of cells as $u_{S\ell}(k_\ell,\mathbf{n})$ does, and yields an output proportional to the weighted root-mean-square of its inputs:

$$v_{C\ell-1}(\mathbf{n}) = \sqrt{\sum_{k_{\ell-1}=1}^{K_{\ell-1}} \sum_{\mathbf{v}\in S_\ell} c_{\ell-1}(\mathbf{v})\cdot u_{C\ell-1}^2(k_{\ell-1},\mathbf{n}+\mathbf{v})} \qquad (2)$$

The efficiencies of the unmodifiable synapses $c_{\ell-1}(\mathbf{v})$ are determined so as to decrease monotonically with respect to $|\mathbf{v}|$, and to satisfy

$$\sum_{k_{\ell-1}=1}^{K_{\ell-1}} \sum_{\mathbf{v}\in S_\ell} c_{\ell-1}(\mathbf{v}) = 1 \qquad (3)$$

The size of the connection area $S_\ell$ of these cells is set to be small in the first module and to increase with the depth $\ell$.

The synaptic connections from S-cells to C-cells are fixed and unmodifiable. As illustrated in Fig. 3, each C-cell has afferent synapses leading from a group of S-cells in the S-plane preceding it (i.e. the S-plane with the same $k_\ell$-number as that of the C-cell). This means that all of the presynaptic S-cells extract the same stimulus feature but from slightly different positions on the input layer. The efficiencies of the synapses are determined in such a way that the C-cell will be activated whenever at least one of its presynaptic S-cells is active. Hence, even if a stimulus pattern which has elicited a large response from the C-cell is shifted a little in position, the C-cell will still keep responding as before, because another presynaptic S-cell will become active instead of the first one. In other words, a C-cell responds to the same stimulus feature as its presynaptic S-cells do, but is less sensitive to the shift in position of the stimulus feature.

Quantitatively, the output of a C-cell of the $k_\ell$-th C-plane in the $\ell$-th module is given by

$$u_{C\ell}(k_\ell,n) = \psi \left[ \frac{1 + \sum_{\nu \in D_\ell} d_\ell(\nu) \cdot u_{S\ell}(k_\ell,n+\nu)}{1 + v_{S\ell}(n)} - 1 \right], \tag{4}$$

where $\psi[x] = \begin{cases} x/(\alpha+x) & (x \geq 0) \\ 0 & (x < 0) \end{cases}.$

The parameter $\alpha$ is a positive constant which determines the degree of saturation of the output. In the computer simulation discussed in Section 5, we choose $\alpha=0.5$ .

The inhibitory cell $v_{S\ell}(n)$, which sends an inhibitory signal to this C-cell and makes up the system of lateral inhibition, yields an output proportional to the (weighted) arithmetic mean of its inputs:

$$u_{S\ell}(n) = \frac{1}{K_\ell} \sum_{k_\ell=1}^{K_\ell} \sum_{\nu \in D_\ell} d_\ell(\nu) \cdot u_{S\ell}(k_\ell,n+\nu) . \tag{5}$$

In Eqs. (4) and (5), the efficiencies of unmodifiable synapses $d_\ell(\nu)$ are determined so as to decrease monotonically with respect to $|\nu|$ similarly as $c_\ell(\nu)$. The size of the connection area $D_\ell$ is set to be small in the first module and to increase with the depth $\ell$.

## 3. SELF-ORGANIZATION OF THE NETWORK

### 3.1. Reinforcement of the Modifiable Synapses

The self-organization of the neocognitron is performed by means of unsupervised learning, that is, "learning without a teacher": During the process of self-organization, the network is repeatedly presented with a set of stimulus patterns to the input layer, but it does not receive any other information about the categories of the stimulus patterns.

As was discussed in Section 2, it is hypothesized that all the S-cells in a single S-plane should have afferent synapses always of the same spatial distribution, and that only the position of the presynaptic cells shift in parallel depending on the positions of the postsynaptic S-cells.

It has not been physiologically demonstrated yet wheather the self-organization of the real neural network can progress always keeping such orderly synaptic connections as hypothesized in the neocognitron. There are, however, some physiological experiments which suggest the validity of this hypothesis: In the amphybian or fish, orderly synaptic connections are formed between retina and optic tectum, not only in the initial development in the embryo, but also in regeneration after severance of the optic nerve. Furthermore, even in the situation where the correspondence between the retinal cells and the target cells is forced to change, orderly synaptic connections can still be formed: In regeneration after removal of half the tectum, it is reported that the whole retina comes to make a compressed orderly projection upon the remaining tectum (e.g., review article by Meyer and Sperry 1974). These experiments suggest that the neural network has some mechanism which makes orderly synaptic connections. Hence, it would not be unreasonable to hypothesize that such a mechanism works even in the process of self-organization of the normal brain.

Anyway, in the neocognitron, in order to develop the self-organization without destroying such orderly synaptic connections in the network, the modifiable synapses are assumed to be reinforced by means of the following procedure.

At first, several "representative" S-cells are chosen from each S-layer every time when a stimulus pattern is presented. The representatives are chosen among the S-cells which have yielded large outputs, but the number of the representatives is so restricted that more than one representative should not be chosen from any single S-plane. The detailed procedure for choosing the representatives is discussed in Section 3.2.

As for a representative S-cell, only the afferent synapses through which non-zero signals are coming are reinforced. With this procedure, the representative S-cell becomes selectively responsive only to the stimulus feature which is now presented. A detailed discussion on the S-cell's response will appear in Section 3.3. All the other S-cells in the S-plane, from which the representative is chosen, have their afferent synapses reinforced by the same amounts as those for their representative. These relations can be quantitatively expressed as follows.

Let cell $u_{S\ell}(\hat{k}_\ell, \hat{n})$ be chosen as a representative. The modifiable synapses $a_\ell(k_{\ell-1}, \nu, \hat{k}_\ell)$ and $b_\ell(\hat{k}_\ell)$, which are afferent to the S-cells of this S-plane, are reinforced by the amount shown below:

$$\Delta a_\ell(k_{\ell-1}, \nu, \hat{k}_\ell) = q_\ell \cdot c_{\ell-1}(\nu) \cdot u_{C\ell-1}(k_{\ell-1}, \hat{n}+\nu) , \tag{6}$$

$$\Delta b_\ell(\hat{k}_\ell) = q_\ell \cdot v_{C\ell-1}(\hat{n}) , \tag{7}$$

where $q_\ell$ is a positive constant which determines the speed of reinforcement.

The cells in the S-planes, from which no representative is chosen, however, do not have their afferent synapses modified at all.

The choice of initial values of these modifiable synapses has little effect on the performance of the neocognitron, provided that they are small and are determined in such a way that each S-plane has a different set of initial values for its afferent synapses. In the computer simulation discussed in Section 5, the initial values of the excitatory modifiable synapses $a_\ell(k_{\ell-1}, \nu, k_\ell)$ are set to be small positive values in such a way that each S-cell has very weak orientation selectivity, and that its preferred orientation differs from S-plane to S-plane. They do not have any randomness. The initial values of inhibitory modifiable synapses $b_\ell(k_\ell)$ are set to be zero.

## 3.2.  Choosing the Representatives

The procedure for choosing the representatives is
as follows:  At first, in an S-layer, we pick up a group
of S-cells whose receptive fields are situated within
a small area on the input layer.  Such a group of S-
cells constitutes a column in an S-layer, if we arrange
the S-planes of an S-layer in a manner shown in Fig. 5.
Accordingly, we call the group an "S-column".  An S-
column contains S-cells from all the S-planes:  That
is, an S-column contains various kinds of feature-
extracting cells in it, but the receptive fields of
these cells are situated almost at the same position.
Hence, the idea of S-columns defined here closely

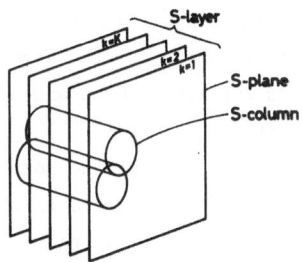

Fig. 5.  Relation between
S-planes and S-columns
within an S-layer.

resembles that of "hypercolumns" proposed by Hubel and Wiesel (1977).  There are a
lot of such S-columns in a single S-layer.  Since S-columns overlap with each other,
there is a possibility that a single S-cell is contained in two or more S-columns.

From each S-column, every time when a stimulus pattern is presented, the S-cell
which is yielding the largest output is chosen as a candidate for the representatives.
Hence, there is a possibility that a number of candidates appear in a single S-plane.
If two or more candidates appear in a single S-plane, only the one which is yielding
the largest output among them is chosen as the representative from that S-plane.  In
case only one candidate appears in an S-plane, the candidate is unconditionally determined as the representative from that S-plane.  If no candidate appears in an S-plane,
no representative is chosen from that S-plane.

Since the representatives are determined in this manner, it is seldom that two
or more representatives are selected among the cells which are activated by one and
the same kind of stimulus feature.  Hence, there is not a possibility of formation
of redundant connections such that two or more S-planes are used for detection of one
and the same stimulus feature.  Incidentally, representatives are chosen only from a
small number of S-planes at a time, and the rest of the S-planes are to send representatives when other stimulus patterns are presented.

## 3.3.  Response of an S-cell

In this section, we discuss how each S-cell comes to respond selectively to differences in stimulus patterns.  Since the structure between two adjoining modules is
similar in all parts of the network, we observe the response of an arbitrary S-cell
$u_{S1}(k_1,\mathbf{n})$ as a typical example.  Fig. 6 shows the synaptic connections converging to
such a cell.  For the sake of simplicity, we will omit the suffixes S and $\ell=1$, and the
arguments $k_\ell$ and $\mathbf{n}$, and represent the response of this cell simply by u.  Similarly,

we will use the notation v for the output of the inhibitory cell $v_{CO}(n)$, which sends an inhibitory signal to cell u. For the other variables, the arguments $k_\ell$ and $n$, and sufixes S, C, $\ell$ and $\ell-1$ will also be omitted.

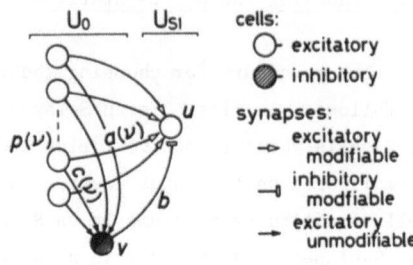

Fig. 6. Synaptic connections converging to an S-cell.

Let $p(v)$ be the response of the cells of layer $U_0$ situated in the connection area of cell u, so that

$$p(v) = u_0(n+v) . \qquad (8)$$

In other words, $p(v)$ is the stimulus pattern (or feature) presented to the receptive field of cell u.

With this notation, Eqs. (1) and (2) can be written:

$$u = r \cdot \phi \left( \frac{1 + \sum_v a(v) \cdot p(v)}{1 + \frac{r}{1+r} \cdot b \cdot v} - 1 \right) , \qquad (9)$$

$$v = \sqrt{\sum_v c(v) \cdot p^2(v)} . \qquad (10)$$

When cell u is chosen as a representative, the amounts of reinforcement of the modifiable synapses are derived from Eqs. (6) and (7), that is,

$$\Delta a(v) = q \cdot c(v) \cdot p(v) , \qquad (11)$$

$$\Delta b = q \cdot v . \qquad (12)$$

Let s be defined by

$$s = \frac{\sum_v a(v) \cdot p(v)}{b \cdot v} . \qquad (13)$$

Then Eq. (9) reduces approximately to

$$u \simeq r \cdot \phi \left( \frac{r+1}{r} \cdot s - 1 \right) , \qquad (14)$$

provided $a(v)$ and b are sufficiently large.

Let us suppose that cell u has been chosen N-times as a representative for the same stimulus pattern $p(v)=P(v)$. We also suppose that no other cell has been chosen as a representative from the S-plane in which cell u is located. Assuming that the initial values of the modifiable synapses are amall enough to be neglected, we obtain

$$a(v) = N \cdot q \cdot c(v) \cdot P(v) , \qquad (15)$$

$$b = N \cdot q \sqrt{\sum_v c(v) \cdot P^2(v)} . \qquad (16)$$

Substituting Eqs. (10), (15) and (16) into Eq. (13), we obtain

$$s = \frac{\sum_v c(v) \cdot P(v) \cdot p(v)}{\sqrt{\sum_v c(v) \cdot P^2(v)} \cdot \sqrt{\sum_v c(v) \cdot p^2(v)}} . \qquad (17)$$

If we regard $p(\nu)$ and $P(\nu)$ as vectors, Eq. (17) can be interpreted as the (weighted) inner product of the two vectors normalized by the norms of both vectors. In other words, s gives the cosine of the angle between the two vectors $p(\nu)$ and $P(\nu)$ in the multidimensional vector space. Therefore, we have s=1 only when $p(\nu)=P(\nu)$, and we have s<1 for all patterns such as $p(\nu)\neq P(\nu)$. This means that s becomes maximum for the learned pattern, and becomes smaller for any other patterns.

Consider the case when N is so large that Nq>>1. In this case, Eq. (14) holds.

When an arbitrary pattern $p(\nu)$ is presented, and if it satisfies s>r/(r+1), we have u>0 by Eq. (14). Conversely, for a pattern which makes s≦r/(r+1), cell u does not respond. We can interpret this by saying that cell u judges the similarity between patterns $p(\nu)$ and $P(\nu)$ using the criterion defined by Eq. (17), and that it responds only to patterns judged to be similar to $P(\nu)$. Incidentally, if $p(\nu)=P(\nu)$, we have s=1 and consequently u≈1.

Roughly speaking, during the self-organization process, patterns $p(\nu)$ which satisfy s>r/(r+1) are taken to be of the same class as $P(\nu)$, and patterns $p(\nu)$ which make s≦r/(r+1) are assumed to be of different classes from $P(\nu)$.

The patterns which are judged to be of different classes will possibly come to be extracted by the cells in the other S-planes.

Since the value r/(r+1) tends to 1 with increase of r, a larger value of r makes the cell's response more selective to one specific pattern or feature. In other words, a large value of r endows the cell with a high ability to discriminate patterns of different classes. However, a higher selectivity of the cell's response is not always desirable, because it decreases the ability to tolerate the deformation of patterns. Hence, the value of r should be determined at a point of compromise between these two contradictory conditions.

In the above analysis, we supposed that cell u is trained only for one particular pattern $P(\nu)$. However, this does not necessarily mean that the training pattern sequence should consist of the repetition of only one pattern $P(\nu)$. Once the self-organization of the network progresses a little, each cell in the network becomes selectively responsive to one particular pattern. Hence, even if pattern other than $P(\nu)$ appear in the training pattern sequence, they usually do not elicit any response from cell u, unless they closely resemble $P(\nu)$. Hence, they do not have any effect on the reinforcement of the synapses afferent to u. Thus, among many patterns in the training pattern sequence, only pattern $P(\nu)$ effectively works for the training of cell u.

In the above discussion, we also assumed that the stimulus pattern is always presesented at the same position in the input layer, and consequently that the representative from an arbitrary S-plane, if any, is always the same cell. Actually, however, this restriction is not necessary for our discussion, either. We have the same results, even if the position of presentation of the stimulus pattern varies each time. We can explain this reason as follows. If the stimulus pattern is presented at a different position, another cell in that S-plane will become the representative, but the

position of the new representative relative to the stimulus pattern will usually be
kept still the same as before. On the other hand, according to the algorithm of self-
organization discussed in Section 3.1, all the S-cells in that S-plane have their input
interconnections modified always in the same manner as their representative of the
moment. Hence, the shift in position of the stimulus pattern does not affect the
result of self-organization of the network.

The above discussion is not restricted to S-cells of layer $U_{S1}$. Each S-cell in
succeeding modules shows a similar type of response, if we regard the response of the
C-cells in its connection area in the preceding layer as its input pattern.


## 4. A ROUGH SKETCH OF THE WORKING OF THE NEOCOGNITRON

In order to help the understanding of the principles with which the neocognitron
performs pattern recognition, we will make a rough sketch of the working of the network
in the state after completion of the self-organization. The description in this section, however, is not so strict, because the purpose of this section is only to show
the outline of the working of the network.

At first, let us assume that the neocognitron has been self-organized with repeated presentations of a set of stimulus patterns like "A", "B", "C" and so on. In the
state when the self-organization has been completed, various feature-extracting cells
are formed in the network as shown in Fig. 7. The upper half of Fig. 7 is a simplified sketch of the whole network, while the lower half of the figure is an enlarged
illustration showing the detailed structure of a part of the network. (It should be
noted that Fig. 7 shows only an example. It does not mean that exactly the same feature extractors as shown in this figure are always formed in the network.)

If pattern "A" is presented to the input layer $U_0$, the cells in the network yield
outputs as shown in Fig. 7. For instance, S-plane with $k_1=1$ in layer $U_{S1}$ consists of
a two-dimensional array of S-cells which extract $\wedge$-shaped feature. Since the stimulus pattern "A" contains $\wedge$-shaped feature at the top, an S-cell near the top of this
S-plane yields a large output as shown in the enlarged illustration in the lower half
of Fig. 7.

A C-cell in the succeeding C-plane (i.e. C-plane in layer $U_{C1}$ with $k_1=1$) has
synaptic connections from a group of S-cells in this S-plane. For example, the C-cell
shown in Fig. 7 has synaptic connections from the S-cells situated within the thin-
lined circle, and it is activated whenever at least one of these presynaptic S-cells
yields a large output. Hence, the C-cell responds to a $\wedge$-shaped feature situated in
a certain area in the input layer, and its response is less affected by the shift in
position of the stimulus pattern than that of the presynaptic S-cells. Since this C-
plane consists of an array of such C-cells, several C-cells which are situated near
the top of this C-plane respond to the $\wedge$-shaped feature of the stimulus pattern "A".
In layer $U_{C1}$, besides this C-plane, we also have C-planes which extract features with

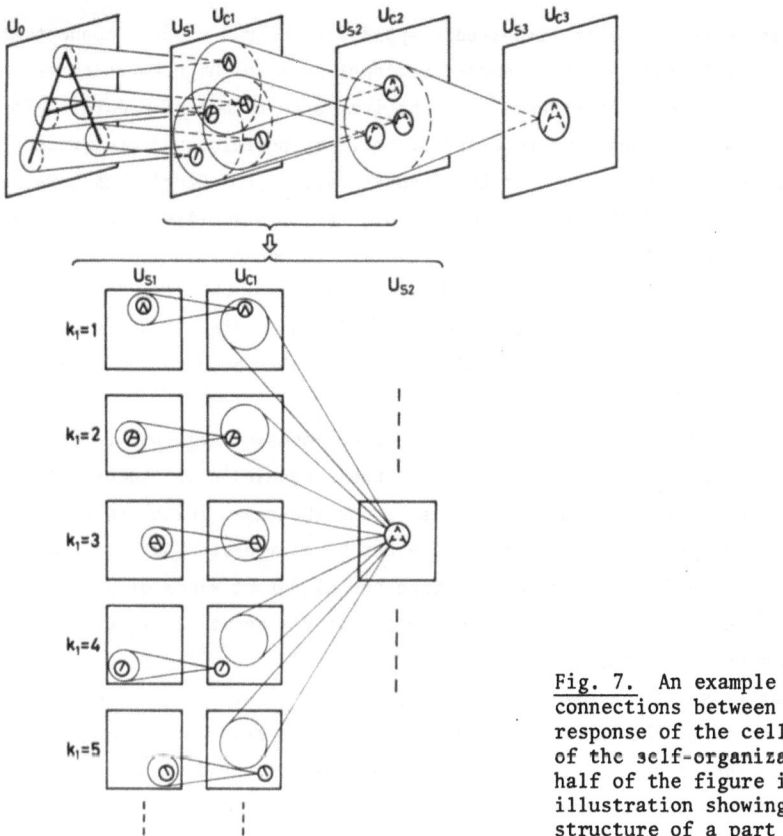

Fig. 7. An example of the synaptic connections between cells, and the response of the cells after completion of the self-organization. The lower half of the figure is an enlarged illustration showing a detailed structure of a part of the network.

shapes like ⊬, ⋏ and so on.

In the next module, each S-cell receives signals from all the C-planes of layer $U_{C1}$. For example, the S-cell of layer $U_{S2}$ shown in Fig. 7 receives signals from C-cells within the thin-lined circles in layer $U_{C1}$. Its afferent synapses have been reinforced in such a way that this S-cell responds only when ⋏-shaped, ⊬-shaped and ⊣-shaped features are presented in its receptive field with configulation like ⋏ₐ. Hence, pattern "A" elicits a large response from this S-cell, which is situated a little above the center of this S-plane. Even if positional relation of these three features is changed a little, this cell will still keep responding, because the preceding C-cells are not so sensitive to the positional error of these features. However, if positional relation of these three features is changed beyond some allowance, this S-cell stops responding. This S-cell also checks the condition that other features such as ends-of-lines, which are to be extracted in S-planes in layer $U_{S1}$ with $k_1=4$, 5 and so on, are not presented in its receptive field. The inhibitory $V_C$-cell, which makes inhibitory synapses to this S-cell of layer $U_{S2}$, plays an important role in checking the absence of such irrelevant features.

Since operations of this kind are repeatedly applied through a cascade connection of modular structures of S- and C-layers, each individual cell in the network comes to have wider receptive field in accordance with the increased number of modules before it, and, at the same time, becomes more tolerant of shift in position and distortion in shape of the stimulus pattern. Thus, in the deepest layer, each C-cell has a receptive field large enough to cover the whole input layer, and is selectively responsive only to one of the stimulus patterns, say to "A", but its response is not affected by shift in position or even by considerable distortions in shape of the stimulus pattern. Although only one cell which responds to pattern "A" is drawn in Fig. 7, cells which respond to other patterns, such as "B", "C" and so on, have also been formed in parallel in the deepest layer.

From these discussions, it might be felt as if an enormously large number of feature-extracting cell-planes become necessary with the increase in the number of stimulus patterns to be recognized. However, this is not the case. With the increase in the number of stimulus patterns, it becomes more and more probable that one and the same feature is contained in common in more than two different kinds of stimulus patterns. Hence, each cell-plane, especially the one near the input layer, will generally be used in common for feature extraction, not from only one pattern, but from numerous kinds of patterns. Therefore, the required number of cell-planes does not increase so much in spite of the increase in the number of patterns to be recognized.

We can summarize the discussion in this chapter as follows. The stimulus pattern is first observed within a narrow range by each of the cells in the first module, and several features of the stimulus pattern are extracted. In the next module, these features are combined by observation over a little larger range, and higher-order features are extracted. Operations of this kind are repeatedly applied through a cascade connection of a number of modules. In each stage of these operations, a small amount of positional error is tolerated. The operation by which positional errors are tolerated little by little, not at a single stage, plays an important role in endowing the network with an ability to recognize even distorted patterns.

## 5. COMPUTER SIMULATION

The neural network proposed here was simulated on a digital computer. In the computer simulation, we considered a seven layered network: $U_0 \to U_{S1} \to U_{C1} \to U_{S2} \to U_{C2} \to U_{S3} \to U_{C3}$. In other words, the network has three stages of modular structures preceded by an input layer. The number of cell-planes $K_\ell$ is equally 24 for every layers $U_{S1} \sim U_{C3}$. The numbers of excitatory cells in these seven layers are: $16 \times 16$ in $U_0$, $16 \times 16 \times 24$ in $U_{S1}$, $10 \times 10 \times 24$ in $U_{C1}$, $8 \times 8 \times 24$ in $U_{S2}$, $6 \times 6 \times 24$ in $U_{C2}$, $2 \times 2 \times 24$ in $U_{S3}$, and 24 in $U_{C3}$. In the deepest layer $U_{C3}$, each of the 24 cell-planes contains only one C-cell.

The number of cells contained in a connection area $S_\ell$ is always $5 \times 5$ for every S-

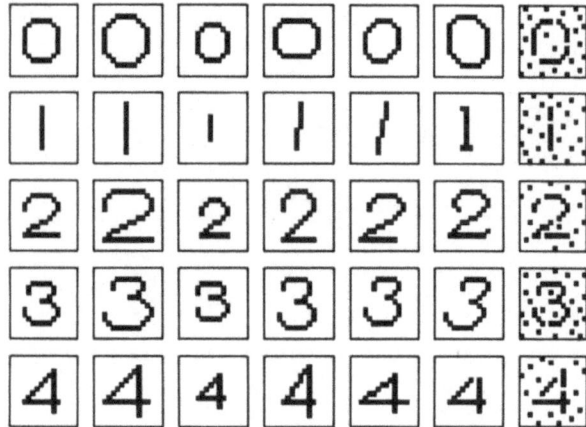

Fig. 8. Some examples of the stimulus patterns which the neocognitron recognized correctly. The neocognitron was first trained with the patterns shown in the leftmost column.

layer. Hence, the number of excitatory afferent synapses to each S-cell is 5×5 in layer $U_{S1}$, and 5×5×24 in layers $U_{S2}$ and $U_{S3}$, because layers $U_{S2}$ and $U_{S3}$ are preceded by C-layers consisting of 24 cell-planes each. Although the number of cells contained in $S_\ell$ is the same for every S-layer, the size of $S_\ell$, which is projected to and observed at layer $U_0$, increases with the depth of the S-layer, because of decrease in density of the cells in a cell-plane.

The number of excitatory input synapses to each C-cell is 5×5 in layers $U_{C1}$ and $U_{C2}$, and is 2×2 in layer $U_{C3}$. Every S-column has a size such that it contains 5×5×24 cells for layers $U_{S1}$ and $U_{S2}$, and 2×2×24 cells for layer $U_{S3}$. That is, it contains 5×5, 5×5 and 2×2 cells from each S-plane, in layers $U_{S1}$, $U_{S2}$ and $U_{S3}$, respectively.

For the other parameters, we choose $r_1=4.0$, $r_2=r_3=1.5$, $q_1=1.0$ and $q_2=q_3=16.0$.

We tested the capability of this network using various kinds of training patterns. Here we show the result of one of these experiments.

During the stage of learning, five training patterns "0", "1", "2", "3" and "4", shown in the leftmost column in Fig. 8 were presented repeatedly to the input layer $U_0$. The positions of presentation of these training patterns were randomly shifted at every presentation. It does not matter, of course, even if the training patterns are presented always at the same position. On the contrary, the self-organization generally becomes easier if the position of pattern presentation is stationary than it is shifted at random. Thus, the experimental result under a more difficult condition is shown here.

The learning was performed "without a teacher", and the neocognitron dit not receive any information about the categories of the training patterns.

After repeated presentations of these five training patterns, the neocognitron by itself gradually acquired the ability to classify these patterns according to the differences in their shape. In this simulation, each of the five training patterns was presented 20 times. By that time, the self-organization of the network was almost completed.

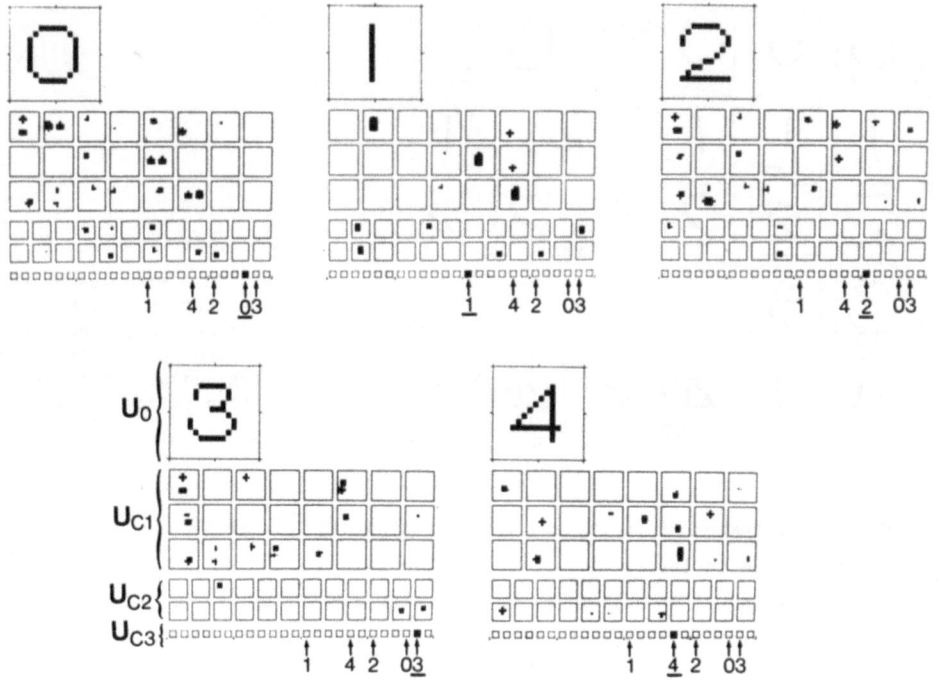

<u>Fig. 9.</u>  Response of the cells of layers $U_0$, $U_{C1}$, $U_{C2}$ and $U_{C3}$ to each of the five stimulus patterns.

Figure 9 collectively shows how the individual cells in the network came to respond to the five training patterns.  In each of the pictures in Fig. 9, out of the seven cell-layers constituting the neocognitron, we display only four layers, namely, the input layer $U_0$ and C-cell-layers $U_{C1}$, $U_{C2}$ and $U_{C3}$ arranging them vertically in order.  In the deepest layer $U_{C3}$, which is displayed in the bottom row of each picture, it is seen that a different cell responds to each stimulus pattern "0", "1", "2", "3" and "4".  This means that the neocognitron correctly recognizes these five stimulus patterns.

We will see next how the response of the neocognitron is changed by shifts in position of presentation of the stimulus patterns.  Figure 10 shows how the individual cells of layers $U_0$, $U_{C1}$, $U_{C2}$ and $U_{C3}$ respond to stimulus pattern "2" presented at four different positions.  As is seen in this figure, the responses of the cells in the intermediate layers, especially the ones near the input layer, vary with the shift in position of the stimulus pattern.  However, the deeper the layer is, that is, the lower in each of the four pictures the layer is, the smaller become the variations in response of the cells in it.  Thus, the cells of the deepest layer $U_{C3}$ are not af-

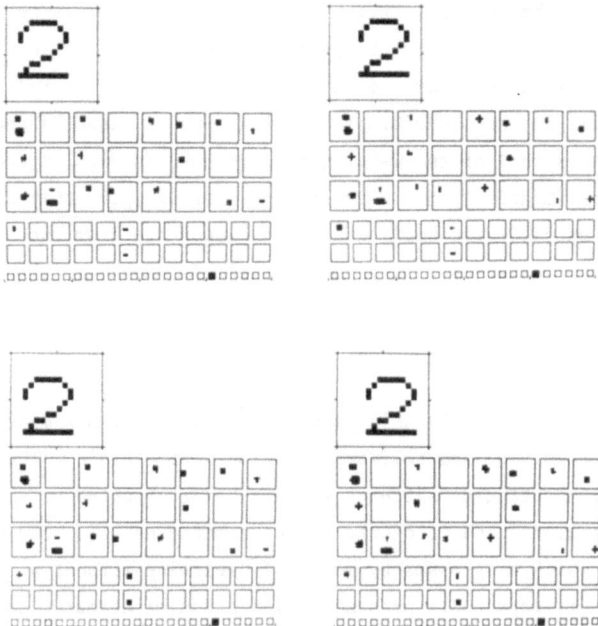

Fig. 10. Response of the cells of layers $U_0$, $U_{C1}$, $U_{C2}$ and $U_{C3}$ to the same stimulus pattern "2" presented at four different positions.

fected at all by shifts in position of the stimulus pattern. This means that the neocognitron correctly recognizes the stimulus patterns irrespective of their positions.

The neocognitron recognizes the stimulus patterns correctly, even if the stimulus patterns are distorted in shape or contaminated with noise. Fig. 8 shows some examples of distorted stimulus patterns which the neocognitron recognized correctly. All the patterns in each row elicited the same response from the cells of the deepest layer $U_{C3}$. Even though a stimulus pattern was increased or diminished in size, or was skewed in shape, the response of the cells of the deepest layer was not affected at all. Similarly, the stimulus patterns were correctly recognized even if they were stained with noise or had some missing parts.

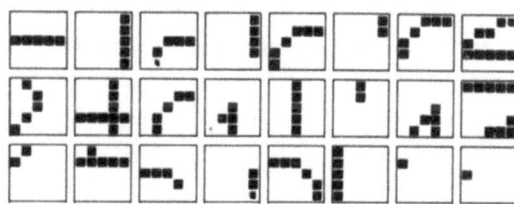

Fig. 11. Receptive fields of the cells of each of the 24 S-planes of layer $U_{S1}$, which has finished learning.

As an example of the feature extracting cells formed in this network, Fig. 11 displays the receptive fields of the cells of each of the 24 S-planes of layer $U_{S1}$. In this figure, the excitatory parts of the receptive fields are indicated by the darkness of the cells in the picture. The arrangement of these 24 receptive fields in Fig. 12 is the same as that of the C-cells of layer $U_{C1}$ in Figs. 9 and 10. Hence, in each display of Figs. 9 and 10, it is understood, for example, that the upper left C-plane of layer $U_{C1}$ extracts the horizontal line components from the stimulus pattern.

## 7. CONCLUSION

The neocognitron proposed in this paper is a neural network model for some mechanisms of pattern recognition and self-organization in the brain. The neocognitron is capable of recognizing stimulus patterns without being affected by shift in position or by a small distortion in shape of the stimulus patterns. It also has a function of self-organization, which progresses by means of "learning-without-a-teacher" process: If the neocognitron is repeatedly presented with a set of stimulus patterns, it gradually acquires the ability to recognize these patterns. It is not necessary to give any instructions about the categories to which these patterns should be classified. The performance of the neocognitron was demonstrated by computer simulation.

We do not advocate that the neocognitron is a complete model for the mechanism of pattern recognition in the brain, but we propose it as a working hypothesis for some neural mechanism of visual pattern recognition.

The structure of the neural network of the neocognitron somewhat resembles that of the hierarchical model for the visual nervous system proposed by Hubel and Wiesel. It is true that recent physiological data show that their hierarchical model does not hold in its original form: At least, there are several synaptic connections other than they initially believed to exist. We agree that our model should be modified in minute details to match the recent physiological data, but we believe that our model is not so misdirected when we observe the main stream of information flow in the network.

It is conjectured that, in the human brain, the process of recognizing familiar patterns such as alphabets of our native language differs from that of recognizing unfamiliar patterns such as foreign alphabets which we have just begun to learn. The neocognitron probably presents a neural network model corresponding to the former case, in which we recognize patterns intuitively and immediately. It would be a future problem to model the neural mechanism which works in deciphering illegible letters.

# REFERENCES

Block, H.D., Knight, B.W., Rosenblatt, F.: "Analysis of a four-layer series-coupled perceptron. II". *Rev. Modern Physics* 34[1], pp. 135-142 (1962)

Fukushima, K.: "Cognitron: a self-organizing multilayered neural network". *Biol. Cybernetics* 20[3/4], pp. 121-136 (1975)

Fukushima, K.: "Neural network model for a mechanism of pattern recognition unaffected by shift in position — neocognitron —", (in Japanese with English abstract). *Trans. IECE Japan (A)* 62-A[10], pp. 658-665 (1979)

Fukushima, K.: "Neocognitron: a self-organizing neural network model for a mechanism of pattern recognition unaffected by shift in position". *Biol. Cybernetics* 36[4], pp. 193-202 (1980)

Fukushima, K.: "Cognitron: a self-organizing multilayered neural network model". *NHK Technical Monograph* No. 30 (1981)

Gross, C.G., Rocha-Miranda, C.E., Bender, D.B.: "Visual properties of neurons in inferotemporal cortex of the macaque". *J. Neurophysiol.* 35[1], pp. 96-111 (1972)

Hubel, D.H., Wiesel, T.N.: "Receptive fields, binocular interaction and functional architecture in cat's visual cortex". *J. Physiol. (Lond.)* 160[1], pp. 106-154 (1962)

Hubel, D.H., Wiesel, T.N.: "Receptive fields and functional architecture in two non-striate visual area (18 and 19) of the cat". *J. Neurophysiol.* 28[2], pp. 229-289 (1965)

Hubel, D.H., Wiesel, T.N.: "Functional architecture of macaque monkey visual cortex". *Proc. Roy. Soc. Lond.* B 198[1130], pp. 1-59 (1977)

Meyer, R.L., Sperry, R.W.: "Explanatory models for neuroplasticity in retinotectal connections". In: *Plasticity and Function in the Central Nervous System*, pp. 45-63, Stein, D.G., Rosen, J.J., Butters, N., eds., New York·San Francisco·London: Academic Press 1974

Rosenblatt, F.: *Principles of Neurodynamics*. Washington, D.C.: Spartan Books 1962

Sato, T., Kawamura, T., Iwai, E.: "Responsiveness of inferotemporal single units to visual pattern stimuli in monkey performing discrimination". *Exp. Brain Res.* 38[3], pp. 313-319 (1980)

# CHAPTER 19

## ON THE SPONTANEOUS EMERGENCE OF

## NEURONAL SCHEMATA

Erich Harth
Physics Department, Syracuse University
Syracuse, NY  13210/USA

Summary

The generation of neural activity in the absence of specific sensory
triggers raises some of the most fundamental and puzzling questions
in neuroscience.  How are spontaneous voluntary actions initiated?
What guides thought processes?  How is endogenous neural activity,
arising from minimal cues or out of noise, fashioned into meaningful
sequences of neural representations?  I believe that two general
principles play a significant role in the processes.  One is the flow
of information from the micro- to the macroscale, analogous to the
dynamics of chaotic or turbulent mechanical systems.  The other is a
stochastic feedback process, called Alopex, which is capable of
generating complex patterns under the guidance of a single scalar
signal.  This process has been applied to a study of visual receptive
fields, but may be involved also in perception and so-called higher
neural processes.

We frequently speak of the machine-brain analogy.  It is instructive to look at some
of the more obvious differences.  A classical machine such as the steam engine has
interacting components (valves, pistons, flywheel, etc.) with sizes in the range of
centimeters to meters.  At scales much below about a centimeter we enter a world of
homogeneous matter with nothing very interesting happening until we get down to
molecular sizes, about $10^{-10}$ meters.  Here the characteristics of the water
molecules and those of the iron atoms are essential to the functioning of the
machine.  They make their presence felt across the largely structureless gap of
almost nine orders of magnitude by determining the macroscopic properties of the
substances.  (The molecular and atomic properties, in turn, are determined by a
range of subatomic phenomena that extend down to the level of quarks.)

Our most sophisticated modern machine is the digital computer.  Its largest
components also have dimensions of meters:  the printers, the tape and disc units,
and other peripherals, and the cables connecting them with the central processor.
We descend through a range of hardware components, the circuitry and microcircuitry
(the integrated 'chips') to the smallest functional units, the semiconductor
elements and the bit-storage locations in the memory units.  Below the scale of
about $10^{-5}$ meters we encounter again a range of virtually homogeneous matter.  Only
when we arrive at the scale of atoms do we again pick up characteristics that are
essential to the functioning of the whole:  the properties of the atoms of the
semiconductor and the impurities.  But the gap, the mechanistic void is now
considerably narrower than in the steam engine.

In the human nervous system distances between central and peripheral components are also of the order of a meter. Large structures in the brain and spinal chord are typically centimeters in size. On the scale of millimeters we have local functional populations of neurons, so-called nuclei, as well as their bundles of interconnecting cables. Gross activity at this level is manifested in the EEG. At smaller scales we have the hypercolumns, columns and minicolumns of cortical organization. Local interactions between neurons extend primarily over distances of a few hundred microns, as do the dendritic trees of individual neurons. The chief functional characteristic of a single neuron is the action potential, but details such as the distribution of postsynaptic potentials over the dendritic surface may play an important role also. Main functional parts of a neuron range from tens of microns (the size of the cell body) down to fractions of a micron for the functional components of the synapse and other organelles. The cell membrane, where the all-important electrochemical transactions of the neurons take place, has a thickness of about 100 angstroms. It contains specific channels and receptor sites and mechanisms involving structure from macromolecular assemblies to small molecules. One distinguishing characteristic of the nervous system is thus the virtually continuous range of scales of tightly intermeshed mechanisms reaching from the macroscopic to the molecular level and beyond (Fig. 1). There are no meaningless gaps of just matter.

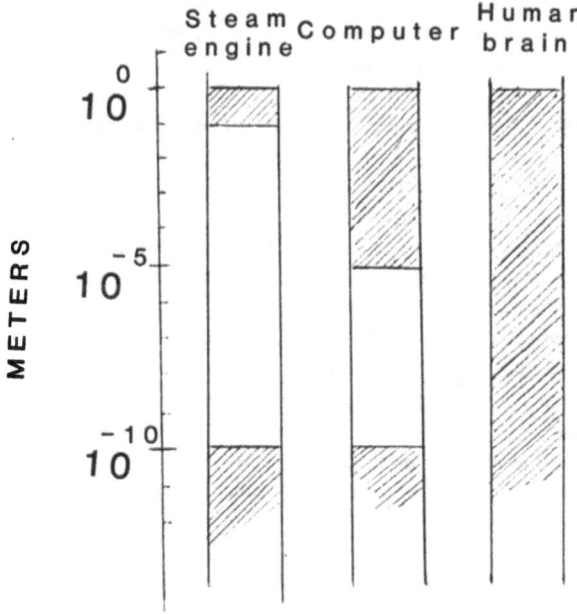

Fig. 1    Scales of mechanisms in machine and brain (see text).

The distinction between macroscale and microscales, so prominent in the classical machine, and somewhat less so in the digital computer, seems of questionable value when we consider the brain. Indeed, this appears to be a characteristic of all biological systems. But in the brain, more so than in any other organ, we are still uncertain about the appropriate level of description of meaningful activity. Thus, it is still a matter of passionate dispute among neuroscientists, how much redundancy there exists in neural activity. Can all relevant information be contained in a statistical description of neuronal population dynamics, or must we know the activity of each neuron in the brain? If the latter, is it enough to know the firing frequencies of each neuron, or is information contained in the precise arrival times and intervals between spikes?

It would seem that features which arise unpredictably, i.e. as a result of noise, cannot carry significant neural information. This is probably true of events carrying energy of the order of kT. But what about others, somewhat higher on our hierarchic scale of structures?

It has recently been pointed out by R. Shaw (1) that the separation of phenomena and energies as belonging either to the microworld of "heat degrees of freedom" or to the macroworld of "large-scale correlated motions of which we claim classically to have complete knowledge" is an artifact that breaks down even classically in so-called chaotic or turbulent systems. Here minute fluctuations become rapidly amplified, by as much as factors of $10^{15}$ according to Helleman (2). The result of knife-edge decisions encountered by phase space trajectories is that the dynamics of the system cannot be predicted in any practical way. When fluctuations of the order of kT can thus affect macroscopic behavior, the trajectories are in principle unpredictable, and the system is turbulent.

In studying the dynamics of neural assemblies one often starts with a difference equation of the type

$$S(n+1) = f[S(n)] \tag{1}$$

which relates a neural state S at the (n+1) time interval to its predecessor S(n). In a series of studies, based in part on computer simulations, Harth et al. (3), Anninos et al. (4), and Wong and Harth (5) have shown that activity in a neural net can be represented by a mapping in one dimension of the unit interval $0 \leqslant \alpha_n \leqslant 1$, where $\alpha_n$ is the fractional number of neurons active during the n-th time interval. The curve of $\alpha_{n+1}$ vs $\alpha_n$ has a rising and a falling phase; the latter is due to the refractoriness of the neurons. Fig. 2, reproduced here from reference (3), shows a family of such curves, computed for different sets of network parameters. Here point A is an unstable stationary state, O and B are stable stationary states or attractors.

In the presence of sustained excitatory and inhibitory input into the network it was shown (5,6,7) that under appropriate conditions the network undergoes phase transitions and hysteresis effects, and can be represented by a 'catastrophe manifold' (Fig. 3). In the case of second order nets, in which $\alpha_{n+1}$ depends on two preceding states

$$\alpha_{n+1} = f_2(\alpha_n, \alpha_{n-1}), \tag{2}$$

it was shown (5) that phase trajectories in the so-called $\Delta$-phase plane (Fig. 4) were generally oblate spirals approaching convergent or divergent focal points. In other experiments (8) network activity was found to fluctuate in an irregular, seemingly unpredictable manner about some stationary value.

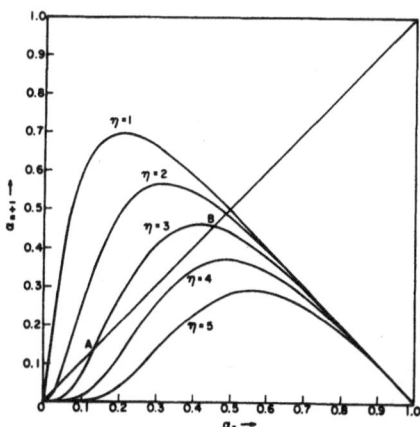

Fig. 2 Level of activity $\alpha_{n+1}$ in a neural net as function of preceding activity $\alpha_n$. (Curves are drawn for different network parameters $\eta$ which denote the minimum number of simultaneous EPSP's required to elicit an action potential). Here points 0 and B are stable stationary states or attractors; A is an unstable stationary state.

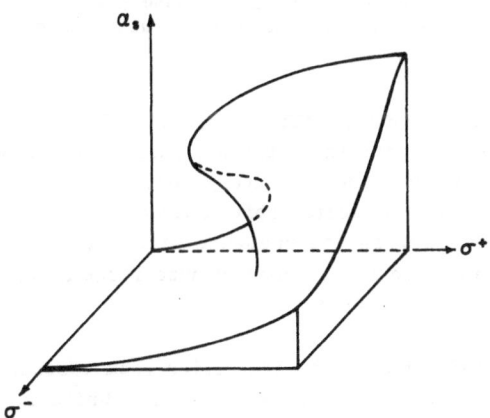

Fig. 3 Surface of stationary states of network activity $\alpha_s$ for different values of sustained excitatory ( $\sigma^+$) and inhibitory inputs ( $\sigma^-$) (Schematic).

One-dimensional maps of the type $x_{n+1} = f(x_n)$ are often 'chaotic' or turbulent. This condition exists when these three conditions are satisfied: 1) trajectories $x_0$, $x_1$, $x_2$... are critically dependent on initial conditions, 2) the average correlation function $C(m)$ between two terms $x_m$ and $x_{n+m}$ tends to zero as $m$ approaches infinity, and 3) the sequence is non-periodic (9). These conditions obtain for functions for which the maps are <u>non-invertible</u>, i.e. the relation $x_n = F(x_{n+1})$ is not single-valued, and for which small intervals in $x_n$ are 'stretched' to give larger intervals in $x_{n+1}$ (1). For the simple parabola $f(x) = 4\lambda x(1-x)$, $0 \leqslant \lambda \leqslant 1$ chaotic behavior occurs upon iterating the equation by substituting $f(x)$ for x, for all values of beyond a critical value of about 0.892. It was shown by Feigenbaum (10) that similar rules hold for other convex one-dimensional maps of the unit interval.

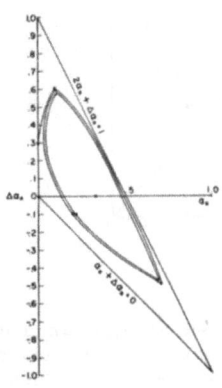

Fig. 4  Trajectory of network activity in ' $\triangle$ -phase plane'.  Here $\triangle \alpha_n = \alpha_{n+1} - \alpha_n$. Points are from computer simulations of a second order net (from Wong and Harth, 1972).

The kinematics graphs (Fig. 2) obtained for neural nets thus appear to be candidates for chaotic behavior, except for the fact that the phase space here is discontinuous for any finite number of neurons.  The trajectory must, therefore, approach either a steady value (vortex point) or a periodic orbit (attractor).  Something like chaotic behavior is possible, however, in the presence of noise.  The situation may be compared with turbulence in fluid flow, where minute fluctuations become amplified into unpredictable macroscopic behavior.

R. Shaw has recently pointed out that the transition from laminar to turbulent flow involves a reversal of the information flow (Fig. 5).  While macroscopic features subside in laminar flow, with information transfer from the macroscale to the world of disordered thermal motions, information in turbulent motion is extracted from the microscale to generate macroscopic features.

I began this discussion with a comparison between the scales on which different mechanisms took place.  In classical physics observations were always made on systems involving large numbers of atoms sharing 'large-scale coordinated motions'. The wide <u>uninteresting</u> gap between the scale of <u>mechanisms</u> and the scale of disordered molecular motions (Fig. 1) gave rise to the distinction between micro-and macroworlds. The continuum of neuronal mechanisms, virtually down to the kT-level, suggests that the information flow to the macroscale, instead of carrying only meaningless thermal noise, may bring out features diffusely stored in micromechanisms.

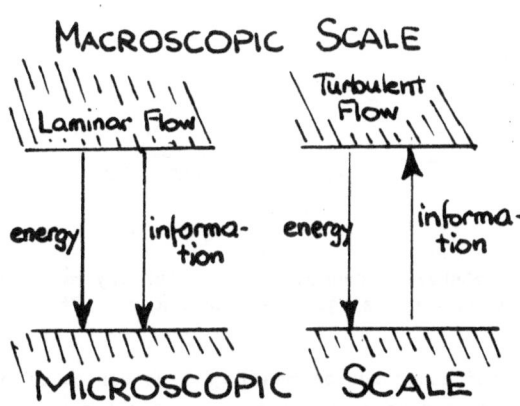

Fig. 5   Energy and information flow in laminar and turbulent flow (After R. Shaw, 1901)

I believe that a useful analogy can be drawn between Shaw's description of laminar and turbulent flow on the one hand, and the storing and retrieving of sensory information on the other.   In the process of sensory perception, as in laminar flow, macroscopic features are _dissipated_, with energy and information flow directed toward the microscales.   But unlike turbulent flow, the brain through a system of closely coupled micromechanisms, is able to prevent the information from being permanently lost in thermal noise.   Similarly, the spontaneous recall of sensory events, and thought procesess in general, may be likened to turbulent flow in which the arrow of information flow becomes reversed (Fig. 5).   We may thus account for the spontaneity and unpredictability of some human behavior.   Macrofeatures of neural activity (e.g. behavior) arise in this picture through the enormous amplification of diffusely stored microfeatures.   If these processes have to pass through critically poised decision _trees_, the 'trajectories' will have the characteristics of chaotic processes.

The foregoing discussion applies primarily to actions that are not _reflexive_, that is not resulting from an immediately preceeding sensory trigger.   When stimuli and actions are separated by times much larger than typical transit times of neural signals through the nervous system, we are dealing with sustained neural processes often referred to as _mental_.   The spontaneous emergence of action, or decision making in general, in the human nervous system is a process whose dynamics are still largely mysterious.   Some important empirical facts have been uncovered in Kornhuber's laboratory (11).   It was found that a spontaneous voluntary act, such a flexing a finger 'at will', has a precursor, called the _readiness potential_: an exponentially rising scalp potential, starting as much as 1.5 seconds before the action (Fig. 6).  It appears that the neural activity associated with the volitional act grows exponentially out of the background of EEG noise, suggesting the characteristics of a positive feedback mechanism.

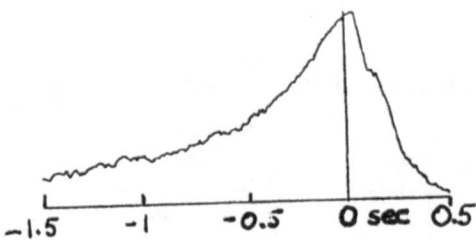

Fig. 6  Precentral scalp potential accompanying a voluntary motor act.  The movement occurs at t=0.  (After Deecke, Grötzinger, and Kornhuber, 1976)

The growth, persistence and coherence of meaningful sequences of neuronal schemata, as in thought processes or in dreams, requires more than the capricous amplification of minute fluctuations in the neural network.  I wish to suggest a second mechanism as responsible for 'higher' processing in the brain.  A process, called _Alopex_, was described in 1974 (12), in which the scalar response of a feature detecting system was used to generate patterns, which in turn provided the inputs to the system. This cyclic process was shown to produce sequences of patterns that converged on the trigger features of the detecting system.  In the case of single neurons in the visual pathway typical _receptive_ _field_ patterns were produced by the system (13). It must be emphasized that the evolution of the patterns is guided by the responses to complete trial patterns, and not by exploratory _scanning_ of the visual field. Thus, the cat cortical receptive field (Fig. 7) obtained by E. Tzanakou in D. A. Pollen's laboratory at Barrow Neurological Institute (14), using the Syracuse Alopex instrumentation, was generated entirely by a single cortical neuron without intervention on the part of the experimenter.  It corresponded exactly in size, shape, position and orientation to the receptive field obtained for the same cell by standard methods.

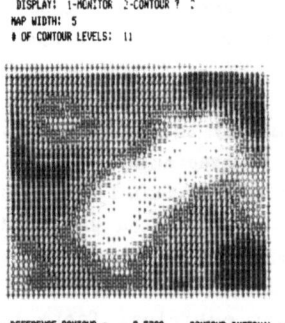

Fig. 7  Visual receptive field of a 'simple' cell in area 17 of cat neocortex.  The field was obtained by E. Tzanakou using the Alopex method.

I have suggested previously (15) that the Alopex process may play a role also in some forms of visual perception. The ubiquitous existence of corticofugal pathways was interpreted as providing mechanisms whereby sensory areas in the neocortex are able to enhance, suppress, or otherwise manipulate their own sources of sensory information, and perhaps generate meaningful patterns when true sensory input is absent. In this way subcortical sensory areas, such as the LGN, may serve as 'internal sketchpads', on which cortical fancy is able to project and develop its images, and test its ideas.

A general system of this type is now being investigated at Syracuse University, using large scale computer simulations. It is initially intended as an exercise in artificial intelligence. A box <u>A</u> (Fig. 8) contains a network of subsystems, each of which responds selectively to a particular input pattern. The subsystems are either independent of one another or affect one another following certain rules of association. All subsystems in A 'see' the same input pattern B. Their responses are combined into a single scalar feedback R which B uses to modify the pattern presented to A. A modified Alopex algorithm is used to determine the next pattern displayed by B.

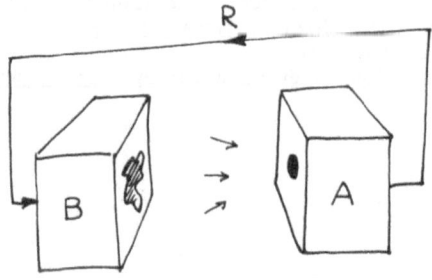

Fig. 8   Schematic diagram of a generalized Alopex process in which the responses R of a system A are used by a device B to modify its stimulus pattern.

The process is stochastic and will, in the simplest case, converge on a single pattern, selected by one of the subsystems in A. A more sophisticated system, employing accommodation (fatigue) and associative connections in A, is expected to lead to sequences of patterns simulating various types of intelligent behavior.

This work was supported in part through grant EY 01215-08 by the National Eye Institute, NIH.

## References

1. R. Shaw: Strange attractors, chaotic behavior, and information flow. Z. Naturforsch. 36a: 80-112 (1981).
2. R. H. G. Helleman: 1978 American Institute of Physics, Conference Proceedings No. 46.
3. E. Harth, T. J. Csermely, B. Beek, and R. D. Lindsay: Brain functions and neural dynamics. J. Theoret. Biol. 26: 93-120 (1970).
4. P. A. Anninos, B. Beek, T. J. Csermely, E. Harth, and G. Pertile: Dynamics of neural structures. J. Theoret. Biol. 26: 121-148 (1970).
5. R. Wong and E. Harth: Stationary states and transients in neural populations. J. Theoret. Biol. 40: 77-106 (1973).
6. E. Harth, N. S. Lewis, and T. J. Csermely: The escape of Tritonia: dynamics of a neuromuscular control mechanism. J. Theoret. Biol. 55: 201-228 (1975).
7. H. R. Wilson and J. D. Cowan: Excitatory and inhibitory interactions in localized populations of model neurons. Biophys. J. 12: 1-24 (1972).
8. S. Finette, E. Harth, and T. J. Csermely: Anisotropic connectivity and cooperative phenomena a a basis for orientation sensitivity in the visual cortex. Biol. Cybernetics 30: 231-240 (1978).
9. E. Ott: Strange attractors and chaotic motions of dynamical systems. Rev. Mod. Phys. 53: 655-671 (1981).
10. M. J. Feigenbaum: Universal behavior in nonlinear systems. Los Alamos Science: 4-27, Summer 1980.
11. L. Deeke, B. Grötzinger, and H. H. Kornhuber: Voluntary finger movement in man: cerebral potentials and theory. Biol. Cybernetics 23: 99-119 (1976).
12. E. Harth and E. Tzanakou: Alopex: a stochastic method for determining visual receptive fields. Vision Res. 14: 1475-1482 (1974).
13. E. Tzanakou, R. Michalak and E. Harth: The Alopex process: visual receptive fields by response feedback. Biol. Cybernetics 35: 161-174 (1979).
14. E. Tzanakou (private communication)
15. E. Harth: Visual perception: a dynamic theory. Biol. Cybernetics 22: 169-180 (1976).

# CHAPTER 20

ASSOCIATIVE AND COMPETITIVE
PRINCIPLES OF LEARNING
AND DEVELOPMENT

The Temporal Unfolding
and Stability of STM and
LTM Patterns

by

Stephen Grossberg
Department of Mathematics
Boston University
Boston, Mass. 02215

## 1. Introduction: Brain, Behavior and Babel.

This article reviews some principles, mechanisms, and theorems from my work over the past twenty-five years. I review these results here to illustrate their inter-connectedness from a recent perspective, to indicate directions for future work, and to reaffirm an approach to theorizing on problems of mind and brain that is still not fashionable despite growing signs that it needs to become so soon.

I say this because, despite the explosive growth of results on the fundamental issues of mind and brain, our science remains organized as a patchwork of experimental and theoretical fiefdoms which rarely interact despite the underlying unity of the scientific problems that they address. The territorial lines that bound these fiefdoms often seem to be as sacrosanct as national boundaries, and for similar cultural and economic reasons. A theorist who succeeds in explaining results from distinct experimental preparations by discovering their unifying mechanistic substrates may, through repeated crossings of these territorial boundaries, start to feel like a traveler without a country, and will often be treated accordingly. My own intellectual travels have repeatedly left me with such a feeling, despite the reassuring belief that theory had provided me with an international passport. To quickly review how some of these territorial passages were imposed by the internal structure of my theory, I will use a personal historical format of exposition, since the familiar territories do not themselves provide a natural descriptive framework.

## 2. From List Learning to Neural Networks: The Self-Organization of Individual Behavior.

My scientific work began unexpectedly in 1957-58 while I was an undergraduate psychology major at Dartmouth College. A great deal of structured data and classical theory about topics like verbal learning, classical and instrumental conditioning, perceptual dynamics, and attitude change were then available. It struck me that the revolutionary meaning of these data centered in issues concerning the self-organization of individual behavior in response to environmental pressures. I was exhilarated by the dual problems of how one could represent the emergence of behavioral units that did not exist before, and how one could represent the environmental interaction that stimulated this emergence even before the units emerged that would ultimately stabilize this interaction. I soon realized that various data which seemed paradoxical when viewed in terms of traditional concepts seemed inevitable when viewed in a network framework wherein certain laws hold. In fact, the same laws seemed to hold, in one version or another, in all the learning data that I studied. This universality suggested an important role for mathematics to quantitatively classify these various cases, which is why I sit in a mathematics department today. Although the laws were derived from psychological ideas, once derived they readily suggested a neurophysiological interpretation. In fact, that is how I learned my first neurophysiology, and crossed my first major experimental boundaries. To a nineteen year-old, these heady experiences were motivationally imprinting, and they supplied enough energy to face the sociological difficulties that my blend of psychology, physiology, and mathematics tends to cause. I might add that this interdisciplinary penetration of boundaries by my laws has prevented them from being widely studied by psychologists to the present time, despite the fact that their manifestations have appeared in a vast array of data and specialized models during the past decade.

## 3. Unitized Nodes, Short Term Memory, and Automatic Activation.

The network framework and the laws themselves can be derived in several ways (Grossberg, 1969a, 1974). My first derivation was based on classical list learning data (Grossberg, 1961, 1964) from the serial verbal learning and paired associate paradigms (Dixon and Horton, 1968; Jung, 1968; McGeoch and Irion, 1952; Osgood, 1953; Underwood, 1966). List learning data force one to confront the fact that new verbal units are continually being synthesized as a result of practice, and need not be the obvious units which the experimentalist is directly manipulating (Young, 1968). All essentially stationary concepts, such as the concept of information itself (Khinchin, 1957) hereby become theoretically useless. I therefore find the recent trend to discuss results about human memory in terms of "information processing" misleading (Klatsky, 1980; Loftus and Loftus, 1976; Norman, 1969). Such approaches either implicitly or explicitly adopt a framework wherein the self-organization of new behavioral units cannot be intrinsically characterized. Because these approaches miss processing constraints that control self-organization, they often construct special-purpose models to explain experiments in which the formation of new units is not too important, or deal indirectly with the self-organization problem by using computer models that would require a humunculus to carry out their operations in a physical setting. I will clarify these assertions as I go along.

By putting the self-organization of individual behavior in center stage, I realized that the phenomenal simplicity of familiar behavioral units, and the evolutionary aggregation of these units into new representations which themselves achieve phenomenal simplicity through experience, should be made a fundamental property of my theory. To express the phenomenal simplicity of familiar behavioral units, I represented them by indecomposable internal representations, or unitized nodes, rather than as composites of phonemes or as individual muscle movements. The problem of how phomenic, syllabic, and word-like representations might all coexist with different importance in different learning contexts was hereby vividly raised.

Once unitized nodes were conceived, it became clear that experimental inputs can activate these nodes via conditionable pathways. A distinction between sensory activation (the input source) and short term memory (the node's reaction) hereby became natural, as well as a concept of "automatic" activation of a node by its input. These network concepts have become popular in psychology during the past decade under the pressure of recent data (e.g., Schneider and Shiffrin, 1976), but they were already needed to analyze classical list learning data that are currently out of fashion.

The following properties of list learning helped to constrain the form of my associative laws. To simplify the discussion, I will only consider associative interactions within a given level in a coding hierarchy, rather than the problem of how coding hierarchies develop and interact between several levels. All of my conclusions can be, and have been, generalized to a hierarchical setting (Grossberg, 1974, 1978a, 1980a).

## 4. Backward Learning and Serial Bowing.

Backward learning effects and, more generally, error gradients between nonadjacent, or remote, list items (Jung, 1968; McGeogh and Irion, 1952; Murdock, 1974; Osgood, 1953; Underwood, 1966) suggest that pairs of nodes $v_i$ and $v_j$ can interact via distinct directed pathways $e_{ij}$ and $e_{ji}$ over which conditionable signals can travel. Indeed, an analysis of how any node $v_i$ can know where to send its signals reveals that no local information exists at the node itself whereby such a decision can be made. By the principle of sufficient reason, the node must therefore send signals towards all possible nodes $v_j$ with which it is connected by directed paths $e_{ij}$. Some other variables must exist which can discriminate which combination of signals should reach their target nodes based on past experience. These auxiliary variables turn out to be the long term memory traces. The concept that each node sends out signals to all possible nodes has recently appeared in models of spreading activation (Collins and Loftus, 1975; Klatsky, 1980) to explain semantic recognition and reaction time data.

The form that the signalling and conditioning laws should take is forced by data about serial verbal learning. A main paradox about serial learning concerns the form of the bowed serial position curve which relates cumulative errors to list positions (Figure 1a). This curve is paradoxical for the following reason. If all that happens during serial learning is a build-up of various types of interference at each list position due to the occurrence of prior list items, then the error curve should be monotone increasing (Figure 1b). Because the error curve is bowed, and the degree of bowing depends on the length of the intertrial interval between successive list presentations, the nonoccurrence of list items after the last item occurs somehow improves learning across several prior list items.

## 5. The Inadequacy of Rehearsal as an Explanatory Concept.

Just saying that rehearsal during the intertrial interval causes this effect does not explain it, because it does not explain why the middle of the list is less rehearsed. Indeed the middle of the list has more time to be rehearsed than does the end of the list before the next learning trial occurs. In the classical literature, one reads that the middle of the list experiences maximal proactive interference (from prior items) and retroactive interference (from future items), but this just labels what we have to explain (Jung, 1953; Osgood, 1953; Underwood, 1966). In the more recent literature, rehearsal is given a primary role in determining the learning rate (Rundus, 1971) although it is believed that only certain types of rehearsal, called elaborative rehearsal, can accomplish this (Bjork, 1975; Craik and Watkins, 1973; Klatsky, 1980). Notwithstanding the type of rehearsal used, one still has to explain why the list middle is rehearsed less than the list end in the serial learning paradigm.

The severity of such difficulties led the serial learning expert Young (1968) to write: "If an investigator is interested in studying verbal learning processes... he would do well to choose some method other than serial learning" (p. 146). Another leading verbal learning expert Underwood (1966) realized the magnitude of the difficulties, but also that they would not go away by ignoring them, when he wrote: "the person who originates a theory that works out to almost everyone's satisfaction will be in line for an award in psychology equivalent to the Nobel prize" (p. 491).

Most recent investigators have followed Young's advice. They have turned to paradigms like free recall (Bower, 1977; Murdock, 1974) wherein single trial presentations minimize self-organizing effects and subject-determined recall strategies sim-

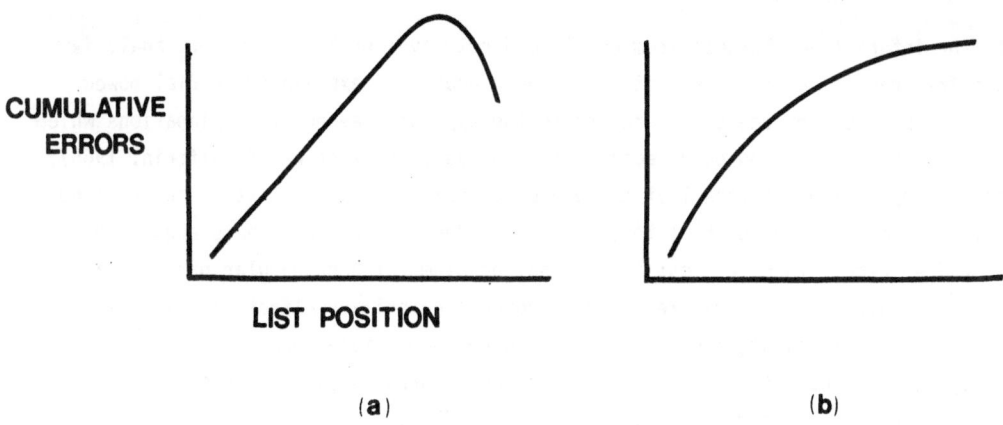

**CUMULATIVE ERRORS**

**LIST POSITION**

(a)                    (b)

FIGURE 1. (a) The cumulative error curve in serial verbal learning is a skewed bowed curve.  Items between the middle and end of the list are hardest to learn.  Items at the beginning of the list are easiest to learn. (b) If position-dependent difficulty of learning were all due to interference from previously presented items, the error curve would be monotone increasing.

plify the interactions between item recall and retrieval probes. However, analysis of the free recall paradigm has not proved deep enough to explain the serial bowed curve. In particular, one cannot resort to the type of free recall explanations which are used to explain the bowed effects in that paradigm (Atkinson and Shiffrin, 1968), since the improvement in recall at the end of a serially learned list is due to long term memory rather than to short term memory. Indeed, I have elsewhere argued that popular free recall theories contain internal problems of a humuncular nature, cannot explain some critical free recall data concerning primacy effects in STM which are not supposed to exist, and cannot even explain how a telephone number can be accurately repeated out of STM, because they do not address various issues which are also raised by serial learning data (Grossberg, 1978b).

## 6. The Inadequacy of Programmatic Time.

The massive backward effect that causes the bowed serial curve forces the use of a real-time theory that can parameterize the temporal unfolding of both the occurrences and the nonoccurrences of events. The bowed curve hereby weakens the foundations of all theories whose time variable is counted in terms of computer program operations, no matter how successful these theories might be in simulating data via humuncular constructions (Anderson and Bower, 1973). The existence of facilitative effects due to nonoccurring items also shows that traces of prior list occurrences must endure beyond the last item's presentation time, so they can be influenced by the future nonoccurrences of items. This fact leads to the concept of stimulus traces, or short term memory (STM) traces, $x_i(t)$ at the nodes $v_i$, $i=1,2,\ldots,n$, which are activated by inputs $I_i(t)$, but which decay at a rate slower than the input presentation rate.

Thus in response to serial inputs, _patterns_ of STM activity are set up across the network's nodes. By sufficient reason, each supraliminally activated node also sends signals along all its directed pathways. The combination of serial inputs, distributed internodal signals, and spontaneous STM changes at each node changes the STM pattern as the experiment proceeds. A major task of learning theory is to characterize the rules whereby these STM patterns evolve through time. Indeed, a major mathematical task is to learn how to think in terms of pattern transformations, rather than just in terms of feature detectors or other local entities.

## 7. Network vs. Computer Parsing: Distinct Error Gradients at Different List Positions.

The general philosophical interest of the bowed curve can be better appreciated by asking: What is the first time a learning subject can possibly know that item $r_n$ is the last list item in a newly presented list $r_1r_2\ldots r_n$, given that a new item is

presented every w time units until $r_n$ occurs?  The answer obviously is: not until at
least w time units <u>after</u> $r_n$ has been presented.  Only after this time passes and no
item $r_{n+1}$ is presented can $r_n$ be correctly reclassified from the list's "middle" to
the list's "end".  Since parameter w is under experimental control and is not a property of the list ordering <u>per se</u>, spatiotemporal network interactions parse a list
in a way that is fundamentally different from the parsing rules that are natural to
apply to a list of symbols in a computer.  Indeed, increasing the intratrial interval
w during serial learning can flatten the entire bowed error curve and minimize the effects of the intertrial interval (Jung, 1968; Osgood, 1953).

To illustrate the difference between computer models and my network approach,
suppose that after a node $v_i$ is excited by an input $I_i$, its STM trace gets smaller
through time due to either internodal competition or to passive trace decay.  Then in
response to a serially presented list, the last item to occur always has the largest
STM trace.  In other words, at every time a <u>recency</u> gradient obtains in STM (Figure
2).  Given this natural assumption - which, however, is not always true (Grossberg,
1978.a,b) - how do the generalization gradients of errors at each list position get
learned (Figure 3)?  In particular, how does a gradient of anticipatory errors occur
at the beginning of the list, a two-sided gradient of anticipatory and perseverative
errors occur near the middle of the list, and a gradient of perseverative errors occur at the end of the list (Osgood, 1953)?  Otherwise expressed, how does a temporal
succession of STM recency gradients generate an LTM <u>primacy</u> gradient at the list beginning but an LTM <u>recency</u> gradient at the list end?  These properties immediately
rule out any linear theory, as well as any theory which restricts itself to nearest
neighbor associative links, unless the theory makes the humuncular assumptions that
the system has absolute knowledge of how to compute the list's beginning, end, and
direction towards its middle (Feigenbaum and Simon, 1962).

## 8.  Graded STM and LTM Patterns: Multiplicative Sampling and Slow Decay by LTM Tra- ces.

Figures 2 and 3 can be reconciled by positing the existence of STM traces and
LTM traces that evolve according to different time scales and rules.  Indeed this reconciliation is one of the strongest arguments that I know for these rules.

Suppose, as above, that each node $v_j$ can send out a sampling signal $S_j$ along each
directed path $e_{jk}$ towards the node $v_k$, $j \neq k$.  Suppose that each path $e_{jk}$ contains a
long term memory (LTM) trace $z_{jk}$ at its terminal point, where $z_{jk}$ can compute, using
only local operations, the product of signal $S_j$ and STM trace $x_k$.  Also suppose that
the LTM trace decays slowly, if at all, during single learning trial.  The simplest
law for $z_{jk}$ that satisfies these constraints is

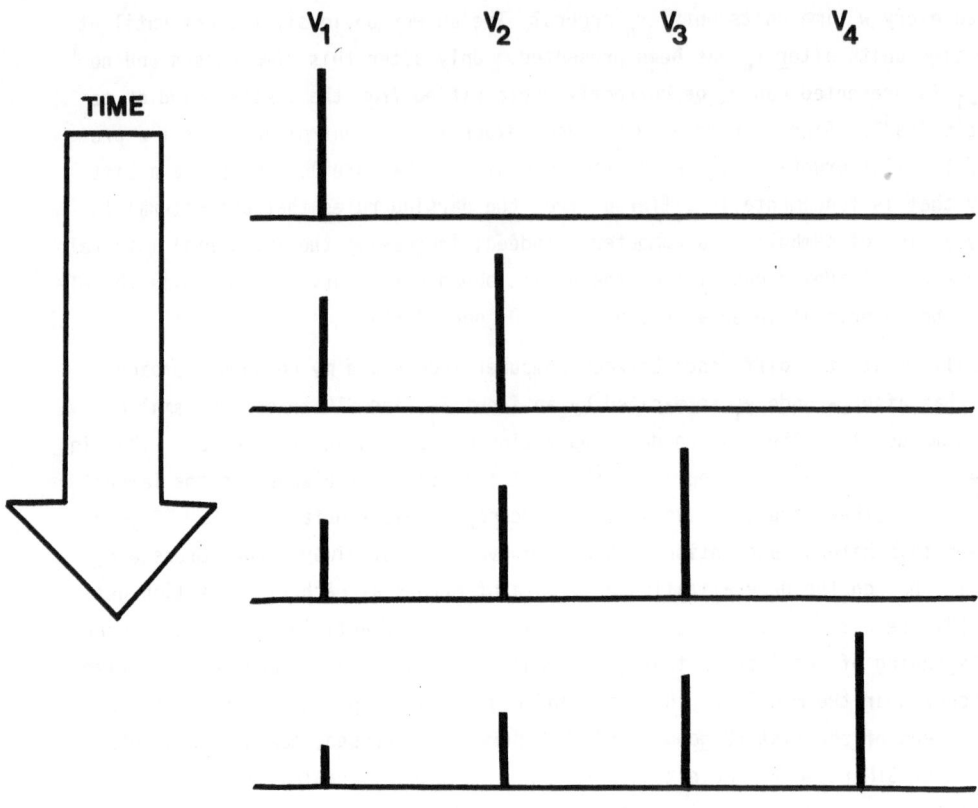

FIGURE 2. Suppose that items $r_1, r_2, r_3, r_4, \ldots$ are presented serially to nodes $v_1, v_2, v_3, v_4, \ldots$, respectively. Let the activity of node $v_i$ at time t be described by the height of the histogram beneath $v_i$ at time t. If each node is initially excited by an equal amount and its excitation decays at a fixed rate, then at every time (each row), the pattern of STM activity across nodes is described by a recency gradient.

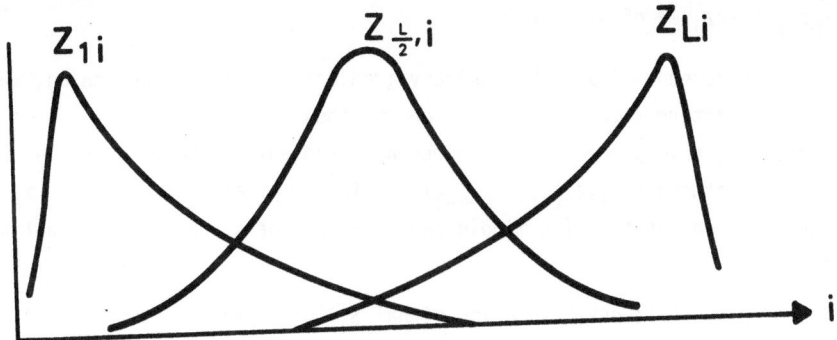

FIGURE 3.  At each node $v_j$, the LTM pattern $z_j = (z_{j1}, z_{j2}, \ldots, z_{jn})$ that evolves
through time is different.  In a list of length $n = L$ whose intertrial interval is
sufficiently long, the LTM pattern at the list beginning ($j \cong 1$) is a primacy gradient.
At the list end ($j \cong L$), a recency gradient evolves. Near the list middle ($j \cong \frac{L}{2}$),
a two-sided gradient is learned.  These gradients are reflected in the distribution
of anticipatory and perseverative errors in response to item probes at different list
positions.

$$\frac{d}{dt} z_{jk} = -c z_{jk} + d S_j x_k, \quad j \neq k. \tag{1}$$

To see how this rule generates an LTM primacy gradient at the list beginning, we need to study the LTM pattern $(z_{12}, z_{13}, \ldots, z_{1n})$ and to show that $z_{12} > z_{13} > \ldots > z_{1n}$. To see how the same rule generates at LTM recency gradient at the list end, we need to study the LTM pattern $(z_{n1}, z_{n2}, \ldots z_{n,n-1})$ and to show that $z_{n1} < z_{n2} < \ldots < z_{n,n-1}$. The two-sided gradient at the list middle can then be understood as a combination of these effects.

By (1), node $v_1$ sends out a sampling signal $S_1$ shortly after item $r_1$ is presented. After rapidly reaching peak size, signal $S_1$ gradually decays as future list items $r_2, r_3, \ldots$ are presented. Thus $S_1$ is largest when trace $x_2$ is maximal, $S_1$ is smaller when both traces $x_2$ and $x_3$ are active, $S_1$ is smaller still when traces $x_2$, $x_3$, and $x_4$ are active, and so on. Consequently, the product $S_1 x_2$ in row 2 of Figure 2 exceeds the product $S_1 x_3$ in row 3 of Figure 2, which in turn exceeds the product $S_1 x_4$ in row 4 of Figure 2, and so on. Due to the slow decay of each LTM trace $z_{1k}$ on each learning trial, $z_{12}$ adds up the products $S_1 x_2$ in successive rows of column 1, $z_{13}$ adds up to the products $S_1 x_3$ in successive rows of column 2, and so on. An LTM primacy gradient $z_{12} > z_{13} \ldots > z_{1n}$ is hereby generated. This gradient is due to the way signal $S_1$ multiplicatively <u>samples</u> the successive STM recency gradients and the LTM traces $z_{1k}$ sum up the sampled STM gradients.

By contrast, the signal $S_n$ of a node $v_n$ at the end of the list samples a different set of STM gradients. This is because $v_n$ starts to sample (viz., $S_n > 0$) only after all past nodes $v_1, v_2, \ldots, v_{n-1}$ have already been activated on that trial. Consequently, the LTM traces $(z_{n1}, z_{n2}, \ldots, z_{n,n-1})$ of node $v_n$ encode a recency gradient $x_1 < x_2 < x_3 < \ldots < x_{n-1}$ at <u>each</u> time. When all the recency gradients are added up through time, the total effect is a recency gradient in $v_n$'s LTM pattern. In summary, nodes at the beginning, middle, and end of the list encode different LTM gradients because they multiplicatively sample and store STM patterns at different times.

Lest the reader who is sensitized to the functional unit issue object to these internodal feedback effects, let me reiterate that similar LTM gradients obtain if the sequences of nodes which are active at any time selectively excite higher-order nodes (chunks) which in turn sample the field of excited nodes via feedback signals (Grossberg, 1974, 1978a).


## 9. Binary vs. Continuous Associative Laws.

The LTM gradient problem illustrates why I have always avoided binary laws for STM and LTM traces. Binary laws have often attracted workers who began with the all-or-none character of individual axonal spikes. However, the continuously fluctuating potentials that receive these spikes often average them in time, thereby yielding graded intercellular signalling effects. For similar reasons, population interactions often obey continuous laws. Workers like Amari (1974, 1977) and Geman (1981) have formally studied how to justify the averaging procedures that can convert binary microscopic rules into continuous macroscopic rules. Because of the psychological derivation of my networks, I have always worked with preaveraged equations from the start.

The example of continuous LTM error gradients is not the only one wherein binary and continuous models yield distinct outcomes. In fact, they usually do. For example, just changing sigmoid feedback signals to binary feedback signals in a competitive network can significantly change network dynamics (Grossberg, 1973, 1978c), notably because sigmoid signals can support infinitely many equilibrium points in competitive geometries wherein binary signals cannot.

## 10. Retrieval Probes and LTM Gating of STM Mediated Signals.

Having shown how STM patterns may be read into LTM patterns, we now need to describe how a retrieval probe $r_m$ can read $v_m$'s LTM pattern back into STM on recall trials, whereupon the STM traces can be transformed into observable behavior. In particular, how can LTM be read into STM without distorting the learned LTM gradients?

The simplest rule generates an STM pattern which is proportional to the LTM pattern that is being read out, and allows distinct probes to each read their LTM patterns into STM in an independent fashion.

To achieve faithful read-out of the LTM pattern $(z_{m1}, z_{m2}, \ldots, z_{mn})$ by a probe $r_m$ that turns on signal $S_m$, I let the product $S_m z_{mi}$ determine the growth rate of $x_i$. In other words, LTM trace $z_{mi}$ <u>gates</u> the signal $S_m$ along $e_{mi}$ before the gated signal reaches $v_i$. The independent action of several probes implies that the gated signals $S_m z_{mi}$ are added, so that the total effect of all gated signals on $v_i$ is $\sum_{m=1}^{n} S_m z_{mi}$. The simplest equation for the STM trace $x_i$ that abides by this rule is

$$\frac{dx_i}{dt} = -ax_i + b \sum_{m=1}^{n} S_m z_{mi} + I_i ,\tag{2}$$

where $-a$ is the STM decay rate that produces Figure 2, $S_m$ is the $m^{th}$ sampling signal, $z_{mi}$ is the LTM trace of pathway $e_{mi}$, and $I_i$ is the $i^{th}$ experimental input.

The reaction of equations (1) and (2) to serial inputs $I_i$ is much more complex than is their response to an isolated retrieval probe $r_m$. Due to the fact that STM traces may decay slower than the input presentation rate, several sampling signals $S_m$ can be simultaneously active, albeit in different phases of their growth and decay.

## 11. Behavioral Choices and Competitive Feedback.

Once one accepts that patterns of STM traces are evolving through time, one also needs a mechanism for choosing those activated nodes which will influence observable behavior. Lateral inhibitory feedback signals are readily implicated as a choice mechanism (Grossberg, 1968, 1969b, 1970). The simplest extension of (2) which includes competitive interactions is

$$\frac{dx_i}{dt} = -ax_i + \sum_{m=1}^{n} S_m^+ b_{mi}^+ z_{mi} - \sum_{m=1}^{n} S_m^- b_{mi}^- + I_i \qquad (3)$$

where $S_m^+ b_{mi}^+$ ($S_m^- b_{mi}^-$) is the excitatory (inhibitory) signal emitted from node $v_m$ along the excitatory (inhibitory) pathway $e_{mi}^+$ ($e_{mi}^-$). Correspondingly, equation (1) is generalized to

$$\frac{dz_{jk}}{dt} = -cz_{jk} + d_{jk} S_j^+ x_k \quad . \qquad (4)$$

The asymmetry between terms $\sum_{m=1}^{n} S_m^+ b_{mi}^+ z_{mi}$ and $\sum_{m=1}^{n} S_m^- b_{mi}^-$ in (3) readily suggests a modification of (3) and a definition of inhibitory LTM traces analogous to (4), where such traces exist (Grossberg, 1969c).

Because lateral inhibition can change the sign of each $x_i$ from positive to negative in (3), and thus the sign of each $z_{jk}$ from positive to negative in (4), some refinements of (3) and (4) are needed to prevent absurdities like the following: $S_m^+ < 0$ and $x_i < 0$ implies $z_{mi} < 0$; and $S_m^+ < 0$ and $z_{mi} < 0$ implies $x_i > 0$. Threshold constraints accomplish this in the simplest way. Letting $[\xi]^+ \equiv \max(\xi, 0)$, these absurdities are prevented if threshold cut-offs are imposed on signals, such as in

$$S_j^+ = [x_j(t - \tau_j^+) - \Gamma_j^+]^+ \qquad (5)$$

and

$$S_j^- = [x_j(t - \tau_j^-) - \Gamma_j^-]^+ , \qquad (6)$$

as well as on sampled STM traces, such as in

$$\frac{dz_{jk}}{dt} = -cz_{jk} + d_{jk} S_j^+ [x_k]^+ . \qquad (7)$$

The equations (3), (5), (6) and (7) have been used by modellers for a variety of purposes. For example, in his seminal article on code development, Malsburg (1973) used these equations, supplemented by his synaptic conservation rule.

## 12. Skewing of the Bow: Symmetry-breaking Between the Future and the Past.

To explain the bowed error curve, we now need to compare the LTM patterns $z_i = (z_{i1}, z_{i2}, \ldots, z_{in})$ which evolve at all list nodes $v_i$. In particular, we need to explain why the bowed curve is <u>skewed</u>; that is, why the list position where learning takes longest occurs nearer to the end of the list than to its beginning (Figure 1a). This skewing effect has routinely demolished learning theories which assume that forward and backward effects are equally strong, or symmetric (Asch and Ebenholtz, 1962; Murdock, 1974). I have elsewhere argued that the symmetry-breaking between the future and the past, by favoring forward over backward associations, makes possible the emergence of a global "arrow in time", or the ultimate learning of long event sequences in their correct order (Grossberg, 1969c, 1974).

Theorem 1 below asserts that a skewed bowed curve does occur in the network, and predicts that the degree of skewing will decrease and the relative learning rate at the beginning and end of the list will reverse as the network's arousal level increases or its excitatory signal thresholds $\Gamma_j^+$ decrease to abnormal levels (Grossberg and Pepe, 1970, 1971). The arousal and threshold predictions have not yet been tested to the best of my knowledge. They are of some conceptual importance because abnormally high arousal or low thresholds can hereby generate a formal network syndrome characterized by contextual collapse, reduced attention span, and fuzzy response categories that resembles some symptoms of simple schizophrenia (Grossberg and Pepe, 1970; Maher, 1977).

To understand what is involved in my explanation of bowing, note that by equation (7), each correct LTM trace $z_{12}, z_{23}, z_{34}, \ldots, z_{n-1,n}$ may grow at a comparable rate, albeit $w$ time units later than the previous correct LTM trace. However, the LTM patterns $z_1, z_2, \ldots, z_n$ will differ no matter when you look at them, as in Figure 3. Thus when a retrieval probe $r_j$ reads its LTM pattern $z_j$ into STM, the entire pattern must influence overt behavior to explain why bowing occurs. The relative size of the correct LTM trace $z_{j,j+1}$ compared to all other LTM traces in $z_j$ will influence its success in eliciting $r_{j+1}$ after competitive STM interactions occur. A larger $z_{j,j+1}$ relative to the sum of all other $z_{jk}$, $k \neq j, j+1$, should yield better performance of $r_{j+1}$ given $r_j$, other things being equal. To measure the distinctiveness of a trace $z_{jk}$ relative to all traces in $z_j$, I therefore define the relative LTM traces, or stimulus sampling probabilities

$$Z_{jk} = z_{jk} \left( \sum_{m \neq j} z_{jm} \right)^{-1} . \tag{8}$$

The appropriateness of definition (8) is strengthened by the following observation. The ordering within the LTM gradients of Figure 3 is unchanged by the relative LTM traces; for example, if $z_{12} > z_{13} > \ldots > z_{1n}$, then $Z_{12} > Z_{13} > \ldots > Z_{1n}$ because all the $Z_{1k}$'s have the same denominator. Thus all conclusions about LTM gradients are valid for relative LTM gradients.

In terms of the relative LTM traces, the issue of bowing can be mathematically formulated as follows. Define the _bowing function_ $B_i(t) = Z_{i,i+1}(t)$. Function $B_i(t)$ measures how distinctive the $i^{th}$ correct association is at time t. After a list of n items is presented with an intratrial interval w and a sufficiently long inter-trial interval W elapses, does the function $B_i((n-1)w + W)$ decrease and then increase as i increases from 1 to n? Does the minimum of the function occur in the latter half of the list? The answer to both of these questions is "yes".

To appreciate the subtlety of the bowing issue, it is necessary to understand how the bow depends upon the ability of a node $v_i$ to sample incorrect future associations, such as $r_i r_{i+2}$, $r_i r_{i+3}$,... in addition to incorrect past associations, such as $r_i r_{i-1}$, $r_i r_{i-2}$,... As soon as $S_i$ becomes positive, $v_i$ can sample the entire past field of STM traces at $v_i, v_2, ..., v_{i-1}$. However, if the sampling threshold is chosen high enough, $S_i$ might shut off before $r_{i+2}$ occurs. Thus the sampling duration has different effects on the sampling of past than future incorrect associations. For example, if the sampling thresholds of all $v_i$ are chosen so high that $S_i$ shuts off before $r_{i+2}$ is presented, then the function $B_i(\infty)$ decreases as i increases from 1 to n. In other words, the monotonic error curve of Figure 1b obtains because no node $v_i$ can encode incorrect future associations.

Even if the thresholds are chosen so that incorrect future associations can be formed, the function $B_i((i+1)w)$ which measures the distinctiveness of $z_{i,i+1}$ just before $r_{i+2}$ occurs is again a decreasing function of i. The bowing effect thus depends on threshold choices which permit sampling durations that are at least 2w in length.

The shape of the bow also depends on the duration of the intertrial interval, because before the intertrial interval occurs, all nodes build up increasing amounts of associative interference as more list items are presented. The first effect of the nonoccurrence of items after $r_n$ is presented is the growth through time of $B_{n-1}(t)$ as t increases beyond the time nw when item $r_{n+1}$ would have occurred in a larger list (Grossberg, 1969c). The last correct association is hereby facilitated by the absence of interfering future items during the intertrial interval. This facilitation effect is a nonlinear property of the network. Indeed, bowing itself is a nonlinear pheno-menon in my theory, because it depends on a comparison of ratios of integrals of sums of products as they evolve through time.

In my review of a bowing theorem below, I will emphasize the effect of the sig-nal threshold $\Gamma$ on the degree of skewing. One can, however, also compute the effect of the intertrial interval W on skewing, as well the role of other network parame-ters, such as STM decay rate and LTM growth rate.

The position of the bow     has not yet been quantitatively computed although it has been qualitatively demonstrated within the full system (3),(5),(6),(7). Complete computations have been made in a related system, the _bare field_, wherein the primary effects of serial inputs on associative formation and competition are preserved (Grossberg, 1969c; Grossberg and Pepe, 1971) on a single learning trial. In the bare field, serial inputs occur with intratrial interval w:

$$I_1(t) = I_2(t+w) = \dots = I_n(t+(n-1)w);$$ (9)

the STM traces decay after they are excited by their inputs:

$$\frac{dx_i}{dt} = -ax_i + I_i \ ;$$ (10)

the LTM traces add up products of signals and STM traces:

$$\frac{dz_{jk}}{dt} = d \ [x_j(t-\tau) - \Gamma \ ]^+ x_k, \ j \neq k;$$ (11)

and the relative LTM traces, or stimulus sampling probabilities, estimate how well a given LTM trace fares after it is read into STM and STM competition takes place:

$$Z_{jk} = z_{jk} \ ( \ \underset{m \neq j}{\Sigma} \ z_{jm} \ )^{-1} \ .$$ (12)

## Theorem 1 (Skewing of the Bowed Curve):

If the bare field
(I) is initially at rest and associatively unbiased; that is, all $x_i(t)=0, -\tau \leq t \leq 0$, and $z_{jk}(0) = \alpha > 0$, $j \neq k$; (9)

(II) the signals $S_i$ and inputs $r_{i+1}$ are well-correlated; that is,

$$w = \tau$$ (10)

(This condition is convenient but not essential);

(III) successive inputs do not overlap in time; that is, $I_1(t)$ is positive only in an interval $(0, \lambda)$ with $\lambda < \tau$ and is zero elsewhere;

(IV) the inputs are not too irregular; that is, $I_1(t)$ is continuous and grows monotonically until it reaches a maximum at time $t = T_{max}$, after which it monotonically decreases to zero at time $t = \lambda$;

(V) at high thresholds, the sampling signals don't last too long; that is, if $\Gamma$ is chosen so large that $v_1$ first emits a signal $S_1$ at the time $T_{max}$, then $S_1$ shuts off before $r_3$ occurs: if

$$\Gamma_0 = \int_0^{T_{max}} e^{-a(T_{max} - v)} I_1(v) dv,$$

then

$$\int_0^\lambda e^{-a(\lambda-v)}I_1(v)dv > \Gamma_0 \geq \int_0^{2\tau} e^{-a(2\tau-v)}I_1(v)dv \tag{11}$$

Under hypotheses (I)-(V), if the intertrial interval is infinite, then the bow occurs ($B_i(\infty)$ is minimized) at the list position closest to $M(\Gamma)$, where

A. (Overaroused Bowing)

$$M(0) = \frac{1}{2}(n-1) \tag{12}$$

B. (Skewing)

$$\frac{dM}{d\Gamma} > 0 \tag{13}$$

C. (No Incorrect Future Associations)

$$M(\Gamma) = n \quad \text{if} \quad \Gamma \geq \Gamma_0 \tag{14}$$

If the intertrial interval is $W < \infty$, then the bow occurs ($B_i((n-1)w+W)$ is minimized) at a list position strictly greater than $M(\Gamma)$.

The function $M(\Gamma)$ can, moreover, be explicitly computed. It satisfies the equation

$$M(\Gamma) = \frac{1}{a\tau} \log\left[\frac{E+\sqrt{E^2+4CD}}{2D}\right] \tag{15}$$

where

$$C = \tau E^{-1}(-\tau,0) \; [AB(\lambda-T_1,T_1)+A\Gamma E(T_2,T_1)+ \frac{A^2}{2}E(2\lambda,2T_2)- \frac{\Gamma^2}{2a}E(-L) \; ], \tag{16}$$

$$D = A\tau E(L)E^{-1}(0,\tau) \; [B(\lambda,0)+ \frac{A}{2a} \; e^{-2a\lambda} \; ] \tag{17}$$

$$E = \Gamma \; [C(\lambda,0) + \frac{A}{a} \; e^{-a\lambda} \; ] \tag{18}$$

with

$$A = \int_0^\lambda e^{av}I_1(v)dv, \tag{19}$$

$$B(t,p) = \int_0^t e^{-2a(v+p)} \int_0^{v+p} e^{aw}I_1(w)dwdv \tag{20}$$

$$C(t,p) = \int_0^t e^{-a(v+p)} \int_0^{v+p} e^{aw}I_1(w)dwdv \tag{21}$$

$$E(x) = e^{-a\tau x} \quad , \tag{22}$$

and

$$E(x,y) = \frac{1}{a}(e^{-ax} - e^{-ay}). \tag{23}$$

## 13. Evolutionary Invariants of Associative Learning: Absolute Stability of Parallel Pattern Learning.

Many features of system (3), (5), (6), (7) are special; for example, the exponential decay of STM and LTM and the signal threshold rule. Because associative processing is ubiquitous throughout phylogeny and within functionally distinct subsystems of each individual, a more general mathematical framework is needed. This framework should distinguish universally occurring associative principles which guarantee essential learning properties from evolutionary variations that adapt these principles to specialized environmental demands. Before we can speak with confidence about variations on an evolutionary theme, we first need to identify the theme.

I approached this problem during the years 1967 to 1972 in a series of articles wherein I gradually realized that the mathematical properties that I used to globally analyze specific learning examples were much more general than the examples themselves. This work culminated in my universal theorems on associative learning (Grossberg, 1969d, 1971a, 1972a).

The theorems are universal in the following sense. They say that if certain associative laws were invented at a prescribed time during evolution, then they could achieve unbiased associative pattern learning in essentially any later evolutionary specialization. To the question: Is it necessary to re-invent a new learning rule to match every perceptual or cognitive refinement, the theorems say "no". More specifically, the universal associative laws enable arbitrary spatial patterns to be learned by arbitrarily many, simultaneously active sampling channels that are activated by arbitrary continuous data preprocessing in an essentially arbitrary anatomy. Arbitrary space-time patterns can also be learned given modest constraints on the temporal regularity of stimulus sampling. The universal theorems thus describe a type of parallel processing whereby unbiased pattern learning can occur despite mutual crosstalk between very complex feedback signals.

Such results cannot be taken for granted. They obtain only if crucial network operations, such as spatial averaging, temporal averaging, preprocessing, gating, and cross-correlation, are computed in a canonical ordering. This canonical ordering constitutes a general purpose design for unbiased parallel pattern learning, as well as a criterion for whether  particular networks are acceptable models for this task. The universality of the design mathematically takes the form of a classification of

oscillatory and limiting possibilities that is invariant under evolutionary special-
izations.

The theorems can also be interpreted in another way that is appropriate in discussions of self-organizing systems. The theorems are <u>absolute stability</u> theorems.
They show that evolutionary invariants obtain no matter how system parameters are
changed within this class of systems. Absolute stability is an important property
in a self-organizing system because parameters may change in ways that cannot be predicted in advance, notably before specialized environments act on the system. Absolute stability guarantees that the onset of self-organization does not subvert the
very properties which make self-organization possible.

The systems which I considered have the form

$$\frac{d}{dt} x_i = A_i x_i + \sum_{k \in J} B_{ki} z_{ki} + C_i(t) \tag{24}$$

$$\frac{d}{dt} z_{ji} = D_{ji} z_{ji} + E_{ji} x_i \tag{25}$$

where $i \in I$, $j \in J$, and I and J parameterize arbitrarily large, not necessarily disjoint,
sets of sampled and sampling cells, respectively. As in my equations for list learning, $A_i$ is a STM decay rate, $B_{ki}$ is a nonnegative performance signal, $C_i(t)$ is an input function, $D_{ji}$ is a LTM decay rate, and $E_{ji}$ is a nonnegative learning signal. Unlike the list learning equations, $A_i$, $B_{ki}$, $D_{ji}$, and $E_{ji}$ are continuous functionals of
the entire history of the system. Equations (24) and (25) are therefore very general,
and include many of the specialized models in the literature.

For example, although (24) does not seem to include inhibitory interactions, such
interactions may be lumped into the STM decay functional $A_i$. The choice

$$A_i = -a_i + (b_i - c_i x_i) G_i(x_i) - (d_i + e_i x_i^{-1}) \sum_{k=1}^{n} H_k(x_k) f_{ki} \tag{26}$$

describes the case wherein system nodes compete via shunting, or membrane equation,
interactions (Cole, 1968; Grossberg, 1973; Kuffler and Nichols, 1976 ; Plonsey, 1969.)
The performance, LTM decay, and learning functionals may include slow threshold changes, nonspecific Now Print signals, signal velocity changes, presynaptic modulating
effects, arbitrary continuous rules of dendritic preprocessing and axonal signalling,
as well as many other possibilities (Grossberg, 1972a, 1974). Of special importance
are the variety of LTM decay choices that satisfy the theorems. For example, an LTM
law like

$$\frac{d}{dt} z_{ji} = [x_j(t-\tau_j) - \Gamma_j(y_t)]^+ (-d_j z_{ji} + e_j x_i) \tag{27}$$

achieves an interference theory of forgetting, rather than exponential forgetting,

since $\frac{d}{dt} z_{ji} = 0$ except when $v_j$ is sampling (Adams, 1967). Equation (27) also allows the vigor of sampling to depend on changes in the threshold $\Gamma_j(y_t)$ that are sensitive to the prior history $y_t = (x_i, z_{ji}: i \in I, j \in J)_t$ of the system before time t.

In this generality, too many possibilities exist to as yet prove absolute stability theorems. One further constraint on system processing paves the way towards such
results. This constraint still admits the above processing possibilities, but it im-
poses some spatiotemporal regularity on the sampling process. Indeed, if the performance signals $B_{ji}$ from a fixed sampling node $v_j$ to all the sampled nodes $v_i$, $i \in I$,
were arbitrary nonnegative and continuous functionals, then the irregularities in each
$B_{ji}$ could override any regularities in $z_{ji}$ within the gated performance signal $B_{ji}z_{ji}$
from $v_j$ to $v_i$.

14. Local Symmetry and Self-Similarity in Pattern Learning and Developmental Invariance.

Absolute stability does obtain even if different functionals $B_j, D_j$, and $E_j$ are
assigned to each node $v_j$, $j \in J$, just so long as the same functional is assigned to
all pathways $e_{ji}$, $i \in I$. Where this is not globally true, one can often partition the
network into the maximal subsets where it is true, and then prove unbiased pattern
learning in each subset. This restriction is called the property of local symmetry
axes since each sampling cell $v_j$ can act as a source of coherent history-dependent
waves of STM and LTM processing. Local symmetry axes still permit (say) each $B_j$ to
obey different history-dependent preprocessing, threshold, time lag, and path strength
laws among arbitrarily many mutually interacting nodes $v_j$.

When local symmetry axes are imposed on (24) and (25), the resulting class of
systems takes the form

$$\frac{d}{dt} x_i = Ax_i + \sum_{k \in J} B_k z_{ki} + C_i(t) \tag{28}$$

and

$$\frac{d}{dt} z_{ji} = D_j z_{ji} + E_j x_i. \tag{29}$$

A simple change of variables shows that constant interaction coefficients $b_{ji}$ between
pairs $v_j$ and $v_i$ of nodes can depend on $i \in I$ without destroying unbiased pattern learning in the systems

$$\frac{d}{dt} x_i = Ax_i + \sum_{k \in J} B_k b_{ki} z_{ki} + C_i(t) \tag{30}$$

and

$$\frac{d}{dt} z_{ji} = D_j z_{ji} + E_j b_{ji}^{-1} x_i. \tag{31}$$

By contrast, the systems (30) and

$$\frac{d}{dt} z_{ji} = D_j z_{ji} + E_j b_{ji} x_i \tag{32}$$

are not capable of unbiased parallel pattern learning (Grossberg, 1972a). A dimensional analysis shows that (30) and (31) hold if action potentials transmit the network's intercelluler signals, whereas (30) and (32) hold if electrotonic propagation is used.

The dimensional analysis hereby suggests that spatial biases in the $b_{ji}$ which are due to differences in axonal diameters can be overcome by an interaction between action potentials and mass action properties of the LTM traces. Temporal biases in time lags that are due to differences in intercellular distances are overcome by the proportionality of action potential velocity to axon diameter (Katz, 1966; Ruch, Patton, Woodbury, and Towe, 1961) in cells whose axon lengths and diameters covary. Such cells are said to be <u>self-similar</u> (Grossberg, 1969f). Self-similar cell populations can preserve the learned meaning of patterns under significant developmental deformations of their mutual distances and sizes. Self-similar rules of network design also permit individual nodes to arrive at globally correct decisions from locally ambiguous data (Grossberg, 1978a). In the developmental biology literature, self-similarity is called self-regulation (Wolpert, 1969).

## 15. The Unit of LTM is a Spatial Pattern: Global Constraints on Local Network Design.

To illustrate the global theorems that have been proved, I consider first the simplest case, wherein only one sampling node exists (Figure 4a). Then the network is called an <u>outstar</u> because it can be drawn with the sampling node at the center of outward-facing conditionable pathways (Figure 4b) such that the LTM trace $z_i$ in the $i$th pathway samples the STM trace $x_i$ of the $i$th sampled cell, $i \in I$. An <u>outstar</u> is thus a functional-differential system of the form

$$\frac{d}{dt} x_i = Ax_i + Bz_i + C_i(t) \tag{33}$$

$$\frac{d}{dt} z_i = Dz_i + Ex_i \tag{34}$$

where A,B,D, and E are continuous functionals such that B and E are nonnegative.

Despite the fact that the functionals A,B,D and E can fluctuate in extremely complex system-dependent ways, and the inputs $C_i(t)$ can also fluctuate wildly through time, an outstar can learn an arbitrary spatial pattern

$$C_i(t) = \theta_i C(t) \quad (\theta_i \geq 0, \ \sum_{k \in I} \theta_k = 1)$$

with a minimum of oscillations in its pattern variables $X_i = x_i \ (\sum_{k \in I} x_k)^{-1}$ and

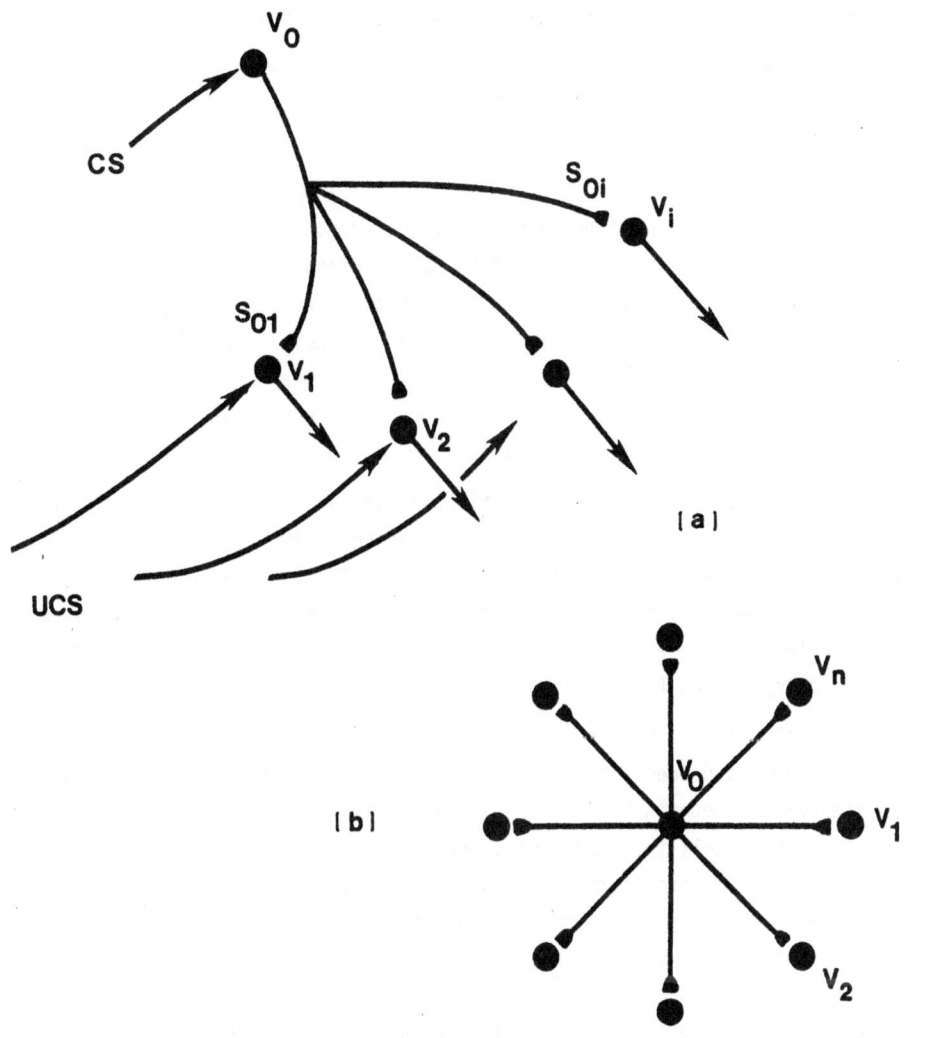

FIGURE 4. (a) In the minimal anatomy capable of associative·learning in a classical conditioning paradigm, a conditioned stimulus (CS) excites a single node, or cell population, $v_0$ which thereupon sends sampling signals to a set of nodes $v_1, v_2, \ldots, v_n$. An input pattern representing the unconditioned stimulus (UCS) excites the nodes $v_1, v_2, \ldots, v_n$, which thereupon elicit output signals that contribute to the unconditioned response (UCR). The sampling signals from $v_0$ activate the LTM traces $z_{0i}$ (which I denote by $z_i$ in the text for brevity) that are computed at the synaptic knobs $S_{0i}$, $i = 1, 2, \ldots, n$. The activated LTM traces can learn the activity pattern across $v_1, v_2, \ldots, v_n$ that represents the UCS. (b) When the sampling structure in (a) is redrawn to emphasize its symmetry, the result is an <u>outstar</u>, whose sampling source is $v_0$ and whose sampled border is the set $\{v_1, v_2, \ldots, v_n\}$.

$Z_i = z_i \left( \sum\limits_{k \in I} z_k \right)^{-1}$. Recall that the $Z_i$'s are the stimulus sampling probabilities
that played such a central role in my explanation of serial bowing. Because the li-
mits and oscillations of the pattern variables have a classification that is independent of particular choices of A,B,C,D, and E, these properties are the evolutionary
invariants of outstar learning.

The following theorem summarizes, albeit not in the most general known form,
some properties of outstar learning. One of the constraints in this theorem is called a local flow condition. This constraint says that a performance signal B can be
large only if its associated learning signal E is large. Local flow prevents the
read-out of old LTM memories from a sampling pathway which has lost its plasticity.
Such a read-out would prevent accurate registration of the UCS in STM and thus accurate LTM encoding of the UCS via STM sampling.

I should immediately remark that a plastic synapse can be dynamically buffered
against recoding by global network interactions (Grossberg, 1976c, 1980a). Such a
synapse can still obey the local flow condition. I should also say that the local
flow condition is needed only if all sampling sources are trying to encode the same
pattern without bias, as in the parallel learning of sensory expectancies (Grossberg,
1980a) or of motor synergies (Grossberg, 1978a).

If the threshold of B is no smaller than the threshold of E, then local flow is
assured. Such a threshold inequality occurs automatically if the LTM trace $z_{ji}$ is
physically interpolated between the axonal signal $S_j b_{ji}$ and the postsynaptic target
cell $v_i$. That is why I call the condition a local flow condition. Such a geometric
interpretation of the location of the LTM trace is not forced by the psychological
derivation of the associative equations, although it is the minimal anatomical realization of this derivation. Local flow gives unexpected support to the minimal realization by showing that pattern learning depends upon a mathematical constraint which
automatically obtains in the minimal realization, but is at best ad hoc and difficult to guarantee in other anatomical interpretations of the associative equations.

## Theorem 2 (Outstar Pattern Learning)

Suppose that

(I) the functionals are chosen to keep system trajectories bounded;

(II) a local flow condition holds:

$$\int_0^\infty B(t)dt = \infty \quad \text{only if} \quad \int_0^\infty E(t)dt = \infty \tag{35}$$

(III) the UCS is practiced sufficiently often:
there exist positive constants $K_1$ and $K_2$ such that for all $T \geq 0$,

$$f(T,T + t) \geq K_1 \quad \text{if } t \geq K_2 \tag{36}$$

where

$$f(U,V) = \int_U^V C(\xi)\exp\left[\int_\xi^V A(\eta)\,d\eta\right]\,d\xi \ . \tag{37}$$

Then, given arbitrary continuous and nonnegative initial data in $t \leq 0$ such that $\sum_{k \in I} z_k(0) > 0$,

(A) practice makes perfect:

The stimulus sampling probabilities $Z_i(t)$ are monotonically attracted to the UCS weights $\theta_i$ if

$$[Z_i(0)-X_i(0)][X_i(0)-\theta_i] \geq 0, \tag{38}$$

or may oscillate at most once due to prior learning if (38) does not hold, no matter how wildly A,B,C,D, and E oscillate;

(B) the UCS is registered in STM and partial learning occurs:

the limits $Q_i = \lim_{t \to \infty} X_i(t)$ and $P_i = \lim_{t \to \infty} Z_i(t)$ exist with

$$Q_i = \theta_i , \quad i \in I. \tag{39}$$

(C) If moreover the CS is practiced sufficiently often, then perfect learning occurs:

$$\text{if } \int_0^\infty E(t)\,dt = \infty, \text{ then } P_i = \theta_i, \ i \in I. \tag{40}$$

Remarkably, similar global theorems hold for systems (28)-(29) wherein arbitrarily many sampling cells can be simultaneously active and mutually signal each other by very complex feedback rules (Geman, 1981; Grossberg, 1969d, 1971a, 1972a, 1980b). This is because all systems of the form (28)-(29) can _factorize_ information about STM and LTM pattern variables from information about how fast energy is being pumped into the system. Pattern variable oscillations can therefore be classified even if wild fluctuations in input and feedback signal energies occur through time. In the best theorems now available, only one hypothesis is not known to be necessary and sufficient (Grossberg, 1972a). It would be most satisfying if this imperfection in the theorems could be overcome.

When many sampling cells $v_j$, $j \in J$, can send sampling signals to each $v_i$, $i \in I$, the outstar property that each stimulus sampling probability $Z_{ji} = z_{ji}(\sum_{k \in I} z_{jk})^{-1}$ oscillates at most once fails to hold. This is so because the $Z_{ji}$ of all active nodes $v_j$ track $X_i = x_i \left( \sum_{k \in I} x_k \right)^{-1}$, while $X_i$ tracks $\theta_i$ and the $Z_{ji}$ of all active nodes $v_j$. The oscillations of the functions $Y_i = \max \{Z_{ji} : j \in J\}$ and $y_i = \min \{Z_{ji} : j \in J\}$ can, however, be classified much as the oscillations of each $Z_i$ can be classified in the outstar case. Since each $Z_{ji}$ depends on all $z_{jk}$, $k \in I$; each $Y_i$ and $y_i$ depends on all $z_{jk}$, $j \in J$, $k \in I$; and each $X_i$ depends on all $x_k$, $k \in I$, the learning at each $v_i$ is influenced by <u>all</u> $x_k$ and $z_{jk}$, $j \in J$, $k \in I$. No local analysis can provide an adequate insight into the learning dynamics of these networks.

Because the oscillations of all $X_i$, $Y_i$, and $y_i$ relative to $\theta_i$ can be classified, the following generalization of the outstar learning theorem holds.

<u>Theorem 3 (Parallel Pattern Learning)</u>.

Suppose that

(I) the functionals are chosen to keep system trajectories bounded;

(II) every sampling cell obeys a local flow condition: for every $j \in J$,

$$\int_0^\infty B_j dt = \infty \text{ only if } \int_0^\infty E_j dt = \infty \tag{41}$$

(III) the UCS is presented sufficiently often:

there exist positive constants $K_1$ and $K_2$ such that (36) holds.
Then given arbitrary nonnegative and continuous initial data in $t \leq 0$ such that $\sum_{k \in I} x_k(0) > 0$ and all $\sum_{k \in I} z_{jk}(0) > 0$,

(A) the UCS is registered in STM and partial learning occurs:

the limits $Q_i = \lim_{t \to \infty} X_i(t)$ and $P_{ji} = \lim_{t \to \infty} Z_{ji}(t)$ exist with

$$Q_i = \theta_i , \quad i \in I . \tag{42}$$

(B) If the $j^{th}$ CS is practiced sufficiently often, then it learns the UCS pattern perfectly:

$$\text{if} \quad \int_0^\infty E_j dt = \infty \quad \text{then} \quad P_{ji} = \theta_i , \quad i \in I \tag{43}$$

Because LTM traces $z_{ji}$ gate the performance signals $B_j$ which are activated by a retrieval probe $r_j$, the theorem enables any and all nodes $v_j$ which sampled the pattern $\theta = (\theta_i, i \in I)$ during learning trials to read it out with perfect accuracy on recall trials. The theorem does not deny that oscillations in overall network activity can occur during learning and recall, but shows that these oscillations merely influence the rates and intensities of learning and recall. Despite the apparent simplicity of these statements, the details of learning, memory, and recall can be dramatically altered by different choices of functionals. As one of many examples, phase transitions in memory can occur, and the nature of the phases can depend on a complex interaction between network rates and geometry (Grossberg, 1974).

Neither Theorem 2 nor Theorem 3 needs to assume that the CS and UCS are presented at correlated times. This is because the UCS condition keeps the baseline STM activity of sampled cells from ever decaying below the positive value $K_1$ in (36). For purposes of space-time pattern learning, this UCS uniformity condition is too strong. In Grossberg (1972a) I show how to replace the UCS uniformity condition by a weaker condition which guarantees that CS-UCS presentations are well enough correlated to guarantee perfect pattern learning of a given spatial pattern by certain cells $v_j$, even if other spatial patterns are presented at irregular times when they are sampled by distinct cells $v_j$.

## 16. The Teleology of the Pattern Calculus: Retina, Command Cell, Reward, Attention, Motor Synergy, Sensory Expectancy, Cerebellum.

Three simple but fundamental facts emerge from the mathematical analysis of pattern learning: the unit of LTM is a spatial pattern $\theta = (\theta_i : i \in I)$; suitably designed neural networks can factorize invariant pattern $\theta$ from fluctuating energy; the size of a node's sampling signal can render it adaptively sensitive or blind to a pattern $\theta$. These concepts helped me to think in terms of pattern transformations, rather than in terms of feature detectors, computer programs, linear systems, or other types of analysis. When I confronted equally simple environmental constraints with these simple pattern learning properties, the teleological pressure that was generated drove me into a wide-ranging series of specialized investigations.

What is the minimal network that can discriminate $\theta$ from background input fluctuations? It looks like a retina, and the $\theta$'s become reflectances. What is the minimal network that can encode and/or perform a space-time pattern or ordered series of spatial patterns? It looks like an invertebrate command cell. How can one synchronize CS-UCS sampling if the time intervals between CS and UCS presentations are unsynchronized? The result leads to psychophysiological mechanisms of reward, punishment, and attention. What are the associative invariants of motor learning? Spatial patterns become motor synergies wherein fixed relative contraction rates across muscles occur, and temporally synchronized performance signals read-out the synergy as a unit. What are the associative invariants of sensory learning? The potential ease of learning and reading-out complex sensory expectancies and spatial representations shows that even eidetic memory is more remarkable as a memory retrieval property than as a learning property. What is the minimal network that can bias the performance of motor acts with learned motor expectancies? It looks like a cerebellum.

An historical review of these investigations is found in the prefaces to a selection of my articles reprinted in Grossberg (1982a). Individually and collectively,
these results add force to the idea that patterns rather than features are the functional units which regulate the neural designs subserving behavioral adaptation.

## 17. The Primacy of Shunting Competitive Networks over Additive Networks.

These specialized investigations repeatedly drove me to consider competitive systems. As just one of many instances, the same competitive normalization property
which arose during my modelling of receptor-bipolar-horizontal cell interactions in
retina (Grossberg, 1970a, 1972b) also arose in studies of the decision rules needed
to release the right amount of incentive motivation in response to interacting drive
inputs and conditioned reinforcer inputs within midbrain reinforcement centers (Grossberg, 1972c,d). Because I approached these problems from a behavioral perspective,
I knew what interactive properties the competition had to have. I have repeatedly
found that shunting competition has all the properties that I need, whereas additive
competition often does not.

As solutions to specialized problems involving competition piled up, networks
capable of normalization, sensitivity changes via automatic gain control, attentional
biases, developmental biases, pattern matching, shift properties, contrast enhancement, edge and curvature detection, tunable filtering, multistable choice behavior,
normative drifts, travelling and standing waves, hysteresis, and resonance began to
be classified within the framework of shunting competitive feedforward and feedback
networks. See Grossberg (1981) for a recent review. As in the case of associative
learning, the abundance of special cases made it seem more and more imperative to
find an intuitive and mathematical framework within which these results could be unified and generalized. I also began to wonder whether many of the pattern transformation and STM storage properties of specialized examples were not instances of an
absolute stability property of a general class of networks.

## 18. The Noise-Saturation Dilemma and Absolute Stability of Competitive Decision-Ma-
king.

A unifying intuitive theme of particular simplicity can be recognized by considering the processing of continuously fluctuating patterns by cellular tissues. This
theme is invisible to theories based on binary codes, feature detectors, or additive
models. All cellular systems need to solve the noise-saturation dilemma which might
cause sensitivity loss in their responses to both low and high input intensities.

Mass action, or shunting, competition enables cells to elegantly solve this problem using automatic gain control by lateral inhibitory signals (Grossberg, 1973, 1980a). Additive competition fails in this task because it does not, by definition, possess an automatic gain control property.

A unifying mathematical theme is that every competitive system induces a decision scheme that can be used to prove global limit and oscillation theorems, notably absolute stability theorems (Grossberg, 1978c,d; 1980c). This decision scheme interpretation is just a vivid way to think about a Liapunov functional that is naturally associated with each competitive system.

A class of competitive systems with absolutely stable decision schemes is the class of adaptation level systems

$$\frac{dx_i}{dt} = a_i(x)[b_i(x_i) - c(x)],$$

(44).

i=1,2,...,n, where $x = (x_1, x_2, ..., x_n)$. These systems include all shunting competitive feedback networks of the form

$$\frac{dx_i}{dt} = -A_i x_i + (B_i - x_i)[I_i + f_i(x_i)]$$

$$-(x_i + C_i)[J_i + \sum_{k \neq i} f_k(x_k)]$$

(45)

which, in turn, are capable of many of the special properties listed above, given suitable choices of parameters and feedback signal functions. A special case of my theorem concerning these systems is the following.

Theorem 4 (Absolute Stability of Adaptation Level Systems).

Suppose that

(I) Smoothness:
the functions $a_i(x)$, $b_i(x_i)$ and $c(x)$ are continuously differentiable;

(II) Positivity:

$$a_i(x) > 0 \text{ if } x_i > 0, x_j \geq 0, j \neq i ;$$

(46)

$$a_i(x) = 0 \text{ if } x_i = 0, x_j \geq 0, j \neq i;$$

(47)

for sufficiently small $\lambda > 0$, there exists a continuous function $\bar{a}_i(x_i)$ such that

$$\bar{a}_i(x_i) \geq a_i(x) \text{ if } x \in [0,\lambda]^n$$

(48)

and

$$\int_0^\lambda \frac{dw}{\bar{a}_i(w)} = \infty$$

(49)

(III) Boundedness:

for each i = 1,2,...,n,

$$\lim_{x_i \to \infty} \sup b_i(x_i) < c(0,0,\ldots,\infty,0,\ldots,0) \tag{50}$$

where $\infty$ is in the $i^{th}$ entry of $(0,0,\ldots,\infty,0,\ldots,0)$;

(IV) Competition:

$$\frac{\partial c(x)}{\partial x_i} > 0, \ x \in \mathbb{R}_+^n \ , \ i = 1,2,\ldots,n \tag{51}$$

(V) Decision Hills:

The graph of each $b_i(x_i)$ possesses at most finitely many maxima in every compact interval.

Then the pattern transformation is stored in STM because all trajectories converge to equilibrium points: given any x(0) > 0, the limit $x(\infty) = \lim_{t \to \infty} x(t)$ exists.

This theorem intuitively means that the decision schemes of adaptation level systems are globally consistent. Globally inconsistent decision schemes can, by contrast, force almost all trajectories to persistently oscillate. This can occur even if n=3 and all feedback signals are linear, as the voting paradox vividly illustrates (Grossberg, 1978c, 1980c; May and Leonard, 1975).

Adaptation level systems exclude distance-dependent interactions. To overcome this gap, Michael Cohen and I (Cohen and Grossberg, 1982) recently studied the absolute stability of the distance-dependent networks

$$\frac{dx_i}{dt} = -A_i x_i + (B_i - C_i x_i)[I_i + f_i(x_i)]$$

$$-(D_i x_i + E_i)[J_i + \sum_{k=1} g_k(x_k)F_{ki}]. \tag{52}$$

Distance-dependence means that $E_{ki} = F_{ik}$. The networks (52) include examples of Volterra-Lotka systems, Hartline-Ratliff networks, Gilpin-Ayala systems, and shunting and additive networks.

In this setting, we constructed a global Liapunov function for these systems and used the LaSalle Invariance Principle, Sard's lemma, and some results about several complex variables to analyze the limiting behavior of (52). Modulo some technical hypotheses, we have proved that almost all systems of the form (52) are absolutely stable, and that systems with polynomial and sigmoid feedback signals can be directly analyzed.

These results show that adaptation level and distance-dependent competitive

networks represent stable neural designs for competitive decision-making.  The fact
that adaptation level systems have been analyzed using Liapunov functionals whereas
distance-dependent networks have been analyzed using Liapunov functions shows that
the absolute stability theory of competitive systems is still incomplete.  Absolute
stability theorems for cooperative systems have also been recently discovered (Hirsch,
1982a,b).  This is an exciting area for intensive mathematical investigation.

The final sections of the article discuss code development issues wherein interactions between associative and competitive rules play a central role.


## 19.  The Babel of Code Development Models.

The experimental interest in geniculocortical and retinotectal development (Gottlieb, 1976; Hubel and Wiesel, 1977; Hunt and Jacobson, 1974) has been paralleled by a
vigorous theoretical interest in these basic phenomena.  Perhaps in no other area of
brain theory is the issue of what constitutes a new model, a new idea, or real progress so badly discussed.  A literal reading of this literature might lead one to
conclude that a one-to-one correspondence between articles and models exists, or at
least between authors and models.  A world of theoretical monads is an anarchy, which
is the antithesis of what a theoretical community should be.  If we are to achieve
the coherence that theory must have to be effective, then the endless numerical and
experimental variations on our laws must not be confused with the invariant structure
of these laws.  A new model is not a change of notation, a use of a discrete instead
of a continuous time variable, a different setting of numerical parameters, or a presentation of the same equations with a different input series.

When Malsburg (1973) adapted the equations which he found in Grossberg (1972b)
for computer simulation and subjected them to a series of input patterns, I was delighted but not surprised by his findings.  I was delighted because here was an interesting new twist in the use of the equations.  I was not surprised because the results are a variant of pattern learning properties which had already been studied.  Now
I will review some of the relationships between code development and pattern learning, state some mathematical results on code development which computer studies missed,
and make some comparative remarks about recent articles in the literature.


## 20.  The Duality Between Code Development and Pattern Learning.

In both pattern learning and code development situations, one often finds two
sets, or fields, $\mathcal{F}^{(1)}$ and $\mathcal{F}^{(2)}$ of cells, which are not necessarily disjoint.  The
sets of sampled cells $v_i$, $i \in J$ and sampling cells $v_j$, $j \in J$ are illustrative.  Condi-

tionable pathways $e_{ji}$ are assumed to exist from one set to the other set of cells, and LTM traces $z_{ji}$ are assigned to the pathways $e_{ji}$. Competitive interactions are assumed to occur within $\mathcal{F}^{(1)}$ and $\mathcal{F}^{(2)}$, if only to solve the noise-saturation dilemma at each level of pattern processing. In what, then, does the difference between a pattern learning and a code development model consist?

In a word, the answer is arrow-reversal, or duality. Whereas the conditionable pathways in a pattern learning example point from sampling cell to sampled cells, the conditionable pathways in a code development example point from sampled cells to sampling cell. Because of arrow-reversal, each sampling cell receives a sum of LTM-gated signals from sampled cells, which in turn influence the activity of the sampling cell and thus whether the sampled cells will be sampled.

If we apply the principle of sufficient reason to the arrow-reversal distinction, it becomes more ambiguous. How, after all, does an individual LTM trace $z_{ji}$ from $v_j$ to $v_i$ know whether $v_j$ is a sampling cell and $v_i$ a sampled cell, or conversely? The answer is that it doesn't. Consequently, similar principles of pattern learning hold in both cases. Only when we ask more global questions about network design do distinctions between the two problems emerge.

For example, how do the fields $\mathcal{F}^{(1)}$ and $\mathcal{F}^{(2)}$ determine whether their cells will be sampling cells, sampled cells, or both? A major part of the answer lies in how sharply $\mathcal{F}^{(1)}$ and $\mathcal{F}^{(2)}$ contrast enhance their input patterns. To fix ideas, suppose that conditionable pathways pass between $\mathcal{F}^{(1)}$ and $\mathcal{F}^{(2)}$ in both directions and that both $\mathcal{F}^{(1)}$ and $\mathcal{F}^{(2)}$ directly receive input patterns. If $\mathcal{F}^{(1)}$ does not sharply contrast enhance its input patterns but $\mathcal{F}^{(2)}$ does, then $\mathcal{F}^{(2)}$ will encode patterns across $\mathcal{F}^{(1)}$ within the $\mathcal{F}^{(1)} \rightarrow \mathcal{F}^{(2)}$ LTM traces, and $\mathcal{F}^{(2)}$ will learn patterns across $\mathcal{F}^{(1)}$ within the $\mathcal{F}^{(2)} \rightarrow \mathcal{F}^{(1)}$ LTM traces. The difference between code development and pattern learning in this example thus resides in an asymmetric choice of competitive parameters within $\mathcal{F}^{(1)}$ and $\mathcal{F}^{(2)}$, not in a choice of new associative or competitive laws.

## 21. Outstars and Instars.

These facts become clearer if we start with the simplest examples of pattern learning and code development, and then build up towards more complex examples. As Section 15 noted, the simplest network capable of pattern learning is an outstar (Figure 5a). By duality, the simplest network capable of code development is an instar (Figure 5b). The main difference between an outstar and an instar is that the source of an outstar excites the outstar border, whereas the border of an instar excites the instar source. The changing efficacy with which practiced border patterns can excite the instar source constitutes code development. Because of the outstar learning theorem, it is no surprise that the LTM traces of an instar can learn a spatial

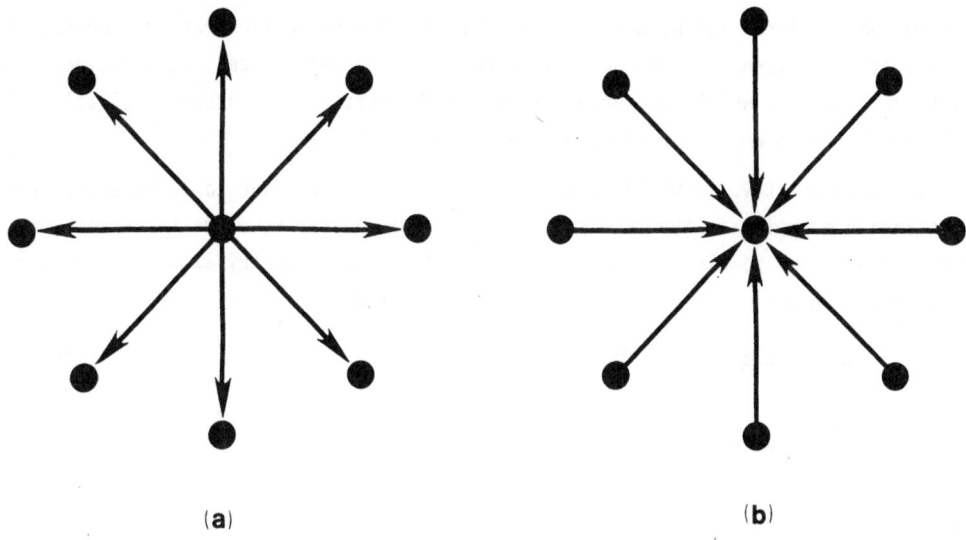

(a)  (b)

FIGURE 5. (a) An <u>outstar</u> is the minimal network capable of associative pattern learning. (b) An <u>instar</u> is the minimal network capable of code development.  The source of
an outstar excites the outstar border, whereas the border of an instar excites the instar source.  In both cases, source activation is necessary to drive the LTM sampling
process.  Since the instar border signals are gated by LTM traces before activating
the instar source, code learning changes the efficacy of source activation and is
changed by it.

pattern that perturbs its border. In an outstar, if a space-time pattern or sequence of spatial patterns plays upon its border while the source cell is sampling, then the source learns a weighted average of the sampled patterns (Grossberg, 1971b). This fact also holds in an instar for the same mathematical reason.

It is instructive to write down equations for an instar and to compare them with illustrative examples in the literature. Because an instar reverses the arrows between sampling and sampled cells, an instar with a local symmetry axis with respect to its sampling cell $v_1$ ($J = \{1\}$) obeys equations such as

$$\frac{d}{dt} x_1 = A_1 x_1 + \sum_{k \in I} B_k z_{k1} + C_1 \tag{53}$$

$$\frac{d}{dt} x_i = A x_i + C_i, \quad i \in I \tag{54}$$

and

$$\frac{d}{dt} z_{i1} = D_1 z_{i1} + E_1 x_i , \tag{55}$$

$i \in I$. In (53), the sampling cell $v_1$ receives LTM gated signals from the sampled cells $v_i$, $i \in I$, in addition to a possible input $C_1$. In (54), the sampled cells $v_i$ share a common STM decay functional A due to the local symmetry axis, but receive distinct inputs $C_i$ from the input patterns ($C_i$: $i \in I$). In (55), the usual LTM trace law holds with a shared LTM decay functional $D_1$ and a shared learning functional $E_1$ due to the local symmetry axis.

The article by Bienenstock, Cooper, and Munro (1982) is devoted to the study of a locally symmetric instar. These authors consider the equation

$$\frac{d}{dt} m_j = - \epsilon m_j + \phi d_j \tag{56}$$

for the $j^{th}$ LTM trace $m_j$ and the $j^{th}$ input $d_j$. They define $\phi$ to be a functional of the past and present values of function

$$c = \sum_j d_j m_j. \tag{57}$$

In particular, they use an average of past values of c as a threshold against which a present value of c is compared. If the present value exceeds threshold, $\phi > 0$, otherwise not. The threshold is assumed to increase as a function of past values of c.

A simple change of notation shows that equations (56) and (57) are a lumped version of an instar. In (55), let i=j, $z_{i1}=m_j$, $D_1=-\epsilon$, and $E_1 = \phi$ to see that (55) subsumes (56). In (54), let A average $C_1$ so fast that

$$x_i \cong C_i, \quad i \in I . \tag{58}$$

In (53), let $C_1 \equiv 0$ and let $A_1$ rapidly average $\sum_{k \in I} B_k z_{k1}$ so fast that

$$x_1 \stackrel{\simeq}{=} \sum_{k \in I} B_k z_{k1} \quad . \tag{59}$$

Letting $B_k = x_k$ shows, by (58), that

$$x_1 \stackrel{\simeq}{=} \sum_{k \in I} C_k z_{ki} \quad , \tag{60}$$

which is the same as c in (57), but in different notation. Now plug $x_1$ into $E_1$ and use a threshold rule as in (27) to complete the reduction.

Despite the obvious nature of this reduction, the authors make a number of claims that illustrate the present fragmentation of the theoretical community. They say that they have introduced in their threshold rule "a new and essential feature" which they call "temporal competition between input patterns". They also write that Cooper, Lieberman, and Oja (1979) were the first to introduce "the idea of such a modification scheme". They note that their equations result "in a form of <u>competition between incoming patterns</u> rather than competition between synapses" which they allege to be the conclusion of alternative theories. They also suggest that "our theory is in agreement with classical experimental results obtained over the last generation." Finally, in 1981 they feel free to "conjecture that some form of correlation modification is a very general organizational principle".

The status of some of these claims is clear from the preceding discussion. I will, however, indicate below how the threshold rule in (56) and (57) generates a temporally unstable code when more than one sampling node exists, and why this threshold rule either cannot explain critical period termination or cannot explain the results of Pettigrew and Kasamatsu (1978). Thus although equations (56) and (57) are a special case of an instar, not all choices of instar functionals are equally good for purposes of stable code development.

## 22. Adaptive Filtering of Spatial Pattern Sequences.

The comparison between pattern learning and code development becomes more interesting when a space-time pattern, or sequence of spatial patterns, is to be parsed by pattern learning or code development mechanisms. In either case, the fact that the LTM unit is a spatial pattern is fundamental, and the task is to show how individual spatial patterns, or subsequences of spatial patterns, can be differentially processed. To do this, one needs to show how distinguishable sampling sources, or subsets of sources, can be sequentially activated by the spatial patterns in the pattern sequence (Figure 6).

In the simplest pattern learning examples, pre-wired sequentially activated sampling nodes can learn an arbitrary space-time pattern (Grossberg, 1969e, 1970b). The price paid for such a ritualistic encoding is that the order of pattern performance, although not its velocity, is rigidly constrained. This <u>avalanche</u> type of anatomy

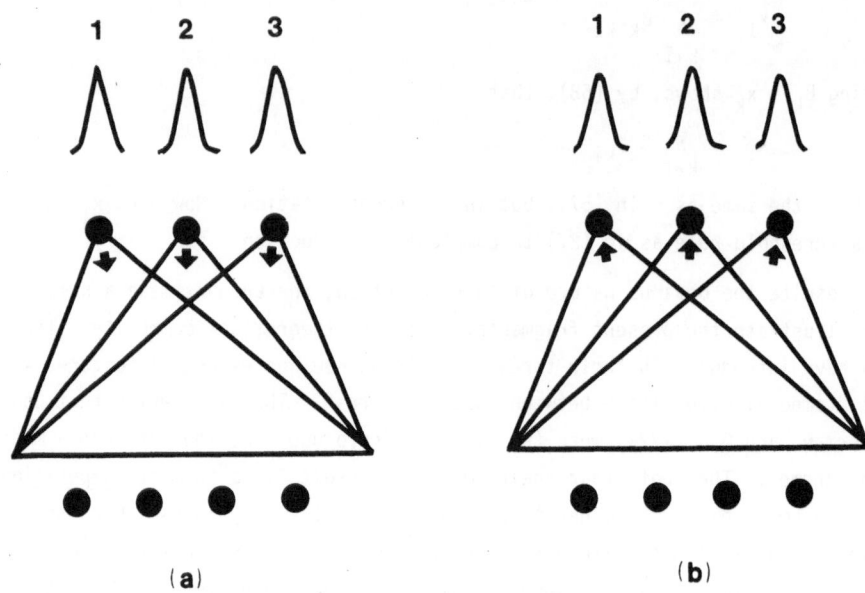

(a)  (b)

FIGURE 6. (a) In the simplest problem of space-time pattern learning, a mechanism is needed to excite discriminable sampling sources in a prescribed order 1,2,3,... during learning trials, and to repeat the same order 1,2,3,... of excitation during recall trials. (b) In the simplest problem of space-time code development, a space-time pattern at the sampled cells initially excites sampling sources in a prescribed order 1,2,3,... due to the a priori biases in the filter from sampled cells to sampling cells. Whether this ordering approaches a stable configuration as development proceeds, or a temporally unstable sequence of coding representations is triggered, depends on details of network design.

Supported in part by the National Science Foundation (NSF IST-80-00257) and the Air Force Office of Scientific Research.

is isomorphic to the anatomies of certain invertebrate command cells (Grossberg, 1974; Stein, 1971), and illustrates that complex acts can be encoded by small numbers of cells if ritualistic performance is acceptable. In examples wherein the order with which sampling nodes will be activated is not prewired into the network, serial learning mechanisms — notably associative and competitive interactions such as those utilized in Section 11 — are needed to learn the correct ordering as practice of the ordering proceeds (Grossberg, 1969c, 1974, 1978a). In examples wherein the filtering rules whereby individual sampling nodes are selected are not prewired into the network, we are confronted with a problem in code development, notably the problem of how sequences of events adaptively select the nodes that will elicit the most accurate predictive commands within their sequential context (Grossberg, 1978a). Most code development models consider special cases of this general problem.

If we generalize the instar equations (53) and (55) to include the possibility that many sampling (encoding) cell indices occur in J, we find equations

$$\frac{d}{dt} x_j = A_j x_j + \sum_{k \in I} B_k z_{kj} + I_j \tag{61}$$

and

$$\frac{d}{dt} z_{ij} = D_j z_{ij} + E_j x_i \tag{62}$$

$i \in I$, $j \in J$. Although all the terms in (61) and (62) work together to achieve code development, perhaps the terms $F_j \equiv \sum_{k \in I} B_k z_{kj}$ should be singled out as the principal ones. We all learn in calculus or linear algebra that the terms $F_j$ are dot products, or inner products, of the vector $B = (B_k: k \in I)$ with the vectors $z_j = (z_{kj}: k \in I)$; viz

$$F_j = B \cdot z_j . \tag{63}$$

If we define the vector $F = (F_j: j \in J)$ of dot products, then we can also recognize that the mapping $B \to F$ is a linear filter. By reversing arrows to go from pattern learning to code development, we hereby convert the property of independent read-out of probed pattern recall (Section 10) into the property of linear filtering — without a change of model.

An elementary formula about dot products underlies basic properties of code development. This formula is the law of cosines:

$$B \cdot z_j = \| B \| \| z_j \| \cos (B, z_j) , \tag{64}$$

where $\| V \|$ is the Euclidean length of vector V. By (64), given a fixed signal pattern B, $F_j$ is maximized among all $z_j$ of fixed length by choosing $z_j$ parallel to B. Thus in response to a fixed pattern B, the nodes $v_j$ for which $z_j$ is most parallel to B will be the ones most highly activated by B, other things being equal. If (62) causes the LTM vectors $z_j$ of highly activated nodes $v_j$ to become more parallel to B due to frequent past B presentations, then on future B presentations these nodes will respond ever more vigorously to B. Let us call alterations in F due to

past experience with B <u>adaptive</u> <u>filtering</u>. Then we can say that an interaction between adaptive filtering from $\{v_i: i \in I\}$ to $\{v_j: j \in J\}$ and competition within
$\{v_j: j \in J\}$ to select active nodes controls at least the most elementary features of
code development.

23.  <u>Synaptic Conservation, Code Invariance, and Code Instability</u>.

These observations about adaptive filtering did not, perhaps surprisingly, appear in Malsburg's original paper. Other important properties also have not been
sharply articulated by computer analysis. For example, presenting a given pattern
several times can recode not only the nodes which respond most vigorously to this pattern but also the responses of inactive nodes to other patterns presented later on,
because each node can encode infinite sets of patterns which it has never before experienced. This has the nice consequence  that the code can learn to recognize certain invariant properties of pattern classes without having to experience all the
patterns in each class (Fukushima, 1980; Grossberg, 1978a).

Another deeper consequence is not so pleasant. If there exist more patterns
than encoding nodes $v_j$, $j \in J$, there need not exist <u>any</u> temporally stable coding rule:
that is, the nodes which respond most vigorously to a given pattern can continually
change through time as the same list is repetitively practiced (Grossberg, 1976b).
I was led to suspect that such a result might hold due to my prior theorems about
absolute stability of parallel pattern learning,which suggested possible destabilizing effects of the STM decay terms $A_j$ in (61). This important instability result
has been missed by all the computer studies that I know because these studies typically present small numbers of patterns to small numbers of cells. Indeed, they usually present small numbers of patterns (e.g., 19) to much larger sets of cells (e.g.,
169), as in the careful analysis of Amari and Takeuchi (1978).

An instability result forces one to ask which properties are essential to code
development and which properties are adventitious. For example, Malsburg supplemented equations (3),(5),(6),(7) with a synaptic conservation rule that requires the
sum $\sum_{k \in I} z_{kj}$ of all the synaptic strengths which converge on a node $v_j$ to be constant
through time. Because I was aware of the duality between pattern learning and code
development, I realized that the synaptic conservation rule is incompatible with the
simplest example of associative learning, namely classical conditioning (Grossberg,
1976a). This is because the UCR pattern must be extinguished in memory as the CR
pattern is learned if synaptic conservation holds. I was therefore reluctant to accept the synaptic conservation rule without an important physical or mathematical
reason.

I could, however, see the need for a type of conservation, or normalization, that would regulate the temporal stability of the sampling process. By the time I read Malsburg's paper, I knew that long-range shunting competition, as opposed to the additive competition which Malsburg inherited from me, can automatically normalize the total suprathreshold STM activity of a network (Grossberg, 1973). The normalized STM activity can, in turn, normalize, or conserve, total synaptic strength across the network via feedback of $E_j$ to $z_{ij}$ in (62). This synaptic conservation mechanism is, moreover, compatible with classical conditioning. I therefore concluded that shunting competition, which can be absorbed into the STM decay terms $A_j$ of (61), should formally replace synaptic conservation until more pressing reasons to the contrary are given. Some experimental tests of synaptic competition vs. STM normalization are suggested in Grossberg (1981).

In their more recent contributions to retinotectal development, Malsburg and Willshaw have continued to use additive interactions, supplemented by the synaptic conservation rule and another rule for matching the similarity between retinal and tectal marker patterns (Malsburg and Willshaw, 1977, 1981; Willshaw and Malsburg, 1976). Since shunting networks automatically have matching properties as well as normalization properties, I take the need for these special assumptions as support for the idea that shunting operations subserve retinotectal development (Grossberg, 1976b, 1981). I have also argued that shunting interactions are operative in a variety of nonneural developmental examples, such as gastrulation in the sea urchin, slime mold aggregation, and regeneration in Hydra (Grossberg, 1978e). In all of these cases, I believe that alternative models have adapted analogies from chemical kinetics that do not incorporate mass action properties of cellular tissues. Notable differences between additive and shunting models occur in their explanations of the self-regulation mechanism that underlies the invariance of form when growth occurs (Gierer and Meinhardt, 1972; Grossberg, 1980c; Meinhardt and Geirer, 1974) and the contrast enhancement mechanism of categorical perception (Anderson, Silverstein, Ritz and Jones, 1977; Grossberg, 1978f).

24. Critical Period Termination, the Stability-Plasticity Dilemma, and Adaptive Resonance.

The fact that no temporally stable code need exist in response to a large family of input patterns, such as occurs in vision, made the problem of terminating those critical periods which are sensitive to behavioral experience seem more severe. This fact suggested that either the critical period is terminated by a chemical switch, but then there is a high likelihood that the code will incorporate adventitious statistical fluctuations of the most recent input sequences, or that the code is stabilized by a gradual process of dynamic buffering in response to network states that signify the behavioral relevance of the coded data. This dilemma led me to build my theory of <u>adaptive</u> <u>resonances</u> (Grossberg, 1976c, 1978a, 1980a, 1982b) which formalizes an answer to what I call the <u>stability-plasticity</u> <u>dilemma</u>.

The stability-plasticity dilemma asks how internal representations can maintain themselves in a stable fashion against the erosive effects of behaviorally irrelevant environmental fluctuations, yet can nonetheless adapt rapidly in response to environmental fluctuations that are crucial to survival. How does a network as a whole know the difference between behaviorally irrelevant and relevant events even though its individual cells do not possess this knowledge? How does a network transmute this knowledge into the difference between slow and fast rates of adaptation, respectively? Classical examples of the stability-plasticity balance are found in the work of Held and his colleagues on rapid visual adaptation in adults to discordant visuomotor data (Held, 1961, 1967; Held and Hein, 1963) and in the work of Wallach and his colleagues on rapid visual adaptation to discordant cues for the kinetic depth effect and cues for retinal disparity (Wallach and Karsh, 1963a,b; Wallach, Moore, and Davidson, 1963). The stability-plasticity issue is raised on a pharmacological level by the experiments of Pettigrew and Kasamatsu (1978) which show that the visual plasticity of normal adult cats can be restored by selectively adding some noradrenaline to cortical tissues which already possess a functioning noradrenaline arousal system.

The adaptive resonance theory which I introduced in Grossberg (1976c) can explain the Pettigrew and Kasamatsu (1978) data; see Grossberg (1982b) for a review. Let me briefly indicate why the Bienenstock, Cooper, and Munro (1982) work cannot.

First note what happens when (56) is embedded in a system such as (61) wherein several sampling nodes can compete for activity. By the threshold rule of (56), a node $v_j$ which has successfully won this competition in the past will acquire a progressively higher threshold due to persistent activation by its own input $\sum_k B_k z_{kj}$. By contrast, other nodes $v_m$ which do not win the STM competition when $\sum_k B_k z_{kj}$ occurs, but which receive significant projections $\sum_k B_k z_{km}$, will maintain a low threshold. Thus, the tradeoff between input size and threshold can ultimately favor a new set of nodes. When this happens, the pattern will be recoded, and a temporally unstable coding cycle will be initiated. This instability does not require a large number of coding patterns to occur. It can occur when only one pattern is repeatedly presented to a network containing more than one encoding node. In fact, the last examples in Grossberg (1976b, p. 132) consider history-dependent threshold changes, much like those in the Bienenstock <u>et</u> <u>al</u> example. I note their instability in a competitive sampling milieu before introducing the adaptive resonance theory in Grossberg (1976c) as a possible way out.

One might object to the above criticism by claiming that the original winning node $v_j$ acquires a high threshold so quickly that only the adaptively enhanced input $\sum_k B_k z_{kj}$ can exceed this threshold. In other words, the parameters may be carefully chosen to quickly shut off the critical period. But then one cannot understand how adding a little noradrenaline can turn it back on. In this example, either the critical period does not shut off, whence temporal instabilities in coding can occur, or it does shut off, whence critical period reversal by noradrenaline application cannot be explained. Of course, quickly raising the threshold might in any case trigger unstable coding by favoring new nodes.

## 25. Stable Coding of Pattern Sequences.

I will end my remarks with two theorems about stable pattern coding (Grossberg, 1976b). These theorems do not even dent the surface of the mathematical challenges raised by the theory of adaptive resonances. The theorems consider the simplest case wherein:

1) The patterns across nodes $v_i$, $i \in I$, are immediately and perfectly normalized. Thus input $C_i(t) = \theta_i C(t)$ generates activity $x_i(t) = \theta_i$.

2) The signals $B_k$ in (61) are linear functions of the activities $x_k$. Choose $B_k = \theta_k$ for definiteness.

3) The competition among nodes $v_j$, $j \in J$, normalizes the total activity (to the value 1 for definiteness) and rapidly chooses the node $v_j$ for STM storage which receives the largest input. In other words,

$$
x_j = \begin{cases} 1 \text{ if } F_j > \max \{\epsilon, F_k : k \neq j\} \\ \\ 0 \text{ if } F_j \leq \max \{\epsilon, F_k : k \neq j\} \end{cases} \tag{65}
$$

where

$$
F_j = \sum_{k \in I} \theta_k z_{kj} \tag{66}
$$

and $\epsilon$ represents the quenching threshold of the competition (Grossberg, 1973).

4) The LTM traces sample the pattern $\theta = (\theta_1, \theta_2, \ldots, \theta_n)$ only when their sampling cell is active. Thus

$$
\frac{d}{dt} z_{ij} = (-z_{ij} + \theta_i) x_j. \tag{67}
$$

Amari and Takeuchi (1978) study essentially identical equations and arrive at related results in the case of one encoding cell. They also study the response of the equations to inputs which simulate experiments on monocular and alternate-monocular deprivation of the kitten visual cortex.

The first result shows that if a single pattern is practiced, it maximizes the input (inner product) to its encoding cell population $v_j$ by making $z_j$ become parallel to $\theta$. Simultaneously, the length of $z_j$ becomes normalized.

## Theorem 5 (Single Pattern Code)

Given a pattern $\theta$, suppose that there exists a unique $j \in J$ such that

$$F_j(0) > \max \{ \epsilon, F_k(0): k \neq j \}. \tag{68}$$

Let $\theta$ be practiced during a sequence of nonoverlapping intervals $[U_k, V_k]$, $k=1,2,\ldots$ Then the angle between $z_j(t)$ and $\theta$ monotonically decreases, the signal $F_j(t)$ is monotonically attracted towards $\| \theta \|^2$, and $\| z_j(t) \|^2$ oscillates at most once as it tracks $F_j(t)$. In particular, if $\| z_j(0) \| \leq \| \theta \|$, then $F_j(t)$ is monotone increasing. Except in the trivial case that $F_j(0) = \| \theta \|^2$, the limiting relations

$$\lim_{t \to \infty} \| z_j(t) \|^2 = \lim_{t \to \infty} F_j(t) = \| \theta \|^2 \tag{69}$$

hold if and only if

$$\sum_{k=1}^{\infty} (V_k - U_k) = \infty . \tag{70}$$

The second result characterizes those sets of input patterns which can generate a temporally stable code, and shows that the classifying vectors $z_j(t)$ approach the convex hull of the patterns which they encode. The latter property shows that the nodes $v_j$ ultimately receive the maximal possible inputs from the pattern sets which they encode.

To state the theorem, the following notation is convenient. A partition $\oplus_{k=1}^{K} P_k$ of a finite set P is a subdivision of P into nonoverlapping and exhaustive subsets $P_j$. The convex hull H(P) of P is the set of all convex combinations of elements in P. Given a set $Q \subset P$, let $R = P \setminus Q$ denote the elements in P that are not in Q. If the classifying vector $z_j(t)$ codes the set of patterns $P_j(t)$, let $P_j^*(t) = P_j(t) \cup \{z_j(t)\}$. The distance between a vector p and a set of vectors Q, denoted by $\| p-Q \|$, is defined by $\| p-Q \| = \inf \{ \| p-Q \| : q \in Q \}$.

## Theorem 6 (Stability of Sparse Pattern Codes).

Let the network practice any finite set $P = \{\theta^{(i)}: i = 1,2,\ldots, M\}$ of patterns for which there exists a partition $P = \oplus_{k=1}^{N} P_k(T)$ at some time $t = T$ such that

$$\min\{u \cdot v : u \in P_j(T), v \in P_j^*(T)\} > \max\{u \cdot v : u \in P_j(T), v \in P^*(T) \setminus P_j^*(T)\} \tag{71}$$

for all $j = 1,2,\ldots,N$. Then

$$P_j(t) = P_j(T) \text{ for } t \geq T, j = 1,2,\ldots, N, \tag{72}$$

and the functions

$$D_j(t) = \| z_j(t) - H(P_j(t)) \| \tag{73}$$

are monotone decreasing for $t \geq T$, $j = 1,2,\ldots,N$. If moreover the patterns $P_j(T)$ are practiced in the time intervals $[U_{jk}, V_{jk}]$, $k = 1, 2, \ldots$ such that

$$\sum_{k=1}^{\infty} (V_{jk} - U_{jk}) = \infty ,\qquad (74)$$

then

$$\lim_{t \to \infty} D_j(t) = 0. \qquad (75)$$

Despite the fact that the code of a sparse pattern class is stable, it is easy to construct examples of pattern sequences which are densely distributed in pattern space for which no temporally stable code exists. To stabilize a behaviorally sensitive developing code in an arbitrary input environment, I have constructed the adaptive resonance theory, which uses the same feedback laws to stabilize infant code development as are needed to analyze data on adult attention. I have therefore elsewhere suggested that adult attention is a continuation on a developmental continuum of the mechanisms needed to solve the stability-plasticity dilemma in infants.

## REFERENCES

Adams, J.A. Human memory. New York: McGraw Hill, 1967.

Amari, S.-I. A method of statistical neurodynamics. Kybernetik, 1974, 14, 201-215.

Amari, S.-I. A mathematical approach to neural systems. In J. Metzler (Ed.) Systems neuroscience. New York: Academic Press, 1977.

Amari, S.-I., and Takeuchi, A. Mathematical theory on formation of category detecting nerve cells. Biological Cybernetics, 1978, 29, 127-136.

Anderson, J.R., and Bower, G.H. Human associative memory. Washington, D.C.: V.H. Winston and Sons, 1973.

Anderson, J.A., Silverstein, J.W., Ritz, S.A., and Jones, R.S. Distinctive features, categorical perception, and probability learning: Some applications of a neural model. Psychological Review, 1977, 84, 413-451.

Asch, S.E., and Ebenholtz, S.M. The principle of associative symmetry. Proceedings of the American Philosophical Society, 1962, 106, 135-163.

Atkinson, R.C., and Shiffrin, R.M. Human memory: A proposed system and its control processes. In K.W. Spence and J.T. Spence (Eds.), Advances in the psychology of learning and motivation research and theory (Vol. 2). New York: Academic Press, 1968.

Bienenstock, E.L., Cooper, L.N., and Munro, P.W.  Theory for the development of neuron selectivity: Orientation specificity and binocular interaction in visual cortex. 1982, preprint.

Bower, G.H. (Ed.).  Human memory: Basic processes.  New York: Academic Press, 1977.

Cohen, M.A., and Grossberg, S.  Global pattern formation by nonlinear networks. Submitted for publication, 1982.

Cole, K.S.  Membranes, ions, and impulses.  Berkeley, Calif.: Univ. of Calif. Press, 1968.

Collins, A.M., and Loftus, E.F.  A spreading-activation theory of semantic memory. Psychological Review, 1975, 82, 407-428.

Cooper, L.N., Lieberman, F., and Oja, E.  A theory for the acquisition and loss of neuron specificity in visual cortex. Biological Cybernetics, 1979, 33, 9.

Dixon, T.R., and Horton, D.L.  Verbal behavior and general behavior theory.  Englewood Cliffs, N.J.: Prentice-Hall, 1968.

Feigenbaum, E.A., and Simon, H.A.  A theory of the serial position effect.  British Journal of Psychology, 1962, 53, 307-320.

Fukushima, K.  Neocognitron: A self-organizing neural network model for a mechanism of pattern recognition unaffected by shift in position.  Biological Cybernetics, 1980, 36, 193-202.

Geman, S.  The law of large numbers in neural modelling.  In S. Grossberg (Ed.) Mathematical psychology and psychophysiology.  Providence, R.I.: American Mathematical Society, 1981.

Gierer, A., and Meinhardt, H.  A theory of biological pattern formation.  Kybernetik, 1972, 12, 30-39.

Gottlieb, G. (Ed.).  Neural and behavioral specificity, Vol. 3. New York: Academic Press, 1976.

Grossberg, S.  Senior Fellowship thesis, Dartmouth College, 1961.

Grossberg, S.  The theory of embedding fields with applications to psychology and neurophysiology.  New York: Rockefeller Institute for Medical Research, 1964.

Grossberg, S.  Some physiological and biochemical consequences of psychological postulates.  Proceedings of the National Academy of Sciences, 1968, 60, 758-765.

Grossberg, S. Embedding fields: a theory of learning with physiological implications. Journal of Mathematical Psychology, 1969, 6, 209-239(a).

Grossberg, S. On learning, information, lateral inhibition, and transmitters. Mathematical Biosciences, 1969, 4, 255-310(b).

Grossberg, S. On the serial learning of lists. Mathematical Biosciences, 1969, 4, 201-253(c).

Grossberg, S. On learning and energy-entropy dependence in recurrent and nonrecurrent signed networks. Journal of Statistical Physics, 1969, 1, 319-350(d).

Grossberg, S. Some networks that can learn, remember, and reproduce any number of complicated space-time patterns, I. Journal of Mathematics and Mechanics, 1969, 19, 53-91(e).

Grossberg, S. On the production and release of chemical transmitters and related topics in cellular control. Journal of Theoretical Biology, 1969, 22, 325-364(f).

Grossberg, S. Neural pattern discrimination. Journal of Theoretical Biology, 1970, 27, 291-337(a).

Grossberg, S. Some networks that can learn, remember, and reproduce any number of complicated space-time patterns, II. Studies in Applied Mathematics, 1970, 49, 135-166(b).

Grossberg, S. Pavlovian pattern learning by nonlinear neural networks. Proceedings of the National Academy of Sciences, 1971, 68, 828-831(a).

Grossberg, S. On the dynamics of operant conditioning. Journal of Theoretical Biology, 1971, 33, 225-255(b).

Grossberg, S. Pattern Learning by Functional-Differential Neural Networks with Arbitrary Path Weights. In K. Schmitt (Ed.). Delay and functional-differential equations and their applications. New York: Academic Press, 1972(a).

Grossberg, S. Neural expectation: Cerebellar and retinal analogs of cells fired by learnable or unlearned pattern classes. Kybernetik, 1972, 10, 49-57(b).

Grossberg, S. A neural theory of punishment and avoidance, I. Qualitative Theory. Mathematical Biosciences, 1972, 15, 39-67(c).

Grossberg, S. A neural theory of punishment and avoidance, II. Quantitative Theory. Mathematical Biosciences, 1972, 15, 253-285(d).

Grossberg, S. Contour enhancement, short-term memory, and constancies in reverberating neural networks. Studies in Applied Mathematics, 1973, 52, 217-257.

Grossberg, S. Classical and instrumental learning by neural networks. In R. Rosen and F. Snell (Eds.), Progress in Theoretical Biology, Vol. 3. New York: Academic Press, 1974.

Grossberg, S. On the development of feature detectors in the visual cortex with applications to learning and reaction-diffusion systems. Biological Cybernetics, 1976, 21, 145-159(a).

Grossberg, S. Adaptive pattern classification and universal recoding I: Parallel development and coding of neural feature detectors. Biological Cybernetics, 1976, 23, 121-134(b).

Grossberg, S. Adaptive pattern classification and universal recoding, II: Feedback, expectation, olfaction, and illusions. Biological Cybernetics, 1976, 23, 187-202(c).

Grossberg, S. A theory of human memory: Self-organization and performance of sensory-motor codes, maps, and plans. In R. Rosen and F. Snell (Eds.), Progress in Theoretical biology, Vol. 5. New York: Academic Press, 1978(a).

Grossberg, S. Behavioral contrast in short-term memory: Serial binary memory models or parallel continuous memory models? Journal of Mathematical Psychology, 1978, 17, 199-219(b).

Grossberg, S. Decisions, patterns, and oscillations in the dynamics of competitive systems with applications to Volterra-Lotka systems. Journal of Theoretical Biology, 1978, 101-130(c).

Grossberg, S. Competition, decision, and consensus. Journal of Mathematical Analysis and Applications, 1978, 66, 470-493(d).

Grossberg, S. Communication, memory, and development. In R. Rosen and F. Snell (Eds.) Progress in theoretical biology (Vol. 5). New York: Academic Press, 1978(e).

Grossberg, S. Do all neural networks really look alike? A comment on Anderson, Silverstein, Ritz, and Jones. Psychological Review, 1978, 85, 592-596(f).

Grossberg, S. How does a brain build a cognitive code? Psychological Review, 1980, 87, 1-51(a).

Grossberg, S. Intracellular mechanisms of adaptation and self-regulation in self-organizing networks: The role of chemical transducers. Bulletin of Mathematical Biology, 1980, 3, 365-396(b).

Grossberg, S. Biological Competition: Decision rules, pattern formation, and oscillations. Proceedings of the National Academy of Sciences, 1980, 77, 2338-2342 (c).

Grossberg, S. Adaptive resonance in development, perception, and cognition. In S. Grossberg (Ed.) Mathematical Psychology and Psychophysiology. Providence, R.I. American Mathematical Society, 1981.

Grossberg, S. Studies of mind and brain. Boston: Reidel Press, 1982(a).

Grossberg, S. Some psychophysiological and pharmacological correlates of a developmental, cognitive, and motivational theory. In J. Cohen, R. Karrer, and P. Tueting (Eds.), Proceedings of the 6th Evoked Potential International Conference, June 21-26,1981, Lake Forest, Illinois. New York: N.Y. Academy of Sciences,1982 (b).

Grossberg, S., and Pepe, J. Schizophrenia: Possible dependence of associational span, bowing, and primacy vs. recency on spiking threshold. Behavioral Science, 1970, 15, 359-362.

Grossberg, S., and Pepe, J. Spiking threshold and overarousal effects in serial learning. Journal of Statistical Physics, 1971, 3, 95-125.

Held, R. Exposure-history as a factor in maintaining stability of perception and coordination. Journal of Nervous and Rental Diseases, 1961, 132, 26-32.

Held, R. Dissociation of visual functions by deprivation and rearrangement. Psychologische Forschung, 1967, 31, 338-348.

Held, R., and Hein, A. Movement-produced stimulation in the development of visually guided behavior. Journal of Comparative and Physiological Psychology, 1963, 56, 872-876.

Hirsch, M.W. Systems of differential equations which are competitive for cooperative, I. Limit sets, 1982, preprint(a).

Hirsch, M.R. Systems of differential equations which are competitive or cooperative, II. Convergence almost everywhere, 1982, preprint(b).

Hubel, D.H., and Wiesel, T.N. Functional architecture of macaque monkey visual cortex. Proceedings of the Royal Society of London (B), 1977, 198, 1-59.

Hunt, R.K., and Jacobson, M. Specification of positional information in retinal ganglion cells of Xenopus laevis: Intraocular control of the time of specification. Proceedings of the National Academy of Sciences, U.S.A., 1974, 71, 3616-3620.

Jung, J. Verbal learning. New York: Holt, Rinehart, and Winston, Inc., 1968.

Katz, B. Nerve, muscle, and synapse. New York: McGraw Hill, 1966.

Khinchin, A.I. Mathematical foundations of information theory. New York: Dover, 1967.

Klatsky, R.L. Human memory: Structures and processes. San Francisco: W.H. Freeman and Co., 1980.

Kuffler, S.W., and Nicholls, J.G. From neuron to brain. Sunderland, Mass.: Sinauer, 1976.

Loftus, G.R., and Loftus, E.F. Human memory: The processing of information. Hillsdale, N.J.: Erbaum, 1976.

Maher, B.A. Contributions to the psychopathology of schizophrenia. New York: Academic Press, 1977.

Malsburg, C. von der. Self-organization of orientation sensitive cells in the striate cortex. Kybernetik, 1973, 14, 85-100.

Malsburg, C. von der., and Willshaw, D.J. How to label nerve cells so that they can interconnect in an ordered fashion. Proceedings of the National Academy of Sciences, U.S.A., 1977, 74, 5176-5178.

Malsburg, C. von der, and Willshaw, D.J. Differential equations for the development of topological nerve fibre projections. In S. Grossberg (Ed.) Mathematical Psychology and Psychophysiology. Providence, R.I.: American Mathematical Society, 1981.

May, R.M. and Leonard, W.J. Nonlinear aspects of competition between three species. SIAM Journal on Applied Mathematics, 1975, 29, 243-253.

McGeogh, J.A. and Irion, A.L. The psychology of human learning, Second Edition. New York: Longmans, Green, 1952.

Meinhardt, H., and Gierer, A. Applications of a theory of biological pattern formation based on lateral inhibition. Journal of Cell Science, 1974, 15, 321-346.

Murdock, B.B. Human memory: Theory and data. Potomac, Maryland: Erlbaum, 1974.

Norman, D.A. Memory and attention: An introduction to human information processing. New York: Wiley, 1969.

Osgood, C.E. Method and theory in experimental psychology. New York: Oxford, 1953.

Pettigrew, J.D., and Kasamatsu, T. Local perfusion of noradrenaline maintains visual cortical plasticity. Nature, 1978, 271, 761-763.

Plonsey, R. Bioelectric phenomena. New York: McGraw-Hill, 1969.

Ruch, T.C., Patton, H.D., Woodbury, J.W., and Towe, A.L. Neurophysiology. Philadelphia, Saunders, 1961.

Schneider, W., and Shiffrin, R.M. Automatic and controlled information processing in vision. In D. La Barge and S.J. Samuels (Eds.), Basic processes in reading: Perception and comprehension. Hillsdale, N.J.: Erlbaum, 1976.

Stein, P.S.G. Intersegmental coordination of swimmeret and motoneuron activity in crayfish. Journal of Neurophysiology, 1971, 34, 310-318.

Underwood, B.J. Experimental psychology, Second edition. New York: Appleton-Century-Crofts, 1966.

Wallach, H. and Karsh, E.B. Why the modification of stereoscopic depth-perception is so rapid. American Journal of Psychology, 1963, 76, 413-420(a).

Wallach, H., and Karsh, E.B. The modification of stereoscopic depth-perception and the kinetic depth-effect. American Journal of Psychology, 1963, 76, 429-435(b).

Wallach, H., Moore, M.E., and Davidson, L. Modification of stereoscopic depth-perception. American Journal of Psychology, 1963, 76, 191-204.

Willshaw, D.J., and Malsburg, C. von der. How patterned neural connections can be set up by self-organization. Proceedings of the Royal Society of London, Series B, 1976, 194, 431-445.

Wolpert, L. Positional information and the spatial pattern of cellular differentiation. Journal of theoretical biology, 1969, 25, 1-47.

Young, R.K. Serial learning. In T.R. Dixon and D.L. Horton (Eds.) Verbal behavior and general behavior theory. Englewood Cliffs, N.J.: Prentice-Hall, 1968.

# CHAPTER 21

## MODELLING NEURAL MECHANISMS OF VISUOMOTOR COORDINATION IN FROG AND TOAD*

Michael A. Arbib
Computer and Information Science Department,
Center for Systems Neuroscience
University of Massachusetts
Amherst, Massachusetts 01003, USA

### ABSTRACT

Frogs and toads provide interesting parallels to the way in which humans can see the world about them, and use what they see in determining their actions. What they lack in subtlety of visually-guided behavior, they make up for in the amenability of their behavior and the underlying neural circuitry to experimental analysis. We provide an overview of problems involved in modelling neural mechanisms of frog and toad visuomotor coordination; and then present a number of background models "in search of the style of the brain." We then review three specific models of neural circuitry underlying visually-guided behavior in frog and toad. They form an 'evolutionary sequence' in that each model incorporates its predecessor as a subsystem in such a way as to explain a wider range of behavioral data in a manner consistent with current neurophysiology and anatomy. The models thus form stages in the evolution of <u>Rana computatrix</u>, an increasingly sophisticated model of neural circuitry underlying the behavior of the frog.+ Finally, we provide a quick tour of a number of studies which have developed from these basic models.

## 1. NEURAL SUBSTRATES FOR VISUALLY-GUIDED BEHAVIOR

Lettvin, Maturana, McCulloch and Pitts [1959] initiated the behaviorally-oriented study of the frog visual system with their classification of retinal ganglion cells into four classes each projecting to a retinotopic map at a different depth in the optic tectum, the four maps in register. In this spirit, we view the analysis of interactions between layers of neurons as a major approach to

------------------------

* The research reported in this paper was supported in part by the National Institutes of Health under grant R01 NS14971-03. My special thanks to Rolando Lara of Universidad Nacional Autonoma de Mexico with whom the recent modelling was conducted during his stay at the University of Massachusetts, 1978-1980. Portions of Sections 1 and 4 appeared in the earlier paper, Arbib (1982).

+ When both models and experiments are further advanced, the time will be ripe for the differential analysis of (different species of) frog and toad. In the present article, however, we conflate data gathered from both frog and toad studies to lay the experimental basis for the models that we discuss.

modelling "the style of the brain". In Section 3, we offer a general view of cooperative computation between neurons within a layer, and between layers within the brain. (The relation of "maps as control surfaces" to the general study of perceptual structures and distributed motor control is given in Arbib [1981].) In following sections, we shall then exemplify these general principles in three specific models of cooperative computation in neural circuitry underlying visuomotor coordination in frog and toad. The final section will then chart directions for further modelling.

Lettvin et al. found that group 2 retinal cells responded best to the movement of a small object within the receptive field; while group 3 cells responded best to the passage of a large object across the receptive field. It became common to speak of these cells as "bug detectors" (following Barlow [1953]) and "enemy detectors", respectively, though subsequent studies make it clear that the likelihood of a given frog behavior will depend on far more than activity of a single class of retinal ganglion cells (Ewert [1976], and Section 4 below). Given the mapping of retinal "feature detectors" to the tectum and the fact that tectal stimulation could elicit a snapping response, it became commonplace to view the tectum as, inter alia, directing the snapping of the animal at small moving objects -- it being known that the frog would ignore stationary objects, and would jump away from large moving objects. However, this notion of a simple stimulus-response chain via the tectum was vitiated by Ewert's observation that after a lesion to PT (pretectum-thalamus) a toad would snap at moving objects of all sizes, even those large enough to elicit escape responses in the normal animal. More detailed neurophysiological studies support the inference that the tectum alone will elicit a response to all (sufficiently) moving objects, and that it is PT-inhibition that blocks this response when the object is large, since tectal cells respond to visual presentation of large moving objects in the PT-lesioned animal [Ingle, 1973].

In Section 4a we present a model of local circuitry in the tectum (a 'tectal column') to explain certain facilitation effects in prey-catching behavior; then in Section 4b we study a linear array of such columns to model certain data on size-dependence of prey-catching activity in toads; and then, in Section 4c, we add PT-inhibition to such an array to model the behavior of an animal confronted with more than one prey-stimulus. These models form three stages in an evolutionary sequence for Rana Computatrix, our developing model of the neural circuitry underlying visuomotor coordination in frog and toad. Tectum and PT are but two of the many brain regions to be incorporated into the model during its further evolution. Section 5 provides a brief perspective of models discussed at greater length in later papers.

## 2. AN OVERVIEW OF MODELLING PROBLEMS

We may determine units in the brain physiologically -- for example, by electrical recording -- and anatomically -- e.g. by staining. In many regions of the brain, we have an excellent correlation between physiological and anatomical units -- we know which anatomical entity yields which physiological response. Unfortunately, this is not yet the case in many studies of visuomotor coordination in frog and toad. We have data on the electrophysiological correlates of animal behavior, and we have anatomical data. Often, though, we do not know which specific cell, defined anatomically, yields an observed electrophysiological response. For example, we have the Golgi anatomy of the frog tectum, shown in Figure 1a, and the physiological responses recorded from tectum during facilitation of prey-catching behavior shown in Figure 6d. However, our identification of the physiological responses with specific anatomically defined cells is still hypothetical. Nonetheless, such choices have to be made in formulating and testing our models.

Another problem that we confront in modelling is that we have both too much and too little anatomical detail: too much in that there are many connections that we cannot put into our model without overloading our capabilities for either mathematical analysis or computer simulation; and too little in that we often do not know which details of synaptology may determine the most important modes of behavior of a particular region of the brain. For example, in starting from the Golgi anatomy of frog tectum shown in Figure 1a, we can either follow Szekely and Lazar into the elaborate synaptology shown in Figure 1b, or we can rather accept their schematic view of a tectal column as the basic unit of structure, as shown in Figure 1c. In the modelling to be described in this paper, we have chosen the latter course, viewing the tectum as an array of interconnected columns each of which has the formal structure shown in Figure 1d, the behavior of the various neurons being described by coupled differential equations. In comparing the Golgi anatomy of Figure 1a with the model of Figure 1d, we see that a number of choices have been made. In Figure 1a we see that there are two types of output cells for the tectum, the pyramidal cell and the large tectal ganglionic neuron. Our model assumes that it is only the output of the former that is relevant to the phenomena that we are considering. Clearly, our models must be of such a kind that they are adaptable when we come to phenomena that in fact can be shown to depend upon the ganglionic output. Note, too, that we have ignored the bipolar neurons and amacrine cells, and that we have made certain assumptions about the connectivity between the neurons that are included in the model. However, an important point of our modelling methodology will be that we set up our simulation in such a way that we can use different connectivity on different simulations. In this fashion, we can generate hypotheses which can then be subjected to further experimental test.

Even if we have made a satisfactory choice of how to correlate physiological units with anatomical units, and of the appropriate connectivity, we still have the

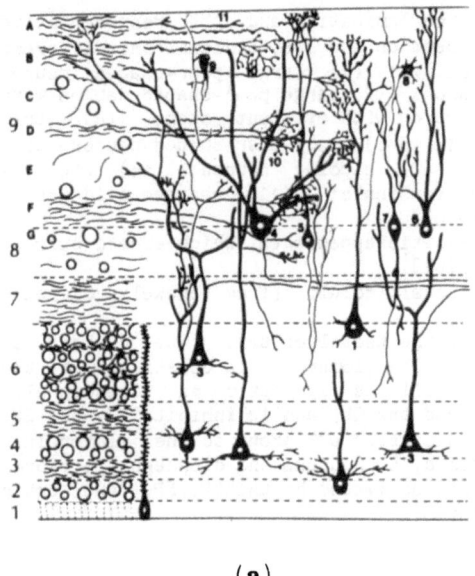

(a)

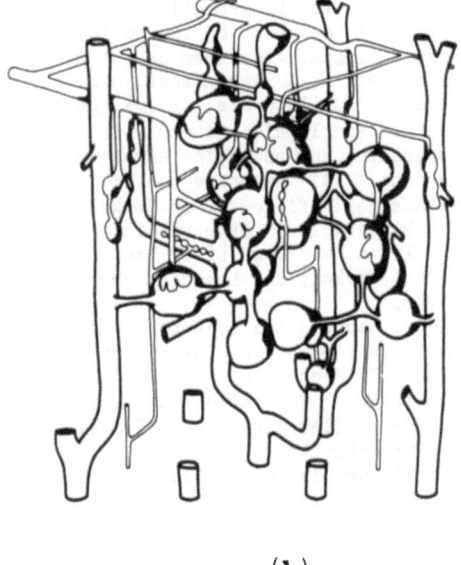

(b)

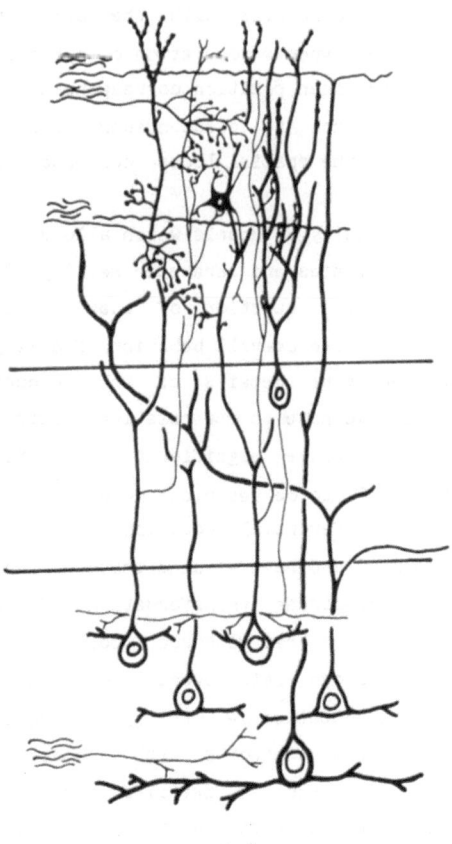

(c)

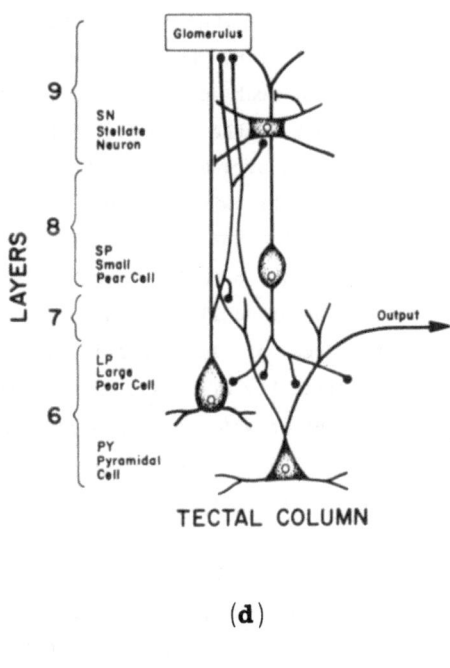

(d)

Figure 1.

Figure 1. (a) Diagramatic representation of the lamination and the representative types of neurons of the optic tectum. Numbers on the left indicate the different tectal layers. Numbered cell-types are as follows: (1) large pear-shaped neuron with dendritic appendages and ascending axon; (2) large pear-shaped neuron with dendritic collaterals; (3) large pyramidal neuron with efferent axon; (4) large tectal ganglionic neuron with efferent axon; (5-6) small pear-shaped neurons with descending and ascending axons respectively; (7) bipolar neuron; (8) stellate neuron; (9) amacrine cell; (10) optic terminals; (11) assumed evidence of diencephalic fibres [from Szekely & Lazar (1976)].

(b) Details of synaptic interaction of dendritic appendages, which exceed current models in intricacy [from Szekely & Lazar (1976)].

(c) Szekely and Lazar's schematic for a tectal column [from Szekely & Lazar (1976)].

(d) Neurons and synaptology of the model of the tectal column. The numbers at the left indicate the different tectal layers. The glomerulus is constituted by the LP and SP dendrites and recurrent axons as well as by optic and diencephalic terminals. The LP excites the PY, the SN, and the GL, and is inhibited by the SN. The SP excites the LP and PY cells, and it sends recurrent axons to the glomerulus; it is inhibited by the SN. The SN is excited by LP neurons and diencephalic fibres and it inhibits the LP and SP cells. The PY is activated by the LP, SP, and optic fibres, and is the efferent neuron of the tectum.

problem of correlating cellular structure and function with the animal's overall behavior. In the modelling to be described in this paper, for example, we have assumed that the activity of the pyramidal cells correlates with the orienting response of the animal. We have also assumed that when a population of pyramidal cells is active, the resultant orientation is to the mean position corresponding to that population, though there is not yet evidence to discountenance further hypotheses, such as that the orientation will be to the spatial locus corresponding to the peak of the pyramidal cell activity.

We have already spoken of the need to have a family of models which allows one to experiment with a number of different connectivities and parameter settings for the cells of the model. There still remains the question of what is the appropriately detailed model. Is it the fact that the overall behavior of a large collection of cells depends critically on the fine details of the response performance of each individual neuron, or can we hope to use relatively simple, computationally efficient, neuron models and still derive significant information about the behavior of the population? In the models to be described below, we have described the behavior of the neuron by a simple differential equation linear in terms of the synaptically weighted input values, and have assumed that the input from one cell to another is given by a simple non-linear transformation of the membrane potential of the source cell. We believe that with such models we can probe whether the neural networks of different kinds can yield overall classes of behavior. Future research will be both less detailed -- trying to provide quantitative analyses correlating classes of neural networks with classes of behaviors--and more detailed, as we try to establish detailed parametric specifications which can be subjected to experimental test in the laboratory. Section 3 will provide a survey of some of the conceptual models that enter into our search for "the style of the brain," while Section 4 will present the first three stages of our attempts to model in some detail the experimental studies of visuomotor coordination in frog and toad.

Our modelling methodology must be based not on a single "take it or leave it" model, but rather on the exploration of a variety of different connectivities within some overall paradigm of brain function. Thus, the models to be described below are dominated by two main considerations: the visual system of the animal must be considered in the context of the ongoing behavior of the animal -- thus the stress on visuomotor coordination, rather than on vision per se; and the analysis will be in terms of the interaction between concurrently active regions of the brain, rather than in terms of any simple one-way flow of information in a hierarchically organized system. We use the term cooperative computation to refer to this style of concurrent neural processing.

In addition to our concern for embedding the brain within the ongoing cycle of the animal's action and perception, and studying the brain itself in terms of the cooperative computation of interacting subsystems, the three models to be exhibited in Section 4 exhibit a style of "evolutionary" modelling. As a first approximation, we continually try to localize the neural processes underlying some overt behavior of the animal within some relatively small portion of the brain. As we come to analyze more functions, though, we find that each function may require activity in many portions of the brain, and that each portion of the brain will be involved in many different activities. Thus, having successfully modelled several phenomena, one should try as far as possible, when modelling a new phenomenon, to do it by minor adaptations of the previous model, preserving the earlier successes, rather than introducing an ad hoc model of a new brain region specifically designed to achieve the new specified task. Thus, in Section 4 we shall start with the model of the single tectal column shown in Figure 1d, and show that it is able to account for certain behavioral and neurophysiological data on facilitation; we shall then show how a linear array of such columns (Fig. 10a) can account for data on worm pattern recognition, without losing the ability of the individual column to exhibit facilitation to a localized stimulus. Finally, as shown in Figure 11a, we shall introduce pretectal cells and newness interneurons in interaction with the linear array of columns of Figure 10a, and see how, again without losing earlier properties of the model, we can account for the certain aspect of prey facilitation. In Section 5, we shall briefly outline further developments which continue to increase the behavioral repertoire of our evolving model, Rana computatrix.

## 3. BACKGROUND MODELS: IN SEARCH OF THE STYLE OF THE BRAIN

Before turning, in Section 4, to the first three stages in the evolution of Rana computatrix, we devote the present section to a number of background models which establish the "style of the brain" with which we approach our modelling of visuomotor coordination in frog and toad.

Since we are concerned with motor control, we of course make use of such concepts as feedback and feedforward. In many treatments of these concepts in the literature on biological control systems, we see the use of lumped models. For example, the direction in which the animal should turn is encoded by a single angle variable. However, since we shall be concerned with the way in which patterns on the retina inpinge upon ongoing activity within the brain, we shall not consider it permissible to regard this angle as explicitly available in the brain as the value of, for example, firing of some neuron. Rather, we must consider it as encoded by the locus of the peak of activity within a neural array. Perhaps the first model of distributed motor control of this kind is that of Pitts and McCulloch (1947).

3a. <u>Distributed</u> <u>Motor</u> <u>Control</u>. Apter (1945, 1946) had shown that each half of the visual field of the cat maps topographically upon the contralateral superior colliculus. In addition to investigating this sensory map, she studied the motor map by strychninizing a single point on the collicular surface, flashing a diffuse light on the retina, and then observing which point in the visual field was affixed by the resultant change in gaze. She found that these sensory and motor maps were almost identical, and this basic finding has been replicated and extended in many recent studies. Starting from these data, Pitts and McCulloch developed the model shown in Figure 2. This outlined the reflex arc that extended from the eyes through the superior colliculus to the ocular-motor nuclei, thereby controlling the muscles that direct the gaze so as to bring the fixation point to the center of gravity of distribution of the visual input's brightness. (Our current knowledge of retinal preprocessing enables us to substitute for the term <u>brightness</u> such a term as <u>contour</u> <u>information</u> or an expression that describes some other feature of the input.) Pitts and McCulloch noted that excitation at a point on the left colliculus corresponds to excitation from the right half of the visual field and so should induce movement of the eye to the right; gaze is centered when excitation from the left is exactly balanced by excitation from the right. Their model is so arranged that each motor neuron controlling muscle fibers in the muscles that contract to move the eyeballs to the right, for example, should receive excitation summing the level of activity in a thin transverse strip of the left colliculus. This process provides all the excitation for muscles turning the eye to the right. Reciprocal inhibition by axons from nuclei of the antagonist eye muscles, which are excited similarly by the other colliculus, performs subtraction. The quasi-center of gravity's vertical coordinate is computed similarly. Eye movement ceases when and only when the fixation point is the center of gravity. Such a model leads to the idea that a plausible subsystem for vertebrate nervous systems may be one in which position of the input on the control surface encodes the target to which the muscular control will be sent. Of course, much remains to be done in turning such a

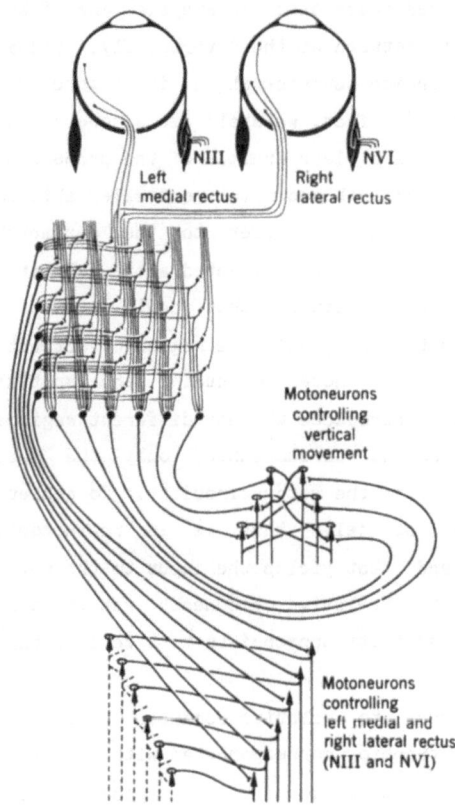

Figure 2. Pitts-McCulloch scheme
for reflex control of eye position
via superior colliculus. Eye can
only be stationary when activity
in two halves of colliculus is
balanced. [Adapted from Pitts &
McCulloch (1947).]

general scheme for distributed motor control into a specific model of a specific
system. For example, the Pitts-McCulloch model does not give an account of
ballistic movements. Again, it does not show us how, for increasing angles of
deviation of the target, visual tracking might first evoke movement of eyes alone,
then of eyes and head, and then of eyes, head, and trunk. It remains an important
task in brain theory to explain how the output of a motor computer would control not
a single pair of antagonist muscles, but rather a whole hierarchy of subcontrollers,
in a distributed way.

3b. A Model of Frog's Snapping. Another problem is that in much visually guided
behavior, the animal does not simply respond to "the center of gravity" of visual
stimulation, but rather is responding to some property of the overall configuration.
Consider, for example, the snapping behavior of frogs confronted with one or more
fly-like stimuli.

Ingle (1968) found that in a certain region around the head of a frog, the
presence of a fly-like stimulus elicits a snap; that is, the frog turns so that its
midline is pointed at the stimulus and zaps it with its tongue. When confronted
with two "flies," either of which is vigorous enough that alone it could elicit a

snapping response, the frog exhibits one of three reactions: it snaps at one of the flies, it does not snap at all, or it snaps in between at the "average fly." Didday (1970, 1976) offered the simple model of this choice behavior shown in Figure 3a. It is presented not as the state of the art -- in fact, we shall see a more recent model built upon it in Section 4c -- but rather as a clear example of the processing of structured stimuli to provide the input to a distributed motor controller akin to that shown in Figure 2. Didday used the term foodness to refer to the parameter representing the extent to which a stimulus could, when presented alone, elicit a snapping response. The task was to design a network that could take a position-tagged "foodness array" and ensure that usually only one region of activity would influence the motor control system. The model maintains the spatial distribution of information, with new circuitry introduced whereby different regions of the tectum compete in such a way that in normal circumstances only the most active region provides an above-threshold input to the motor circuitry. To achieve this effect we first introduce a new layer of cells that is in retinotopic correspondence to the "foodness layer," and that yields the input to the motor circuitry. In some sense, then, it is to be "relative foodness" rather than foodness that describes the receptive field activity appropriate to a cell of this layer.

Didday's transformation scheme from foodness to relative-foodness employs a population of "S-cells" that are in topographic correspondence with the other layers. Each S-cell inhibits the activity that cells in its region of the relative-foodness layer receive from the corresponding cells in the foodness layer by an amount that augments with increasing activity outside its particular region.

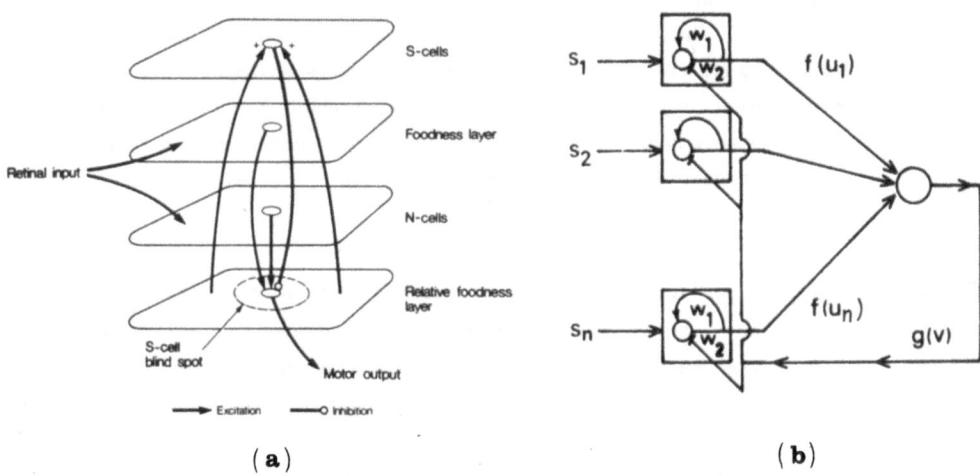

( a )          ( b )

Figure 3. (a) Schematic view of Didday's model of interacting layers of neurons subserving prey-selection. (b) Primitive cooperation model in which the layer of S-cells of (a) is replaced by a single inhibitory neuron [from Amari & Arbib (1977)].

This ensures that high activity in a region of the foodness layer penetrates only if the surrounding areas do not contain sufficiently high activity to block it.

When we examine the behavior of such a network, we find that plausible interconnection schemes yield the following properties:

1. If the activity in one region far exceeds the activity in any other region, then this region eventually overwhelms all other regions, and the animal snaps at the corresponding space.

2. If two regions have sufficiently close activity then a) they may both (providing they are very active) overwhelm the other regions and simultaneously take command, with the result that the frog snaps between the regions; or b) the two active regions may simply turn down each other's activity, as well as activity in other regions, to the point that neither are sufficient to take command. In this case the frog remains immobile, ignoring the two "flies."

One trouble with the circuitry as so far described is that the buildup of inhibition on the S-cells precludes the system's quick response to new stimuli. If in case 2b above, for example, one of those two very active regions were to suddenly become more active, then the deadlock should be broken quickly. In the network so far described, however, the new activity cannot easily break through the inhibition built up on the S-cell in its region. In other words there is hysteresis. Didday thus introduced an "N-cell" for each S-cell. The job of an N-cell is to monitor temporal changes in the activity of its region. Should it detect sufficiently dramatic increase in the region's activity, it then overrides the inhibition on the S-cell and permits this new level of activity to enter the relative foodness layer. With this scheme the inertia of the old model is overcome, and the system can respond rapidly to significant new stimuli. Didday hypothesized that the S-cells and N-cells modelled the "sameness" and "newness" cells, respectively, that had been observed in the frog tectum. Regrettably, no experiments have been done to test this hypothesis.

3c. <u>Competition and Cooperation in Neural Nets</u>. The above model of prey selection is an example of a broad class of models dealing with competition and cooperation in neural nets. As one example of a model of such a kind, let us consider the problem of stereopsis, or segmentation on depth cues. Julesz (1971) has designed "random-dot stereograms" in which each eye receives a totally random pattern, but in which there are correlations between the inputs to the two eyes. Specifically, the different regions in the two inputs are identical save for a shift in position, yielding a different disparity in the two retina (Fig. 4a). Although such a pattern for a naive subject can initially appear to be nothing but visual noise, eventually disparity matching takes place and the subject perceives surfaces at different depths. Barlow, Blakemore and Pettigrew (1967) and Pettigrew, Nikara, and Bishop (1968) have found that cells in cat visual cortex are tuned for retinal disparity, and similar cells are posited in the human. What presumably causes the initial

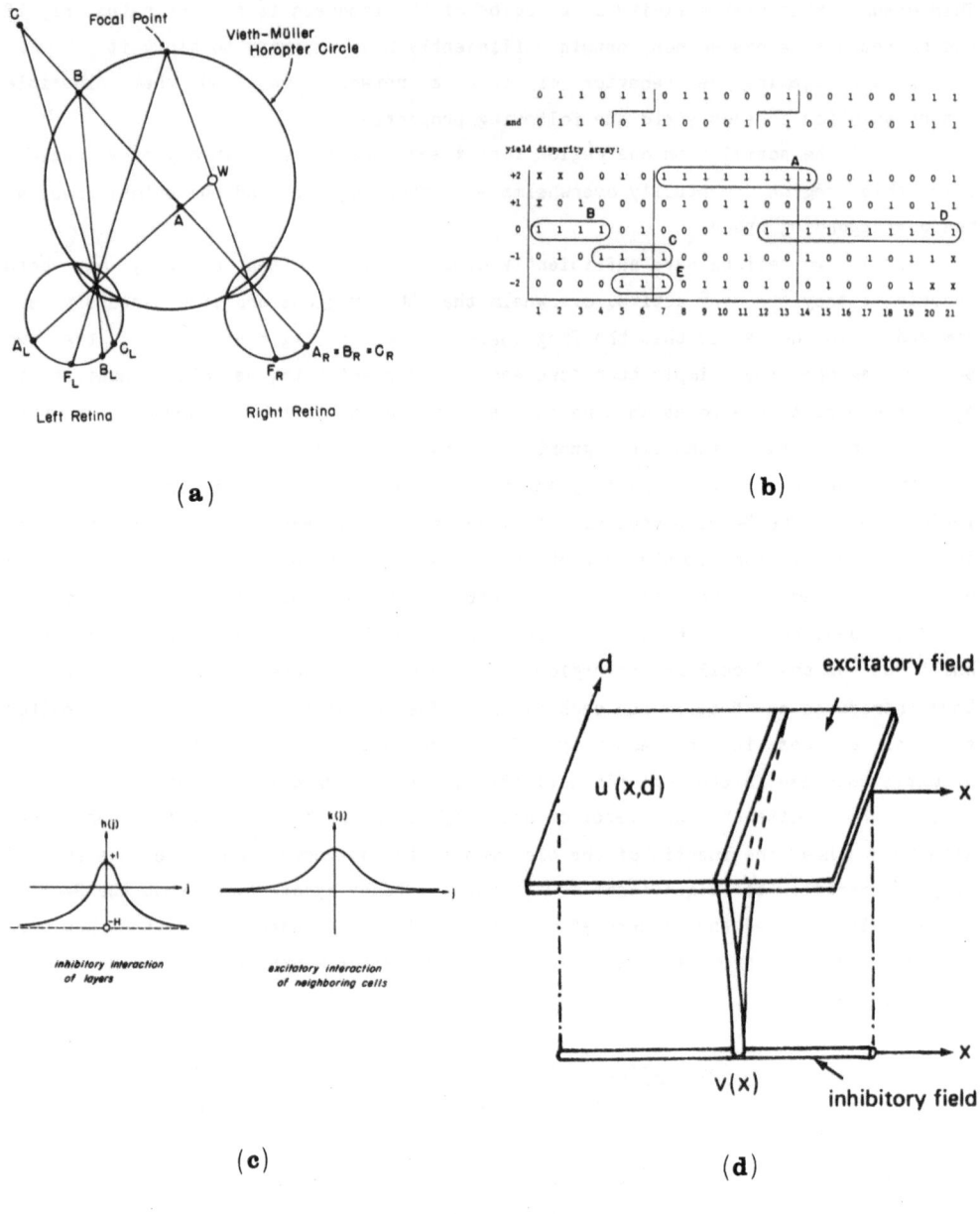

**Figure 4.** (a) Points projecting to the same point of one retina are projected to points with different disparities on the other retina. (b) The problem of resolving ambiguity: We conceptualize "layers" of cells (they are really in "columns"), one for each gross disparity. The aim is to segment the activity into connected regions. (c) Coupling coefficients for one approach to the problem: moderate local cross-excitation within layers; increasing inhibition between layers as difference in disparity increases. [From Arbib, Boylls & Dev (1974).] (d) The full model of competition and cooperation which allows the idea shown in (c) to be subject to mathematical analysis [from Amari & Arbib (1977)].

perception of visual noise is that in addition to the correct correlation of points in the two retinas, there are many spurious correlations, and computation is required to reduce them (Fig. 4b).

Dev (1975) [see also Sperling (1970), Arbib, Boylls and Dev (1974), Nelson (1975), and Marr and Poggio (1977)] has proposed that the cells of a given disparity be imagined as forming a population arrayed in a spatial map corresponding to the map of visual direction. Connections between cells could then be arranged so that nearby cells of a given disparity would be mutually excitatory, whereas cells nearby in visual direction but different in disparity would have inhibitory interaction (Fig. 4c). In this way, the activity of the array would organize into a pattern where in each region of visual direction, cells of only one disparity type would be highly active. As a result the visual input would eventually be segmented into a number of distinct surfaces.

In the stereopsis model, then, we have competition in the disparity dimension and cooperation in the other dimensions. The Didday model (Fig. 3a) can be regarded as the limiting case where there is only a competition dimension, namely that of prey location. Such informal observations have laid the basis for rigorous mathematical analysis of competition and cooperation in neural nets. For example, Amari and Arbib (1977) both offer the "primitive cooperation model" of Figure 3b which allows us to gain a mathematical handle on Didday's results, as well as a more sophisticated model, shown in Figure 4d, which allows us to provide a stability analysis of a model of the kind studied by Dev for stereopsis. Amari (1982) gives an up-to-date perspective on such models.

3d. <u>Motor Schemas</u>. We owe to the Russian school founded by Bernstein the general strategy which views the control of movement in terms of selecting one of a relatively short list of modes of activity, and then within each mode specifying the few parameters required to tune the movement. Where the Russians used the term synergy, we will use the term <u>motor schema</u>. The problem of motor control is thus one of sequencing and coordinating such motor schemas, rather than directly controlling the vast number of degrees of freedom offered by the independent activity of all the motor units. We have, to use the language of Greene, to get the system "into the right ballpark," and then to tune activity within that ballpark —— the dual problems of activation and tuning.

In the familiar realm of feedback control theory, a controller (which we will now think of as a motor schema) compares feedback signals from the controlled system with a statement of the desired performance of the system to determine control signals which will move the controlled system into ever greater conformity with the given plan. However, the appropriate choice of control signal must depend upon having a reasonably accurate model of the controlled system —— for example, the appropriate thrust to apply must depend upon an estimate of the mass of the object that is to be moved. Moreover, there are many cases in which the controlled system will change over time in such a way that no a priori estimate of the system's parameters can be reliably made. To that end, it is a useful practice to interpose an identification algorithm which can update the parametric description of the controlled system in such a way that the observed response of the system to its control signals comes into greater and greater conformity with that projected on the basis of the parametric description. We see that when a motor schema is equipped with an identification algorithm (Fig. 5a) and when the controlled system is of the class whose parameters the algorithm is designed to identify, and when, finally, the changes in parameters of the controlled system are not too rapid, then in fact the combination of controller and identification algorithm within the motor schema provides an adaptive control system, which is able to function effectively despite continual changes in the environment.

3e. A Model of the Cerebellum. We have suggested that the problem of motor control is one of sequencing and coordinating motor schemas, rather than directly controlling the vast number of degrees of freedom offered by the independent activity of all the muscles. We have suggested that an "identification algorithm" can adapt a motor schema to changing conditions within some overall motor task.   To see how this analysis can make contact with an interacting layers approach to neural circuitry, we now examine a model of the cerebellum (Arbib, Boylls, and Dev, 1974; Boylls, 1975, 1976).   The model brings together the notion of a motor schema with the notion of maps as control surfaces, and is important in that it exhibits neural layers acting as control surfaces representing levels of activation for the coordination of muscles, complementing our study of retinotopic representations of visual input.

To provide neurophysiological data for the model, we consider cerebellar function in locomotion of the high decerebrate cat (Shik et al., 1966). Where Sherrington had noticed that stimulation of Deiter's nucleus in the standing animal would lead to extension of all the limbs, Orlovskii found that in the high decerebrate cat, stimulation of Deiter's nucleus during locomotion would not affect extension during the swing phase, but would increase extension during the support phase. Since the locomotory "motor schema" has been shown to be available even in the spinal cat (both in classical work by Sherrington (1910) and in modern studies (compare Herman et.al. 1976)), it seems reasonable to view the system in which the cerebellum and Deiter's nucleus are involved as providing an identification algorithm for the parametric adjustment of the spinal schema (Fig. 5b). We now turn to Boylls' model which shows how the adjustment of these parameters might be computed within the cerebellar environs.

As is well known (Eccles et al., 1967), the only output of the cerebellar cortex is provided by the Purkinje cells, which provide inhibitory input to the cerebellar nuclei. Each Purkinje cell has two input systems. One input is via a single climbing fiber which ramifies and synapses all over the Purkinje cell's

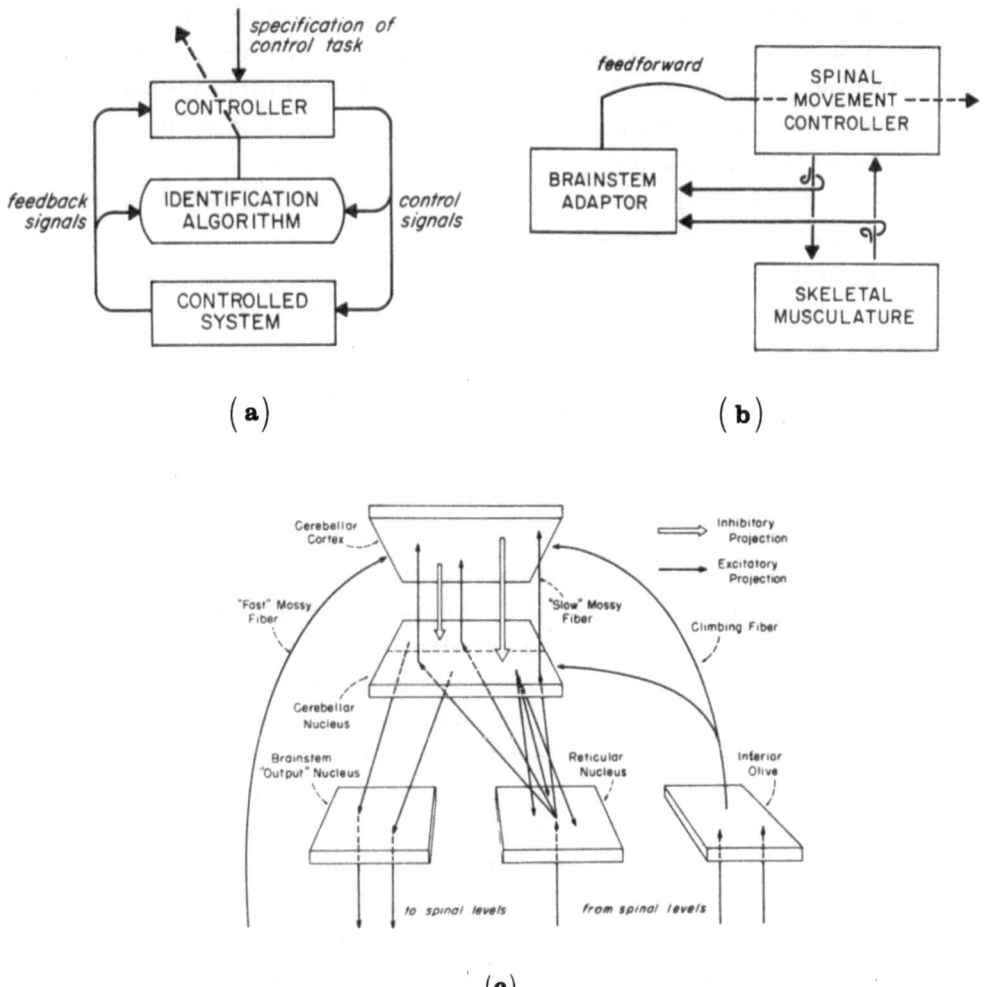

Figure 5. (a) An identification algorithm updates the parameters used to match the controller to the current properties of the system being controlled. (b) It is posited that the brainstem (cerebellum in interaction with various nuclei) serves as an identification algorithm for spinal movement controllers. (c) Schematic of the interacting control surfaces in the Boylls model of the tuning of motor schemas by cerebellum and related nuclei [from Arbib, Boylls & Dev (1974)].

dendritic tree. The other input system is via the mossy fibers, which activate granule cells whose axons rise up into the layer of Purkinje cell dendrites (which are flat, with the planes of all their dendritic trees parallel to one another) to form T's, whose crossbars run parallel to one another at right angles to the planes of the Purkinje dendritic trees. (There are a number of interneurons in the cerebellar cortex, but we shall not model these here, but shall instead concentrate on the basic cerebellar circuit of mossy and climbing fibers, and of granule and Purkinje cells.)

The climbing fiber input to a Purkinje cell is so strong that, when its climbing fiber is fired, a Purkinje cell responds with a sharp burst of four or five spikes, known as the climbing fiber response (CFR). Many authors have thought that the 'secret' of the climbing fiber is this sharp series of bursts, but Boylls suggested below that the true role of the climbing fiber input is to provide the suppression of Purkinje cell activity for as much as 100 milliseconds which has been found to follow the CFR (Murphy and Sabah, 1970).

The overall architecture of Boylls' model as played over an array of interacting control surfaces is shown in Figure 5c, which is an anatomical template of circuitry ubiquitous in cerebellar transactions. That is, specific labels could be given to, say, the 'brainstem output nucleus' as red or Deiters nucleus, the 'reticular nucleus' could be reticularis tegmenti pontis or paramedian, etc. From this architecture we gather that the output from the cerebellar nuclei via the brainstem 'output' nucleus results from the interaction between cerebellar cortical inhibition as supplied by the Purkinje cells and between drives from the reticular nucleus. Tsukahara (1972) has demonstrated the possibility of intense reverberation between the reticular and cerebellar nuclei following removal of Purkinje inhibition, and Brodal and Szikla (1972) and others have demonstrated the anatomical substrate for such loops, with a somatotopic mapping in both directions. We thus postulate that there will be explosively excitatory driving of the cerebellar nucleus by reticulo-cerebellar reverberation unless blocked by Purkinje inhibition.

The output of cerebellar tuning is expressed as a spatio-temporal neuronal activity pattern in a cerebellar nucleus, which can then be played out via the brainstem nuclei to spinal levels. A careful analysis of the anatomy enabled Boylls to predict that the agonists of a motor schema would be 'represented' along a saggital strip of the cerebellar cortex, while its antagonists will lie orthogonal to that strip (in the medio-lateral plane). Applications of this formula to cortical topography of the anterior lobe, as developed by Voogdt (1969) and Oscarsson (1973), allowed Boylls to identify particular cortical regions as associated with equally particular types of hindlimb-forelimb, flexor-extensor synergic groupings. This led to conclusions which are experimentally testable.

The Boylls model suggests that activity within the cerebellar nucleus is initiated through topically precise climbing fiber activity; the mechanism involves their direct cerebellar nuclear activation coupled with the suppression of the target Purkinje cell activity in the cortex via the above-mentioned 'inactivation response'. Once activity is installed in cortico-nuclear interactions via climbing fiber intervention, the underlying reverberatory excitation helps to retain or 'store' it. At the same time, this activity is transmitted to the cerebellar cortex on mossy fibers, eventually altering the inhibitory pattern in the nuclear region surrounding the active locus. The spread of parallel fibers yields a form of lateral inhibition which provides spatial 'sculpting' in a way depending on the elaborate geometry of cerebellar cortex and cortico-nuclear projections. Mossy inputs of various types tune the resultant patterns to the demand of the periphery; and the program is spinally 'read out' as appropriate. Testing of the various hypotheses has required computer simulation of this neuronal apparatus. Simulation results corroborated the conjecture that cerebellar related circuitry could support the short-term storage of motor schema parameters initiated (and periodically refreshed) by climbing fiber activity.

## 4. THE FIRST THREE STAGES OF RANA COMPUTATRIX

4a. Facilitation of Prey-Catching Behavior. Frogs and toads take a surprisingly long time to respond to a worm. Presenting a worm to a frog for 0.3 sec may yield no response, whereas orientation is highly likely to result from a 0.6 sec presentation. Ingle [1975] observed a facilitation effect: if a worm were presented initially for 0.3 sec, then removed, and then restored for only 0.3 sec, the second presentation would suffice to elicit a response, so long as the intervening delay was at most a few seconds. Ingle observed tectal cells whose time course of firing accorded well with this facilitation effect (Fig. 6d). This leads us to a model [Lara, Arbib and Cromarty, in press] in which the "short-term memory" is in terms of reverberatory neural activity rather than in terms of the short-term plastic changes in synaptic efficacy demonstrated, for example, by Kandel [1978] in Aplysia. Our model is by no means the simplest model of facilitation -- rather, it provides a reverberatory mechanism for facilitation consistent with Ingle's neurophysiology and the known local neuroanatomy of the tectum. Unfortunately, the current knowledge of tectal circuitry is scanty, and much of the structure of the tectal column to be postulated below is hypothetical, and is in great need of confrontation with new and detailed anatomy and neurophysiology.

The model described in this section addresses facilitation at a single locus of tectum. Further developments address the interaction of a number of columns, and we shall discuss these in Sections 4b and 4c.

The anatomical study of frog optic tectum by Szekely and Lazar [1976] provides the basis for our model of the tectal column (Fig. 1a). In the superficial sublayers of tectum we see the thalamic input (which may also ramify in deeper layers), below which are the retinal type 1 and 2 inputs, with the retinal type 3 and 4 inputs deeper in turn. Deeper still, in layer 7, are the tectal efferents, which come from two cell types, the pyramidal cells and the so-called tectal ganglion cells. Our model of prey-catching will use only the pyramidal cells as efferents; we shall ignore the tectal ganglion cells which may (this is speculative) provide the output path for avoidance behavior. We incorporate the stellate cells as inhibitory interneurons, and ignore the amacrine interneurons. The other major components to be incorporated in our model are the large and small pear-shaped cells. Little of the anatomical connectivity of these cells is known,

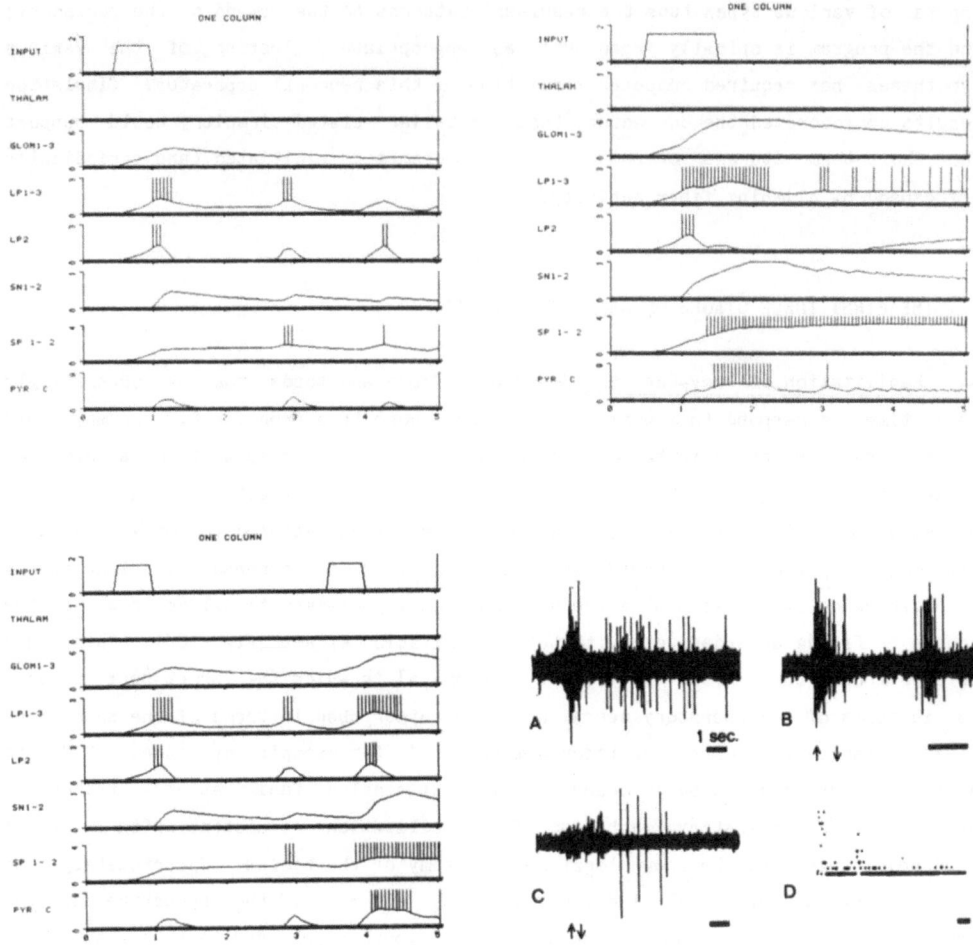

Figure 6. (a) Computer simulation of tectal cells response when a brief stimulus is presented. The onset of the stimulus produces a long lasting depolarizatin in the glomerulus which then fires the large pear-shaped cell (SP). This neuron in turn sends recurrent axons to the glomerulus and the stellate cell (SN) which acts as the inhibitory neuron in the column. When the inhibitory effect of SN releases the LP cell, a rebounding excitation occurs. The small pear-shaped cell is integrating the activity of GL, LP, and SN neurons to give a delayed short response. (b) If in the above situation we present a stimulus of longer duration then we show that now the pyramidal neuron fires. In (c) we show that when a second stimulus of the 'subthreshold duration' used in (a) is presented, the pyramidal cell (PY) responds. (The frequency of the spikes are a graphical convention. The spikes are drawn simply to highlight when the membrane potential of a cell is above threshold.) [From Lara, Arbib & Cromary (in press).]

(d) Physiological behavior of cells related to prey catching facilitation. A shows a brief class 2 burst followed by a delayed response of a tectal cell. In B the behavior of a tectal cell is shown, responding to the presentation of the stimulus and again with a delay. C shows a tectal neuron that produces a delayed response to the presentation of the stimulus. Finally, D shows the postimulus histogram of a tectal cell showing a delayed peak at 3 to 4 seconds. [From Ingle (1975).]

let alone the physiological parameters of their connections.

As we discussed in Section 2, the tectal column model (Fig. 1d) is abstracted somewhat crudely from the anatomy of Szekely and Lazar. It comprises one pyramidal cell (PY) as sole output cell, one large pear-shaped cell (LP), one small pear-shaped cell (SP), and one stellate interneuron (SN). (The simulation results of Figs. 6 and 7 were actually based on a larger column with 1 PY, 3 LP, 2 SP and 2 SN, but the results for the column of Fig. 1d are essentially the same.) All cells are modelled as excitatory, save for the stellates. The retinal input to the model is a lumped "foodness" measure, and activates the column through glomeruli with the dendrites of the LP cell. LP axons return to the glomerulus, providing a positive feedback loop. A branch of LP axons also goes to the SN cell. There is thus competition between "runaway positive feedback" and the stellate inhibition. (For a full presentation of the differential equations used in the simulation, see Appendix 1 of Lara, Arbib and Cromarty [in press].)

The role of SN in our tectum model is reminiscent of Purkinje inhibition of the positive feedback between cerebellar nuclei and reticular nuclei, a basic component of Boylls' model of cerebellar modulation of motor synergies outlined in Section 3. As mentioned above, Tsukahara [1972] found that reverberatory activity was indeed established in the subcerebellar loop when picrotoxin abolished the Purkinje inhibition from the cerebellar cortex. It would be interesting to conduct an analogous experiment by blocking inhibitory transmitters in the tectum.

Returning to the tectal model: glomerular activity also excites the SP cell which also sends its axon back to the glomerulus. The SP cell also excites the LP cell to recruit the activity of the column. The PY cell is excited by both SP cell and LP cell. Clearly, the overall dynamics will depend upon the actual choice of excitatory and inhibitory weights and of membrane time constants. It required considerable computer experimentation to find the weights that yielded the neural patterns discussed below. Further study was devoted to a sensitivity analysis of how weighting patterns affect overall behavior. It is our hope that our hypotheses on the ranges of the parameters involved in the model will stimulate more detailed anatomical and physiological studies of tectal activity.

Excitation of the input does not lead to runaway reverberation between the LP and its glomerulus; rather, this activity is "chopped" by stellate inhibition and we see a period of alternating LP and SN activity. The SP cell has a longer time constant, and is recruited only if this alternating activity continues long enough.

In one simulation experiment, we graphed the activity of the pyramidal cell as a function of the time for which a single stimulus is applied (Fig. 7a). There is, as in the experimental data, a critical presentation length below which there is no pyramidal response. Input activity activates the LP, which re-excites the glomerulus but also excites the SN, which reduces LP activity. But if input continues, it builds on a larger base of glomerular activity, and so over time there is a build-up of LP-SN alternating firing. If the input is removed too soon, the

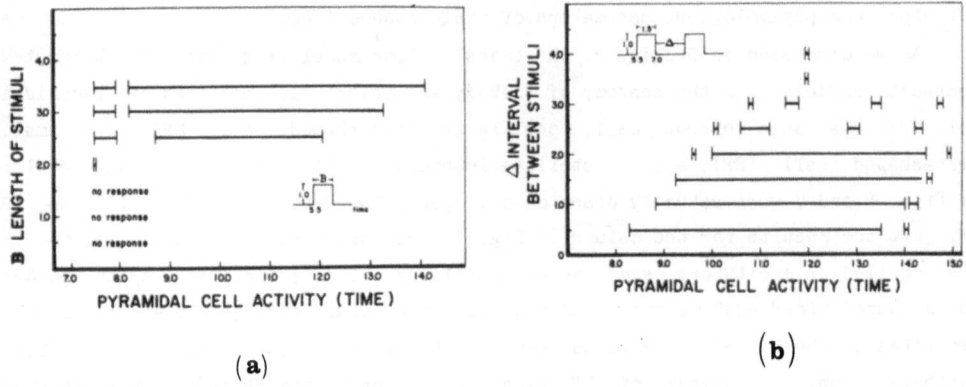

Figure 7. (a) Computer simulation of the PY behavior when stimuli are presented for different intervals. (b) Computer simulation of the temporal pattern of the facilitation process after the presentation of a brief stimulus.

reverberation will die out without activating the SP cells enough for their activity to combine with the LP activity and trigger the pyramidal output. However, if input is maintained long enough, the reverberation may continue, though not at a level sufficiently high to trigger output. However, a second simulation experiment (Fig. 7b) shows that re-introduction of input within a short time after cessation of this "subthreshold" length of input presentation can indeed "ride upon" the residual reverberatory activity to build up to pyramidal input after a presentation time too short to yield output activity on an initial presentation.

4b. A Simple Model of Pattern Recognition in the Toad. The facilitation model was 'local' in that it analyzed activity in a small patch of tectum rather than activity distributed across entire brain regions. We now outline Ewert's [1976, for a review] study of pattern recognition in the toad, analyzing what features of a single moving pattern will increase the animal's snapping responses. We then show how a one-dimensional array of tectal columns, of the type studied in the previous section, can model certain of these data. In Section 5, we briefly discuss our use of a two-dimensional array of such columns to model the whole range of Ewert's data on pattern recognition.

The toad is placed in a transparent cylinder. An object moves around a circular track concentric with, and on the floor outside, the cylinder. Some objects elicit no response. Other objects do elicit an orienting response (though the cylinder wall prevents the toad from actually snapping). Since the object keeps moving along its track, it can elicit a second response, and a third, and so on. Ewert's suggestion, then, is that the more 'attractive' is the object, the more frequently will the toad orient to it, so that the response rate is a measure of foodness. (Note a paradox here. The less attractive the object, the greater the

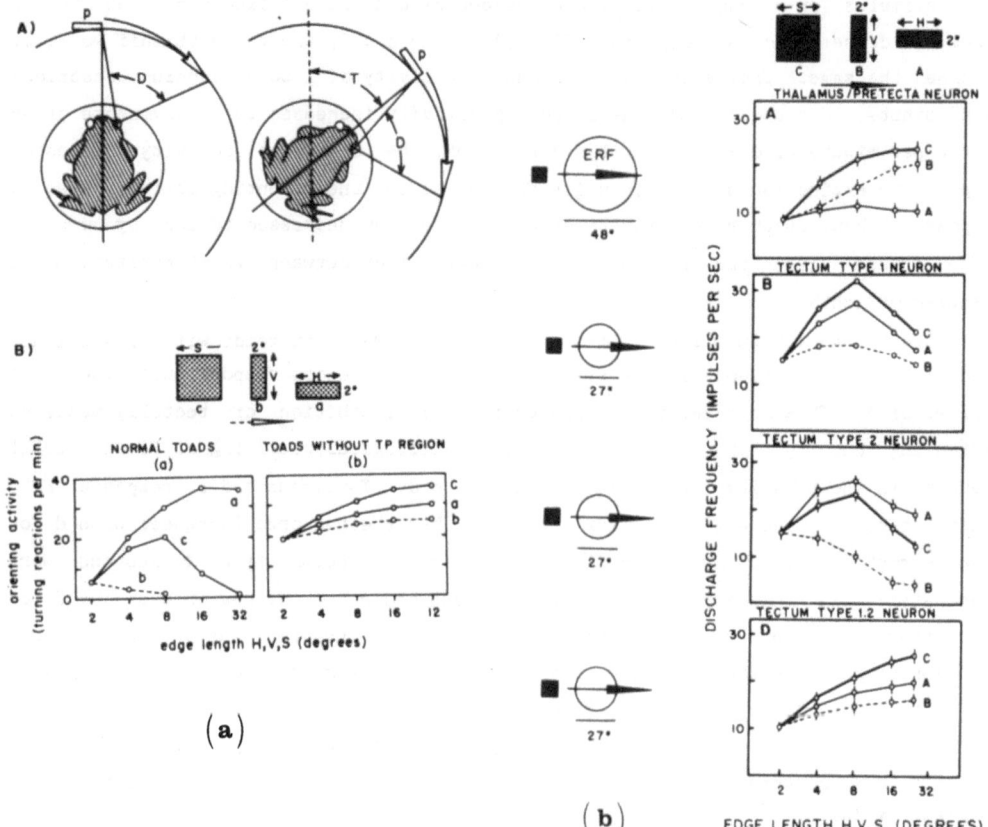

Figure 8. (a) Prey orienting behavior to different configuration of the stimulus.
A) Turning reaction to the stimulus presentation. B) Orienting activity to three
configurations (a,b,c): facilitation to stimulus a, inhibition to stimulus b, and
an initial facilitation and then an inhibition to stimulus c. When pretectum
ablation occurs this discrimination disappears.

(b) Tectal and pretectal cell activity to different configurations of the
stimulus (a,b,c). A) Response of a pretectal neuron which is mostly sensitive to
stimulus b and c. B and C show the response of two tectal cells to the three types
of stimuli. Neuron C response is mostly sensitive to stimuli type a and c, and its
response is greatly reduced for stimulus type b. This response is similar to the
observed behavioral response. D shows the response of both tectal cells (B and C)
without pretectum and how the discriminative abilities of these cells are lost.
[From Ewert (1976).]

integration time to a response, and thus the greater the distance the animal has to
move to orient towards the object if it orients at all.)

Ewert presented three types of rectangular · stimuli: a "worm" subtending 2
degrees in the direction normal to the motion, and some d degrees in the direction
of motion; an "antiworm" subtending some d degrees in the direction orthogonal to
motion, and 2 degrees in the direction of motion; and a "square" subtending d
degrees in both directions. The prey dummy was moved at 20 degrees per second at a
distance of about 7 cm from the toad. Ewert studied the toad's response rate for

each stimulus for a range of different choices of d degrees (fixed for each trial) from 2 degrees to 32 degrees (Fig. 8). For d = 2, the three stimuli were, of course, the same. They elicited an orienting activity of 2 to 3 turning reactions per minute. For the "worm", the orienting activity increased to an asymptote of 35 turns per minute at d = 16; for the "antiworm", the orienting activity decreased rapidly to extinction at d = 8; while for the square the orienting activity reached a peak of about 20 turns per minute at d = 8, and then decreased to zero by d = 32. (The square gives the impression of a competition between "worm" excitation and "antiworm" inhibition.)

Ewert repeated this series of behavioral experiments in toads with PT-lesions, and found that for none of the stimuli was there decreased response with increased values of d. This more detailed evidence for PT inhibition of tectally-mediated orienting was further elaborated by neurophysiological recording of PT and tectal neurons in the behaving toads. In the intact toad, PT-neurons had a response rate insensitive to increasing d for "worms", but the response increased with d for "antiworms", and even more rapidly for squares. Tectum type 1 neurons were insensitive to changing d for "antiworms", but had a peak of response at d = 8 for both "worms" and squares; while the firing rate of tectum type 2 neurons was similar to the orienting activity of the intact toad -- monotonically declining with d for "antiworms", peaking at d = 8 for squares, and declining slightly after d = 8 for "worms". (Note the slight discrepancy here -- one would expect the response to "worms" to be non-decreasing if, as Ewert does, one takes tectal type 2 activity as the neural correlate of orienting behavior.)

On this basis, Ewert postulated a simple model: A filter in PT responds best to an antiworm stimulus; a tectum type 1 cell responds as a filter tuned to a worm

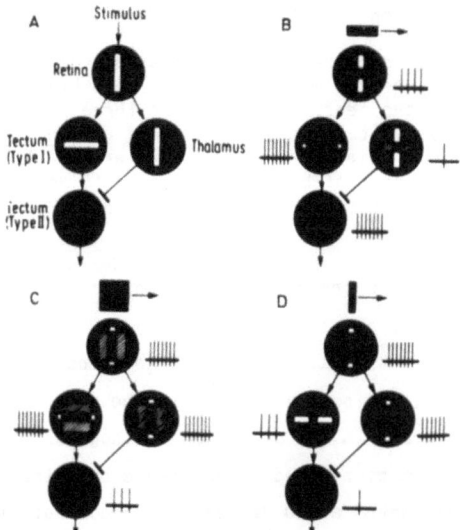

Figure 9. Schematic representation of the Ewert-von Seelen model of worm-'antiworm' discrimination. The tectum type II cell is excited by a tectal 'worm filter' and a thalamic-pretectal 'antiworm filter'.

stimulus; and a tectum type 2 cell is excited by the tectal type 1 cell and inhibited by a PT-cell (Fig. 9). Thus the type 2 cell responds with increased activity to increasing d for a worm stimulus; with decreased activity to increasing d for an antiworm stimulus; and with some tradeoff (dependent upon the actual parameters of the filters and the connectivity) for a square. Ewert and von Seelen [1974] fitted parameters to a linear formulation of this model to fit (part of) the response curves observed by Ewert. Note, however, that the domain of linearity is strictly limited; and that the model yields the average firing rate of the neuron: the model is thus lumped over time, and says nothing about the temporal pattern of neuronal interactions. Arbib and Lara [in press] have studied a one-dimensional array of tectal columns as in Figure 10a (without PT interaction) to provide a model of spatiotemporal neural interactions possibly underlying Ewert's "worm" phenomena. For example, in the Ewert study of the toad's response to an object moving along a track, we may regard the object's movement at one position as facilitating the animal's orientation to the object in a later position. The key question here is "How does the facilitation build up in the right place?" Part of the answer lies in noting the large receptive fields of the tectal columns; and analyzing how activity in a population of tectal columns can yield orientation in a particular direction. Thus, rather than analyzing activity in a single column, Arbib and Lara [in press] study the evolution of a waveform of activity in a one-dimensional array of columns (Fig. 10a). We show in Figs. 10b, c, d the response to a moving stimulus of various lengths. These reproduce Ewert's observations on the increasing attraction of a 'worm' with increasing length; Arbib and Lara also report a number of other computational experiments. The elaboration of this model to a two-dimensional array of columns (Lara, Cervantes and Arbib, 1982) is integrated with our model (Section 3c) of tectal-pretectal interactions in prey-selection to yield a model rich enough to extend an explanation of Ewert's data on pattern recognition into the temporal domain in a way which addresses the antiworm and square data, as well as the worm data.

4c. A Model of Prey-Selection. We saw in Section 3b that Ingle [1968] had studied the response of frogs to pairs of fly-like stimuli, each of which was such that when presented alone it would elicit a snapping response, and found that, under differing conditions, the animal would snap at one of the stimuli, snap between them, or not snap at all. We now turn to a model of such prey-selection which refines the Didday model discussed above (Fig. 3a) but differs in that -- in view of Ewert's study of PT-lesions -- it uses PT-tectal interactions, rather than positing that all the necessary circuitry is embedded in the tectum. Moreover, the new model extends the 'array of tectal columns' model of Section 4b to provide yet a third stage in the evolution of Rana Computatrix.

To make that transition from the Didday model, we now identify the "foodness layer" of Figure 3 with the retinal outflow to tectum and pretectum, and identify

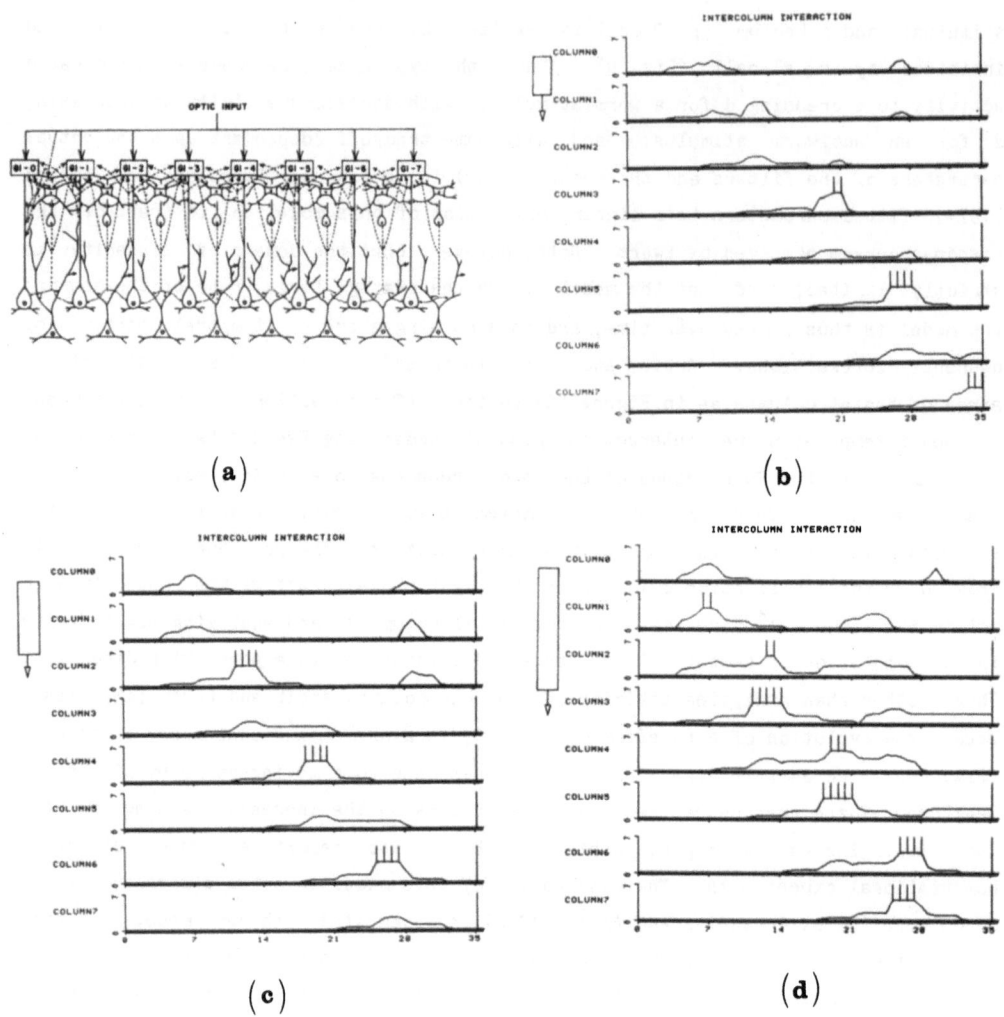

(a)

(b)

(c)

(d)

Figure 10. (a) Architecture of the model of the tectum. Each column is constituted by one GL (glomerulus), one LP (large pear-shaped) cell, one SP (small pear-shaped) neuron, one SN (stellate neuron), and one PY (pyramidal cell). The afferents are the optic fibres that arrive at the GL, LP, SP, and PY cells, and the efferents are the PY axons. LP cells are activated by the GL and the optic input, and they send recurrent axons to their own as well as neighboring glomeruli. The SN are activated by the LP cells, and they inhibit LP and SP neurons of their own as well as neighboring columns. The SP receive excitation from GL and are inhibited by SN; finally, PY receives afferents from the retina, the LP and SP neurons.

(b), (c) and (d) present a computer simulation of tectal response to a moving stimulus of different sizes. The graphs show the behavior of the 8 PY neurons of the tectal model of (a) to a moving stimulus. (b) Notice that in this case an alternate response is given in columns 3, 5, and 7 when the stimulus size only covers one glomerulus. (c) Here the stimulus covers 2 glomeruli simultaneously. The results show that the strength of activation increases when the size of the object is elongated. The latency of response is also shorter (column 2). (d) In this figure the stimulus covers 3 GL simultaneously. It can be seen that the latency of response is shorter and the total activity is greater than in (b) and (c). Notice that all columns fire with this stimulus. [From Arbib & Lara (in press).]

the "relative foodness layer" with the pyramidal cells of tectum. We now see that Figure 3a is too simple because it does not include other cells of the tectal column. The new model [Lara and Arbib, in press] interconnects a one-dimensional array of tectal columns with a layer of cells called S-cells, in retinotopic correspondence with the columns, which represent cells of the pretectum-thalamus (Fig. 11a). (In the 1970 model, the S-cells were identified with the sameness cells reported in the tectum by Maturana, Lettvin et al.) Each S-cell is excited by activity in the relative foodness layer, save for a blind spot centered at the locus corresponding to that of the S-cell. In the Didday model, the S-cell then provides an inhibitory input to cells within its blind spot on the relative foodness layer. Lara and Arbib [to appear], however, do not make the corresponding assumption that an S-cell must inhibit the PY cell in the corresponding tectal column. Rather they conduct a number of experiments on the dynamic consequences of choosing different sites for pretectal inhibition of columnar activity. The reader is referred to their paper for details.

The system described so far exhibits hysteresis. Should a new peak be introduced in the input array, it may not affect the output activity even if it is rather large, for it may not be able to overcome the considerable inhibition that has built up on the S-cells. The model thus follows Didday in postulating a further array of NE-cells (representing the newness cells of Lettvin et al.) which register sudden changes in input, and uses these to interrupt the ongoing computation to enable new input to affect the outcome.

Clearly, the detailed dynamics of the model will depend on the size of the blind spot, and the relative parameters of excitation and inhibition. We were able to adjust the coefficients in such a way that with several peaks in the foodness input array, the activity passed through to the tectal column would excite the S-cells in such a way that they would lower the corresponding peaks in tectal activity. However, if one peak were stronger than the others, it would be less inhibited, and would begin to recover; in doing so, it would suppress the other peak more, and thus be inhibited less; the process continuing until the stronger peak recovered sufficiently to control a "snap" in the corresponding direction (Fig. 11c). However, there were cases in which the mutual suppression between two peaks sufficed to hold each below a level sufficient to release behavior (Fig. 11b). We also showed that if the tectum became habituated to one of the stimuli, a standoff would be resolved in favor of the novel stimulus (Fig. 11d).


## 5. A PERSPECTIVE FOR FURTHER MODELLING

In Section 4, we exhibited an evolutionary sequence of models -- tectal column, one-dimensional array of columns; array with pretectal inhibition -- which explains an increasingly broad range of behavioral data on visuomotor coordination in frog

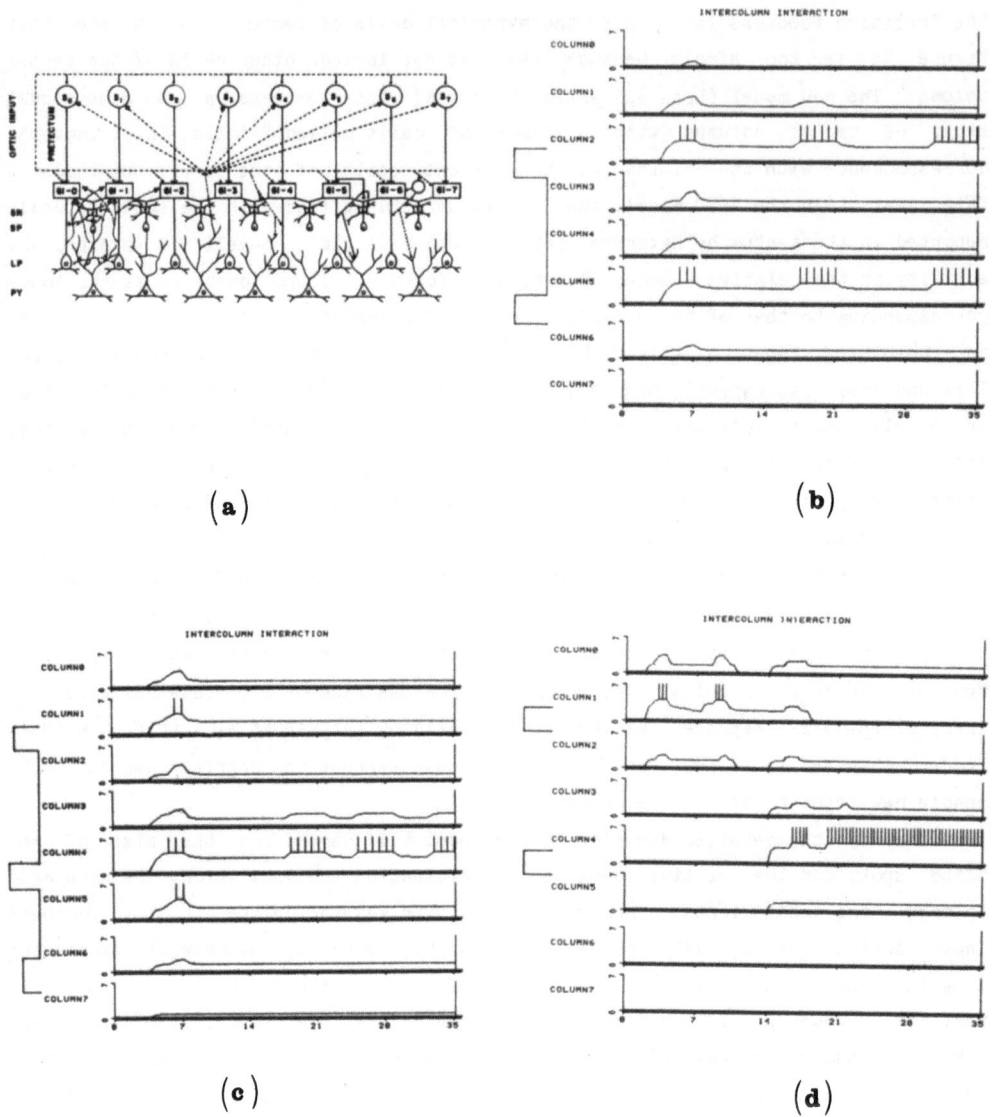

(a)  (b)

(c)  (d)

Figure 11.  (a) Architecture of the model for the interactions between  tectum  and  pretectum  in  prey selection.  Each column receives the afferents from one sameness neuron;  each PY (pyramidal) neuron excites all pretectal cells except the one whose blind  spot  is in its receptive field.  The NE (newness) neurons arrive at the same site as the corresponding optic fibres.  (b) Computer simulation of the behavior  of  PY  neurons  to two equally intense stimuli.  The stimuli are presented in columns 2 and 5.  Notice that an alternation of excitation and inhibition is  present  without  convergence  to  any  of the stimuli.  (c) Computer simulation of the behavior of PY neurons to two equally intense stimuli to columns 2 and 5 biased by  a  third  one. When  the  third stimulus is applied in column 7, then the response converges to the stimulus presented in column 5.  (d) Computer simulation of habituation  effects  on  PY  activity.  We first present a stimulus in column 1.  After a period of rest, we present two equally intense stimuli in columns 1 and 4, the response  converging  to  column 4, because the pathway of column 1 is habituated.

and toad. We note three important features of the style of modelling developed here.

1. New phenomena are addressed not by the creation of ad hoc models but by the orderly refinement and expansion of models already created. Of course, we expect that future development along this line will lead to redefinition and refinement of earlier models, rather than simple addition of new circuitry in each case. On the other hand, we would expect that the model, once sufficiently developed, will explain many data beyond those which specifically entered into its design.

2. Each 'model' in the sequence is in fact a 'model-family'. We design a family of overall models, and then conduct simulation experiments to see which choices -- of connectivity, synaptic weights, time constants -- yield neural dynamics, and input-output relations, compatible with available data.

3. The choices mentioned above are only loosely constrained by the experimental data presently available. To carry out simulations, we make choices which often must, perforce, go beyond these data. In making such choices, we form explicit hypotheses (whose details are spelt out in our papers cited in Section 4) which may serve to stimulate new experiments. These experiments in turn will stimulate more refined modelling. The continuing cycle will lead to an increasingly sophisticated understanding of the neural mechanisms of visuomotor coordination.

We close with a brief discussion of future directions for this modelling effort. We have already mentioned the transition from a one-dimensional to a two-dimensional array of tectal columns (and corresponding pretectal elements) as a further development of Rana Computatrix in Lara, Cervantes and Arbib (1982).

The models of Section 4 have nothing to say about the control of avoidance behavior, nor does the basic version described here address more than a few of the prey-predator discrimination phenomena discussed in Section 4b. A two-dimensional array of columns will allow us to study the full range of these phenomena.

Other developments in the modelling of frog brain visuomotor coordination will come as we try to take more and more regions of the brain into account -- for example, the cerebellum, the retina, and the forebrain. We will also want to look at more complex behaviors of the animal, not only prey-selection and predator avoidance, but also behaviors which require the integration of a number of motor schemas. For example, we are currently considering the way in which an animal will approach a worm when a vertical paling barrier is interposed. In this case, the animal's behavior can be analyzed in terms of the coordinated activation of three motor schemas: one for side-stepping, one for orienting, and one for snapping. The understanding of this behavior, then, reinforces our need to model the animal's behavior in terms of the cooperative computation of a number of brain regions. We have now adapted the Dev conceptual model of stereopsis, described above in Section 3c, to a model of depth perception in the frog, in which we take account not only of the disparity cues available in the binocular field of the animal, but also of accommodation cues available in the monocular field (the animal can still strike with an accurate depth estimation if it has only one eye, and the worm is presented in the monocular field corresponding to that eye).

Clearly, then, developments in modelling will require both the generation of general concepts for vision and motor control, as well as specific studies which try to provide a variety of detailed models adapted to experimentation on different kinds of animals and different kinds of situations. We shall also need to get a better understanding of how regions of the brain are coordinated in complex behaviors. Finally, it will not be enough to understand how the adult brain behaves in any given situation; we must also understand the development of the brain (for example, by modelling the development of retinal-tectal connections), and by studying learning mechanisms.

There are further refinements not incorporated into the basic model. Increased motivation (due, e.g., to food odor or to hunger) will cause the animal to snap at larger moving objects than it would otherwise approach. Such an effect might be modelled by direct excitation of tectal columns, or by diffuse inhibition of the S-cells, probably under the control of telencephalic regions. Forebrain mechanisms allow the animal to learn simple discriminations. And there are habituation phenomena which we have begun to model (Fig. 11d). Habituation disappears when there is PT ablation. Moreover, the habituation is stimulus specific, and it appears that pattern recognition is necessary both for habituation and dishabituation to occur. For example, Ewert has studied habituation of a toad's snapping response to simple moving patterns and has discovered a hierarchy -- an ordering $A \leq B$ of patterns, such that if the toad habituates to A it will automatically be habituated to B, but not vice versa. Such data provide a continuing challenge to the theory-experiment interaction that will drive the future evolution of Rana Computatrix.

## REFERENCES

Amari, S., 1982, Competitive and Cooperative Aspects in Dynamics of Neural Excitation and Self-Organization. In: Competition and Cooperation in Neural Nets (S. Amari and M.A. Arbib, Eds.), Lecture Notes in Biomathematics, Springer-Verlag (this volume).

Amari, S., and Arbib, M.A., 1977, Competition and cooperation in neural nets. In: Systems Neuroscience, (J. Metzler, Ed.), New York: Academic, p. 119-165.

Apter, J.T., 1946, Eye movements following strychninization of the superior colliculus of cats. J. Neurophysiol. 9: 73-85.

Apter, J.T., 1945, Projection of the retina on the superior colliculus of cats. J. Neurophysiol. 8: 123-134.

Arbib, M.A., 1981, Perceptual structures and distributed motor control. In: Handbook of Physiology: The Nervous System II. (V.B. Brooks, Ed.), Bethesda, Md.: Amer. Physiological Society, 1449-1480.

Arbib, M.A., 1982, Rana Computatrix: An Evolving Model of Visuomotor Coordination in Frog and Toad. In: Machine Intelligence 10 (J. Hayes and D. Michie, Eds.), Ellis Horwood.

Arbib, M.A., Boylls, C.C., and Dev., P., 1974, Neural models of spatial perception and the control of movement. In: Cybernetics and Bionics, (W.D. Keidel, W. Handler, and M. Spreng, Eds.), Munich: Oldenbourg, 216-231.

Arbib, M.A., and Lara, R. (in press), A neural model of the interaction of tectal columns in prey-catching behavior. Cognition and Brain Theory, 5.

Barlow, H., 1953, Summation and inhibition in the frog's retina. J. Physiol. (Lond.) 119: 69-88.

Barlow, H.B., Blakemore, C., and Pettigrew, J.D., 1967, The neural mechanism of binocular depth discrimination. J. Physiol. 193: 327-342.

Boylls, C.C., 1974, A Theory of Cerebellar Function with Applications to Locomotion. Ph.D. Thesis, Stanford University.

Dev, P., 1975, Computer simulation of a dynamic visual perception model. Int. J. Man-Mach. Stud., 7: 511-528.

Didday, R. L., 1970, The Simulation and Modelling of Distributed Information Processing in the Frog Visual System. Ph.D. Thesis, Stanford University.

Didday, R. L., 1976, A model of visuomotor mechanisms in the frog optic tectum. Math. Biosci. 30: 169-180.

Ewert J. P., 1976, The visual system of the toad: behavioral and physiological studies in a pattern recognition system. In: The Amphibian Visual System: A Multidisciplinary Approach. (K. Fite, Ed.), Academic Press, pp. 142-202.

Ewert, J. P. and von Seelen, W., 1974, Neurobiologie and System-Theorie eines visuellen Muster-Erkennungsmechanismus bei Kroten. Kybernetic 14: 167-183.

Ingle, D., 1968, Visual releasers of prey-catching behavior in frogs and toads. Brain Behav. Evol. 1: 500-518.

Ingle, D., 1973, Disinhibition of tectal neurons by pretectal lesions in the frog. Science 180: 422-424.

Ingle, D., 1975, Focal attention in the frog: behavioral and physiological correlates. Science 188: 1033-1035.

Ingle, D., 1976, Spatial visions in anurans. In: The Amphibian Visual System. (K. Fite, Ed.), Academic Press, New York, pp. 119-140.

Julesz, B., 1971, Foundations of Cyclopean Perception. Chicago: Univ. of Chicago Press.

Kandel, E.R., 1978, A Cell Biological Approach to Learning. Grass Lecture No. 1. Society for Neuroscience: Bethesda, MD.

Lara, R., and Arbib, M.A. (to appear), A neural model of interaction between tectum and pretectum in preytectum in prey selection.

Lara, R., Arbib, M.A., and Cromarty, A.S. (in press), The role of the tectal column in facilitation of amphibian prey-catching behavior: a neural model. J. Neuroscience.

Lara, R., Cervantes, F., and Arbib, M.A., 1982, Two-dimensional Model of Retinal-Tectal-Pretectal Interactions for the Control of Prey-Predator Recognition and Size Preference in Amphibia. In: Competition and Cooperation in Neural Nets (S. Amari and M.A. Arbib, Eds.), Lecture Notes in Biomathematics, Springer-Verlag (this volume).

Lettvin, J. Y., Maturana, H., McCulloch, W. S. and Pitts, W. H., 1959, What the frog's eye tells the frog brain. Proc. IRE. 47: 1940-1951.

Marr, D., and Poggio, T., 1977, Cooperative computation of stereo disparity. Science. 194: 283-287.

Nelson, J. I., 1975, Globality and stereoscopic fusion in binocular vision. J. Theor. Biol. 49: 1-88.

Pettigrew, J.D., Nikara, T., and Bishop, P.O., Binocular interaction on single units in cat striate cortex. Exp. Brain Res. 6: 391-410.

Pitts, W.H., and McCulloch, W.S., 1947, How we know universals, the perception of auditory and visual forms. Bull. Math. Biophys. 9: 127-147.

Rosenfeld, A., Hummel, R.A. and Zucker, S.W., 1976, Scene labelling by relaxation operations. IEEE Trans. Syst. Man Cybern. 6: 420-433.

Sperling, G., 1970, Binocular vision: a physical and a neural theory. Am. J. Psych. 83: 461-534.

Szentagothai, J. and Arbib, M. A., 1974, Conceptual Models of Neural Organization. NRP Bulletin vol. 12, no. 3: 310-479. (Also: The MIT Press, 1975.)

Tsukahara, N., 1972, The properties of the cerebello-pontine reverberating circuit, Brain Res. 40: 67-71.

Waltz, D.L., 1978, A parallel model for low-level vision. In: Computer Vision Systems (A.R. Hanson and E.M. Riseman, Eds.), New York: Academic, p.175-186.

# CHAPTER 22

Two-Dimensional Model of Retinal-Tectal-Pretectal Interactions for the Control
of Prey-Predator Recognition and Size Preference in Amphibia

Rolando Lara, Francisco Cervantes and Michael Arbib*
Centro de Investigaciones en Fisiología Celular
Universidad Nacional Autónoma de México
Apartado Postal 70-600
04510 México, D.F.
*Center for Systems Neuroscience, and Department of
Computer and Information Science
University of Massachusetts
Amherst, MA    01003/USA
Work supported in part by NIH under grant NS14971-03.

I.  Introduction

In the present paper, we propose a model of the interactions among retina, tectum and pretectum in the amphibian brain which simulates prey-predator recognition,
direction invariance of prey-predator recognition as a consequence of tectal architecture, size preference and latency of response of the animal depending on its motivational state.  The model is an extension of the one dimensional model of the tectum,
described elsewhere, [1-3] which takes into consideration the anatomical, physiological
and behavioral studies of the tectum.  With this model we have been able to study the
different experimental hypotheses described below with the aim that a single model
could explain the observed results, reproduce the experimental observations, and predict new experiments for future research.

Amphibia (and we emphasize frog and toad) have been considered a good biological
model for the study of visuomotor coordination because much of their behavior is guided
by visual stimuli, they are almost static animals, their nervous system is not as complex as those of higher vertebrates, although some of the complexity of their behavior
and processing of information is present in these animals, and they have been extensively studied from anatomical, physiological, behavioral and theoretical points of
view.

Ethological studies [4,5,6] have shown that these animals have innate mechanisms to
recognize different stimuli in their environment to elicit the proper response.  It has
been shown that the geometry of the visual stimulus in relation to the direction of motion plays a prominent role in the orienting response of the animal:  objects whose
longest axis moves in the direction of motion are considered as prey, while objects
whose longest axis moves perpendicular to the direction of motion are considered as
predators, because they do not elicit prey orienting behavior or they give an avoidance response [4] (see Fig. 1).  It has also been shown that this prey/non-prey recognition is invariant to both the direction and speed of the stimulus. [7]

Ingle [8,9] has shown that the size, color preference and latency of response to
prey stimuli can change depending on the motivational state of the animal.  He showed

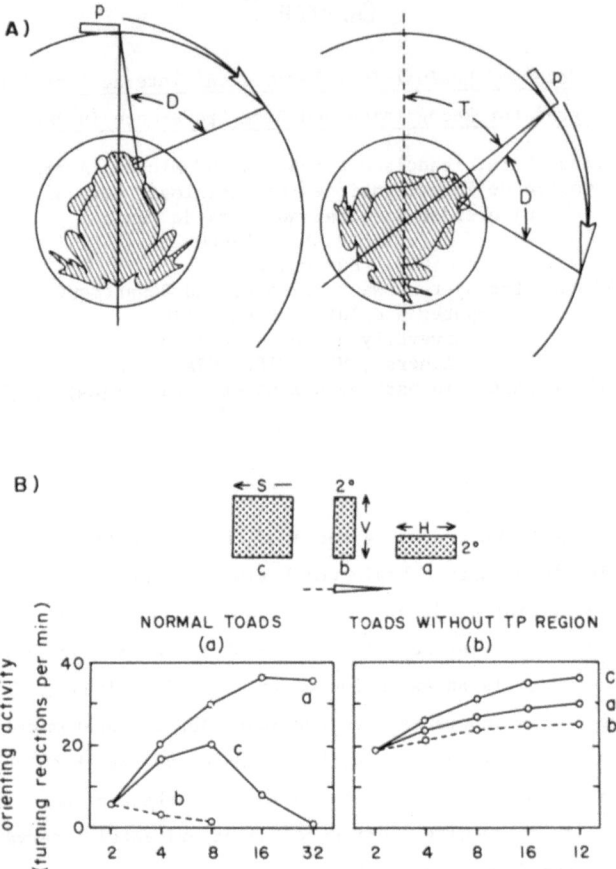

Fig. 1. Prey orienting behavior to different configurations of the stimulus. A) Turning reaction to the stimulus presentation. B) Orienting activity to three configuration (a,b,c): facilitation to stimulus a, inhibition to stimulus b, and an initial facilitation and then an inhibition to stimulus c. When pretectum ablation occurs this discrimination disappears. (From Ewert, 1976)

that animals which were good feeders had low response thresholds and preferred $16^{\circ}$ long worms to worms of $6^{\circ}$ length, while slow feeders had the opposite preference; thus response readiness is correlated with size preference. Ingle suggests that these mechanisms of prey preference may be associated with the known size constancy capacity that amphibia have for a limited distance.

A great deal of research has been aimed at trying to find the neuronal mechanisms responsible for these processes. Ewert has shown[4] that prey-orienting behavior is disrupted when the tectum is destroyed. Moreover, the tectum receives information from the retina in a retinotopic way and electrical stimulation of a specific tectal region

elicits the orienting response to the corresponding retinal projection. This suggests
that the tectum plays a prominent role in prey orienting behavior. Ewert has also shown
that lesions of the pretectum, another brain region which receives retinotopic information from the retina and establishes closed loop interactions with the tectum,[10],[11]
disrupts the recognition abilities of the animal to the different configurations of the
stimuli.[4] (see Fig. 1b) Furthermore, he observed that toads with pretectum ablation
snap indiscriminately to any object, they switched their preference from black to white
stimuli, and they lost size selectivity. This suggests that the interaction among retina, tectum and pretectum may be responsible for the processes of prey-predator recognition, size preference and size constancy.

Trying to establish the role that each one of these brain regions may play in the
control of these behaviors, Ewert studied the neuronal responses in the retina, tectum
and pretectum to the different configurations of the stimulus.[4],[12-14] He showed that
in toads and frogs ganglion retinal cells of types II, III and IV do not change considerably their rate of response when a worm-like stimulus of different sizes was presented; whereas when an anti-worm-like stimulus whose longest axis moves perpendicularly
to the direction of motion was presented, ganglion cells of types II and III initially
increased their rate of response up to the size of their respective excitatory receptive field and then the rate of response decreased when the object was larger than the
excitatory receptive field (ERF). The inhibitory effect is stronger in ganglion type
II cells than ganglion type III cells. Class IV ganglion cells increase their rate of
response depending on the size of the object. (see Fig. 2) Ganglion cells of types
II, III and IV increase their rate of response depending on the speed of the object,
but they respond independently of the direction of motion.[12],[13]

From the above results, Ewert concluded that the observed behavioral responses
could not be explained simply by the retinal responses; thus he continued studying the
response of tectal and pretectal cells to the different configurations of the stimuli.
He found that some tectal cells responded in a strikingly similar way to the behavior
of the animal when the different stimuli were presented: facilitation of the rate of
firing when the stimulus was elongated along the direction of motion, inhibition when
the stimulus was elongated perpendicularly to the direction of motion, and an initial
facilitation and then an inhibition when the stimulus was expanded in both directions
(Fig. 3). Moreover when pretectal ablation occurs this tectal cell responds similarly
to the behavioral response when the different stimuli are presented. Ewert also showed
that some of the tectal cells could discriminate between prey and non-prey stimuli independently of the direction of motion while other tectal neurons were directionally
selective or responded more strongly to a predator-like stimulus.[7] For the above reasons, Ewert suggests that this neuron, tectal type T5(2), may be responsible both for
the discrimination between prey and non-prey stimuli and for indicating the position
to which the animal should orient. The tectal neuron performs this through combined
activity with pretectal cells, possibly through an inhibitory effect. Studies of the

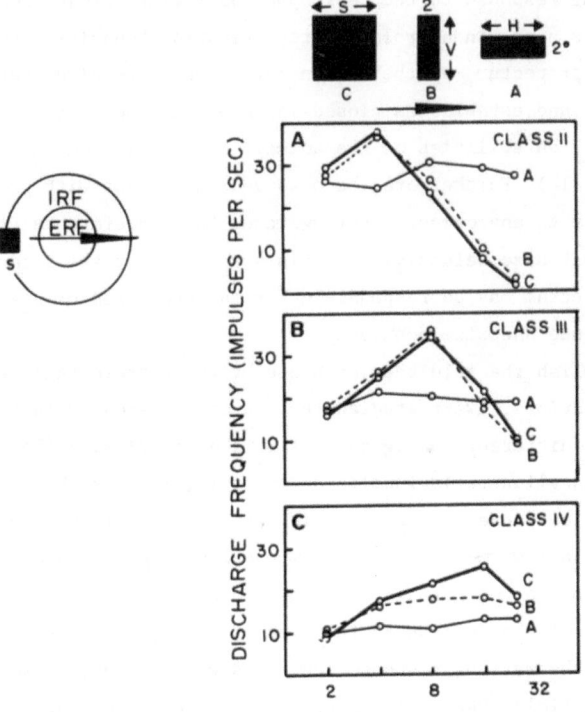

Fig. 2. Retinal ganglion cell response (II, III, IV) to different configurations of the stimulus (a,b,c). For stimulus type a the response of the three ganglion neurons is almost invariant for the different sizes of the stimulus. For stimulus type b and c, ganglion type II and III increase their rate of response up to their respective receptive field and then the response is reduced. For type IV ganglion cells the rate of response is proportional to the size of the stimulus. (From Ewert, 1976)

response of the pretectal cells[14] showed that most of these neurons had large receptive fields and that they were more sensitive to a predator-like stimulus (see Fig. 3). One of these pretectal cells TH5 with a relatively small receptive field responded mostly to non-prey like stimuli; for this reason, Ewert postulates that this cell inhibits the activity of tectal cells when a predator-like stimulus is present, thus allowing the animal to orient to the proper prey stimulus. In this way, Ewert suggests that the combined activity of retina-tectum and pretectum may control prey-predator recognition. With respect to the direction invariance recognition, Ewert simply says that it must be a consequence of the tectal architecture rather than of a sophisticated "software"-like processing of information.

In relation to size preference, Ingle, following Ewert's hypothesis of pretectal inhibition over tectal activity, suggests that the changes in size, color and latency of response could also be modulated by the pretectum.[6,8,9] He postulates that in normal

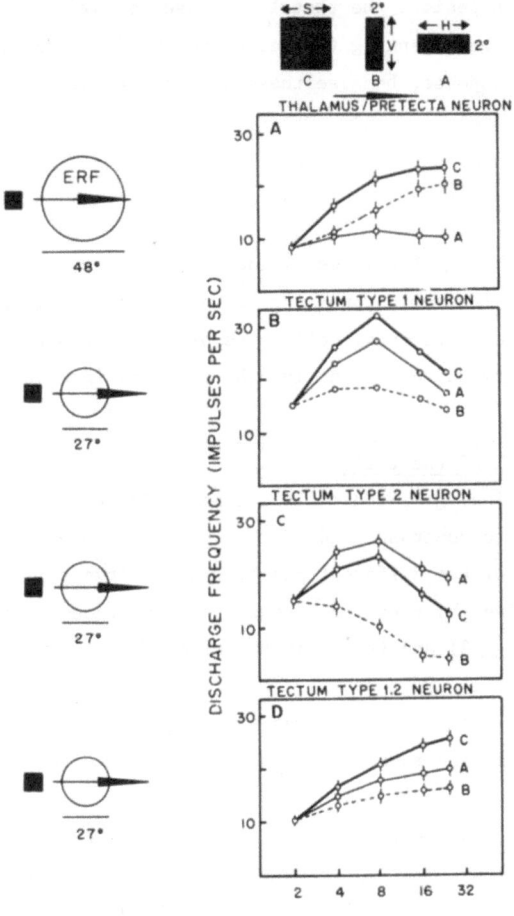

Fig. 3. Tectal and pretectal cell activity to different configurations of the stimulus (a,b,c). A) Response of a pretectal neuron which is mostly sensitive to stimulus
b and c. B and C show the response of two tectal cells to the three types of stimuli.
Neuron C response is mostly sensitive to stimuli type a and c and its response is greatly reduced for stimulus type b. This response is similar to the observed behavioral
response. D shows the response of both tectal cells (B and C) without pretectum and
how the discriminative abilities of these cells are lost. (From Ewert, 1976)

conditions tectal cells are mostly guided by type II ganglion neurons, and the afferents

from ganglion type III and IV cells are normally inhibited by the pretectum. Ingle

suggests that retinal type II cells can overcome the pretectal inhibition through a

facilitatory effect as a consequence of recurrent excitatory activity but the response

has a long latency. In this way he explains why animals normally prefer small size

stimuli. Whenever the pretectal inhibition is decreased, either by an increased motivational state or by a lesion, the tectal response is now controlled by tectal ganglion type III cells, thus changing the receptive field, color size preference, and reducing the latency of response, because these neurons arrive closer to the soma of tectal neurons.

## II.  The Model

The description of the model will be divided in four parts; a brief description of the black box model of the retina, waiting for a more realistic retinal model, the description of the different pretectal cells, the proposed architecture for the two-dimensional model of the tectum and, finally, the proposed interaction among retina, tectum and pretectum.

### II.1  Black box model of the retina

As we have seen, retinal ganglion cells are sensitive to geometry in relation to the direction of motion, to contrast, and to the speed of the stimulus.[4],[15]

Our black box model of the different ganglion cells (types II, III and IV) is based on the curves obtained by Ewert for the response of these cells to prey and non-prey like stimuli (see Fig. 2) and the speed function obtained by Grüsser and Grüsser-Cornehls[15] and Ewert.[4]

The model simply defines the rate of response of type II, III and IV ganglion cells depending on the size and speed of motion:  the first with Ewert's graphs and the second with the following equation.

$$R = kv^\delta;$$ where k and $\delta$ are constants and v is the speed of the

object.  R is the frequency response of the retinal cell.

Each type of ganglion cell projects point to point to each tectal and pretectal column.  In the present model we have not considered the spatial representation of the different retinal receptive fields; we have only considered that the center of each type of retinal cell projects to the corresponding point either in the tectum or pre-tectum.  Each retinal ganglion axon projects to a specific column, and excites the surrounding neighbors with less intensity.

Each time a stimulus arrives at the receptive field of a group of ganglion cells, they will generate a response frequency R depending on the size, speed, and direction of motion of the object.  The parameters of the stimulus are specified by the modeller. We simulated the presence of a stimulus simply by a variable which defines when the stimulus should be present in a given zone and for how long it should rest there, de-pending on the speed and size of the object.

### II.2  Pretectal cells

Because of the limited data about the anatomy of the pretectal region, in this model we simply considered single units to simulate the postulated behavior of pretec-tal cells.  We proposed two types of pretectal neurons:  The first which represents the pretectal TH-3 cell of Ewert which is mostly sensitive to predator-like stimuli and which Ewert postulates to play a very important role, through its inhibitory action, for prey-predator recognition of tectal neurons. This neuron receives retinal afferents from ganglion cells of types III and IV. The second pretectal neuron is related to prey selection and has already been described elsewhere.[3] Briefly, this neuron is called a sameness cell, and receives the excitatory effects of all tectal pyramidal cells except the one or the ones of its blind spot. These cells in their turn inhibit retinotopically the corresponding tectal column. (see Fig. 6)

II.3 Two-dimensional model of the tectum

The two-dimensional model of the tectum is shown in Fig. 4 and is an expansion of the one-dimensional model of the tectum described elsewhere.[2,3] The model is now constituted of 64 columns. For the two-dimensional model the number of cells and their

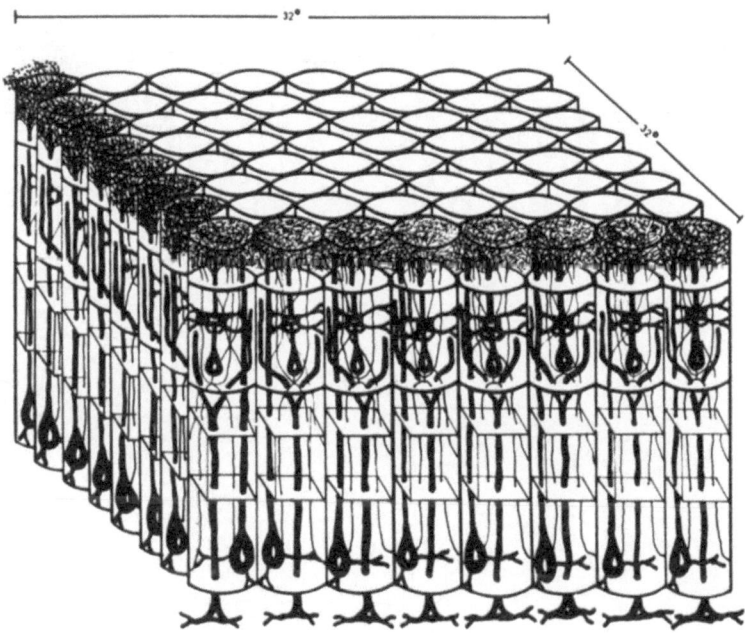

Fig. 4.   Representation of the two-dimensional model of the tectum constituted of 64 columns.

interactions is greatly increased. We will give a brief description of the most important considerations. We have elsewhere[2,3] modelled the tectum as constituted of tectal columns. Briefly, each column is constituted of one large (LP) and one small (SP) pear shaped cell, one stellate neuron (SN), one pyramidal (PY) cell (the efferent cell of the column) and one glomerulus (GL) where the optic fibres arrive. The connections are indicated in Figure 5:

a) The glomerulus is constituted of the dendrites of LP and SP cells of its own column, the afferent optic fibres from ganglion type II cells, and the recurrent

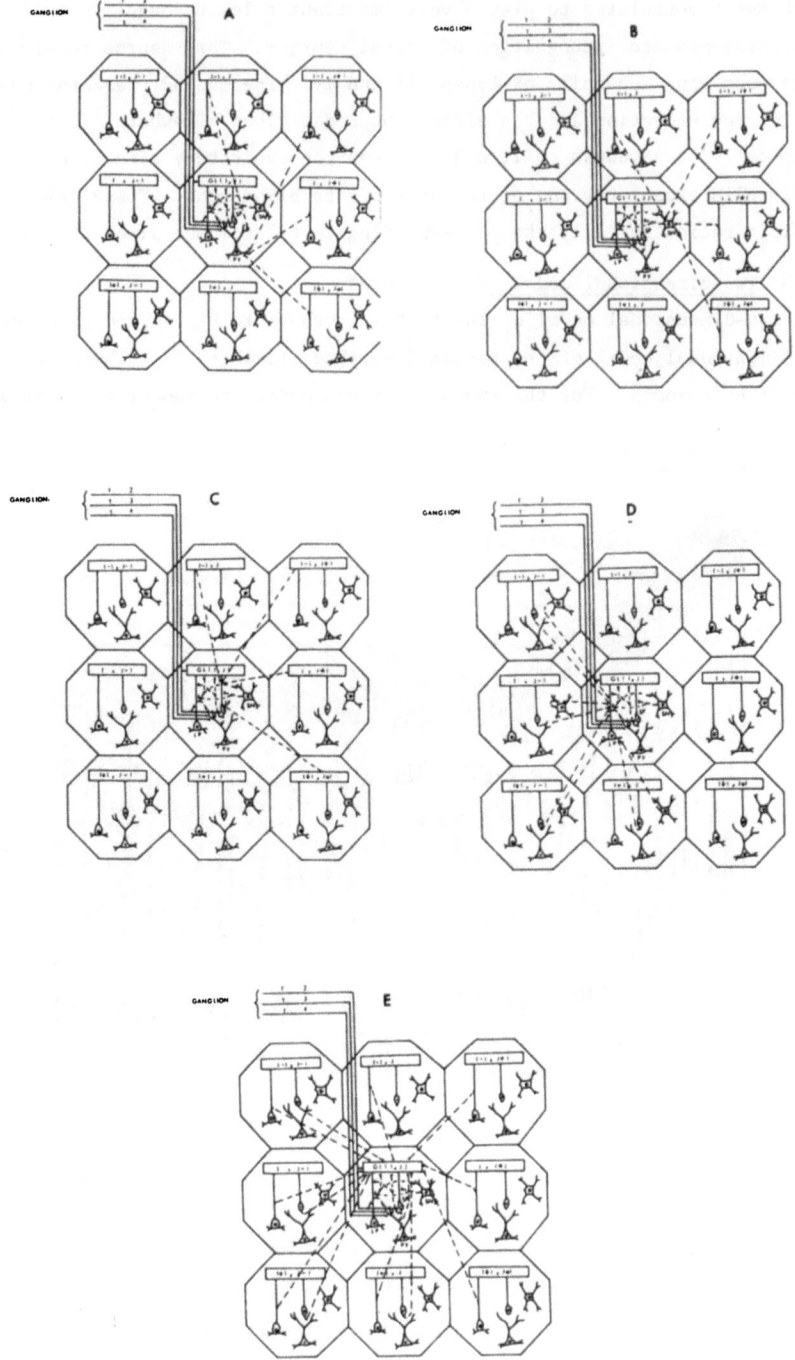

Fig. 5. Interactions among the tectal columns in the two-dimensional model of the tectum. (caption continued, next page)

A) Cellular afferents to PY cells. This cell is activated by LP, SP of its own column and by LP's of neighboring columns and it also receives afferents from retinal ganglion cells type II, III and IV. B) Cellular afferents to the SN neuron. This cell is excited both by LP's of its own and neighboring columns. C) Cellular afferents to SP. This cell receives excitatory afferents from the ganglion cells type II and from neighboring glomerulus and it is inhibited by SN of its column. D) Cellular afferents to LP. This neuron is excited by both retinal ganglion type II cells and by SP cells both from neighboring and its own column and it is inhibited by SN cells both from neighboring as well as its own column. E) Cellular afferents to the glomerulus. It receives afferents from retinal ganglion cells type II and is excited by recurrent axons from LP and SP cells both from its column and neighboring columns.

axons of LP and SP cells both from its own column as well as from neighbor ones (see Fig. 5e).

b) Each LP cell receives afferents from retinal ganglion type II cells both through the glomerulus and through its dendrites along the length between glomerulus and cell body, from SP cells from its own as well as from neighboring columns, and they are inhibited by the SN of its own column as well as by neighboring ones (see Fig. 5d).

c) Each SP neuron is also activated by retinal ganglion type II cells through the glomerulus and interglomerular dendrites both from its own column and from right neighboring columns; this cell is inhibited by SN neurons from its own column (see Fig. 5c).

d) The SN receives afferents from LP cells both from its column and from right neighboring columns (see Fig. 5b).

e) The PY cell, the efferent cell of the column, receives afferents from optic fibres of type II, III and IV, from LP and SP cells of its own column and from LP cells of neighboring columns (see Fig. 5a).

### II.4  Interactions among retina, tectum and pretectum

The diagram that shows the interactions among retina, tectum and pretectum is shown in Fig. 6. This figure shows that the retina sends retinotopically its fibres to both tectum (II, III and IV) and pretectum (III and IV). The two pretectal cells inhibit the activity of LP, SP and PY neurons of its corresponding column (see inset Fig. 6). One of these pretectal cells receives its afferents from ganglion retinal cells type III and IV, while the other receives its afferents from all the PY tectal neurons except from those of their blind spot. Finally the PY cell activity defines the direction of the orienting response and the discriminative abilities of tectal neurons.

The mathematical description of the two-dimensional model of the interactions among retina, tectum and pretectum can be seen in the appendix.

It is important to notice that the final architecture proposed for the interactions among retina, tectum and pretectum is the result of testing several alternatives from which this architecture is the one that best reproduces the physiological and behavioral results. For a detailed analysis of the architecture proposed for the tectum see Lara et al.[1-3]

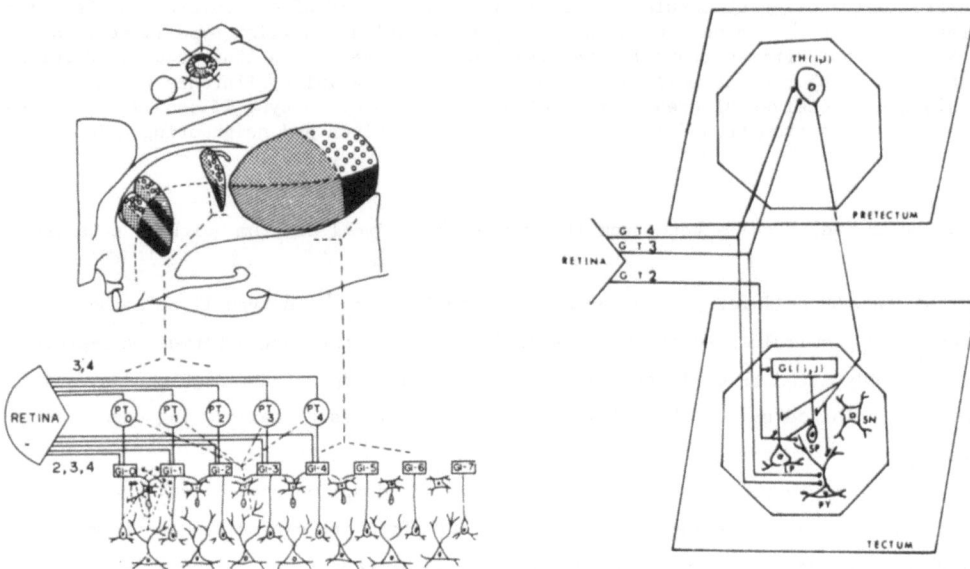

Fig. 6. Interactions among retina, tectum and pretectum. The retina sends fibres in a retinotopical way to both tectum (2, 3, and 4) and pretectum, TH3 (3, 4). The tectum PY cell excites all sameness cells (shown as PT for simplicity) except the one of its own column. Both pretectal neurons inhibit LP, SP and PY of the tectal column (right section).

### III. Computer Simulation

As mentioned above, the aim of the present model is to test the hypothesis both of prey–predator recognition independent of the direction of motion and size preference through a model based on anatomical and physiological grounds. We will present the results of the computer simulation in two ways: pyramidal tectal response of each of the 64 columns of the two–dimensional model of the tectum and through graphs that show several computer experiments that could be directly compared with experimental results. In the first case we divided the tectum in 64 sections each one representing in the horizontal axis the period of simulation and the vertical axis the PY activity. The PY activity is shown through its membrane potential and whenever the membrane potential reaches the threshold value we indicate it by an action potential. We did not simulate the generation of the action potential in our neurons but it is simply a way of showing the results that could be easily compared with experimental results. Both ways of showing the actual behavior of the model allow us to make analogies and comparisons with experimental observations.

For the different simulations we used three types of stimuli:  rectangles whose longest axis moves in the direction of motion (type a); rectangles whose longest axis moves perpendicular to the direction of motion (type b); and squares of different sizes (type c).

### III.1  Behavior of pretectal cell TH3

Our purpose in the simulation of the behavior of this neuron to the different types of stimuli is to show how the interaction of ganglion type cells III and IV could generate their properties.  Trying different weights (see table for the final values) the behavior of this cell to the different types of stimuli is shown in Fig. 7A.  As can be seen in this figure, the response of this cell to the different types of stimuli is very similar to the behavior of the pretectal cell that Ewert suggests is related to prey-predator recognition:  see Fig. 3 for comparison.  This neuron responds more strongly to stimuli of type c, then to those of type b, while the response to stimuli type a does not change very much for different sizes of the object presented.

### III.2  Behavior of tectal cells to different configurations of the stimulus without the inhibitory effect of pretectal cells

It has been shown that tectal neurons without the inhibitory effect of pretectal cells respond better to stimuli of type c, then to stimuli of type a, while they give a slow response to stimuli of type b.[4]  (see Fig. 3)

It is also known that ganglion retinal cells of types II, III and IV arrive at the tectum.  It has been suggested that the facilitation effect for prey-catching activity is mainly controlled by type II ganglion neurons[4,6] but anatomical studies and changes in the receptive field of the animal, latency of response etc. also suggest that tectal cells controlling prey-orienting behavior are also stimulated by ganglion cells of types III and IV.

For the above reasons, we tested different configurations in our model and the configuration that best fitted the physiological results is shown in Fig. 5.  This figure shows that ganglion type II cells arrive at the GL, LP, SP and PY, as already described in detail elsewhere,[1,2,3] while neurons of type III and IV only arrive at the PY neuron.  Thus the column activity is mainly controlled by type II ganglion afferents while the PY response is the combination of all three ganglion cell types.

The response of this neuron to the different types of stimuli is shown in Fig. 7D where it can be seen that it responds best to stimuli type c, then to those of type a, and lowest to those of type b.  This behavior reproduces in general the experimentally observed behavior of tectal cells without pretectum (see Fig. 3D for comparison).

### III.3  Model of the interaction among retina, tectum and pretectum for prey-predator recognition

As we mentioned above, there is a tectal neuron whose behavior closely matches the behavioral response of the animal[4]:  its response is facilitated for stimulus

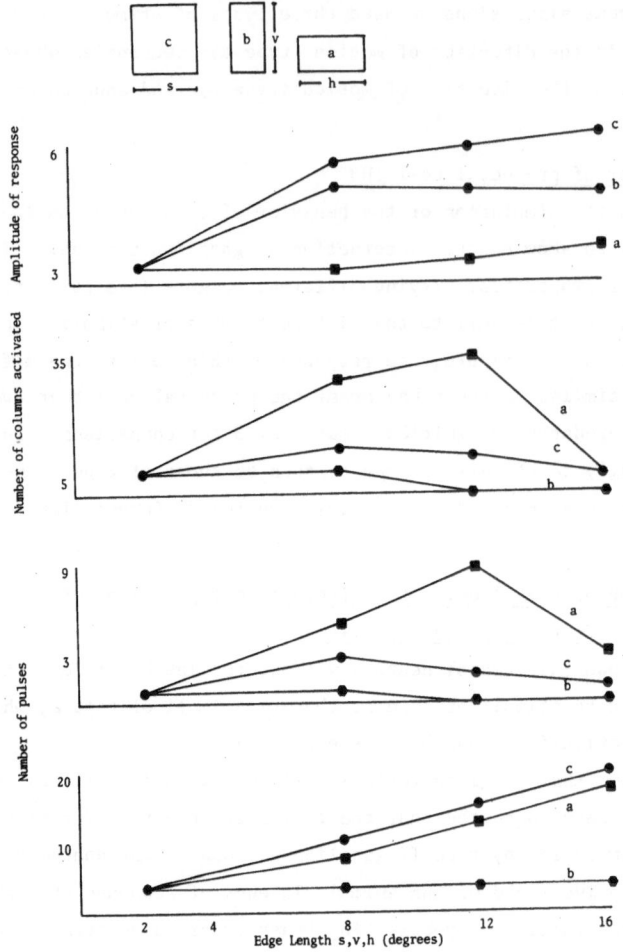

Fig. 7. Computer simulation of the response of pretectal and tectal cells to the different configurations of the stimulus (a,b,c). A) Pretectal cell response: it is mostly sensitive to stimulus type b and c. B) Response of the 64 PY cells to the three types of stimuli (a,b,c): the tectum is mostly sensitive to stimulus type a and it is inhibited by stimulus type b. C) PY response to the three configurations of the stimulus type a. D) PY response when pretectum ablation occurs: PY cells are mostly sensitive to stimulus type c, then to those of type a.

type a, inhibited by stimulus type b, while a combination of an initial facilitation and then an inhibition is observed for stimuli type c.

The interaction among retinal cells, pretectum and tectum in our model, is shown in Fig. 6. In this figure it can be seen that retinal ganglion cells arrive both at the pretectum and to the pyramidal cell of the tectum. The pretectal cell then inhibits LP, SP and PY in the tectal column (see Lara & Arbib[1-3] for this configuration between tectum and pretectum).

In Fig. 7C we show the response of the tectal columns through the activity of

the pyramidal neuron to the three types of stimuli. It can be seen that the PY greatly increases its response for stimulus type a, while for stimulus type b the response
is greatly inhibited and for stimulus type c there is an initial facilitation and then
an inhibition. These results reproduce in a general way the observed physiological
and behavioral observations (see Fig. 2 for comparison). Fig. 7B shows a graph of the
number of times the tectum is activated when a stimulus is presented, thus it is a
better measure of the possible control by the tectum of the orienting response. This
figure is equivalent to Fig. 7C.

In Fig. 8 we show the response of the 64 columns of the tectum to the three types
of stimuli. It can be seen that the tectal activity is stronger for stimulus type a
(Fig. 8a), then type c (Fig. 8b) and the lowest activity is for stimulus type b (Fig.
8c).

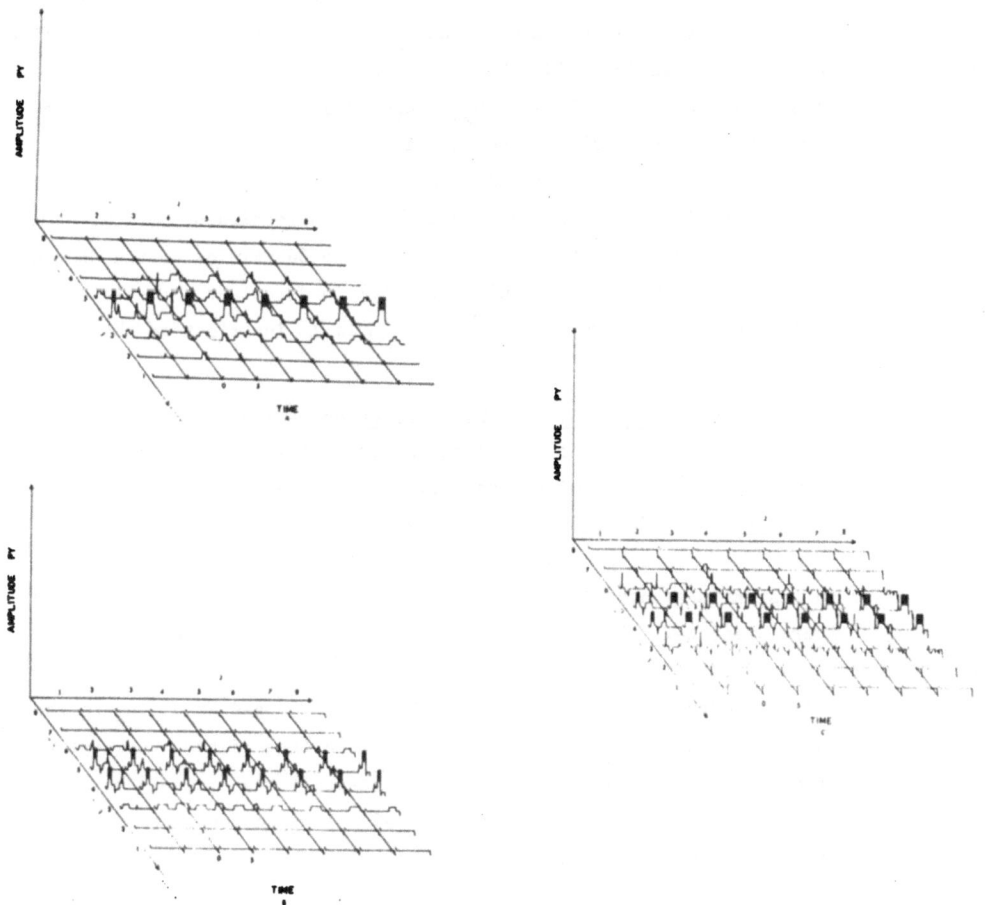

Fig. 8. Computer simulation of the PY response of the 64 columns of the tectum to
the different configurations of the stimulus: (A = type a, B = type b, C = type C).
The horizontal axis shows the temporal response of each of the 64 columns, while the
vertical axis shows the PY cell response. Dark parts of the trace show the action
potentials of PY cells. The response of the PY cells is stronger for stimulus type
c, then to type a, and finally to type b.

III.4  Directional invariance for prey-predator recognition

It has been shown both behaviorally and physiologically[7] that prey-predator recognition is independent of the direction of motion of the stimulus, the only relevant
factor being the relationship between geometry and direction of motion. We test the
behavior of our model for prey-predator recognition in eight different directions.
We used a 12 x 2° stimulus which we know produces a very weak (almost no response in
most tectal cells) response as a stimulus of type b and a strong response as a stimulus of type a. We represent the results of this simulation in the same way that Ewert
represented his experimental observations. We used the contrast formula:

$$D_{wa} = \frac{R_w - R_a}{R_w + R_a}$$

where $D_{wa}$ is a measure of the discrimination between worm (w) and antiworm (a) stimuli. $R_w$ is the response to worm-like stimuli and $R_a$ is the response to antiworm-like
stimuli. We then formed a circle divided by eight lines each one representing a given
direction of the stimulus and the values of $D_{wa}$ ranges between -1 (the origin) and +1
(the outer circle). The inner circle is when $D_{wa}$ is equal to 0.

In Fig. 9 we show the response of tectal columns for prey and non-prey like stimuli showing that the discrimination is independent to the direction of motion.

Fig. 10 and 11 show the actual tectal response of the 64 columns to the presentation of prey (Fig. 10) and non-prey stimuli (Fig. 11) in different directions. It
can be seen that the response to prey or non-prey stimuli is invariant to the direction of motion.

III.5  Size preference as a result of the interaction between tectum and pretectum

It has been observed that prey selection and latency of response can be modulated
depending on the motivational state of the animal.[9] Ingle has suggested that these
changes are the result of a reduced inhibitory effect from pretectal neurons to tectal activity.

In order to test this hypothesis we used our model of prey selection described
elsewhere,[2,3] where pretectal cells receive afferents from all tectal PY neurons except those of a given region which constitute their blind spot. In our present version of this model we postulate that the blind spot can be changed according to the
motivational state of the animal simply by an inhibitory effect of the pretectal neurons, possibly from the telencephalon, increasing or reducing the size of the blind
spot according to the motivational state of the animal. Thus we postulate that when
the animal is greatly motivated for prey catching behavior the following phenomena
occur:

The inhibitory effect of pretectal cells of type 2 is greatly reduced (possibly
increasing the threshold value of this neuron), and the blind spot of the sameness
cells is increased.

Based on the above postulates we studied prey selection of the animal in the normal and motivated state. We represent our results in the same way that Ingle and

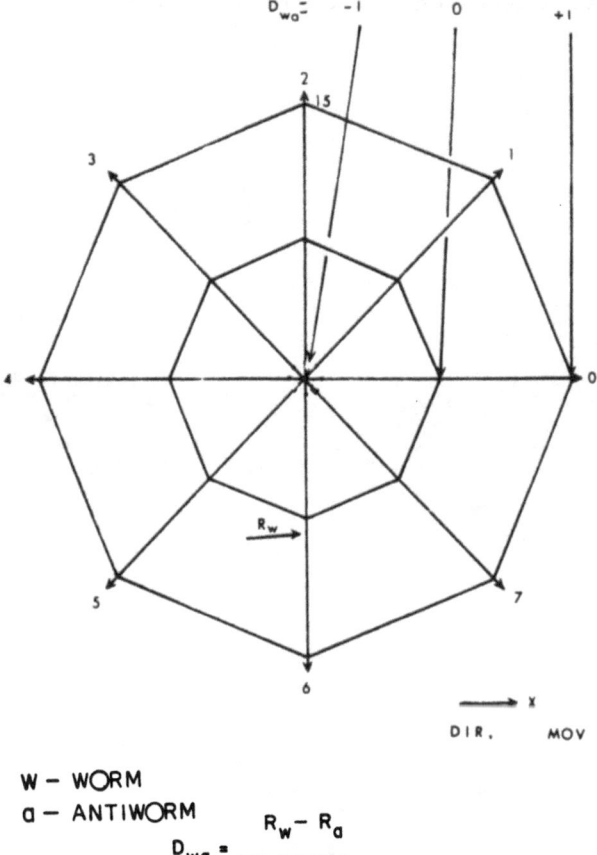

W — WORM

a — ANTIWORM

$$D_{wa} = \frac{R_w - R_a}{R_w + R_a}$$

Fig. 9. Computer simulation of PY cell response of the two-dimensional model of retina-tectum-pretectum for prey-predator recognition in eight different directions. The graph shows that the PY response is direction invariant. The formula shows how it has been quantified for the invariance of prey-predator recognition: $R_w$ is the number of responses when a worm-like stimulus is presented and $R_a$ is the response when an antiworm-like stimulus is shown. The outer circle is the value of $D_{wa}$ of the PY cell response.

Ewert did in their experimental observations[8]: we used a stimulus of a given size as a point of comparison; we then used stimuli of different sizes and studied which one of them was chosen when the animal was both in a normal or in motivated state. We also studied the latency of response in both cases.

Fig. 12 shows that a normal animal prefers smaller stimuli of 4° while the motivated animal always preferred larger stimuli. This figure also shows that the latency of response is greatly reduced in the motivated animal. These results reproduce in general terms the observed experimental results (see inset for comparison).

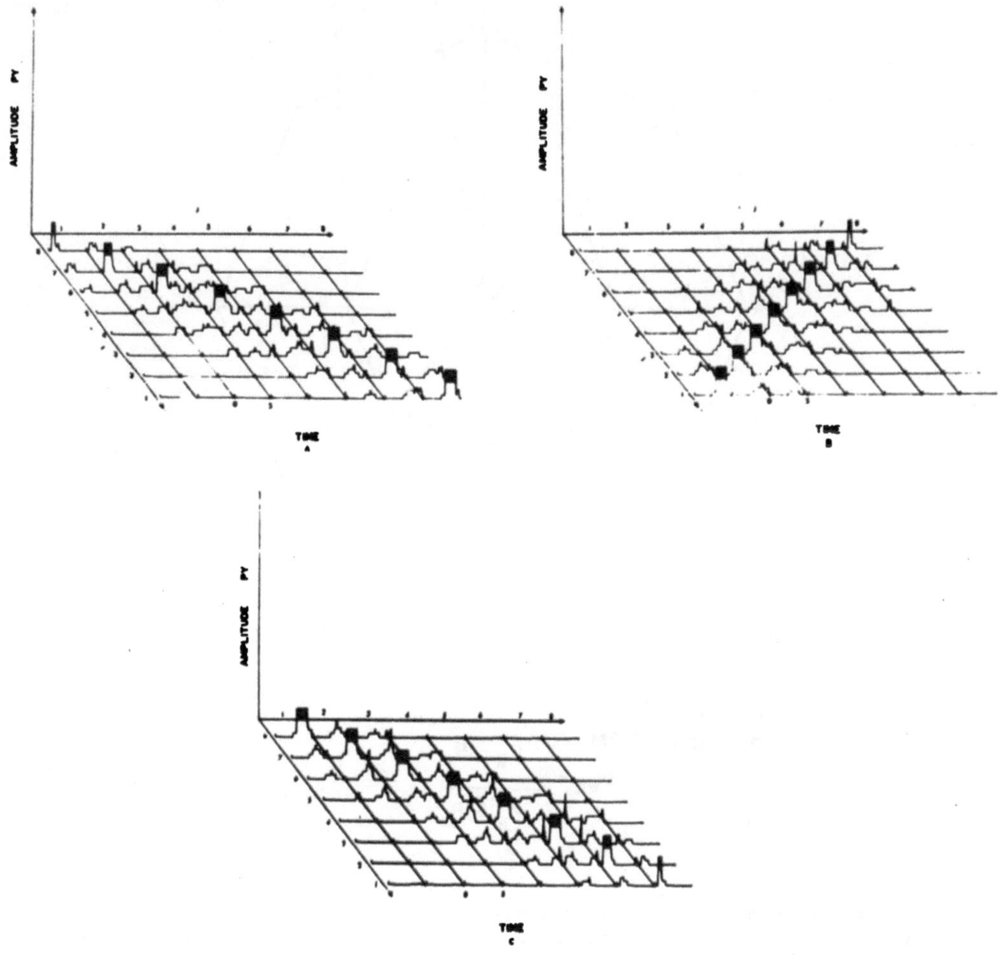

Fig. 10.  Computer simulation of the 64 columns for prey-like stimulus in three different directions.  The response is direction invariant.

IV.  Discussion

With the present model we have been able to simulate a great range of physiological, anatomical and behavioral observations.  When we simulated a tectal column we reproduced the behavioral and physiological results obtained for prey-catching facilitation.[1]  This model was based on the anatomical and physiological studies of this region.  We then expanded our model of the tectal column to a one dimensional model of the tectum where we reproduced the facilitation of tectal response when the stimulus is elongated along the direction of motion as well as the facilitation to double stimuli moved along the direction of motion, with the preference of the animal being

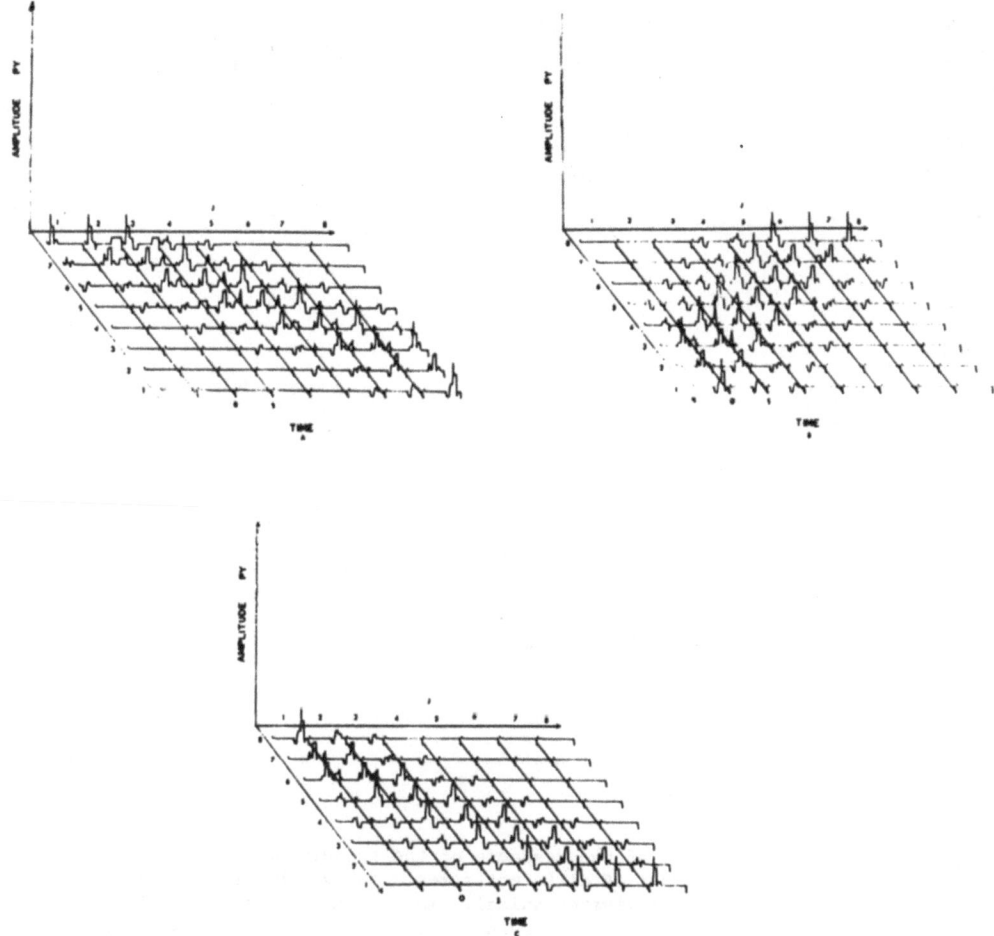

Fig. 11. Computer simulation of the 64 columns for predator-like stimulus in three different directions showing that it is direction invariant.

to orient to the leading of the two objects.[2] We then proposed a one-dimensional model of the interactions among retina, tectum and pretectum for prey selection.[3] With the expansion to two dimensions we have here been able to reproduce prey-predator recognition independently of the direction of motion, and size preference depending on the motivational state of the animal. With this method we thus have been able to integrate step by step, modelling with an increased hierarchical complexity, a theory of how the processing of information performed by the different brain regions of amphibia may control their behavior.

Our modelling studies in combination with the experimental evidence we have used suggest that the behavior of amphibia is controlled by the cooperative activity among

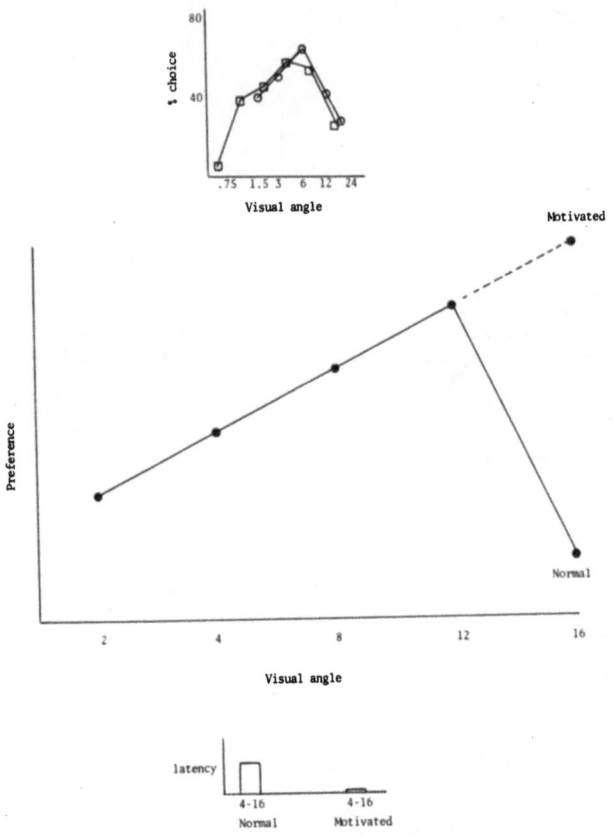

Fig. 12. Computer simulation of the response of the two-dimensional model of retina-tectum-pretectum to two stimuli of different sizes. The graph shows the hierarchy of preference in a normal and a motivated animal. In a normal animal the animal prefers the 4° high stimulus to the 16° high stimulus, while this preference is inverted in the motivated animal. In the lower part of the figure it is shown that the latency of response is also reduced in the motivated animal. See inset for comparison with experimental results. (From Ingle, 1977)

different brain regions. Each region itself has functional units for the processing of information of specific properties of the stimulus. In this way the retina has different features of the stimulus. Then the pretectal neurons, or maybe other thalamic cells, process features of the stimuli related to predator-like stimuli or static objects.[6,15] We know that in the tectum there are cells which seem to be very sensitive to prey-like stimuli; however, there is also evidence of cells which are mostly sensitive to predators as well as to other variables such as distance of the object, position in the visual field, etc.[5,7] This suggests that each region has different functional groups processing different properties of the stimulus whose integrative activity give the desired response. Our present modelling has thus neglected the existence of other functional units with specific processing of information which will generate the proper response of the animal. Our simulation has only considered a part of the integrative activity of the tectum which controls the size, orientation and the recognition of prey stimuli. Further modelling should integrate other functional units in the tectum and in other regions so we may have a clearer idea of how these function- al units interact among each other to produce the proper response. Similar ideas of the type of processing of information that the nervous system is doing has been men- tioned by Ewert[5,7] and Pellionisz.[16]

The specific hypotheses of the present model can be listed as follows:

1) The tectal columns controlling prey orienting behavior receive afferents from retinal ganglion cells of type II, III and IV. The tectal column facilitates the re- sponse to retinal type II afferents; while retinal ganglion cells of type III and IV control prey orienting behavior when the animal is in a motivated state or for the regulation of size constancy in the animal in combination with pretectal cells.

2) The inhibitory effect of pretectal cells gives tectal neurons the capacity for prey-predator recognition. The inhibitory effect of pretectal cells is mainly directed to the PY cell although a small effect can also be seen either on LP and SP or in the SN.

3) The directional invariance of prey-predator recognition is the result of tec- tal architecture.

4) Size preference is the result of a combined effect of a reduced pretectal in- hibition of pretectal cell TH3 over the tectum and an increased blind spot of the same- ness cell for prey selection. The change in the latency of response is also the result of this interaction. Both effects may be controlled by the telencephalon.

The present model can be considered as a different way of simulating the ideas of Ewert and von Seelen for the relations among retina, tectum and pretectum for prey- predator recognition. Ewert and von Seelen[17] proposed a model of prey-predator recog- nition in which the retina, tectum and pretectum acted as filters for specific config- urations of the stimulus. Specifically, they proposed that the tectum was mostly sen- sitive to predator stimuli. The inhibitory effect of the pretectum to the tectum en- abled the latter to discriminate between prey and predator. The limits of this model are: 1) they do not show how the architecture of the different brain regions will give rise to the properties of their postulated filters; 2) they only simulate prey- predator recognition with neither the possibility to reproduce other phenomena nor the capacity for expansion; 3) because of the linear nature of the model, it is only restricted to a given range of values; and 4) because the model is lumped both in space and time, it cannot be tested against the time course of response of specific cell types with specific retinotopic coordinates.

Our model, on the other hand, because of its anatomical and physiological bases can be tested against such experiments.

1. Lara, R., Arbib, M.A. and Cromarty, A.S. The role of the tectal column in facilitation of amphibian prey-catching behavior: a neural model. J. Neuroscience (to appear).

2. Arbib, M.A. and Lara, R. A neural model of the interaction of tectal columns in prey-catching behavior. Submitted to Biological Cybernetics.

3. Lara, R. and Arbib, M.A. A neural model of interaction between pretectum and tectum in prey selection. Cognition & Brain Theory (in press).

4. Ewert, J.P. The visual system of the toad: Behavioral and physiological studies on a pattern recognition system. In The Amphibian Visual System. (Fite, K. ed.) Academic Press, New York, pp. 142-202, 1976.

5. Schurg-Pfeiffer, E. and Ewert, J.P. Investigations of neurons involved in the analysis of gestalt prey features in the frog rana temporaria. J. Comp. Physiol. 141: 139-152 (1981).

6. Ingle, D. Spatial vision in anurans. In The Amphibian Visual System. (Fite, K. ed.) Academic Press, New York, pp. 119-141, 1976.

7. Ewert, J.P., Borchers, H.W., Wieterschein, von. Directional sensitivity, invariance and variability of tectal T5 neurons in response to moving configurational stimuli in the toad Bufo bufo (L). J. Comp. Physiol. 132:191-201, 1979.

8. Ingle, D. and Cook, J. The effect of viewing distance upon size preference of frogs for prey. Vision Res. Vol. 17, 1009-1013, 1977.

9. Ingle, D. Size preference for prey-catching in frogs: relationship to motivational state. Behav. Biol. 9:485-491, 1973.

10. Trachtenberg, M.C. and Ingle D. Thalamo-tectal projections in the frog. Brain Res. 79:419-430, 1974.

11. Scalia, F. The optic pathways of the frog: nuclear organization and connections. In Frog Neurobiology (Llinás, R. and Precht, W. eds.) Springer Verlag, 1976.

12. Ewert, J.P. and Hack, F.J. Movement sensitive neurons in the toad's retina. Exp. Brain Res. 16:41-59, 1972.

13. Ewert, J.P., Krug, H., Schnitz, G. Activity of retinal R3 ganglion cells in the toad. Bufo bufo (L.) in response to moving configurational stimuli: Influence of the movement direction. J. Comp. Physiol. 1979.

14. Ewert, J.P. Single unit response of the toad (Bufo americanus) caudal thalamus to visual objects. Z. Vergl. Physiol. 74: 81-102, 1971.

15. Grüsser, O.J., Grüsser-Cornehls, V. Neurophysiology of the anuran visual system. In Frog Neurobiology. (Llinás, R. and Precht, W. eds.). Springer Verlag, 1976, pp. 297-385.

16. Pellionisz, A. Modelling of neurons and neuronal networks. In The Neurosciences: Fourth Study Program (eds. Schmitt, F.O. & Worden, F.G.) MIT Press, Cambridge, pp. 525-546.

17. Ewert, J.P. and von Seelen, W. Neurobiologie und System-Theorie eines Visuellen Muster-Erkennungsmechanismus bei Kroten. Kybernetik. 14:167-183, 1974.

## Appendix

We provide the mathematical definition of the two-dimensional model of the interactions between tectum and pretectum which complements the description of the one dimensional model of the interaction between tectum and pretectum given in Lara and Arbib.[3]

The specifications of threshold functions, membrane constants and weights is given in Tables 1, 2 and 3 respectively.

Glomerulus:

The equation defining the behavior of the glomerulus of the ith, jth unit column is given as follows:

$$\tau_{gl} \; \dot{gl}_{ij}(t) = -kl \, gl_{ij}(t) + U2_{ij}(t) + I_{ij}(t)$$

where $\tau_{g1}$ and k1 are constants, U2 is the optic input from retinal ganglion cells type II, and $I_{ij}$ are the recurrent inputs from LP and SP cells of the unit as well as those of neighboring columns, and are defined as:

$$I1(t)=w_{g1.sp}(SP_{i-1,\ j-1}(t)+SP_{i,\ j-1}(t)+SP_{ij}(t)+SP_{i+1,\ j-1}(t)+SP_{i+1,\ j}(t))$$

$$I2(t)=w_{g1.lp}(LP_{i-1,j-1}(t)+LP_{i-1,j}(t)+LP_{i-1,j+1}(t)+LP_{i,j-1}(t)+LP_{ij}(t)+LP_{i.j+1}(t))$$

$$I_{ij}(t)=I1(t)+I2(t)+LP_{i+1,j-1}(t)+LP_{i+1,j}(t)+LP_{i+1,j+1}(t)$$

where the values of w are given in Table 3.

## Stellate Neurons: (SN)

The ith jth stellate neuron can be defined as follows:

$$A = LP_{i-1,j}(t)+LP_{i-1,j+1}(t)+LP_{ij}(t)+LP_{i,j+1}(t)+LP_{it1,j+1}(t)$$

$$\tau_{sn}\dot{sn}_{ij}(t)=-K2\ sn_{ij}(t)+w_{sn-1p}A$$

where $\tau_{sn}$ is the membrane constant of these neurons, K2 and $w_{sn.1p}$ are constants and can be seen in Tables 2 and 3 respectively.

## Large pear shaped cells: (LP)

The behavior of the ith jth LP neuron can be defined as follows:

$$A1 = SN_{i-1,\ j-1}(t)+SN_{i,\ j-1}(t)+SN_{i\ j}(t)+SN_{i+1,\ j-1}(t)+SN_{i+1,\ j}(t)$$

$$A2 = SP_{i-1,\ j-1}(t)+SP_{i,\ j-1}(t)+SP_{i\ j}(t)+SP_{i+1,\ j-1}(t)+SP_{i+1,\ j}(t)$$

$$\tau_{1p}\dot{1p}_{ij}(t) = -1p_{ij}(t)-w_{1p.sn}A1+w_{1p.th}TH_{ij}(t)+w_{1p.sp}A2+g1_{ij}(t)+U2_{ij}(t)$$

where $\tau_{1p}$ is the membrane constant of these neurons; gl is the glomerulus input; U2 is the optic input from retinal ganglion type II cells; TH is the thalamic input; and w's are the weight factors shown in Table 3.

## Small pear shaped cells: (SP)

The behavior of the ith jth SP neuron is defined as follows:

$$A3=g1_{i-1,j}(t)+g1_{i-1\ j+1}(t)+g1_{i,j+1}(t)+g1_{i+1,j+1}(t)$$

$$\tau_{sp}\dot{sp}_{ij}(t)=sp_{ij}(t)-w_{sp.sn}sn_{ij}+A3\ w_{sp.th}(IH_{ij}(t))+U2_{ij}(t)$$

where $\tau_{sp}$ is the membrane constant of these neurons; SN is the inhibitory effect of the stellate cells; TH is the inhibitory effect from thalamic neurons; and U2 is the optic input from fibres type II from the retina; w's are the weighting factors that can be seen in Table 3.

## Pyramidal Neurons: (PY)

The ith jth PY cell is defined as follows:

$$A4=LP_{i-1,\ j}(t)+LP_{i-1,\ j+1}(t)+LP_{ij}(t)+LP_{i,j+1}(t)+LP_{i+1,j+1}(t)$$

$$A5=w_{py.u2}U2_{ij}(t)+w_{py.u3}U3_{ij}(t)+w_{py.u4}U4_{ij}(t)$$

$$\tau_{py}\dot{py}_{ij}(t)=-py_{ij}(t)+w_{py.sp}.SP_{ij}(t)+w_{py.1p}A4-w_{py.th}TH_{ij}(t)+A5$$

where $\tau_{py}$ is the membrane constant of these neurons; SP is the excitatory effect of small pear cells; TH is the inhibitory effect of both pretectal neurons; and U2, U3 and U4 is the optic input from retinal ganglion cells type II, III and IV respectively.

The w's are the different weight factors shown in Table 3.

## Table 1

### Threshold Functions

LP = f ( lp - 1.0 )
SP = f ( sp - 2.0 )
SN = h ( sn - 0.2 )
PY = h ( py - 5.559 )
TH = g ( th - 3.7 )

## Table 2

### Membrane Constants

$\tau_{gl} = 0.35$, k1 = 0.15
$\tau_{sn} = 0.65$, k2 = 0.4
$\tau_{lp} = 0.3$
$\tau_{sp} = 0.9$
$\tau_{py} = 0.12$
$\tau_{th} = 1$, k3 = 7

## Table 3

### Weights

| | | |
|---|---|---|
| $w_{gl.lp}$ | = 1.0 | LP to GL |
| $w_{gl.sp}$ | = 0.1 | SP to GL |
| $w_{lp.sp}$ | = 0.8 | SP to LP |
| $w_{lp.sn}$ | = 8.0 | SN to LP |
| $w_{lp.th,s}$ | = 0.1, 0.4 | TH to LP |
| $w_{lp.s}$ | = 0.2 | S to LP |
| $w_{sp.sn}$ | = 20.0 | SN to SP |
| $w_{sp.th,s}$ | = .1, 0.4 | TH to SP |
| $w_{sn.lp}$ | = 2.1 | LP to SN |
| $w_{sp.s}$ | = 0.2 | S to SP |
| $w_{py.lp}$ | = 0.8 | LP to PY |

Table 3:  Weights (continued)

| | | |
|---|---|---|
| $w_{py.sp}$ | = 1.0 | SP to PY |
| $w_{py.th,s}$ | = 0.9 | TH to PY |
| $w_{py.u3}$ | = 0.3 | U3 to PY |
| $w_{py.u4}$ | = 6.0 | U4 to PY |
| $w_{py.u2}$ | = 4.5 | U2 to PY |
| $w_{th.u3}$ | = 0.3 | U3 to TH |
| $w_{th.u4}$ | = 5.0 | U4 to TH |
| $w_{py.s}$ | = 0.4 | S to PY |

# CHAPTER 23

## TENSOR THEORY OF BRAIN FUNCTION.

## THE CEREBELLUM AS A SPACE-TIME METRIC

Andràs PELLIONISZ & Rodolfo LLINÁS

Department of Physiology & Biophysics
New York Univeversity Medical Center
550 First Ave, New York, 10016, USA

## 1. EXPERIMENTAL NEUROSCIENCE AND BRAIN THEORY
### -TO KNOW AND TO UNDERSTAND THE BRAIN-

**1.1.** **THE GOAL OF BRAIN THEORY** is to transform knowledge relating to the properties of the central nervous system into an understanding of brain function. As in other fields of science using theories as heuristic tools of understanding, not only the quality of a particular theory has to be carefully measured but its ultimate usefulness must also be considered. Usefulness serves then as an independent test of the degree of understanding provided by a given abstraction. An understanding, e.g. of the cerebellum, may be gauged by the ability of a theory to provide a "blueprint" of a device capable of accomplishing the described function, in this case motor coordination. Such an understanding of brain function has great potential for robotics, an ultimate beneficiary of neuroscience. In this paper we wish to make the point that the Tensor Network Theory of Central Nervous System (a geometrical abstraction and generalization of the known vector-matrix approaches) is capable of yielding a usable understanding. We offer in support of this assertion a tensorial network model capable of generating motor coordination via cerebellar-type neuronal circuits.

**1.2.** **PRINCIPAL METHODS OF BRAIN THEORY: CONCEPTUALIZATION AND ABSTRACTION.** To achieve its goals, brain theory relies chiefly on conceptualization and abstraction, using model-making and mathematics as its basic tools. Such enterprise in all fields of science commences by collecting data that are perceived relevant for the goal of research. Having gathered a set of fragmented facts, collectively addressed as KNOWLEDGE, abstraction is introduced to explore the SYSTEM which relates these facts to a given global function. Revealing a system of relations among facts, addressed as UNDERSTANDING, is similar to the establishing of the geometry that may exist among a set of points. Therefore, in our view, building an understanding implies the construction in our minds of an internal geometrical representation of the relations in the external world. The above approach may be implemented by collaborative efforts among scientists. For instance, in our particular case, we found that our interest and expertise were complementary and could be integrated into formulating a theory by combining the functional view, by R.L., indicating that CNS function is to be understood in the manner of geometry, with the implementation of such view via tensor analysis, by A.P., and by continuing a dialogue for close to a decade so far.

In this paper we summarize our view that the fundamental concept of brain function is the establishment of related geometries (that the brain is a geometric object in the KRONIAN sense, PELLIONISZ & LLINAS, 1979b). We stress that, as of today, the best abstract language for describing geometrical properties in a formal manner appears to be tensor analysis (PELLIONISZ & LLINAS, 1979a, 1980, 1982). For an overview of broader aspects of this approach see, e.g. MELNECHUK, 1979.

1.3. THE VARIETY OF CONCEPTS OF ORGANIZATION OF CNS FUNCTION. Firstly, the most significant initial step in the process of abstraction is the introduction of AN APPROPRIATE FUNDAMENTAL CONCEPT. Model-making (a conceptual representation of data) becomes thereby possible. The next step of abstraction is then the rigorously self-consistent formalization of the concept that we may call a theory.

Secondly, finding the most SUITABLE METHOD OF ABSTRACTION is equally important. Indeed, no matter how sound and powerful the mathematical apparatus may be, the results of the abstraction can only be as good as the quality of the basic concept it relies on. Metaphorically speaking, -c.f. ARBIB, 1972-, the need for having both a proper concept and a suitable method of abstraction is comparable to the need, when fishing, of both knowing where to fish and also of having a proper net to entrap them securely. In these terms, formulating a brain theory without a suitable mathematical system of abstraction (e.g. as in ECCLES, 1981) is comparable to fishing in the proper spot but with bare hands. On the other hand, having the finest and strongest net and casting it into empty waters (e.g. -as pointed out later- in McCULLOCH & PITTS, 1943) is, again, futile.

Indeed, THE PRIMARY TASK OF CONCEPTUALIZATION IS THE SEARCH FOR ADEQUATE FUNDAMENTAL ORGANIZING PRINCIPLES OF BRAIN FUNCTION. Historically one of the most significant views relating to brain was the concept of REFLEXES (SHERRINGTON, 1906). The next major change was to consider brain function as embodying LOGICAL CALCULUS (McCULLOCH & PITTS, 1943). However, it is quite evident today that this approach missed its mark, as the control paradigm used by the brain in say, a coordinated goal-oriented sensorimotor action, is not BOOLEAN. A different, but somewhat primitive concept of motor actions is that of assuming the existence of a "LOOKUP-TABLE" (RAYBERT, 1977, 1978), i.e. to suppose that a motor action relies on the selection of a movement-sequence from a stored set of solutions. Yet another concept, borrowed from engineering, is the one of LINEAR SYSTEM THEORY (e.g. for the representation of eye movements, c.f. ROBINSON, 1968) which provides a highly practical framework for an abstract mathematical interpretation of special sensorimotor systems. In highly nonlinear visuomotor systems (such as the visual control of a fly's navigation) NONLINEAR SYSTEM THEORY has provided algorithms for the movement-control (POGGIO & REICHARDT, 1981). However, these authors do admit that "...the visual control of flight certainly relies on other algorithms as well...", e.g. Taylor-series expansion (ibid). Recently BIZZI (1981) proposed the view that regards the control paradigm used in the visuomotor system primarily as a COMPUTATIONAL problem. For a review of various other organizing concepts proposed for brain function, see SZENTAGOTHAI & ARBIB (1975).

1.4. THE GRADUALLY INCREASING LEVEL OF ABSTRACTION IN NEUROSCIENCE. At the level of simple description of the morphology and physiology of neuronal networks, abstraction (relating the detailed knowledge to the ultimate function) was avoided altogether. Such a traditional view of the brain was directed at the structural geometry, aiming at establishing the system of spatial relations among the different types of cells in neuronal networks. This approach is best known from circuitry drawings by RAMON Y CAJAL (1911). While such a descriptive method is intuitively clear, it has serious limitations as it does not explain how the network provides the global function, nor offers a way to handle its complexity in a quantitative manner. A more detached way to relate the structure to function is the use of computer models (e.g. PELLIONISZ, 1970, PELLIONISZ et al., 1977), the latter based on quantitative histological data (LLINAS, 1971). While such models make it possible to address quantitative properties of large networks, they still do not explain their function.

A somewhat more abstract handling of the functioning of neuronal networks relies on the notion of "patterns" of activities over a field of neurons. It is of historical interest to note that the idea of patterns was first used in a poetic manner by SHERRINGTON (1906) who envisioned the brain as an "enchanted loom". Later, such patterns were handled either by hand-drawn sketches (e.g. ECCLES, 1981), or by mathematical abstraction (e.g. BEURLE, 1962, KATCHALSKY et al., 1974) or by computer-simulation graphics (e.g. PELLIONISZ, 1970). For exact mathematical studies of the intrinsic dynamical features of excitation-inhibition patterns, AMARI & ARBIB (1977) developed the concept of competition and cooperation in neuronal nets.

The concept of "patterns" of activities raises the deep question of whether nervous function is to be understood at an abstract level as a fundamentally deterministic or rather, as a stochastic process. While there appears to be a consensus on the deterministic nature of brain function, it is clear that stochastic methods may still be applied both for investigation (e.g. HARTH & TZANAKOU, 1974) and even for functional analysis of neuronal systems, given that the level of description is close to the QUANTA inherent in the described phenomena (c.f. HOLDEN, 1976). An important step towards a more abstract mathematical analysis of local neuronal mechanisms in networks is represented by the work of MALSBURG & WILLSHAW, 1976, and LEGENDY, 1978, who already used some aspects of the next most important step of abstraction, the vector and matrix approach.

## 2. THE VECTOR-MATRIX APPROACH IN BRAIN THEORY

While the vector-matrix approach opens the way for tremendous advances in mathematical brain theory, technically it appears almost trivial. Having N input and M output neurons, the morphological system of interconnections can be mathematically characterized by an array of N x M quantities (a matrix). The activity of the input can also be formally described by an ordered set of N quantities (a vector, where the components can be, for example, the firing frequencies of the N or M neurons). However, as we elaborate below, the vector-matrix approach has also proved to be treacherous. Indeed, it generated an enormous impetus for the mathematically-minded to build up the apparatus of the abstraction, but the biological relevance of the application was not always beyond question. While the formalism utilized by many workers was usually mathematically appropriate this MEANS OF ANALYSIS occasionally became a GOAL in itself, with its advancement the vector-matrix representation not always remaining relevant to biological problems. However, before pointing out the deficiencies that we perceive in the applications on record, a brief overview of the works in question seems appropriate.

Even though vectorial notation in brain theory was introduced quite early (PITTS & McCULLOCH, 1947 and WIENER, 1948), the profound reason for its applicability was not pointed out. This is not altogether surprising since at that time the overwhelming view was that neuronal nets perform logical operations (McCULLOCH & PITTS, 1943). Accordingly, the conceptual difference between computers and brains was hardly distinguishable, and thus computer science overshadowed brain theory for more than a decade. Not until von NEUMANN (1957) stated the important theoretical difference between the two information processing systems did these two efforts gradually grow distinct. An interesting paper in the transition to the more recent, non-BOOLEAN approaches is that of von FOERSTER (1967). His work was technically a vector-matrix approach, but the conceptology was mixed: the networks were in part assumed to perform operations of mathematical logic, but at other times the network was characterized as a computational machine.

Modern vector-matrix approaches to brain theory can be dated back to the PERCEPTRON (ROSENBLATT, 1962, MINSKY & PAPERT, 1969). Perceptron is basically a computational geometric theory that was applied, unfortunately, to conjectural circuits rather than to real neuronal systems. Therefore, this line of research gradually moved away from neuroscience into the realm of artificial intelligence which aims at developing brain-like machines independent of the modes of operation of the brain. More directly related to real brains was the approach taken by ANDERSON (see, e.g. 1981). His early work (ANDERSON et al., 1972) led to an independent line of work by COOPER (1974). In the field of associative memory, one of the highest brain functions, pioneering work was done by KOHONEN. His approach, while using extensive vector-matrix calculations, reveals a common conceptual limitation of the vector-matrix approach quite explicitly: "considerations of the network should be understood as a SYSTEM-THEORETICAL APPROACH only; no assumptions will be made at this stage about the actual data represented by the patterns of activity" (KOHONEN et al., 1981).

An altogether different approach was represented by papers that focused on the physical geometrical properties of the circuitry (e.g. MALSBURG & WILLSHAW, 1976, SCHWARTZ, 1977, COWAN, 1981). While the central concept in these papers appears to be geometry, the relation of the physical and functional geometries of the networks has not been established by a conceptually homogeneous treatment merging the two together.

2.1. <u>THE VECTOR-MATRIX APPROACH IN CEREBELLAR RESEARCH</u>. An early notion from one of us (A.P., mentioned in SZENTAGOTHAI, 1968) related to a vector-matrix approach to compare cerebellar circuitries with the STEINBUCH-matrix (1961). The extensive mathematical analyses by GROSSBERG have relevance to the cerebellar studies as well, having featured Purkinje cells as non-recurrent temporal discriminators (GROSSBERG, 1970). Conceptually quite interesting is the note by GREENE regarding motor generators as devices "mapping...various mathematical spaces". While this was clearly a geometrical consideration, he stopped short of a full analysis: "this method will not be derived from a general model or an abstract study of these spaces..." (GREENE, 1972). The reason may be that throughout the seventies a rather serious problem existed in the conceptualization of cerebellar function. This organ was considered by many workers mainly as a device that learns motor patterns. Thus, it was explicitly regarded as a special kind of PERCEPTRON for more than a decade by some theorists (e.g. ALBUS) and even by experimentalists (e.g. ECCLES). Once the tensor approach had been introduced (PELLIONISZ & LLINAS, 1978), the perceptron-concept of cerebellar function was formally expanded into a vector-matrix approach (ALBUS, 1979). Lately, the rather different "synergy" concept of cerebellar function (BERNSTEIN, 1947) has also been vectorially restated by BOYLLS (1981).

It is instructive to quote an experimentalist's recent view on cerebellar modeling: "...the models so far proposed are essentially a pattern recognizer device which accounts for space-related, but not for time-related, operation of the cerebellum. Time-related operations such as prediction or programming are now often considered in connection with cerebellar functions. It seems to be urgent to build a new model of the cerebellum which allows experimentalists to investigate entire spatio-temporal features of cerebellar neuronal circuitry." (ITO, 1981).

2.2. INADEQUACIES OF THE VECTOR-MATRIX APPROACH. As a result of cerebellar and non-cerebellar studies, today's literature on brain theory is rich in vector-matrix studies. This holds in theoretical neurobiology (c.f. REICHARDT & POGGIO, 1981), in neuronal communication and control theory (c.f. SZEKELY, LABOS & DAMJANOVICH, 1981) and also in modeling of such higher brain functions as associative memory (c.f. HINTON & ANDERSON, 1981). In our opinion, however, all of the works cited seem to miss a rather profound aspect of vector-matrix representations, as we shall elaborate below.

The single main deficiency of the VECTORIAL treatment of nervous activity in the mentioned papers can be summarized by the following argument. It is a fact that activities over a set of neurons can be mathematically represented by a vector. VECTORS, HOWEVER, SHOULD NOT BE TREATED SEPARATELY FROM THE MATHEMATICAL SPACE IN WHICH THEY REPRESENT A MATHEMATICAL POINT, AND WHEN CONSIDERING BRAIN FUNCTION SHOULD NOT BE CONSIDERED AS SEPARATE FROM THE FRAME OF REFERENCE BY WHICH THEY ARE RELATED TO INVARIANTS OF THE PHYSICAL REALITY.

A resulting technical deficiency can be easily spotted considering the following. THERE IS NO REASON AT ALL TO ASSUME THAT THE BRAIN IS LIMITED TO THE USE OF ORTHOGONAL FRAMES OF REFERENCE WHEN IT ASSIGNS INTERNAL VECTORIAL EXPRESSIONS TO THE PHYSICAL INVARIANTS OF THE EXTERNAL WORLD. SINCE IN NON-ORTHOGONAL (OBLIQUE) FRAMES OF REFERENCE THERE ARE TWO DIFFERENT VERSIONS OF A VECTOR (A COVARIANT AND A CONTRAVARIANT), INTRODUCING A VECTORIAL NOTATION WITHOUT SPECIFYING THE VERSION OF THE VECTOR IS MEANINGLESS UNLESS THE ORTHOGONAL NATURE OF THE FRAME OF REFERENCE IS FIRST DEMONSTRATED. In none of the papers mentioned was it explicit whether the "vectors" are of covariant- or contravariant character, or proven that the reference-frame implied was truly orthogonal.

In fact, in most applications it is implicitly or explicitly assumed that the vectors describing CNS function are Euclidean, although this is much more of a dogma than a fact. Indeed, it is often implied that a vector "automatically" belongs to an Euclidean space. Since this assumption is incorrect, a central problem of the use of vectors in brain theory can be depicted as in Fig. 1.

Another problem with the MATRICES of the vector-matrix approach relates to their functional purpose. Ultimately any given brain theory is only as good as the understanding that it provides about the functional purpose of neuronal networks. If one does not grasp the common functional purpose of individually different

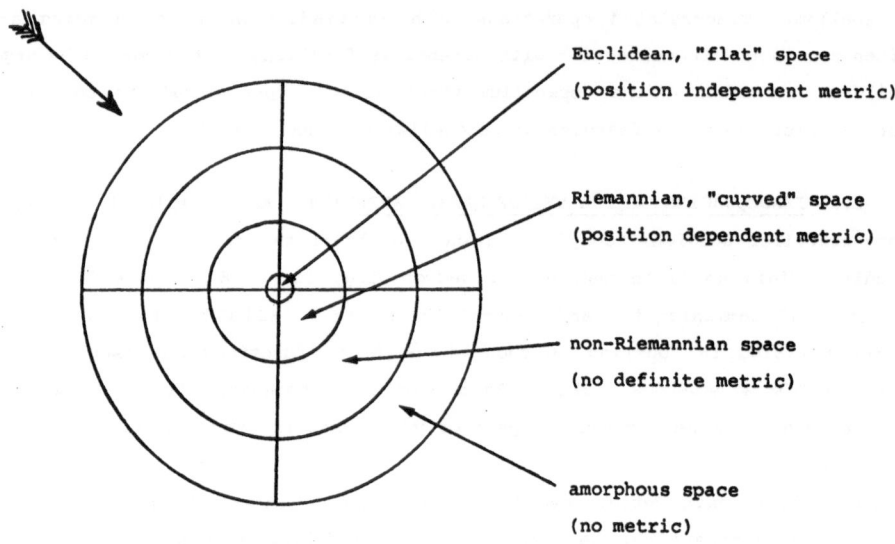

"brain vector"

Euclidean, "flat" space
(position independent metric)

Riemannian, "curved" space
(position dependent metric)

non-Riemannian space
(no definite metric)

amorphous space
(no metric)

FIG. 1. Symbolic depiction of the principal problem with "brain vectors". An ordered set of quantities that represents nervous function is, by definition, a vector. However, a central issue is that of identifying the abstract CNS-space in which this vector represents a mathematical point.

neuronal networks all belonging to a given class (e.g. the functional purpose of cerebellar networks in all specimens and species), one can hardly provide a good model of the function of e.g. the part of the brain that the cerebellum is. Formally, any particular system of interconnections between multiple input and output elements can be mathematically described as a matrix. However, the usefulness of a network theory can only be determined by its ability to state the functional purpose of the general features of the transformation that the whole class of all such network-matrices implements. The existence of an underlying entity capable of representing the transformation by a class of neuronal networks is implicitly assumed in neuroscience: i.e. data derived from particular neuronal networks are generalized to the set of neuronal networks (from A brain to THE brain). Our answer to this problem is to regard the brain as a geometric object in the KRONIAN sense where every individual connectivity-matrix is considered a particular expression of the same reference-frame invariant vector-relationship: the network tensor.

2.3. THE ORIGIN OF THE PROBLEMS IN THE VECTOR-MATRIX APPROACH. Vectorial representations pose problems for two basic reasons. Firstly, there is no single definition of the term "vector" that would be both generally accepted as well as mathematically impeccable. Instead, several imprecise and/or overly strict "definitions" of this word coexist, creating contradictions (for details, see HOFFMANN, 1966). Secondly, "vector" is a collective term, relating many things at once (relating components to one another and also the components to an invariant). As it can be seen all too often in science, dealing with holistic features is frequently avoided, for it is easier to single out a certain aspect and ignore the rest. This problem occurs in the use of vectors, too, and at more than one level: i) vectorial components are often treated as detached from one another (it is easier not to be bothered with too many things at once), and ii) sometimes the full set of vectorial components is totally detached from the physical invariant that it was originally assigned to. This separate existence of the set of components, detached from reference frames and the geometry of the space of the physical invariant, simplifies the mathematical abstraction. As a result, one can operate within a "known" abstract domain of how to handle vectors, without having to bother simultaneously with a largely unknown system through which such vectors are related to the reality of external invariants.

In experimental approaches the first kind of detachment is most common since the investigation of a part of a system (e.g. only of the horizontal vestibulo-ocular reflex arc) greatly simplifies the methodology. In pure mathematical studies the second danger is obvious, since the intricacies of the supposedly known mathematical behavior of the vectorial components themselves distracts the worker from the unexplored properties of the vector in being related to physical invariants. HEAVISIDE pointed out that such differences in the way of thinking of the physically- or mathematically minded used to exist also in the so-called "exact" natural sciences: "...people, like FARADAY, naturally think in vectors ... if they think about vectors, they think of them as vectors ... for general purposes of reasoning the manipulation of the scalar components instead of the vector itself is entirely wrong" (quotation in CROWE, 1967).

A particularly unfortunate combination of the above kinds of detachments occurs when one implies the definition of a "vector" as a directed magnitude (a physical entity, such as a displacement), and simultaneously means by the same definition a Cartesian mathematical vector (x,y,z), which is assigned to the physical entity. In such a special case (but only in such a case) a separation of the vectorial components from one another is mathematically justifiable, since the orthogonal x,y,z components are independent of each other both in the way they are established from a displacement and in the way they generate such displacement. Also, with an x,y,z vector usually no consideration is given to the frame of reference used since an orthogonal frame is always implied. Likewise, the properties of the vectorspace can be totally ignored since in that space the "trivial" Euclidean geometry governs. In the simple case of the vectorial treatment of physical displacements there is not much room for confusion in spite of the facts that a) the vector-definition is ambiguous and unclear, b) this kind of vector represents only a very special case where the components may truly be separable form one another, c) the geometrical properties of the space can only be ignored because the particular space involved is well known. THE CONCEPT OF VECTOR, HOWEVER, IS NOT LIMITED TO THE PRIMITIVE CASE OF CARTESIAN, THREE-DIMENSIONAL ORTHOGONAL VECTORS WITH EUCLIDEAN GEOMETRY IN THE VECTORSPACE. THEREFORE, WHEN DEALING WITH "BRAIN VECTORS", NONE OF THE ABOVE SIMPLIFYING GENEROSITIES CAN BE PERMITTED.

2.4. THE ABSTRACTION AND GENERALIZATION OF THE VECTOR-MATRIX TREATMENT INTO A TENSORIAL APPROACH. Fundamentally the vectorial description of the activity of N neurons is only permissible if the general definition of the vector, as a mathematical point of an abstract space, is used: "...it is essential to adopt the fundamental convention of using the term point of an abstract N-dimensional manifold (N being any positive integer) to denote a set of N values assigned to any N variables $x_1$, $x_2, \ldots, x_n$. This is an obvious extension of the use of the term in the one-to-one correspondence which can be established between pairs or triplets of co-ordinates and the points of a plane or space..." (LEVI-CIVITA, 1926).

Thus, a pivotal point of our tensor-argument is that in the CNS the same event-point (an invariant) that can be externally represented, e.g. by a Cartesian vector of M(x,y,z,t) may internally be represented in the CNS by another mathematical vector $F(f_1, f_2, \ldots, f_n)$, another ordered set of numbers, such as the firing frequencies of a set of N neurons. The tensor concept hinges on the fact that the two vectorial descriptions (M and F) are equally appropriate; i.e., neither is, a priori, preeminent. When using such vectors (as F) in a generalized sense, it is clear that one has both a mathematical instrument, an ordered set of quantities, (i.e. the mathematical vector) and a physical entity (the latter, by its nature, being invariant to the existence of any set of quantities that may be arbitrarily assigned to it). The co-ordination of an invariant with a mathematical vector is implemented by means of a reference-frame, through which the invariant and the mathematical vector become related entities. Given that an abstract relation exists between a single invariant and any mathematical vector that is assigned to it, it follows that if several different vectors are assigned to an invariant (e.g. by using different frames of reference) then all such vectors will also be related to one another. Thus, the abstraction that is implied in the generalized concept of vectors invokes a further abstraction: the generalized concept expressing their relation. THE GENERAL CONCEPTUAL DEFINITION OF A TENSOR: A MATHEMATICAL DEVICE THAT EXPRESSES THE RELATIONS AMONG MATHEMATICAL VECTORS BELONGING TO INVARIANT ENTITIES. The relation of the invariant itself to the mathematical vector is formally expressed by a tensor of rank one (the generalized concept of a "vector"), while the relation of two mathematical vectors, belonging to the same invariant, is expressed by a tensor of rank two (a generalization of the concept of "matrix"). Given that both mathematical vectors are assigned to the same physical entity, it follows that M and F are tensorially related to one another. This consideration provides a basis for a tensorial treatment of the internal representation of external invariants by the brain.

The exact mathematical qualities of any particular type of tensor depend both on the properties of the physical invariants in question as well as on the properties of the mathematical vectors attributed to them by particular kinds of reference frames. For example, if one deals with invariants such as a location in the physical space and assigns three-dimensional orthogonal frames of reference to them, then the relations among the resulting mathematical vectors will be expressed by so-called Cartesian tensors (of the second rank). Such Cartesian tensors are always, by definition, three-by-three matrices, with components that transform according to both covariant and contravariant rules upon rotating the frame of reference (c.f. TEMPLE, 1960). However, if one assigns to the same invariant both a covariant N-dimensional- and a contravariant M-dimensional mathematical vector (by using two different frames of reference, having N and M axes respectively, and using a different method of assigning the components in each), then the relation between these two mathematical vectors will be expressed by a tensor which has N x M components, transforming differently than those of the previous tensor, since the latter is obviously non-Cartesian. The lesson from this example is, that for an application in which fundamental features of the applied vectors are unknown, one has to be very careful not to define tensors overly strictly, just as vectors must not be limited to Cartesian vectors if one deals with a RIEMANNIAN space. For further reading of tensors in general unrestricted coordinate systems consult, e.g. SYNGE & SCHILD, 1949 or WREDE, 1972.

In defining a vector as mathematical point, a vector itself becomes the least interesting entity: an ordered set of quantities. More significantly, such a general definition clearly distinguishes the set of numbers from the physical invariant to which it was assigned. However, in the case of "brain vectors" it is not a trivial task to identify the particular physical invariant to which a particular vector in the CNS is assigned. The subsequent task may be even more difficult: to determine what frame of reference is implied in such a co-ordination of an invariant with a mathematical point. A further task is to establish whether the way of assigning the components to the invariant is a covariant- or contravariant procedure. These tasks of relating sensorimotor "brain vectors" to external invariants is probably the easiest in the so-called vestibulo-ocular reflex, ( c.f. PELLIONISZ & LLINAS, 1980 a). In such a primitive system it is evident that the brain-vectors at both ends of the reflex arc are directly related to an obvious physical invariant, the head-displacement and the (optimally) identical eye-displacement, respectively. Another extreme, when the exact system of relations of the internal brain vectors and external physical entities is rather unclear, may be represented by a linguistic sensorimotor transformation performed by the CNS: e.g. when an interpreter, whose acoustic sensory input vector is a Japanese word, transforms this to a corresponding vocal motor output vector, an English word, where both vectors are assigned to the same invariant.

General vectors necessitate even more profound considerations than the particular ways and means by which a mathematical vector is assigned to the invariant. These are THE QUESTIONS CONCERNING THE PROPERTIES OF THE N-DIMENSIONAL ABSTRACT MATHEMATICAL SPACE, (that which we call a "hyperspace" in order to distinguish it from the three-dimensional physical space) to which the particular mathematical point belongs. Thus, the basic question concerns the system of relations among the points of the abstract space: and indirectly, it raises the question of the type of geometry of the CNS hyperspace. This internal geometry is related to another geometry over the invariants themselves (the latter being the "net" of Euclidean geometry that can be used to "envelop" the locations in a physical space). Therefore, the process of assigning mathematical points (vectors) to invariants, by using a frame of reference, can be seen as invoking an abstract geometry that is related to the geometry of the invariants.

In this view, the intrinsic geometry of the CNS hyperspace (the multidimensional space over the points F) is an internal representation of the external geometry (the latter existing over the set of points M). Thus the adequate mathematical approach is that of related geometries: one in the four-dimensional PHYSICAL space externally (usually represented by Euclidean geometry), and the other a FUNCTIONAL geometry in the CNS hyperspace (a largely unknown, but certainly not Euclidean geometry). In addition to these two different geometries, the structure of the network (form and shape of neuronal elements and connectivities) represents yet another, a STRUCTURAL geometry. A fundamental problem in neurobiology is therefore to establish formally the relation of these three geometries; physical within the external world, functional inside the CNS and structural over the neuronal elements. In fact, even before the tensor concept had been applied to CNS function, it was already felt that the cerebellar structure represented a kind of functional geometry, acting as "a continuously modified mirror of the motor functional state" (LLINAS, 1974). By introducing tensor analysis we feel that a proper language was found by which the relation of geometries can be formally stated.

Some technical and terminological requirements of the tensorial approach can be summarized as follows. Technically, starting from a particular mathematical "brain vector", first the invariant to which the vector is assigned must be identified. Second, the means and ways of this coordination must be established. The frame of reference, by which the set of coordinates was assigned to an invariant, cannot be taken for granted; it must be explicitly revealed (e.g. SIMPSON et al., 1981). Next, since it is clear that there are two different ways of arriving at a pair of mathematical vectors belonging to the same invariant (in any case where a not-necessarily orthogonal vector is considered) A DISTINCTION WHETHER THE VECTOR IN QUESTION IS OF COVARIANT OR CONTRAVARIANT TYPE MUST ALWAYS BE EXPLICITLY MADE.

As far as terminology is concerned, in our tensor approach we intend to make a clear distinction between two separate classes and three categories instead of lumping them under an ill-defined name of "vector". One class is that of the physical entities, while another is their mathematical representation. The three categories used in our tensorial approach, therefore, are: INVARIANTS, COVARIANTS and CONTRAVARIANTS. Adapting such terminology would be a revival of the classic original concepts and terminology first introduced by J.J. SYLVESTER (early 19th century), and it would also reestablish a harmonious UNIFIED TREATMENT OF REALITY (INVARIANTS) TOGETHER WITH ITS ABSTRACTION (THE COVARIANTS AND CONTRAVARIANTS). Moreover, since tensor analysis thus becomes the theory of relations among invariants and the differently transforming covariants and contravariants, this terminology expresses the geometrical nature of tensor analysis in the basic sense that geometry is the theory of invariants in a transformation group (KLEIN, 1939).

### 3. TENSOR NETWORK THEORY OF THE CENTRAL NERVOUS SYSTEM

3.1. THE EMERGENCE OF THE TENSOR NETWORK THEORY OF THE CENTRAL NERVOUS SYSTEM. Our initial approach was based on the tensor idea, stating in its most rudimentary form (PELLIONISZ & LLINAS, 1978, 1979 a,b), that any network-matrix is a particular representation of a more general class: a tensor. Having read our completed initial manuscript (PELLIONISZ & LLINAS, 1979b) Dr. D. FINKELSTEIN was kind enough to bring to our attention the parallelism between our approach and that of KRON (1939), who developed a tensorial theory of electrical networks, based on the notion that electrical circuits can be interpreted as representations of a general tensor. This led us to the "grandfather" of KRON's tensor theory, Dr. Banesh HOFFMANN, who communicated our concept to KRON's followers (PELLIONISZ & LLINAS, 1979 a). It was also Dr. FINKELSTEIN who last year learned and brought to our attention the fact that KRON himself (although in total ignorance of the realities of neural systems) had attempted to conjecture an "artificial brain" in the form of a generalized rotating electrical machinery, which was different from our approach: "Two spatially-orthogonal polyhedra are immersed into the plasma to crystallize the amorphous field into a large number of N-dimensional magnetohydrodynamic generators (or 'generalized' rotating electrical machines) connected to 2N multidimensional transmission networks. The generators play the role of 'neurons' of the projected 'artificial brain'." (KRON, 1960).

3.3. A CONCISE EXAMPLE OF THE TENSOR APPROACH YIELDLNG A NETWORK THEORY. Tensor Network Theory is contrasted in Fig. 2. with the traditional view of representing the structural geometry of networks: The tensor concept is aimed at describing and understanding their functional geometry.

TENSOR NETWORK THEORY: LEVELS OF ABSTRACTION DIRECTED TOWARDS FUNCTIONAL GEOMETRY

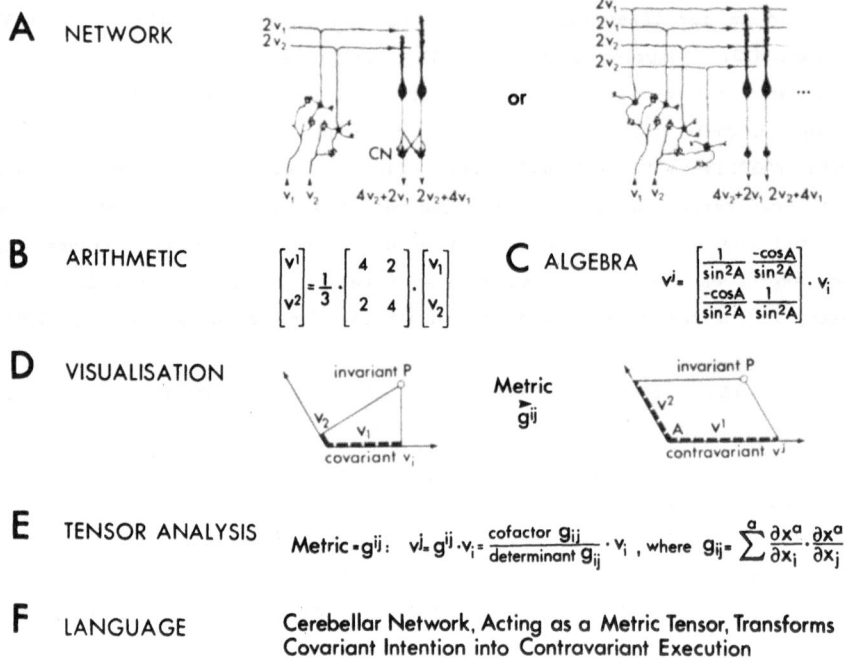

FIG.2. Networks, vectors, matrices and tensors. A concise demonstration of how a set of neuronal networks (A) can perform a transformation to be described by a general tensor, such as the metric (F). The arithmetical (B) or algebraical (C) vector-matrix approaches represent intermediate levels of abstraction from the level of network representations (A) to the full generality of coordinate system free description by tensor analysis (E,F); from PELLIONISZ & LLINAS, 1982.

Tensor theory considers the network as a parallel system with multiple inputs and outputs. Fig. 2A presents networks similar to the cerebellar neuronal circuits drawn by RAMON Y CAJAL (1911). The depicted two networks (and an infinite number of other networks) may perform the same numerical transformation from a multiple input to a multiple output. Mathematically such transformation can be represented by a matrix. The matrix shown in Fig. 2B is an arithmetical abstraction of the particular networks shown in Fig. 2A. Expressed by the arithmetic formula of a matrix, the function of a network is to transform an input vector into an output vector, both given in a particular frame of reference.

At the higher level of abstraction (corresponding to the vector-matrix applications mentioned above), such transformations can be expressed in a manner which applies to ALL two-dimensional, rectilinear, non-orthogonal, oblique frames of reference with any angle A. Such a multilinear algebraic expression is shown in Fig. 2C. As pointed out above, general (non-orthogonal) frames of reference are used throughout Tensor Network Theory, since the CNS is not limited to orthogonal coordinate systems. In such algebraic expressions as Fig. 2C the matrix does not reveal the functional advantage gained from the vectorial transformation it implements. The functional role becomes evident by a visualization of the vector transformation (Fig. 2D). Here it is intuitively clear that the input and output of the matrix transformation (using any of the possible network-variations shown e.g. in Fig. 2A) are the TWO DIFFERENT KINDS OF VECTORIAL EXPRESSIONS OF THE SAME PHYSICAL ENTITY P.

The components $v_i$ of the input vector are covariant (they are obtained by the orthogonal projection method) while the components $v^j$ of the output vector are contravariant (obtained by the parallelogram method). Fig. 2D illustrates the significance of defining the covariant or contravariant nature of a vector when using a non-orthogonal coordinate system. It is apparent that the transformation from $v_i$ to $v^j$ in Fig. 2D expresses a general relationship that exists not only for the depicted $v_i$ and $v^j$ vectors, but for all co- and contravariant pairs of vectors. Such a relation exists in every system of coordinates, regardless of the directions or number of axes in the coordinate system.

The general covariant-contravariant vectorial relationship may be expressed in a totally coordinate-system free manner by the mathematical device $g^{ij}$, the so-called metric tensor. The metric transforms a covariant vector into its contravariant counterpart. This operation is not restricted to the two dimensional space illustrated in Fig. 2D, but applies to any N-dimensional hyperspace. In fact, two-dimensional Euclidean geometry is only utilized in our theory to visualize vectorial relations that can exist in multidimensional, non-Euclidean CNS hyperspaces, where non-rectilinear (or even non-linear) non-orthogonal reference frames may be used.

The contravariant metric is formally expressed with the use of the notation of tensor analysis in Fig 2E. This expression, while not intuitively obvious, is immensely powerful because it is devoid of the inherent limitations incurred by the use of particular reference frames. The metric tensor, as given in general (coordinate system-free) tensor notation: $g^{ij}$ encompasses all expressions in any possible frame of reference. It is the ultimate expression for the class of all neuronal networks implementing such transformation.

As far as NOTATION is concerned, some remarks are necessary here. The indexing system of vectors and matrices can be selected either to conform with the one used in multilinear algebra (where upper and lower indices refer to the row or column-elements, respectively), or to conform with the convention used in tensor analysis, where upper or lower indices denote contravariant or covariant tensors, respectively. Both indexing systems are of great usefulness, the former in the particular arithmetic handling of vector-matrix operations, the latter in general conceptual analysis. It is an unfortunate, but easily demonstrable fact, however, that the two conventions cannot be fully consistent with each other. Following KRON, we have previously used both conventions (the first one, suitable to computational details, mostly in PELLIONISZ & LLINAS, 1979, the latter for generalized coordinates e.g. in PELLIONISZ & LLINAS, 1980 b). For the future we suggest the usage of the general tensor notation (EINSTEIN convention), where the upper and lower indices denote contravariant and covariant entities respectively, and in which a summation is implied whenever the same index occurs twice, once as a superscript and once as a subscript.

The tensorial approach can also be put into "plain English", as in Fig. 3F. Unfamiliar as it may sound at first, the concise definition that the cerebellum acts as a metric tensor of the motor hyperspace is an exact mathematical statement. As an additional advantage provided by the concept and formalism of tensor analysis capable of grasping both the particular and the general, this abstract expression can be made concrete, when necessary, at any of the various levels of abstraction.

3.3. THE TENSOR THEORY OF THE CEREBELLUM: THE CEREBELLAR NEURONAL NETWORK ACTING AS A SPACE-TIME METRIC TENSOR. By relating the invariants to their different kinds of vectorial representations in the CNS, and by treating neuronal networks in their particular structure as well as in their general function, tensor network theory provides a scheme that is capable of explaining how motor coordination emerges from the activities of cerebellar networks. Fundamentally, MOTOR COORDINATION IS DEFINED AS THE TRANSFORMATION OF MOTOR INTENTION INTO MOTOR EXECUTION. This is a necessary operation since intention, as a general rule, cannot be used directly to achieve the desired goal.

Mathematically, motor coordination is a transformation of a covariant vector into its contravariant counterpart, where both CNS vectors express the same physical invariant (the motor displacement of the part of the body in question). Tensorially, this transformation is achieved by the contravariant metric tensor of the CNS motor hyperspace.  It was suggested that this tensorial function is implemented by the cerebellar neuronal connectivity-matrix.

Such a concise abstract definition of motor coordination is a departure from traditional verbalizations of this well-known concept.  However, such tensorial description may be considered the only unambiguous, short and formal definition of motor coordination even in a volume totally devoted to this subject (c.f. LLINAS & SIMPSON, 1981).  The functional properties of covariant intention- and contravariant execution vectors are demonstrated in Fig. 3.

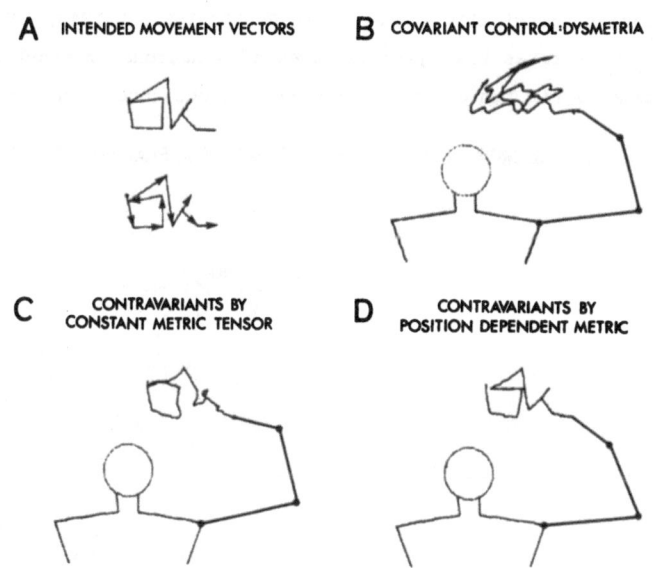

**A** INTENDED MOVEMENT VECTORS   **B** COVARIANT CONTROL:DYSMETRIA

**C** CONTRAVARIANTS BY CONSTANT METRIC TENSOR   **D** CONTRAVARIANTS BY POSITION DEPENDENT METRIC

FIG. 3.   Computer simulation of the tensor-paradigm of cerebellar coordination: covariant-contravariant transformation of motor intention vectors into motor execution vectors via a cerebellar metric tensor. (from PELLIONISZ & LLINAS, 1980 a)

A more detailed explanation of motor coordination must expand the basic conceptual scheme in two directions.  One is the actual elaboration of the neuronal networks involved in such cerebellar operations, the other is a fundamental reconsideration of cerebellar functioning as concerning not only space but time functions (c.f. ITO, 1981).

Indeed, the latter question, while being of great interest over two decades (BRAITENBERG & ONESTO, 1961), has hitherto remained unresolved.  In our view, the reason for this has been the deep-seated dogma that CNS functioning utilizes a NEWTONIAN frame of reference, a separable space-time representation. In physical sciences it has long been accepted that such a frame can only be used when a practically instantaneous synchronizing signal (e.g. light) is available to establish simultaneity (c.f. MINKOWSKI, 1908, WEYL, 1952, or the general overview of this subject by REICHENBACH, 1958). We have pointed out (PELLIONISZ & LLINAS, 1980 b and 1982) that CNS functioning utilizes an inseparable representation of the space-time continuum. Tensor theory is eminently capable of an exact abstract description of such space-time processes; basically stating that THE CEREBELLUM ACTS NOT JUST AS A SPACE METRIC, BUT AS A SPACE-TIME METRIC TENSOR OF THE INTERNAL CNS HYPERSPACE. This concept can be demonstrated to conform with the existing neuronal circuitries of the cerebellum (PELLIONISZ & LLINAS, 1982) relying on the functioning of arrangements of Purkinje cells that act as temporal predictor-modules (originally suggested in PELLIONISZ & LLINAS, 1978 and 1979).

Fig. 4. demonstrates the basic features of a neuronal network capable of acting as a space-time metric tensor, transforming an input intention vector (composed of

### CEREBELLAR NEURONAL NETWORK ACTING AS A SPACE-TIME METRIC TENSOR

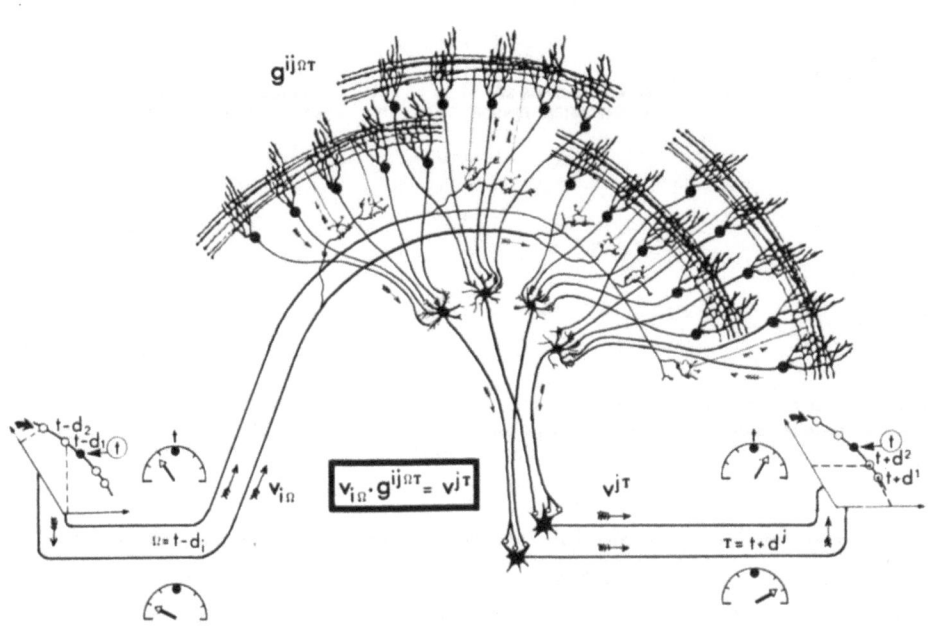

FIG. 4. Circuitry layout of the cerebellar neuronal network acting as a space-time metric tensor. The transformation through the network converts an asynchronous covariant motor intention vector into an asynchronous contravariant motor execution vector. "Stacks" of Purkinje cells serve as temporal lookahead-modules. (For a detailed explanation see PELLIONISZ & LLINAS, 1982)

asynchronous, temporally lagging covariant components) into an output execution vector (composed of asynchronous, temporally leading contravariant components). While the climbing fibers are not shown in this simplified scheme, their action is congruent with the tensor model, being interpreted as altering the metric (changing the curvature of the CNS hyperspace, PELLIONISZ & LLINAS 1980 a). This function is required since the physical geometry of the musculoskeletal system invokes an internal geometry of the CNS hyperspace in which the metric is not constant, but is position-dependent. Thus, the short-lasting alteration, via the climbing fibers, of Purkinje cell firing properties (found by ITO, 1981) is interpreted here not as a learning phenomenon, but as an active interference with the curvature of the motor hyperspace.

The above scheme (for a thorough explanation see PELLIONISZ & LLINAS, 1982) is expandable, both conceptually and in its network-details, to CNS functions much beyond cerebellar operations. Basically, Fig. 4. already suggests, that whenever the intention is expressed in a DIFFERENT frame of reference from that of the execution, then further vectorial transformations are necessitated.

3.4. <u>TENSORIAL SCHEME FOR AN EXPANDED SENSORIMOTOR SYSTEM.</u> In a general case, a sensorimotor action involves a different frame of reference in the sensory- and in the motor systems. Since both such frames are most probably non-orthogonal, there are two different vectorial expressions in each frame. Thus, a full system typically contains four different vectorial expressions (belonging to the same external physical invariant), where all expressions correspond to distinct neurological functions. These are the sensory RECEPTION and PERCEPTION, and the motor INTENTION and EXECUTION vectors.

The scheme of Fig.5. shows these basic vectorial expressions involved in a sensorimotor CNS system, commencing with a lower dimensional non-orthogonal sensory frame in which the primary sensory reception information is derived covariantly from the invariant. At the motor end, the same physical invariant emerges as the physical sum of the component-vectors of the contravariant expression, given in a higher dimensional different (motor) frame of reference. To make the mathematically non-unique transformation from e.g. a two-dimensional sensory system to a three-dimensional motor frame, the sensorimotor transformation utilizes a covariant embedding procedure. This starts from a contravariant-type sensory perception vector and yields a covariant-type motor intention vector. (In this operation, a further covariant-type proprioception vector is required.) Thus, such a full sensorimotor system contains alternating covariant-contravariant vectorial expressions made possible by containing both sensory- and motor metrics. Since both the input and output vectors are asynchronous, such metrics operate in space-time domain; i.e. they must contain temporal lookahead-modules.

# SENSORIMOTOR TENSOR NETWORK WITH DIFFERENT SENSORY AND MOTOR FRAMES

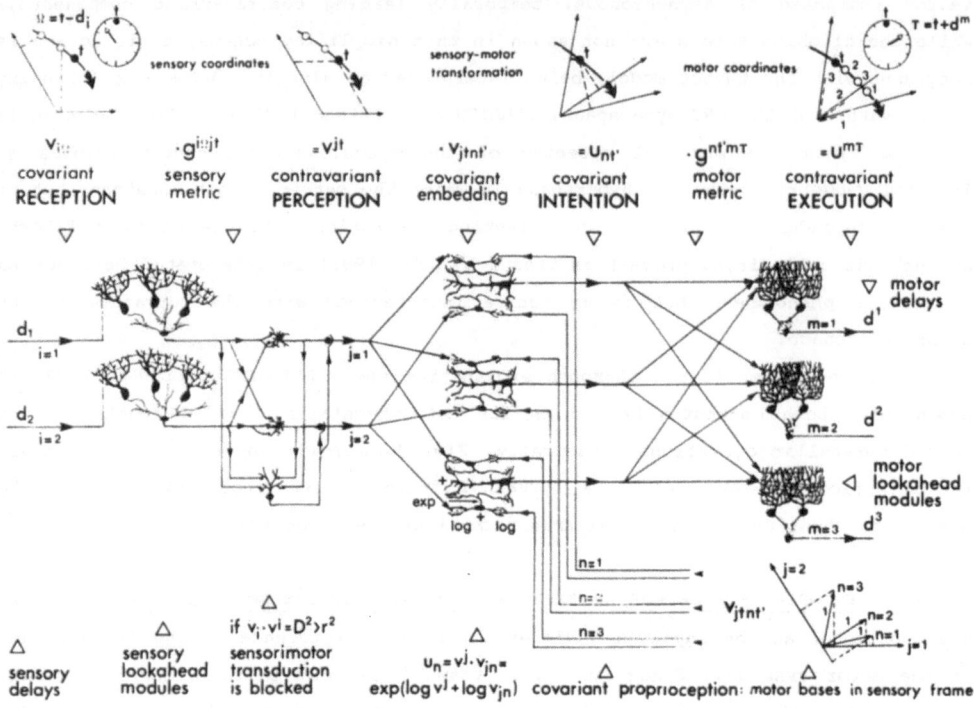

FIG. 5. Tensorial scheme of a sensorimotor system. Two different (sensory and motor) frames of reference are used with two different vectorial expression in each. Thus, the transformation from asynchronous covariant (sensory reception) vector to asynchronous contravariant (motor execution) vector is implemented through a contravariant sensory vector (perception) and a covariant motor vector (motor intention). At the levels of visualization, verbal description, network representation and tensor notation those minimum necessary transformations are shown, necessary for a two-to-three dimensional system. (PELLIONISZ & LLINAS, 1982)

3.5. <u>TENSORIAL SCHEMES FOR GEOMETRIC BRAIN-LIKE MACHINES.</u> The circuitry schematics of Fig. 5. can be presented in a circuitry layout that conforms to existing neuroanatomical realities of the CNS in lower species (see Fig. 6.)

This tensorial "blueprint" for an amphibian brain is an obvious forerunner of a more elaborate set of schemes for the CNS function, providing abstract geometrical explanations of CNS operations beyond sensory and motor tasks. Since such schemes can be implemented by appropriate software and/or hardware techniques, it is expected that a new class of geometric brain-like devices will emerge from such studies. Here, however, only a few characteristic features of such tensorial schemes are pointed out.

## TENSORIAL SCHEME OF A SENSORIMOTOR NEURONAL NETWORK

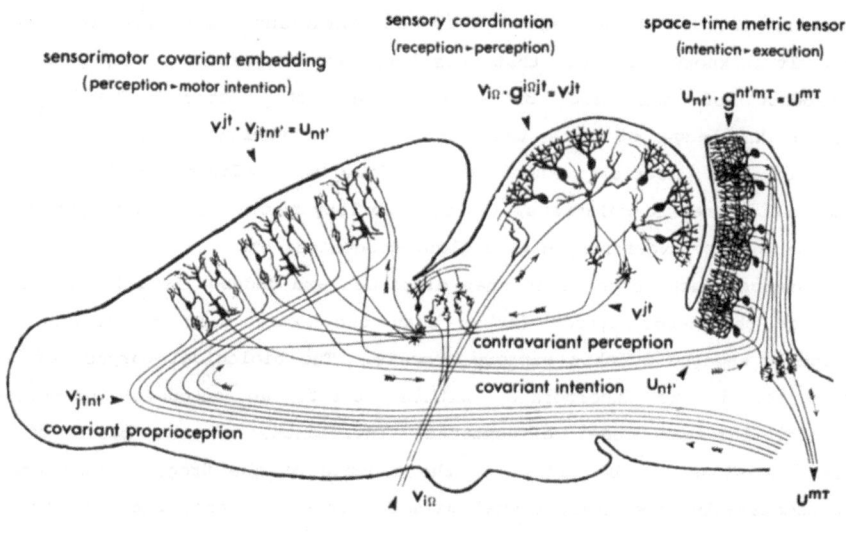

sensory coordination
(reception → perception)

sensorimotor covariant embedding
(perception → motor intention)

space-time metric tensor
(intention → execution)

$$v_{i\Omega} \cdot g^{i\Omega jt} = v^{jt}$$

$$v^{jt} \cdot v_{jtnt'} = u_{nt'}$$

$$u_{nt'} \cdot g^{nt'mT} = u^{mT}$$

contravariant perception

$v^{jt}$

covariant intention $u_{nt'}$

$v_{jtnt'}$ ►

covariant proprioception

$v_{i\Omega}$

$u^{mT}$

covariant sensory reception            contravariant motor execution

FIG. 6. Tensorial "blueprint" of the amphibian brain. The neuronal networks shown implement tensor transformations from a two-dimensional covariant sensory reception vector into a higher, three-dimensional contravariant motor execution vector. The cerebellum is featured as the motor metric, the superior colliculus as the sensory metric, while the cortical network is a tensor-transformer implementing a sensorimotor covariant embedding. (from PELLIONISZ & LLINAS, 1982)

Interestingly, the covariant-contravariant scheme of a covariant embedding followed by a metric transformation reconceptualizes the important notion (NEUMANN, 1956) of functional reliability of neuronal networks achieved by means of structural redundancy (for a more detailed treatment of this question, see PELLIONISZ & LLINAS, 1982). The fact that covariant components may be established independently from each other together with the overcompleteness of the embedding hyperspace, yield an increased reliability of the contravariant expression. While conceptually different, an overcomplete tensorial system resembles the classical scheme of redundant organizations in the sense that in both there is a trade-off between reliability and structural economy.

Further, regarding the schemes of Figs. 5. & 6. it is worth pointing out that to have separate sensory and motor space-time metrics may be a biologically desirable goal in itself, not just a necessity because of the differences in reference frames. An important consideration is that any subsystem of the brain that has to reach realistic conclusions about the external invariant itself (as opposed to only dealing with the coordinate-REPRESENTATION of the invariant) must incorporate a separate metric. This is because neither a covariant nor a contravariant vector alone can express an invariant such as, for example, distance. It is known, however, that this invariant can easily be expressed by the inner product of the two different types of vectors: $D^2 = v^i \cdot v_i$. The simultaneous knowledge of both the co- and contravariant vectors implies the knowledge of the metric, since $v^i \cdot v_i = (g^{ij} \cdot v_j) v_i$. Given that the metric summarizes all geometric relations, its availability permits realistic judgments to be made about other invariants, e.g. angles, geodesics, etc.

For instance, such is the case when distances are judged visually, without the necessity of the motor system (other than the extraocular musculature) being simultaneously engaged in this sensory process. The biological purpose of such a sensory system is to incorporate, within itself, an adequate (geometrically homeomorphic) internal model of external invariants, on the basis of which behavioral decisions can be made. In the example of the frog, a jump does not follow automatically from every visual message; it occurs only when the the fly's location is within range. Thus, the CNS must judge that this physical invariant is within the range of possible jumps. As pointed out in PELLIONISZ & LLINAS 1982, the function of the superior colliculus (optic tectum) is interpreted as a sensory metric, conceptually similar to the motor metric of the cerebellum (PELLIONISZ & LLINAS 1980 a). Thus, a possibility seems to emerge that -at least from some basic conceptual points of view- separate cerebellar and tectal models could be merged into a consistent sensorimotor model. Such an interrelation of tectal function with other parts of CNS has also been considered by LARA & ARBIB (in press).

The schemes of Figs. 5. and 6. show that having available both the covariant reception vector and contravariant perception vector, a simple pre-tectal network (that takes the inner product of the covariant and contravariant sensory vectors) is capable of reaching geometrical judgements relating to the external invariant itself and thus either blocking or permitting a sensorimotor transduction. This example serves as a reminder that higher CNS actions (e.g. pattern recognition, association, etc.) can be conceptualized as requiring such geometrical decision-making processes; this can establish, e.g., "how closely" certain objects resemble each other in the external world, or, for that matter, at "what distance" one IDEA (a point in a CNS hyperspace) lies from another. While it is reassuring that, given an internal metric in the CNS hyperspace that embeds external reality, such geometrical decisions can be very easily implemented by neuronal networks, it is yet to be determined how such CNS metrics could arise and how the CNS could learn to use them (c.f. PELLIONISZ & LLINAS, 1981).

Much further work is needed to achieve the goals outlined in this paper, both in experimental neuroscience and brain theory. We believe, however, that the concepts presented here can begin to provide a framework within which the two fields of research can eventually be unified, leading to results of use to other fields as well. These efforts could generate the impetus for development of significant novel mathematical devices as well as provide a source for usable control paradigms for new types of intelligent machines.

ACKNOWLEDGEMENT: This research was supported by a USPHS Grant NS-13742 from the NINCDS. We thank Dr. J.I. SIMPSON for reading the manuscript and for his valuable criticism.

## REFERENCES

ALBUS, J. (1979) A model of the brain for robot control. I. _Byte_, June, pp. 10-34.

AMARI, S. & ARBIB, M.A. (1977) Competition and cooperation in neural nets. In: _Systems Neurosciences_ (Metzler, J. ed.), pp. 119-165. Academic Press, New York.

ARBIB, M. (1972) _The Metaphorical Brain_. Wiley, New York

ANDERSON, J. A. & HINTON, G. E. (1981) Models of information processing in the brain. In: _Parallel Models of Associative Memory_ (Hinton, G.E. & Anderson, J.A. eds), pp. 9-44. Lawrence Erlbaum Assoc., New Jersey.

ANDERSON, J. A., COOPER, L., NASS, M., FREIBERGER, W., & GRENANDER, U. (1972). _AAAS Symposium, Theoretical Biology and Bio-mathematics_. AAAS, Washington, D.C.

BERNSTEIN, N. A (1947) _O Postroyenii Dvizheniy (On the Construction of Movements)_ Moscow, Medgiz

BEURLE, R. L. (1962) Functional organization in random networks. In: _Principles of Self-Organization_ (Foerster, H. von & Zopf, G. W. eds), pp. 291. Pergamon, N.Y.

BRAITENBERG, V. & ONESTO, N. (1961) The cerebellar cortex as a timing organ. Discussion of an hypothesis. _Proc. 1st Int. Conf. Med. Cybernet._, pp. 1-19. Giannini, Naples.

BIZZI, E. (1981) Visuomotor control as a computational problem. In: _Theoretical Approaches in Neurobiology_ (Reichardt, W.E. and Poggio, T. eds), pp.177-184. MIT Press, Cambridge

BOYLLS, C. C. (1981) Synopsis given in LLINAS, 1981

COOPER, L. N. (1974) A possible organization of animal memory and learning. In: _Proceedings of Nobel Symposium on Collective Properties of Physical Systems_ (Lundquist, B. & Lundquist, S. eds), pp. 252-264. Academic Press, New York.

COWAN, J. (1981) The role of development in the specification of the afferent geometry of the primate visual cortex. In: _Theoretical Approaches in Neurobiology_ (Reichardt, W.E. and Poggio, T. eds), pp.116-133. MIT Press, Cambridge.

CROWE, M. J. (1967) _A History of Vector Analysis_, Univ. of Notre Dame Press, London.

ECCLES, J. C. (1981) The modular operation of the cerebral neocortex considered as the material basis of mental events. _Neuroscience_ 6, 1839-1856.

FARLEY, B. G. & CLARKE, W. A. (1961) Activity in networks in neuronlike elements. _Information Theory (Fourth London Symposium)_ (Cherry, C. ed), Butterworth Scientific Publications, London.

FINKELSTEIN, D. Personal communication.

FOERSTER, H. von (1967) Computation in neural nets. In: _Currents in Modern Biology_, vol. 1, pp. 47-93. North-Holland Publishing Co., Amsterdam.

GREENE, P. H. (1972) Problems of organization of motor system. In: _Progress in Theoretical Biology_ (Rosen, R. & Snell, F.M. eds), vol 2, pp. 303-338. Academic Press, New York

HARTH, E. & TZANAKOU, E. (1974) Alopex: A stochastic method for determining visual receptive fields. _Vision Res._ 14, 1475-1482.

HOLDEN, A. V. (1976) _Models of the Stochastic Activity of Neurones._ Springer Verlag, Berlin-Heidelberg-New York.

GROSSBERG, S. (1970) Neural pattern discrimination. J. Theor. Biol. 27, 291-337.

HINTON, G. E. & ANDERSON J. A. (eds) (1981) Parallel Models of Associative Memory: Lawrence Erlbaum Assoc., Hillsdale, NJ.

HOFFMANN, B. (1966) About Vectors. Dover, New York.

ITO, M. (1981) Experimental tests of constructive models of the cerebellum. In: Neural Communication and Control, Advances in Physiological Sciences, Vol 30 (Szekely, Gy., Labos, E., & Damjanovich, S. eds), pp.155-261. Pergamon Press/Akademiai Kiado.

KATCHALSKY, A. K., ROWLAND, V. & BLUMENTHAL, R. (1974) Dynamic patterns of brain cell assemblies. Neurosci. Res. Prog. Bull. 12, 1-187.

KLEIN, F. (1939) Elementary Mathematics form an Advanced Viewpoint - Geometry. Dover, New York.

KOHONEN, T, OJA, E. & LEHTIO, P. (1981) Storage and processing of information in distributed associative memory systems. In: Parallel Models of Associative Memory (Hinton, G.E. & Anderson, J.A. eds), pp.105-141. Lawrence Erlbaum Associates, Hillsdale, NJ.

KRON, G. (1939) Tensor Analysis of Networks. John Wiley, London.

KRON, G. (1960) Building-blocks of self-organizing automata-I: Outline of a "dynamo-type "artificial brain". General Electric Technical Information Series, 60GL164, pp. 1-24. Schenectady, NY.

LARA, R. & ARBIB, M. A. (in press) A neural model of interaction between pretectum and tectum in prey selection. Cognition & Brain Theory.

LEGENDY, C. (1978) Cortical columns and the tendency of neighboring neurons to act similarly. Brain Res. 158, 89-107.

LEVI-CIVITA, T. (1926) The Absolute Differential Calculus (Calculus of Tensors) (Persico, E., ed). Dover, NY.

LLINAS, R. (1971) Frog cerebellum: Biological basis for a computer model. Math. Biosci. 11: 137-151.

LLINAS, R. (1974) Eighteenth Bowditch Lecture: Motor aspects of cerebellar control. Physiologist 17, 19-46.

LLINAS, R. (1981) Microphysiology of the cerebellum. Chapter 17 in: Handbook of Physiology, vol II: The Nervous System. Part II. (Brooks, V. B. ed), pp. 831-976. Amer. Physiol. Soc., Bethesda, MD.

LLINAS, R. R. & SIMPSON, J. I. (1981) Cerebellar control of movement. In: Motor Coordination, Handbook of Behavioral Neurobiology, Vol.5. (Towe, A.L. & Luschei, E.S. eds), pp. 231-302. Plenum Press, NY.

MALSBURG, C., von der & WILLSHAW, D. J. (1976) A mechanism for producing continuous neural mappings: Ocularity dominance stripes and ordered retinotectal projections. Exp. Brain Res. Suppl. 1, 453-469.

McCULLOCH, W. S. & PITTS, W. (1943) A logical calculus of the ideas imminent in nervous activity. Bull. Math. Biophys. 5, 115-133.

MELNECHUK, T. (1979) Network notes. Trends in NeuroSciences, April. pp. 6-7.

MINKOWSKI, H. (1908) Space and time. In: The Principle of Relativity, by Lorentz H.A., Einstein, A., Minkowski, A. and Weyl, H., pp.75-91. Dover, Toronto.

MINSKY, M. & PAPERT, S. (1969) Perceptrons. MIT Press, Cambridge, MA.

NEUMANN, J. von (1957) The computer and the brain. Yale Univ. Press, New Haven, CT

NEUMANN, J. von (1956) Probabilistic logics and the synthesis of reliable organisms from unreliable components. In Automata Studies (Shannon, C.E. & McCarthy J. eds), pp. 43-98. Princeton Univ. Press, Princeton, NJ.

PELLIONISZ, A. (1970) Computer simulation of the pattern transfer of large cerebellar neuronal fields. Acta biochim. biophys. Acad. Sci. Hung. 5, 71-79.

PELLIONISZ, A. & LLINAS, R. (1978) A formal theory for cerebellar function: The predictive distributed property of the cortico-nuclear cerebellar system as described by tensor network theory and computer simulation. Soc. Neuroscience Abst 4, 68.

PELLIONISZ, A. & LLINAS, R. (1979 a) A note on a general approach to the problem of distributed brain function. Matrix and Tensor Quarterly of the Tensor Society of Great Britain, pp.48-51.

PELLIONISZ, A. & LLINAS, R. (1979 b)  Brain modeling by tensor network theory and computer simulation. The cerebellum: Distributed processor for predictive coordination. Neuroscience 4, 323-348.

PELLIONISZ, A. & LLINAS, R. (1980 a)  Tensorial approach to the geometry of brain function: Cerebellar coordination via a metric tensor. Neuroscience 5, 1125-1136.

PELLIONISZ, A. & LLINAS, R. (1980 b)  Tensorial representation of space-time in CNS: Sensory-motor coordination via distributed cerebellar space-time metric. Soc. Neuroscience Abst 6, 510.

PELLIONISZ, A. & LLINAS, R. (1981)  Genesis and modification of the geometry of CNS hyperspace.  Cerebellar space-time metric tensor and "motor learning".  Soc. Neuroscience Abst 7, 641.

PELLIONISZ, A. & LLINAS, R. (1982)  Space-time representation in the brain. The cerebellum as a predictive space-time metric tensor. Neuroscience (submitted).

PELLIONISZ, A., LLINAS, R. & PERKEL, D.H. (1977) A computer model of the cerebellar cortex of the frog. Neuroscience 2, 19-36.

PITTS, W. H & McCULLOCH, W.S. (1947)  How we know universals: The perception of auditory and visual forms. Bull. Math. Biophys. 9, 127-147.

POGGIO, T. & REICHARDT, W. E. (1981)  Characterization of nonlinear interactions in the fly's visual system. In: Theoretical Approaches in Neurobiology (Reichardt, W.E. & Poggio, T. eds), pp.64-84. MIT Press, Cambridge, MA.

RAMON Y CAJAL, S. (1911)  Histologie du Systeme Nerveux de l'Homme et des Vertebres, vol. 1-2. Maloine, Paris.

RAYBERT, M. H. (1977)  Analytical equations vs table lookup.  In: IEEE Proc. on Decision and Control, New Orleans, 566-569.

RAYBERT, M. H. (1978)  A model for sensorimotor control in learning.  Biol. Cybern. 29, 29-36.

REICHARDT, W. E. & POGGIO, T. (eds) (1981)  Theoretical Approaches in Neurobiology.  MIT Press, Cambridge

REICHENBACH, H. (1958)  The Philosophy of Space & Time.  Dover, New York.

ROBINSON, D. A. (1968)  The oculomotor control system: A review.  Proc. IEEE 56, 1032-1049.

ROSENBLATT, F. (1962)  Principles of Neurodynamics. Spartan Books, N.Y.

SCHWARTZ, E.L. (1977)  Afferent geometry in the primate visual cortex and the generation of neuronal trigger features. Biol. Cybern. 28, pp.1-14.

SHERRINGTON, C. (1906)  The Integrative Action of the Nervous System. Scribner, New York.

SIMPSON, J. I., GRAF, W. & LEONARD, C. (1981) The coordinate system of visual climbing fibers to the flocculus. In: Progress in Oculomotor Research (Fuchs, A. & Becker, eds), pp 475-484. Elsevier/North Holland, Amsterdam.

STEINBUCH, K. (1961) Die Lernmatrix. Kybernetik 1, 36-45.

SYNGE, J. L. & SCHILD, A. (1949)  Tensor Calculus. Dover, New York.

SZENTAGOTHAI, J. (1968)  Structuro-functional Considerations of the Cerebellar Neuronal Network.  Proc. IEEE 56 (6), 960-968.

SZENTAGOTHAI, J. & ARBIB, M.A. (1975)  Conceptual Models of Neural Organization. MIT Press, Cambridge, MA.

SZEKELY, GY., LABOS, E. & DAMJANOVICH, S. (eds) (1981)  Neural Communication and Control, Advances in Physiological Sciences, Vol. 30.  Pergamon/Akademiai Kiado.

TEMPLE, G. (1960)  Cartesian Tensors. John Wiley, New York.

WEYL, R. (1952)  Space-Time-Matter. Dover, New York.

WIENER, N. (1948)  Cybernetics. MIT Press, Cambridge, MA.

WREDE, R. C. (1972)  Introduction to Vector and Tensor Analysis. Dover, New York.

***

# CHAPTER 24

MECHANISMS OF MOTOR LEARNING

M. Ito
Department of Physiology, Faculty of Medicine,
University of Tokyo, Bunkyoku, Tokyo 113/Japan

INTRODUCTION

Since classic works by Flourens (18-2) and Luciani (1891), it has generally been assumed that the cerebellum has capabilities of learning. A decade ago, theoretical exploration of neuronal network structures of the cerebellum pointed to the possibility that the learning is based on self-organization of the cerebellar cortical network which is effected through a certain type of synaptic plasticity (Marr, 1969; Albus, 1971). At that time, there was no evidence for the postulated synaptic plasticity, and experimental verification of this has become a central problem of cerebellar physiology. The question how the neuronal network of the cerebellum reorganizes itself is supplemented by another question how a piece of the cerebellum is incorporated in control of diverse bodily functions to form a learning control system. Recently, cerebellar physiology has provided several concrete examples of adaptive phenomena representing a simple form of learning such as adaptive modification of the vestibulo-ocular reflex under circumstances of vestibular-visual interaction. This presentation outlines the outcome of recent investigations in my laboratory of rabbit's cerebello-vestibulo-ocular system and examines how the experimental data support the Marr-Albus model of the cerebellum.

MARR-ALBUS MODEL OF THE CEREBELLUM

The cerebellar cortex receives two distinctively different inputs, mossy fibers and climbing fibers. Mossy fibers make excitatory synaptic contact with granule cells of the cerebellar cortex whose axons (parallel fibers) in turn supply excitatory synapses onto Purkinje cells, the output cells of the cerebellar cortex, while climbing fibers make direct excitatory contact with dendrites of Purkinje cells (Fig. 1). Marr (1969) postulated that mossy fibers provide the major inputs to the cerebellar cortex which are eventually converted to output signals of Purkinje cells, while climbing fibers convey "instruction" signals which have an action of reorganizing the relationship between mossy fiber inputs and Purkinje cell outputs (Fig. 1). This arrangement enables us to reconstruct a learning machine like a Simple Perceptron (Albus, 1971). Marr-Albus model of the cerebellum is based on the plasticity assumption that the transmission efficacy of the parallel fiber synapse mediating mossy fiber signals to a Purkinje cell is modified when impulses of this parallel fiber attain at the Purkinje cell dendrite with impulses of a climbing fiber (Fig. 1). Marr (1969) suggested that this heterosynaptic interaction from a climbing fiber to a parallel fiber enhances the transmission efficacy, but albus (1971) preferred the opposite, i.e., a depression instead of enhancement for some practical reasons.

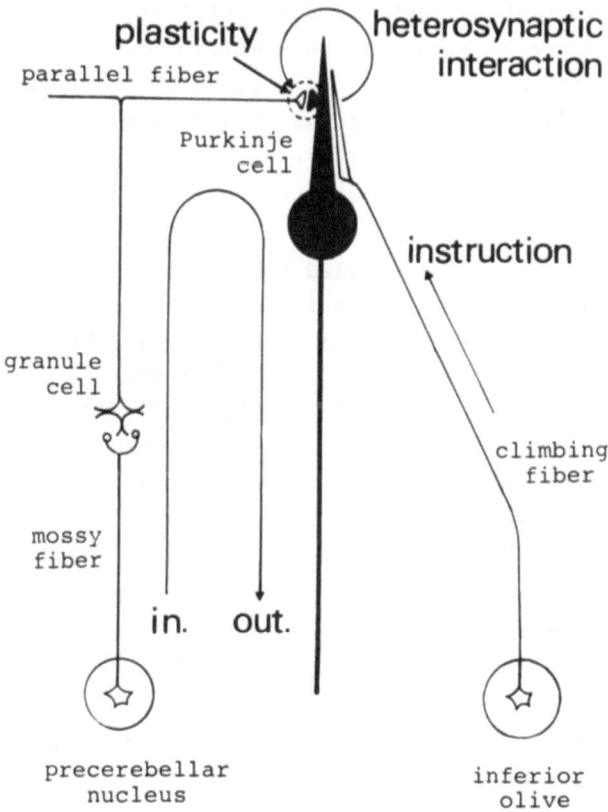

Fig. 1 Basic assumption in Marr-Albus model of the cerebellum.

EXPERIMENTAL EVIDENCE FOR MARR-ALBUS' PLATICITY ASSUMPTION

Support for Marr-Albus' plasticity assumption was derived from observations of Purkinje cell responses under two experimental situations where the cerebellar cortical network is supposed to undergo a reorganization: 1) in cerebellar lobules VI and VII during adaptive alteration of monkey's hand movements to a suddenly imposed load change (Gilbert and Thach, 1977), and 2) in the cerebellar flocculus during adaptive modification of rabbit's vestibulo-ocular reflex to sustained vestibular-visual interaction (Ito, 1977; Dufossé, Ito, Jastreboff and Miyashita, 1978). However, these lines of evidence provide no more than an indirect support, as responses of Purkinje cells to natural stimuli in alert animals involve a number of complicating factors.

A more direct test of Marr-Albus's plasticity assumption has recently been performed by substituting electric pulse stimulation of a vestibular nerve and the inferior olive in decerebrate rabbits for natural vestibular and visual stimulation in alert rabbits (Ito, Sakurai and Tongroach, 1981, 1982). In this experiment, Purkinje cells were sampled from the rostral flocculus and identified by their characteristic responses to stimulation of the contralateral inferior olive. Basket cells were also sampled and identified by absence of olivary responses and also by their location in the molecular layer adjacent to identified Purkinje cells. Single pulse stimulation of a vestibular nerve, either ipsilateral or contralateral, at a rate of 2/sec excited Purkinje cells with a latency of 3-6 msec. This early excitation represents activation through vestibular mossy fibers, granule cells and their axons (Fig. 2). Similar excitation was seen commonly among putative basket cells.

Conjunctive stimulation of a vestibular nerve at 20/sec and the inferior olive at 4/sec, for 25 sec per trial, effectively depressed the early excitation of Purkinje cells by that nerve, without an associated change in spontaneous discharge. The depression recovered in about ten minutes, but it was followed by the onset of a slow depression lasting for an hour (Fig. 3). No such depression occurred in the early excitation by the vestibular nerve not involved in the conjunctive stimulation. No depression was detected in early excitation of putative basket cells from either vestibular nerve, nor in the inhibition or rebound facilitation in Purkinje cells following the early excitation. Vestibular nerve-evoked field potentials in the granular layer and white matter of the flocculus were not affected.

These responses represent impulse transmission in mossy fibers and cortical networks except for the parallel fiber-Purkinje cell synapses. These observations lead to the conclusion that the signal transmission at a parallel fiber-Purkinje cell synapse undergoes a sustained depression after conjunctive activation with the climbing fiber impinging on the same Purkinje cell, in the manner postulated by Albus (1971). However, because of technical difficulties, it is not yet certain whether the depression of parallel fiber-Purkinje cell synapses has an even slower phase corresponding to permanent memory.

Intrinsic mechanisms of the long-lasting depression at parallel fiber-Purkinje cell synapses have further been examined using ionto-phoretic application of glutamate which is the putative neurotrans-mitter of granule cells. Application of glutamate in conjunction with 4/sec olivary stimulation was found to depress very effectively the glutamate sensitivity of Purkinje cells; aspartate sensitivity, tested as control, was depressed to a much less degree. The depression diminished in about ten minutes, but this recovery was followed by a slow depression lasting for an hour. This observation suggests that subsynaptic chemosensitivity of Purkinje cells to the putative neuro-transmitter of parallel fibers is involved in the depression observed after conjunctive stimulation of a vestibular nerve and the inferior olive.

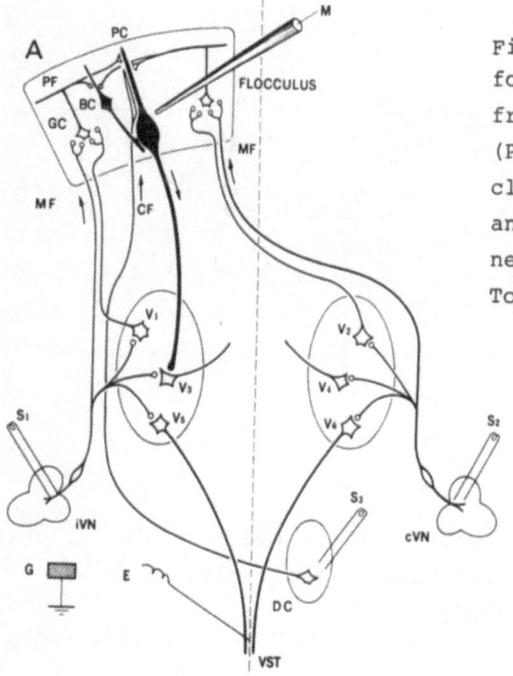

Fig. 2 Experimental arrangement for stimulating and recording from flocculus Purkinje cells (PC). MF, mossy fiber. CF, climbing fiber. iVN, cVN, ipsi- and contra-lateral vestibular nerve. (Ito, Sakurai and Tongroach, 1982).

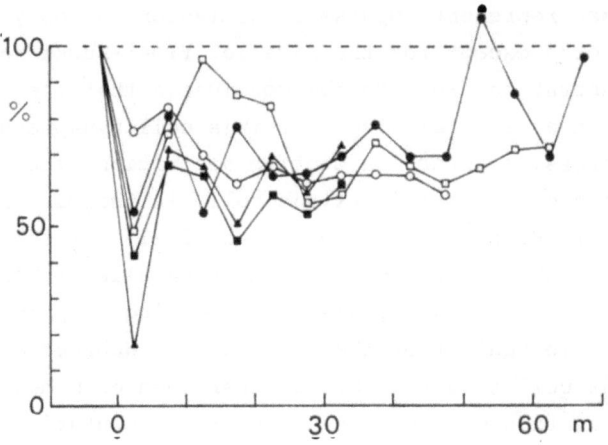

Fig. 3 Recovery time course after conjunctive vestibular-olivary stimulation. Ordinate, values of the firing index of Purkinje cells of rabbit's flocculus activated by electric supramaximal stimulation of a vestibular nerve at 2/sec. The plotted values of firing index are normalized by control values before conjunctive stimulation which happened at the zero time (Ito, Sakurai and Tongroach, 1982).

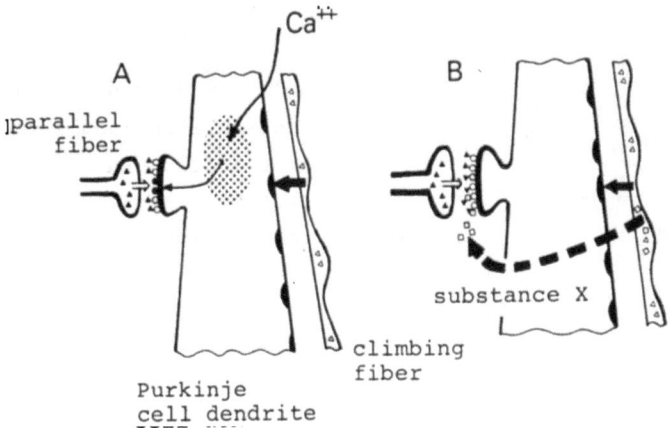

Fig. 4  Diagrammatical illustration of the two possible mechanisms of the heterosynaptic interaction between a climbing fiber and a parallel fiber-Purkinje cell synapse.

POSSIBLE MECHANISMS OF THE PLASTICITY AT PARALLEL FIBER-PURKINJE CELL
SYNAPSES

Marr (1969) adopted Hebb's idea of a plasticity condition that
postsynaptic membrane excitation should coincide with presynaptic
activity. According to this hypothesis, climbing fiber impulses act
upon parallel fiber-Purkinje cell synapses through membrane excitation
in Purkinje cells. However, this is unlikely because 1) vestibular
volleys which excite flocculus Purkinje cells does not produce any
sign of plastic changes in parallel fiber-Purkinje cell synapses, 2)
iontophoretc application of glutamate which by itself excite Purkinje
cells does not produce any plastic change in the gultamate sensitivity
of Purkinje cells, and 3) climbing fiber impulses are effective at
a rate as low as 4/sec, implying that impulse discharges evoked by
climbing fiber activation do not play a role in the plasticity of
parallel fiber-Purkinje cell synapses. Therefore, climbing fiber
impulses should interact with parallel fiber impulses in the manner
different from Hebb's synapse.

There are two possibilties to be tested expcrimentally. First,
since climbing fiber activation of Purkinje cells appears to involve
a voltage-dependent increase of calcium permeability of the dendritic
membrane (Llinás and Sugimori, 1980), Ekerot and Oscarsson (1981)
suggest that an increased intradendritic calcium concentration affects
subsynaptic receptors of Purkinje cell dendrites, just as intracellular calcium desensitizes acetylcholine receptors in muscle endplates
(Miledi, 1980). Second, climbing fibers may liberate a chemical sub-
stance(s) which reacts with subsynaptic receptor molecules at parallel fiber neurotransmitter, thereby rendering the receptors insensitive to the parallel fiber neurotransmitter. An analogous phenomenon
has been reported in cerebral cortical neurons that application of
thyrotropin-releasing hormon causes a reduction in glutamate sensitivity without affecting aspartate sensitivity (Renaud, Blume, Pittman, Lamour and Tan, 1980). These possibile two ways for climbing
fiber-parallel fiber interaction are illustrated diagrammatically in
Fig. 4.

FLOCCULUS CONTROL OF THE VESTIBULO-OCULAR REFLEX

Purkinje cells of the flocculus project to vestibular nuclei and supply inhibitory synapses to relay cells of the vestibulo-ocular reflex (Ito, Highstein and Fukuda, 1970; Fukuda, Highstein and Ito, 1972; Baker, Precht and Llinás, 1972). The flocculus receives vestibular afferents as a mossy fiber input and visual afferents as a climbing fiber input (Maekawa and Simpson, 1973). From these structural aspects (Fig. 5), the flocculus hypothesis of the vestibulo-ocular reflex control has been formulated to propose that the flocculus adaptively controls the vestibulo-ocular reflex, referring to retinal error signals (Ito, 1970, 1972, 1974). Since retinal error signals imply "instruction" concerning the performance of the reflex, the hypothesis is in accordance with Marr-Albus model of the cerebellum. It is supposed that retinal error signals conveyed by visual climbing fiber pathway modify parallel fiber-Purkinje cell synapses in the flocculus and thereby alter the signal transfer characteristics across the flocculus sidepath, which eventually leads to adaptive modification of the vestibulo-ocular reflex .

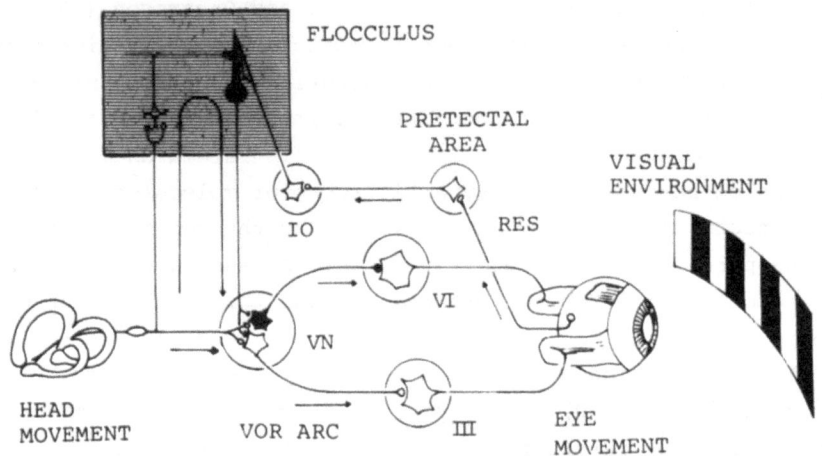

Fig. 5  Structure of the flocculo-vestibulo-ocular system in rabbit. VN, vestibular nuclei. IO, inferior olive. RES, retinal error signals. III, VI, oculomotor and abducens nuclei.

Finding of a microzonal structure of the flocculus has an important meaning in designing an experimental test for the flocculus hypothesis of the vestibulo-ocular reflex control. The flocculus of rabbits has been separated into at least five longitudinal zones projecting differentially to the vestibular nuclear complex and the lateral cerebellar nucleus (Yamamoto and Shimoyama, 1977; Yamamoto, 1978). Only one zone of these five (II in Fig. 6) is involved in control of the horizontal vestibulo-ocular reflex which is usually tested in adaptation experiments. Other zones (I, III-V in Fig. 6) are involved in other types of vestibulo-ocular reflex (vertical and rotatory) or in other functions than vestibulo-ocular reflex.

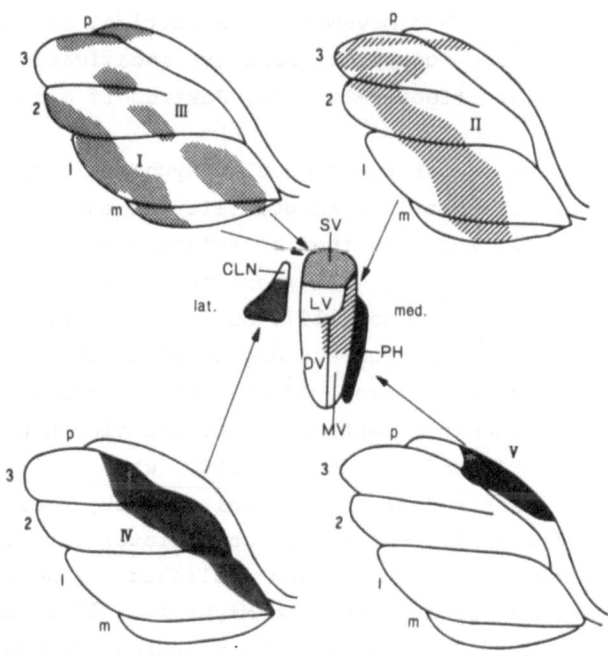

Fig. 6 Microzones of rabbit's flocculus. (from Yamamoto, 1978). Sides views of the left flocculus. 1-3, major folia of the flocculus. m and p, subsidiary folia. Central diagram shows the dorsal view of the left vestibular and lateral cerebellar nuclei. SV, LV, MV, DV, superior, lateral, medial and descending vestibular nuclei. PH, nucleus prepositus hypoglossi. CLN, cerebellar lateral nucleus. lat., lateral. med., medial.

EXPERIMENTAL SUPPORT FOR THE FLOCCULUS HYPOTHESIS

Melvill-Jones and Gonshor's (1966a, b) demonstration of adaptive modification of the horizontal vestibulo-ocular reflex offered the way of testing the flocculus hypothesis experimtnally. The gain of the vestibulo-ocular reflex readily changes under the vestibular-visual interaction which can be created by means of dove prism goggles or X2 telescopic lenses, or by combined optokinetic and rotatory stimulation. In pigmented rabbits, sinusoidal movement of a conventional striped drum in combination of sinuosidal rotation of the animal is effective (Nagao, unpublished). In albino rabbits, this is ineffective presumably because of visual anomaly (Collewijn, Winterson and Dubois, 1978), but movement of a vertical slit light in stead of a striped drum causes an adaptive modification of the vestibulo-ocular reflex effectively (Ito, Jastreboff and Miyashita, 1980).

One line of evidence for the flocculus hypothesis has been obtained by lesion experiments. The adaptability of the horizontal vestibulo-ocular reflex was abolsihed by ablation of the flocculus (Ito, Shiida, Yagi and Yamamoto, 1974) or the vestibulocerebellum including the flocculus (Robinson, 1976 on cat). Ablation of the flocculus caused a serious retrograde degeneration of the inferior olive neurons (Barmack and Simpson, 1980). This complication, however, was avoided by using chemical destruction of the flocculus with kainic acid (Ito, Jastreboff and Miyashita, 1980), which still abolished the adaptation (Ito, Jastreboff and Miyashita, 1982). Destruction of the dorsal cap of the inferior olive which mediates the visual climbing fiber signals to the flocculus also abolished the adaptability of the vestibulo-ocular reflex (Ito and Miyashita, 1975; Haddad, Demer and Robinson, 1980). Death of olivary neurons caused a complication, i.e., attenuation of the inhibitory action of flocculus Purkinje cells on vestibular nuclei cells after deprivation of climbing fibers (Dufossé, Ito and Miyashita, 1978). This complication was avoided by placing lesions in the visual pathway entering to the dorsal cap, but not directly on the dorsal cap, which still abolished the vestibulo-ocular adaptability (Ito and Miyashita, 1975).

Another line of evidence supporting the flocculus hypothesis has been derived from recording of impulse discharges from flocculus Purkinje cells. In rabbits, responsiveness of flocculus Purkinje cells to vestibular mossy fiber inputs during head rotation was altered in amplitude in parallel with the vestibulo-ocular adaptation (Dufossé,

Ito, Jastreboff and Miyashita, 1978). In monkeys, however, changes observed in Purkinje cell responsiveness to vestibular stimuli were in the opposite direction to the theoretical prediction, and therefore were ascribed to secondary effects of the vestibulo-ocular adaptation (Miles, Braitman and Dow, 1980). Microzonal structure have not yet been worked out in monkey flocculus, and there is no firm basis for assuming that flocculus Purkinje cells sensitive to horizontal gaze velocity, sampled by Miles et al. (1980), are really involved in control of the vestibulo-ocular reflex. This situation makes difficult to interprete the monkey data in connection with the flocculus hypothesis.

## Comment

Marr-Albus model of the cerebellum has now gained a sound experimental support for its plasticity assumption. Applicability of this model to the flocculus control of the vestibulo-ocular reflex is supported by neuronal circuit analysis, lesion experiments and unit recording from Purkinje cells. However, it is not yet clear whether the plasticity of parallel fiber-Purkinje cell synapses have a slow phase comparable with the permanent memory which the cerebellum may have. Technical difficulties for long-term stable recording should be overcome for answering this question. Transmitter chemosensitivity is shown to be involved in the plasticity of parallel fiber-Purkinje cell synapses. Molecular mechanisms of this plasticity provides an intersting subject of future investigations. In view of these positive evidence, further expansion of the model may be encouraged so that it has capabilities of not only spatial pattern discrimination but also processing of temporal signals. Eventual goal of the model may be that it handles the motor programs which the cerebellum is assumed to manage. Such a model will facilitate greatly the efforts for investigating the physiological meaning of the cerebro-cerebellar communication loop and its contribution to voluntary movement control.

REFERENCES

Albus, J.S. (1971) A theory of cerebellar function. Math. Biosci. 10, 25-61.

Baker,R.G., Precht, W. and Llinás, R. (1972) Cerebellar modulatory action on the vestibulo-trochlear pathway in the cat. Exp. Brain Res. 15, 364-385.

Dufossé, M., Ito, M. and Miyashita, Y. (1978) Diminution and reversal of eye movements induced by local stimulation of rabbit cerebellar flocculus after partial destruction of the inferior olive. Exp. Brain Res. 33, 139-141.

Dufossé, M., Ito, M., Jastreboff, P.J. and Miyashita, Y. (1978) A neuronal correlate in rabbit's cerebellum to adaptive modification of the vestibulo-ocular reflex. Brain Res. 150, 511-616.

Ekerot, C. -F. and Oscarsson, O. (1981) Prolonged depolarization elicited in Purkinje cell dendrites by climbing fibre impulses in the cat. J. Physiol. 318, 207-221.

Flourens, P. (1842) Recherches experimentales sur les proprietes et les fonctions due systeme nerveux dans les animaux vertebres. Paris: Bailliere.

Fukuda, J., Highstein, S.M. and Ito, M. (1972) Cerebellar inhibitory control of the vestibulo-ocular reflex investigated in rabbit IIIrd nucleus. Exp. Brain Res. 14, 511-526.

Gonshor, A. and Melvill-Jones, G. (1976a) Short-term adaptive changes in the human vestibulo-ocular reflex arc. J. Physiol. 265, 361-379.

Gonshor, A. and Melvill-Jones, G. (1976b) Extreme vestibulo-ocular adaptation induced by prolonged optical reversal of vision. J. Physiol. 256, 381-414.

Haddad, G.M., Demerm J.L. and Robinson, A.D. (1980) The effect of lesions of the dorsal cap of the inferior olive on the vestibulo-ocular and optokinetic system of the cats. Brain Res. 185, 265-275.

Ito, M. (1970) Neurophysiological aspects of the cerbellar motor control system. Int. J. Neurol. 7, 162-176.

Ito, M. (1972) Neural design of the cerebellar motor control system. Brain Res.40, 81-84.

Ito, M. (1974) The control mechanisms of cerebellar motor system. In: The Neurosciences, Third Study Program, ed. F.O. Schmitt, F.G. Worden, 293-303.

Ito, M. (1977) Neural events in the cerebellar flocculus associated with an adaptive modification of the vestibulo-ocular reflex of the rabbit. In: Baker, R, Berthoz, A. (eds.) Control of Gaze by Brain Stem Neurons.Elsevier, Amsterdam (Developments in Neuroscience, vol. 1) 391-398.

Ito, M. and Miyashita, Y. (1975) The effects of chronic destruction of inferior olive upon visual modification of the horizontal vestibulo-ocular reflex of rabbits. Proc. Jpn Acad. 51, 716-760.

Ito, M., Highstein, S.M. and Fukuda, J. (1970) Cerebellar inhibition of the vestibulo-ocular reflex in rabbit amd cat and its blockage by picrotoxin. Brain Res. 17, 524-526.

Ito, M. Jastreboff, P.J. and Miyashita, Y. (1982) Retrograde influence of surgical and chemical flocculectomy upon dorsal cap neurons of the inferior olive. Neurosci. Lett. 20, 45-48.

Ito, M., Jastreboff, J.I. and Miyashita, Y. (1981) Specific effects of unilateral lesions in the flocculus upon eye movements in albino rabbits. Exp. Brain Res. in the press.

Ito, M., Sakurai, M. and Tongroach, P. (1981) Evidence for modifiabity of parallel fiber-Purkinje cell synapses. In: Advances in Physiological Sciences 2, 97-105. Oxford: Pergamon.

Ito, M., Sakurai, M. and Tongroach, P. (1982) Climbing fibre induced depression of both mossy fibre responsiveness and glutamate sensitivity of cerebellar Purkinje cells. J. Physiol. in press.

Ito, M., Shiida, T., Yagi, N. and Yamamoto, M. (1974) The cerebellar modification of rabbit's horizontal vestibulo-ocular reflex induced by sustained head rotation combined with visual stimulation. Proc. Jpn. Acad.50, 85-89.

Llinás, R. and Sugimori, M. (1980) Electrophysiological properties of in vitro Purkinje cell dendrites in mammalian cerebellar slices. J. Physiol. 305, 197-213.

Luciani, L. (1891) Il cerevelletto: Nuovi Studi di Fisiologia normale et pathologica. Flourens: Le Monnier.

Maekawa, K. and Simpson, J.I. (1973) Climbing fiber responses evoked in vestibulocerebellum of rabbit from visual system. J. Neurophysiol. 36, 649-666.

Marr, D. (1969) A theory of cerebellar cortex. J. Physiol.202, 437-470.

Miledi, R. (1980) Intracellular calcium and desensitization of acetylcholine receptors. Proc. Roy. Soc. B 209, 447-452.

Miles, F.A., Braitman, D.J. and Dow, B.M. (1980) Long-term adaptive changes in primate vestibuloocular reflex. IV. Electrophysiological observations in flocculus of adpated monkeys. J. Neurophysiol. 43, 1477-1493.

Renaud, L.P., Blume, H.W., Pittman, Q.J., Lamour, Y. and Tan, A.T. (1979) Thyotropin-releasing hormone selectively depressed glutamate excitation of cerebral cortical neurons. Science N.Y. 205, 1275-1277.

Robinson, D.A. (1976) Adaptive gain control of vestibulo-ocular reflex by the cerebellum. J. Neurophsyiol. 39, 954-969.

Yamamoto, M. (1978) Localization of rabbit's flocculus Purkinje cells projecting to the cerebellar lateral nucleus and the nucleus prepositus hypoglossi, investigated by means of the horseradish peroxidase retrograde axonal transport. Neurosci. Lett. 7, 197-202.

Yamamoto, M. and Shimoyama, I. (1977) Differential localization of rabbit's flocculus Purkinje cells projecting to the medial and superior vestibular nuclei, investigated by means of the horseradish peroxidase retrograde axonal transport. Neurosci. Lett. 5, 279-283.

# CHAPTER 25

# DYNAMIC AND PLASTIC PROPERTIES OF THE BRAIN STEM NEURONAL NETWORKS AS THE POSSIBLE NEURONAL BASIS OF LEARNING AND MEMORY

Nakaakira Tsukahara* and  Mitsuo Kawato
Department of Biophysical Engineering, Faculty of Engineering
Sciences, Osaka University, Toyonaka-shi Osaka 560 JAPAN

*and National Institute for Physiological Sciences,
Okazaki-shi Aichi 444 JAPAN

It has long been believed that nerve connections in the brain are stable and rigid after their formation at an early developmental stage. However, although this rigidity of the nerve connections probably provides an important basis for instinctive behaviors, as the built-in generator of fixed action pattern, it cannot readily account for any kinds of adaptive behaviors, such as learning and memory. In this context it has recently been recognized that some of the nerve connections of the brain have plasticity. That is, they can be modified to some extent. This "plasticity" is thought to be the neural basis for adaptive behaviors. In other words, the "plasticity" and "rigidity" of the neural networks may relate to the environmental versus genetic factors which underlie behavior.

The most remarkable example of plasticity in the brain is the formation of new synaptic connections, a process called "sprouting", which is now recognized as a wide-spread phenomenon in the central nervous system (Tsukahara, 1981, see for review).

It is generally agreed that the degree and extent of sprouting is more remarkable after denervation at the neonatal stage than at the adult stage. Furthermore, sprouting phenomenon is not limited to cases of removal of direct synaptic inputs in some peripheral and central synapses.

## Modifiable synapses in the motor system -- which synapses are modifiable and in what circumstances?

The red nucleus (RN) attracted our attension because of its remarkable plasticity. So far, three kinds of sprouting phenomena have been found in RN (Tsukahara, 1981). (1) lesion-induced sprouting of the corticorubral synapses in neonates; (2) lesion-induced sprouting of the same synapse in the adult; (3) sprouting of corticorubral synapses without lesions of the synaptic inputs. Especially (3) is important because it supports the idea that sprouting can occur in the absence of lesions and can be the neuronal basis of the more physiological phenomenon such as learning. Furthermore, by using the classical conditioning paradigm in which the corticorubral synapses are contained in the pathway responsible for the conditioned response, it is possible to correlate the synaptic plasticity to the behavioral plasticity and to clarify the way the synaptic modification takes place. It turns out that pairing of the conditioned stimulus to the corticorubral input and the unconditioned stimulus to the forelimb skin which most probably activates the cerebello-rubral input. Therefore, contingency of the cerebrorubral and cerebellorubral excitation in the red nucleus should result in the modification of the synaptic transmission of the cerebrorubral input.


## Plasticity of neuronal networks

Not all plasticity can be attributed to changes at the cellular level. There may be circumstances where a certain form of plasticity appears at the network level. It has been frequently postulated that short-term memory is attributable to properties of neuronal networks such as reverberating circuits. There are some experimental results that suggest the possibility of impulse reverberation on cerebellar circuits. Mutually excitatory connections exist between cerebellar nuclei and precerebellar nuclei, such as the paramedian reticular nucleus, pontine tegmental reticular nucleus, lateral reticular nucleus and the inferior olive (Allen and Tsukahara, 1974). It has been shown that prolonged depolarization can be produced in the red nucleus neurons after removal of Purkinje cell inhibition by injecting picrotoxin intravenously or by surgical ablation of the cerebellar cortex. Because transection of the spinal cord or ablation of the cerebral sensorimotor cortex did not abolish this prolonged depolarization, it is probably not mediated by loops via the spinal cord or cerebral sensorimotor cortex. The prolonged depolarization exhibited regenerative properties with a threshold, an important property of positive feedback systems. The prolonged depolarization could be reversibly abolished by local cooling of the fibers connecting the precerebellar nuclei and the cerebellar nuclei at the inferior and middle cerebellar peduncles. This indicates that the prolonged depolarization is not due to pacemaker properties of neurons of the cerebellar nuclei, but to a loop passing through the inferior and middle cerebellar peduncles.

Systematic investigations suggest that two-neuronal group reverberating circuits (pontine tegmental reticular nucleus or paramedian reticular nucleus as the one arm and the nucleus interpositus as the other arm) or three-neuronal group reverberating circuits (rubro-inferior olive-dentate-rubral loop via the parvocellular red nucleus or rubrolateral reticular nucleus-interposito-rubral loop via the magnocellular red nucleus) can be the substrate for the production of the prolonged depolarization (Tsukahara, 1981, for review).

Based on these experimental findings, it is now possible to consider the cerebro-cerebello-rubral learning system in the theoretical context.


## Organization of the cerebro-cerebellar motor control system

In the previous review article (Allen and Tsukahara, 1974), a general model for motor control was proposed as shown in Fig. 1. The association cortex may participate in the translation of the idea to move into a patterned activation of certain motor cortical columns and their elemental movements. Since the hemisphere of the cerebellum appears to perform its function without the aid of direct peripheral inputs, the hemisphere would appear to be more suited for participation in planning the movement than in actual execution and updating of the movement as was proposed for pars intermedia of the cerebellar cortex.

As suggested in Fig. 1, the most reasonable possibility for the lateral cerebellum is that it participates in the programing or longrange planning of the movement. Its function is largely anticipatory, based on learning and previous experience and also on preliminary, highly digested sensory information that some of the association area receive. Once the movement has been planned within the association cortex, with the help of the cerebellar hemisphere and basal ganglia, the motor cortex issues the command for movement. At this point the pars intermedia makes an important contribution by updating the movement based on the sensory description of the limb position and velocity on which the intended movement is to be superimposed. This is a kind of short-range planning as opposed to the long-range planning of the association cortex and lateral cerebellum.

In learning a movement, we first execute the movement very slowly because it cannot be adequately preprogramed. Instead, it is performed largely by cerebral interventions as well as the constant updating

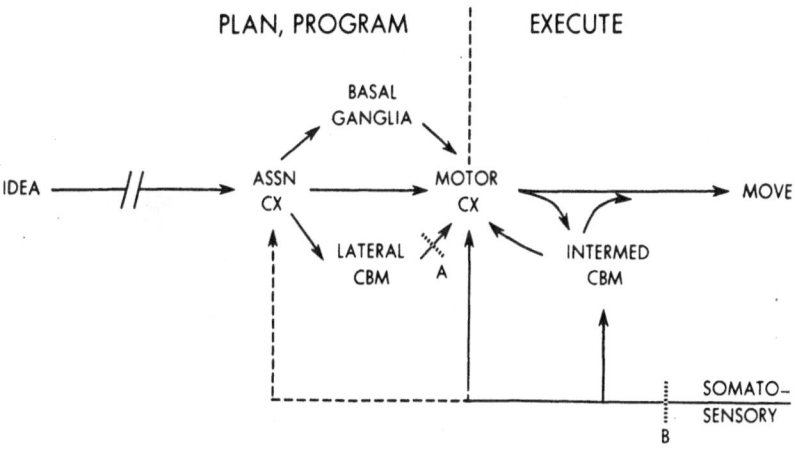

Fig. 1: Scheme showing proposed roles of several brain structures in
movement.  Thin dashed line represents a pathway of unknown importance.  Heavy dashed lines at A and B represent lesions described in text.  It is proposed that basal ganglia and cerebellar
hemisphere are involved with association cortex in programing
of volitional movements.  At the time that the motor command
descends to motorneurons, engaging the movement, the pars intermedia updates the intended movement, based on the motor command
and somatosensory description of limb position and velocity on
which the movement is to be superimposed.  Follow-up correction
can be performed by motor cortex when cerebellar hemisphere and
pars intermedia do not effectively perform their functions (Ref.
1).

of the pars intermedia.  With practice, a greater amount of the movement
can be preprogramed and the movement can be executed more rapidly.
Eccles and Ito view the cerebellum as providing an internal substitute
for the external world.  This eliminates the need for peripheral sensory
input and allows one to increase the speed of the learned movement by
programing.  This cerebellar operation we consider to take place in the
lateral zone.  Thus, we can consider that many of the trained movements,
which constitute a large fraction of our movement repertorie, are not
completely preprogramed but are provisional, subject to continuous re-
vision.  The role of the cerebellum, presumably the pars intermedia,
in our untrained or exploratory movement is attested to be the clumsiness
and slowness with which they must be performed when the cerebellum is
ablated and we must rely on cerebral control alone.
         Fig. 2 illustrates the neuronal circuitry relating the cerebellar

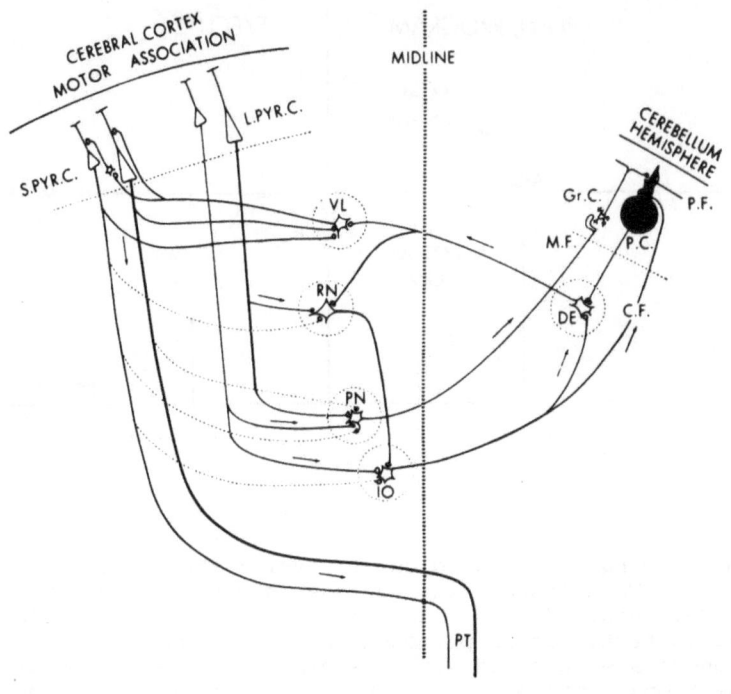

Fig. 2: Cerebrocerebellar loop through cerebellar hemisphere. Pyramidal
tract is shown with dashed lines to precerebellar relay nuclei
to emphasize the fact that primarily corticobulbar fibers from
association cortex carry information to cerebellar hemisphere.
Outflow of cerebellar hemisphere feeds back to motor cortex,
which does send its signals down the pyramidal tract. Only
anterior, parvocellular portion of red nucleus is shown here.
RN=red nucleus; PN=pontine nuclei; IO=inferior olivary nucleus;
PT=pyramidal tract; VL=ventro-lateral nucleus of thalamus;
S.PYR.C.=small pyramidal cell; L.PYR.C.=large pyramidal cell.
(Ref. 1)

hemisphere. The parvocellular part of the red nucleus, which develops
extensively in primates especially in man, constitutes a part of the
lateral cerebellar system in that it receives input from the association
cortex and sends messages to the cerebellar hemisphere via the inferior
olive (Sasaki, 1980). Since a considerable amount of new data on the
plasticity of the red nucleus are at hand, a more concrete model of the
"internal model" can be realized by the circuit of the cerebro-rubro-
olivo-cerebellar loop.

## Model of the cerebro-rubro-olivo-cerebellar system

Fig. 3 illustrates a modification of Fig. 1 in the light of the model proposed here. Let us consider the red nucleus neuron (parvocellular part) which receives input x(t) from the association cortex and gives output y(t). The neuron has n inputs whose instantaneous firing frequencies are designated by $x_1(t)$, $x_2(t)$, $---$ , $x_n(t)$. Let $w_i$ be the synaptic efficacy of the i-th input $x_i(t)$. Membrane potential y(t) of the neuron is a sum of n postsynaptic potential. For convenience, we assume that the output signal of the neuron, u(t) is equal to its membrane potential. The input vector is

$$X(t) = [x_1(t), x_2(t), --- , x_n(t)]^T. \tag{1}$$

The set of synaptic weights is designated by the vector $W = (w_1, w_2, -- - , w_n)^T$. The output signal of the neuron is

$$y(t) = \sum_{i=1}^{n} w_i x_i(t) = W^T X(t) = X(t)^T W . \tag{2}$$

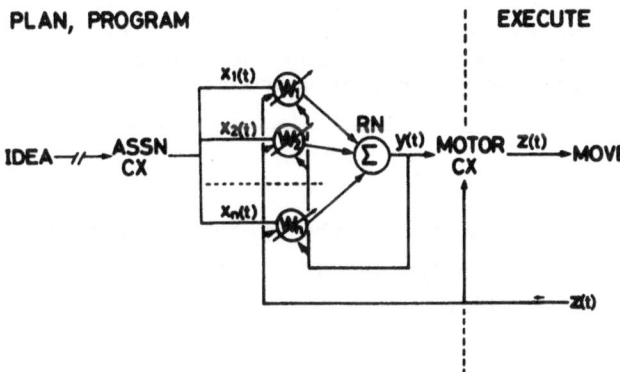

**PLAN, PROGRAM**      **EXECUTE**

Fig. 3: Scheme showing proposed roles of several brain structures in move ment with emphasis on an adaptive role of the red nucleus proposed in the model.

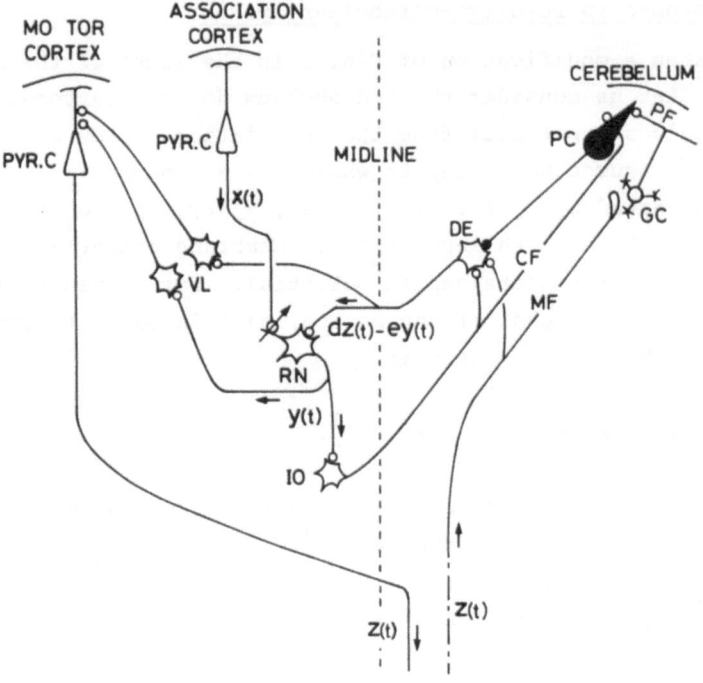

Fig. 4: Cerebrocerebellar communication through cerebellar hemisphere
and the red nucleus. A concrete "neural model" is proposed to
resides in the red nucleus whose output is transmitted to the
cerebral motor cortex via the rubro-thalamic pathway and can
be utilized for the control of movements.

We introduce a learning principle of the synaptic efficacy $w_i(t)$. Denoting
the desired response which is the output of the whole control system as
$z(t)$, the error signal at time t is

$$es(t) = dz(t) - ey(t) = dz(t) - ew^T X(t) = dz(t) - eX(t)^T w. \qquad (3)$$

We assume that rate of change for the i-th synaptic weight is proportional to a product of the i-th input $x_i(t)$ and the error signal $es(t)$.
Suppose that the output from the cerebellar nucleus computes subtraction
of the desired input $z(t)$ from real input $y(t)$ for a "neural model" of
the external world in the parvocellular RN. The impulse frequency of
the cerebellar input to the red nucleus is $dz(t) - ey(t)$. If the i-th

synaptic weight of a neuron in the parvocellular RN changes according to crosscorelation between cerebral input $x_i(t)$ and cerebellar input $dz(t)-ey(t)$,

$$\tau \dot{w}_i(t)=cx_i(t)[dz(t)-ey(t)] \tag{4}$$

where $\tau$ is a time constant of synaptic modification and c, d and e are positive constants, then the output of the neuron tends to approximate $(d/e)z(t)$ (see below). If this output signal, $y(t)$ is transmitted, as shown in Fig. 4, to the cerebral motor cortex via the thalamus and utilized as the signals controlling the corticospinal motor control system, we have a concrete "neural model" or "program" which is acquired by learning and can be utilized for the control of movements.

It has been questioned in the previous review article (Allen and Tsukahara, 1974) about the existence of the rubro-thalamic pathway. However, recent studies by Condé and Condé (Condé and Condé, 1980) supplied evidence in favour of the presence of the rubro-thalamo-cortical projection. Furthermore, it has been reported that there is a three neuron reverberating circuit via the parvocellular red nucleus (Murakami et al., 1981).

## Model prediction

This model predicts that if the rubro-olivo-cerebellar loop is destroyed, the internal model would be destroyed also and the motor performance would be severely disturbed. This prediction is in accord with the frequently reported symptoms of "tremor" after lesions of anywhere in the rubro-olivary pathway (Poirier et al., 1969). This may be interpreted as the oscillation due to the delay of the feedback. Due to the longer delay of the peripheral feedback than that of the internal feedback, the motor control system may be unstable and this may be manifested as the "tremor". This symptom of the lesions of the rubro-olivo-cerebello-rubral loop has long been known without an adequate explanation of this mechanisms. The present model may provide the first theoretical explanation of this phenomenon.

Another prediction of the model would be that after lesions of the rubro-olivo-cerebellar loop, motor learning does not take place. This, of course, is deficits of motor learning such as reported by Ito and also by Llinas, although the interpretation of this deficit has been related Marr's hypothesis in the case of Ito's paradigm.

## On learning equation

Amari (1977) introduced new version of learning called orthogonal learning as a model of association for static patterns. His model was originally for static pattern association, but we will show that a slightly modified model is appropriate for neural identifier (that is, "internal model").

In vector notation the synaptic modification rule (4) is represented as follows.

$$\tau \dot{W}(t) = cX(t)es(t)$$
$$= cX(t)[dz(t)-ey(t)]$$
$$= cX(t)[dz(t)-eX(t)^{T}W(t)].$$

The input u(t) to the control object is assumed to be a continuous random process. Because $z(t)$ and $x_i(t)$ are functionals of $u(t)$, they are also regarded as continuous random processes. We represent the above equation in more explicit form as follows.

$$\tau \dot{W}(t,\omega) = cdX(t,\omega)z(t,\omega)-ceX(t,\omega)X(t,\omega)^{T}W(t,\omega), \tag{5}$$

here $\omega$ is a sample point in a probability space. The input signal $X(t,\omega)$ and desired response $z(t,\omega)$ are assumed to be strongly mixing processes (stochastic processes for which the "past" and the "future" are asymptotically independent). Geman (1979) has shown that when the "rate" of mixing is rapid relative to the rate of change of solution processes, the averaged deterministic equation of (5) is a good approximation of (5). We take the average (take the expected value) of the right-hand side of the stochastic equation (5).

$$\tau \dot{M}(t) = cdE[X(t,\omega)z(t,\omega)]-ceE[X(t,\omega)X(t,\omega)^{T}]M(t) \tag{6}$$

Geman proved the following theorem in a more general fashion.

Theorem (Geman, 1979)

For all $\tau$ sufficiently large, $\sup_{t \geq 0} E[W^2(t)] < \infty$, and $\lim_{\tau \to \infty} \sup_{t \geq 0} E[\{M(t)-W(t)\}^2] = 0$.

Consequently we only need to study (6) when the time constant of synaptic modification $\tau$ is sufficiently long. The cross correlation vector between the input signals $X(t,\omega)$ and the desired response $z(t,\omega)$ is defined as

$$E[X(t,\omega)z(t,\omega)]=E\begin{bmatrix} x_1(t,\omega)z(t,\omega) \\ x_2(t,\omega)z(t,\omega) \\ - \\ - \\ - \\ x_n(t,\omega)z(t,\omega) \end{bmatrix} \equiv P.$$

The symmetric and positive definite input correlation matrix R of the x-input signals is defined as

$$E[X(t,\omega)X(t,\omega)^T]=E\begin{bmatrix} x_1(t,\omega)x_1(t,\omega) & x_1(t,\omega)x_2(t,\omega) & -- \\ x_2(t,\omega)x_1(t,\omega) & x_2(t,\omega)x_2(t,\omega) & -- \\ - & - & -- \\ - & - & -- \\ - & & - & x_n(t,\omega)x_n(t,\omega) \end{bmatrix}$$

$$\equiv R.$$

In this notation the averaged equation (6) is rewritten as follows.

$$\tau\dot{M}(t)=cdP-ceRM(t) \tag{6'}$$

Because the matrix R is positive definite and the constant c and e are positive, M(t) converges asymptotically to $(d/e)R^{-1}P$. From the above theorem, the synaptic weight vector W(t,$\omega$) also converges to $(d/e)R^{-1}P$ in mean if $\tau$ is sufficiently long.

We show that $(d/e)R^{-1}P$ is optimal for approximating $(d/e)z(t)$ by y(t) in the mean square sense. The square of the error es(t) is

$$es(t)^2=[dz(t)-ey(t)]^2=d^2z^2-2dezX^TW+e^2W^TXX^TW.$$

The mean square error $\xi$, the expected value of es(t)$^2$, is

$$\xi=E[es(t)^2]=d^2E[z^2]-2deE[zX^T]W+e^2W^TE[XX^T]W$$
$$=d^2E[z^2]-2deP^TW+e^2W^TRW. \tag{7}$$

It may be observed from (7) that the mean square error (mse) performance function is a quadratic function of the weights, a "bowl-shaped" surface. The gradient at any point on the performance surface may be obtained by differentiating the mse function, equation (7), with respect to the weight vector. The gradient vector is

$$\nabla = -2deP + 2e^2 RW$$

Set the gradient to zero to find the optimal weight vector $W^*$:

$$W^* = (d/e) R^{-1} P \tag{8}$$

The minimum mse is obtained from (8) and (7):

$$\xi_{min} = d^2 E[z^2] - (d^2/e^2) P^T R^{-1} P \tag{9}$$

The optimal $W^*$ is exactly same as the asymptotic value of $W(t,\omega)$ which is obtained from the equation (5), and the gradient vector $\nabla$ is proportional to the right-hand side of the averaged equation (6'). Therefore the modified, temporal, orthogonal learning (4) is closely related to the method of steepst descent.

Because the output $y(t)$ of the neuron adapts to immitate the desired output $z(t)$ which is output of the underlying physiological process by synaptic modification obeying the learning rule (4), the neuron under consideration is regarded as a neural identifier for the unknown motor system.

If we feed the delayed input $u(t-\Delta t)$ to the n known systems of neural identifier and feed the desired output $z(t)$ to the modifiable synapses, then the learning rule (4) attains optimal weighting coefficients $W^*_{\Delta t}$, with which $y(t) = W^T X(t-\Delta t)$ best approximates $(d/e)z(t)$. If we feed input $u(t)$ to the neural identifier after $W^*_{\Delta t}$ is obtained, $y(t)$ clearly approximates $z(t+\Delta t)$. This means the neural machine works as a predictor. These two steps are simultaneously realized by the following slight change of the learning rule (4).

$$\tau \dot{W}(t) = cX(t-\Delta t)[dz(t) - ey(t-\Delta t)]$$
$$= cX(t-\Delta t)[dz(t) - eX(t-\Delta t)^T W(t-\Delta t)]$$

The modified learning rule means that changing rate of synaptic weights is proportional to a product of delayed input signal $X(t-\Delta t)$ and the error signal between desired output $z(t)$ and delayed output signal $y(t-\Delta t)$ of the neuron.

## References

1) Allen, G.I. and Tsukahara, N.:Cerebrocerebellar communication
   systems. Physiol. Rev. 54, 957-1006 (1974)
2) Amari, S.:Neural theory of association and concept-formation. Biol.
   Cybern. 26, 175-185 (1977)
3) Condé, F. and Condé, H.:Demonstration of a rubrothalamic projection
   in the cat, with some comments on the origin of the rubrospinal
   tract. Neuroscience, 5, 789-802 (1980)
4) Geman, S.:Some averaging and stability results for random differential equations. SIAM J. Appl. Math. 36, 86-105 (1979)
5) Ito, M. In:Integrative Control Functions of the Brain. Vol. 3,
   (Ito, M., Tsukahara, N., Kubota, K. and Yagi, K. eds). pp.351-
   367, Kodansha/Elsevier, Tokyo, Amsterdam, 1981
6) Marr, D.:A theory of the cerebellar cortex. J. Physiol. (Lond.)
   202, 437-470 (1969)
7) Murakami,F. and Ozawa, N., Katsumaru, H. and Tsukahara, N.:Reciprocal
   connections between the nucleus interpositus of the cerebellum
   and precerebellar nuclei. Neuroscience Lett. 25, 209-213 (1981)
8) Poirier, L.J., Bouvier, G., Bédard, P., Bouchard, R., Larochelle, L.,
   Olivier, A. et Singh, P.:Essai sur les circuits neuronaux
   mipliqués dans le tremblement postural et l'hypokinesie. Rev.
   Neurol. 120, 15-40 (1969)
9) Sasaki, K. In:Integrative Control Functions of the Brain. Vol. 2,
   (Ito, M., Tsukahara, N., Kubota, K. and Yagi, K. eds). pp.123-
   138, Kodansha/Elsevier, Tokyo, Amsterdam, 1980
10) Tsukahara, N.:Synaptic plasticity in the red nucleus. In:Neuronal
    Plasticity (Cotman, C.W. ed). pp.113-130, Raven Press, New
    York, 1978
11) Tsukahara, N.:Synaptic plasticity in the mammalian central nervous
    system. Ann. Rev. Neurosci. 4, 351-379 (1981)
12) Tsukahara, N.:Sprouting and the neuronal basis of learning. Trends
    in Neuroscience, 4, 234-237 (1981)

# Journal of
# Mathematical
# Biology

ISSN 0303-6812                                    Title No. 285

**Editorial Board:**
H. T. Banks, Providence, RI; H. J. Bremermann, Berkeley,
CA; J. D. Cowan, Chicago, IL; J. Gani, Lexington, KY;
K. P. Hadeler (Managing Editor), Tübingen;
F. C. Hoppensteadt, Salt Lake City, UT; S. A. Levin
(Managing Editor), Ithaca, NY; D. Ludwig, Vancouver;
L. A. Segel, Rehovot; D. Varjú, Tübingen in cooperation
with a distinguished advisory board.

The **Journal of Mathematical Biology** publishes papers in
which mathematics leads to a better understanding of
biological phenomena, mathematical papers inspired by
biological reaserch and papers which yield new experimental data bearing on mathematical models. The scope
is broad, both mathematically and biologically and
extends to relevant interfaces with medicine, chemistry,
physics, and sociology. The editors aim to reach an
audience of both mathematicians and biologists.

*A selection of articles published:*
**A. Hastings:** Multiple Limit Cycles in Predator-Prey
Models
**R. E. Plant:** Bifurcation and Resonance in a Model for
Bursting Nerve Cells
**G. Bard Ermentrout, J. Rinzel:** Waves in a Simple, Exci-
table or Oscillatory, Reaction-Diffusion Model
**M. Brill, G. West:** Contributions to the Theory of Inva-
riance of Color Under the Condition of Varying Illumina-
tion
**J. P. Keener:** On Cardiac Arrythmias: AV Conduction
Block
**B. Türke:** Analysis of Pattern Recognition by Man using
Detection Experiments
**R. M. Miura:** Accurate Computation of the Stable Solitary
Wave for the Fritz Hugh-Nagumo Equations
**A. H. Cohen, P. J. Holmes, R. H. Rand:** The Nature of the
Coupling Between Segmental Oscillators of the Lamprey
Spinal Generator for Locomotion: A Mathematical Model

Subscription information and sample copy upon request

Springer-Verlag
Berlin
Heidelberg
New York

# Bio-mathematics

Managing Editor: S. A. Levin

Springer-Verlag
Berlin
Heidelberg
New York

Volume 8

A. T. Winfree

## The Geometry of Biological Time

1979. 290 figures. XIV, 530 pages
ISBN 3-540-09373-7

The widespread appearance of periodic patterns in nature reveals that many living organisms are communities of biological clocks. This landmark text investigates, and explains in mathematical terms, periodic processes in living systems and in their non-living analogues. Its lively presentation (including many drawings), timely perspective and unique bibliography will make it rewarding reading for students and researchers in many disciplines.

Volume 9

W. J. Ewens

## Mathematical Population Genetics

1979. 4 figures, 17 tables. XII, 325 pages
ISBN 3-540-09577-2

This graduate level monograph considers the mathematical theory of population genetics, emphasizing aspects relevant to evolutionary studies. It contains a definitive and comprehensive discussion of relevant areas with references to the essential literature. The sound presentation and excellent exposition make this book a standard for population geneticists interested in the mathematical foundations of their subject as well as for mathematicians involved with genetic evolutionary processes.

Volume 10

A. Okubo

## Diffusion and Ecological Problems: Mathematical Models

1980. 114 figures, 6 tables. XIII, 254 pages
ISBN 3-540-09620-5

This is the first comprehensive book on mathematical models of diffusion in an ecological context. Directed towards applied mathematicians, physicists and biologists, it gives a sound, biologically oriented treatment of the mathematics and physics of diffusion.

# Lecture Notes in Biomathematics